OXFORD STATISTICAL SCIENCE SERIES

SERIES EDITORS

A. C. ATKINSON   R. J. CARROLL   D. J. HAND
D. A. PIERCE   M. J. SCHERVISH
D. M. TITTERINGTON

D1584944

OXFORD STATISTICAL SCIENCE SERIES

Statistical Modelling in GLIM 4

SECOND EDITION

MURRAY AITKIN
Department of Psychology, University of Melbourne

BRIAN FRANCIS
Centre for Applied Statistics, Lancaster University

JOHN HINDE
Department of Mathematics, National University of Ireland

WITHDRAWN

OXFORD
UNIVERSITY PRESS

OXFORD
UNIVERSITY PRESS

Great Clarendon Street, Oxford OX2 6DP

Oxford University Press is a department of the University of Oxford.
It furthers the University's objective of excellence in research, scholarship,
and education by publishing worldwide in

Oxford New York

Auckland Cape Town Dar es Salaam Hong Kong Karachi
Kuala Lumpur Madrid Melbourne Mexico City Nairobi New Delhi
Shanghai Taipei Toronto

With of☐cesin

Argentina Austria Brazil Chile Czech Republic France Greece
Guatemala Hungary Italy Japan South Korea Poland Portugal
Singapore Switzerland Thailand Turkey Ukraine Vietnam

Oxford is a registered trade mark of Oxford University Press
in the UK and in certain other countries

Published in the United States
by Oxford University Press Inc., New York

© Murray Aitkin, Brian Francis and John Hinde 2005

The moral rights of the authors have been asserted
Database right Oxford University Press (maker)

First published 2005

All rights reserved. No part of this publication may be reproduced,
stored in a retrieval system, or transmitted, in any form or by any means,
without the prior permission in writing of Oxford University Press,
or as expressly permitted by law, or under terms agreed with the appropriate
reprographics rights organization. Enquiries concerning reproduction
outside the scope of the above should be sent to the Rights Department,
Oxford University Press, at the address above

You must not circulate this book in any other binding or cover
and you must impose this same condition on any acquirer

A catalogue record for this title is available from the British Library

Library of Congress Cataloging-in-Publication Data

Aitkin, Murray A.
   Statistical modelling in GLIM 4, Second edition/
Murray Aitkin, Brian Francis, John Hinde.—2nd ed.
      p. cm.
   Summary: "This text examines the theory of statistical modelling with generalized
linear models. It also looks at applications of the theory to practical problems,
using the GLIM4 package"—Provided by publisher.
   Includes bibliographical references and index.
   ISBN 0–19–852413–7 (alk. paper)
   1. Linear models (Statistics) I. Francis, Brian. II. Hinde, John, 1952– III. Title.
   QA276.A397 2005
   519.5—dc22   2004024956

ISBN 0 19 852413 7 (Hbk)

10 9 8 7 6 5 4 3 2 1

Typeset by Newgen Imaging Systems (P) Ltd., Chennai, India
Printed in Great Britain
on acid-free paper by
Biddles Ltd., Kings Lynn

Preface to the second edition

The second edition of this book is extended in scope and restructured in content, though as before we have made some compromises to keep the length of the book within reasonable bounds. We have added three new chapters on finite mixtures and random effect models, with a detailed coverage of Gaussian quadrature and nonparametric maximum likelihood analysis of these models.

The content is restructured to correspond to the modern computing environment, and also to make the book more generally useful. We have expanded the presentation of output so that the book is more self-contained without needing collateral use of GLIM4, though the extended features of the package are a substantial improvement over those of GLIM3.

Chapter 1 is reduced and now discusses only those features of GLIM4 of which we make particular use, especially the presentation graphics. Chapter 2 now gives a detailed discussion of the population modelling process, with a full discussion of frequentist, Bayesian and likelihood inferential approaches to simple models, though in later chapters we present mainly the frequentist analysis. Chapters 3–6 discuss as before the normal, binomial, Poisson/multinomial and survival models. Chapter 7 discusses continuous and finite mixtures, Chapter 8 overdispersed models and Chapter 9 variance component models. The data sets and macros used in this edition can be downloaded from the Oxford Web site.

The second edition was written over the period 1995–2004. Dorothy Anderson was not involved in this edition. We thank many students, course participants and colleagues for helpful comments and corrections, and the Australian Research Council and the American Institutes for Research for support of Murray Aitkin. We are grateful to Ross Darnell for detailed checking of the text and examples; all remaining errors are entirely our responsibility.

Melbourne, Lancaster, and Galway
2004

Preface to the first edition

The aim of this book is twofold: to give an exposition of the principles of statistical modelling with the necessary statistical theory, and to describe the application of these principles to the analysis of a wide range of practical examples using GLIM3, the statistical package for Generalized Linear Interactive Modelling developed by the Working Party on Statistical Computing of the Royal Statistical Society.

There are few published accounts of the principles and practice of statistical modelling. The recent book by McCullagh and Nelder (1983), to which we make frequent reference, provides an authoritative account of much of the theory but does not give details of computing systems used for model fitting. Our book gives a full description of the use of GLIM3 for model fitting, and detailed discussions of many examples. The theoretical orientation of our book is slightly different, as we place greater emphasis on the direct use of the likelihood function for inference.

This book was written over the period 1982–1987, as part of an Economic and Social Research Council research programme at the Centre for Applied Statistics in the analysis of complex social data, which supported Dorothy Anderson and John Hinde. Drafts of chapters were developed and revised from regular intensive courses in statistical modelling in GLIM3 given at the Centre and elsewhere over this period. We are grateful to many of the students on these courses, some of them experienced statisticians, whose comments and criticisms improved these early drafts and our own understanding of statistical modelling and GLIM3.

The book can be used in several ways. It is written in a sequence which we have found appropriate for intensive courses. Chapter 1 gives a gentle introduction to GLIM3 for novices, and Chapter 2 a general introduction to the principles of statistical modelling, with two simple examples. This chapter also develops the necessary theory of maximum likelihood estimation and likelihood ratio testing. Chapter 3 discusses the normal model, Chapter 4 binomial data, Chapter 5 multinomial and Poisson data, and Chapter 6 survival data.

The book is designed to be used for graduate and advanced undergraduate courses in statistics, in conjunction with interactive use of GLIM, either in lectures or in practical/tutorial sessions. We have included relevant parts of the actual display from GLIM sessions (mostly plots and parameter estimates) to make the exposition self-contained as far as possible, so that it can be used as a lecture text without access to GLIM if necessary. The book is equally suitable as a self-teaching manual for professional statisticians and research workers. The data sets and macros in Appendix 4 can be bought separately from the University of Lancaster in machine-readable form.

We have made some compromises in coverage in keeping the book to a reasonable length. Our aim has been to concentrate on the central issues of modelling and data analysis. Many GLIM macros for specialized problems have been published in the GLIM Newsletters, in the GLIM82 and GLIM85 conference proceedings, and in several journals in the field of statistical computing.

We are grateful for the generous support of the Economic and Social Research Council and the University of Lancaster. We particularly thank Doreen Hough who processed the manuscript through numerous drafts with great patience and prodigious speed and accuracy. We would like to thank Mick Green, whose macros for the piecewise exponential distribution are included in Chapter 6, and to other colleagues who commented on earlier drafts of this book. We are also grateful to Alan Airth and Ron McKay for their careful checking of the examples: any remaining errors are entirely our responsibility.

1988   MA
Tel Aviv, Melbourne, Lancaster,   DA
and Exeter   BF
                                                          JH

Contents

1
Introducing GLIM4

1.1   Statistical packages and statistical modelling

Statistical modelling is an iterative process. We have a research question of interest, the experimental or observational context, the sampling scheme and the data. Based on these data and the other information, we develop an initial model, fit it to the data and examine the model output. We then plot the residuals, check the model assumptions and examine the parameter estimates. Based on this model output, we may modify the first model possibly through some transformation or the addition or subtraction of model terms. Fitting the second model may in turn lead to further models for the data, with the form of the current model being based on the information provided by the previous models. This *interactivity* between the data and the data modeller is a crucial feature of the process of statistical modelling. While automatic model selection procedures provided in many software packages can provide guidance in large problems, we take a different route in this book.

We focus on generalized linear models and extensions of these models, which can be fitted using an iterative reweighted least squares (IRLS) algorithm. The computational extensions which we cover here (into such areas as survival analysis and random effects models) are developed by building on the standard IRLS algorithm, incorporating it into more complex *procedures* or *macros*.

There are numerous interactive and programmable statistical packages on the market which can be used to fit generalized linear models. In no particular order, these include R, S-PLUS, GENSTAT, GLIM, SAS, STATA and XLISPSTAT, and, with appropriate procedure development, all could be used to fit the models introduced in this book. All can be used interactively and allow the user to develop models in the way outlined above. In addition, there are other packages such as SPSS and MINITAB which can fit specific types of generalized linear model although not in a unified framework. In this book we have chosen to use GLIM, which is a compact program with a single weighted least squares algorithm for generalized linear models.

We have written this book around GLIM in preference to the other packages mentioned for two reasons. The first is historical; the first edition of this book used GLIM, and two of the authors of this book helped to develop the current version of GLIM. Second, while other packages such as R now provide most of the facilities of GLIM, the command set of GLIM is compact and can be learnt relatively quickly. We therefore begin this book by providing an introduction to the main facilities of GLIM4 (Francis *et al.*, 1993), the most recent release of GLIM.

GLIM is a command driven package. Commands for data manipulation, data transformation, data display and model fitting may be entered, with a few restrictions, in any order. In addition, sequences of frequently used commands (known in GLIM as *macros*) may be named and saved for later use, providing a basic facility for users to define their own procedures.

We have assumed that GLIM is being used interactively. GLIM can also be run in batch mode, that is, by submitting a file of previously assembled commands. Batch use of the package may be appropriate for time-consuming tasks such as the fitting of complex models to large sets of data, but GLIM is primarily designed for interactive use.

1.2   Getting started in GLIM4

On loading GLIM, an introductory message appears:

```
GLIM 4, update 9 for Microsoft Windows Intel  on date at time
      (copyright) 1992 Royal Statistical Society, London
?
```

The question mark which follows the message is the GLIM prompt, which tells you that GLIM is waiting to receive instructions or commands from you. We use the convention of representing *input* to and *output* from GLIM in `typewriter` typefaces, to distinguish them from ordinary text. Instructions are issued to GLIM by using *directives*. Directives all begin with a dollar ($). (If the dollar symbol is not available on your keyboard, try another currency symbol such as £.) The dollar sign provides the means by which GLIM recognizes directive names. Notice that commands are terminated by a $—the terminating $ can be the start of the next directive.

All GLIM directive names can be abbreviated—always to the first four characters, and sometimes further. For example, `$print %pi $` will print the value of π; `$print` can be abbreviated to `$pri` and also to `$pr`. Unlike many other packages, directives do not need to start on a new-line; there can be many directives on one line. If we omit the terminating dollar, then GLIM responds with a prompt, but preceding the prompt symbol is the abbreviated directive name.

```
$PRI ?
```

GLIM is indicating that the end of the `$print` directive has not been reached. Nothing has been printed as the list of items following the `$print` directive has not been terminated by a dollar. In response to this prompt, we can either enter a dollar (to indicate the end of the list of items to be printed) or enter further items.

An on-line help system exists in GLIM through the $manual $ directive. So, for example,

```
$manual print $
```

provides help on the $print directive. The $manual $ directive on its own provides a general help screen giving various help options.

1.3   Reading data into GLIM

We start by introducing a simple set of data. Table 1.1 presents a small set of data, given in Erickson and Nosanchuk (1979), extracted from a larger set collected by Atkinson and Polivy (1976) to examine the effects of unprovoked verbal attack. Ten male and nine female subjects were asked to fill out a questionnaire which mixed innocuous questions with questions attempting to assess the subject's self-reported hostility. A hostility score for each individual was calculated from these responses. After completing the questionnaire, the subjects were then left waiting for a long time, and were subjected to insults and verbal abuse by the experimenter when the questionnaire was eventually collected. All subjects were told that they had filled out the questionnaire incorrectly, and were instructed to fill it out again. A second hostility score was then calculated from these later responses.

For each of the 19 individuals involved in this psychological experiment we have three items of information, or variables: (1) the hostility score of the individual before being insulted; (2) the hostility score of the individual after being insulted; (3) the sex of the individual.

One of the variables (the sex of the individual) is a categorical variable with only two values: male and female. The remaining two variables are continuous—that is, variables which are measured on an underlying continuous scale to a certain measurement accuracy (Section 2.2 contains a further discussion of types

Table 1.1. The insult data set

Hostility		Sex	Hostility		Sex
Before	After		Before	After	
51	58	Female	86	82	Male
54	65	Female	28	37	Male
61	86	Female	45	51	Male
54	77	Female	59	56	Male
49	74	Female	49	53	Male
54	59	Female	56	90	Male
46	46	Female	69	80	Male
47	50	Female	51	71	Male
43	37	Female	74	88	Male
			42	43	Male

of variables). The data are arranged in a data matrix, with the rows of the matrix representing individuals, units or cases and the columns representing the variables recorded for the individuals. There are no missing data; we have complete information on all the variables for each individual.

We define and read this set of data into GLIM by setting up three *variates* to hold the data columns. Any variate, once defined, has a fixed length (that is, it can hold a fixed number of data values). A variate also has a name or identifier, and this name may be chosen by the user. The data values stored in a variate must be numerical; GLIM cannot read non-numeric characters as data. The two continuous variables will cause no problem as their values are numeric, but the categorical variable will need to be coded, associating a different number with each category. There are many ways of coding categorical values. For the categorical variable, we could code males as 0 and females as 1, or females as −1 and males as 1: any two different numbers will do. Here, we code females as 1 and males as 2; the reason for using 1 and 2 will become clear later in this section.

We need to issue directives to GLIM to define the length and names of the three vectors that we wish to define. Once this is done, we need to instruct GLIM to read the data values into these vectors. The directives to do this are as follows:

```
$slength 19 $data hbefore hafter sex $read
51 58 1    54 65 1    61 86 1    54 77 1
49 74 1    54 59 1    46 46 1    47 50 1
43 37 1    86 82 2    28 37 2    45 51 2
59 56 2    49 53 2    56 90 2    69 80 2
51 71 2    74 88 2    42 43 2 $
```

The directive $slength defines the standard or default length of all variates—this does not however prohibit other vectors being defined with lengths other than the standard length.

The $data directive has two functions—it defines the names of the three variates, and tells GLIM that data values following the $read directive should be assigned to these variates. The three variates are called hbefore, hafter and sex. Variate names (and names for all user structures in GLIM) start with a letter and can have up to eight characters; truncation to eight characters occurs if a name is longer. In addition, names are not case-sensitive—HBEFORE is the same as hbefore.

The reading of the data values is initiated by the $read directive. In our example, the $read directive name will be followed by 57 numbers; the data values are read in parallel, one case at a time. For each case, the data values for the vectors in the data list are read in order. Thus, the first data value of hbefore is followed by the first data value of hafter, the first value of sex, the second value of hbefore, and so on. We order the data with the female cases first and the male cases second. Spaces, tabs, commas or new-lines separate the data values, and there is no need to start a new-line for a new case. Note that we do not enter a dollar immediately after the $read directive, as it takes as items the data values of the variates being read in. The terminating dollar follows the last data value.

For the fitting of statistical models and certain other applications, GLIM makes a distinction between two kinds of vectors: *variates* and *factors*. All vectors are initially defined as variates; vectors representing categorical variables may be explicitly redefined as factors by using the $factor directive. For a categorical variable with k categories to be represented as a factor, the categories must be coded using the integers $1, 2, \ldots, k$. We declare sex to be a factor:

```
$factor sex 2 $
```

The $factor directive defines a factor sex with two values (1 and 2) referred to in GLIM as *levels*. Here, we have coded female subjects as level 1 and male subjects as level 2.

We have so far assumed that the data have been typed in interactively, and this method is obviously suitable only for small data sets. In most applications, the data would be entered into a file and stored on disk, allowing the dataset to be edited.

The simplest way of inputting data from a file is to use the $input directive. This simply reads a sequence of GLIM directives stored in a file into GLIM. So a file insult might contain the above sequence of directives for reading the insult data, with the addition of a $return directive as the last directive in the file. This file would be read into GLIM by

```
$input 'insult' $
```

The input file may contain data, or more generally can contain any sequence of GLIM directives terminated with a $return directive. This is useful for developing code and running it interactively. To view the GLIM code on the screen as it is being read in, we precede the $input command by $echo on $. This will echo commands and data from the file to the screen.

1.4   Assignment and data generation

An alternative to the $read directive is the $assign directive, which is used to assign data values to a variate. $assign takes the length of the variate from the number of data values entered and this is often a more flexible way of inputting data. If the variate already exists, then its contents are overwritten and its length reset.

```
$assign hbefore = 51,54,61,54,49,54,46,47,43,86,28,45,
59,49,56,69,51,74,42 $
$assign hafter = 58,65,86,77,74,59,46,50,37,82,37,51,
56,53,90,80,71,88,43 $
$assign sex = 1,1,1,1,1,1,1,1,2,2,2,2,2,2,2,2,2,2,2 $
$factor sex 2 $
```

Here, as before, we have declared the vector identifier `sex` to be a factor after the values have been assigned. `sex` is declared by default to be a variate by the `$assign` directive, then redeclared as a factor by the `$factor` directive.

The `$assign` directive is in fact a general command to assign values and the contents of variates to a variate. On the left of the equals sign (=) is the destination variate; on the right hand side are elements separated by commas or spaces; the elements can be lists of source variates, numbers and number sequences. The destination variate can be the same as one of the source variates, allowing the length of variates to be changed.

For example,

```
$assign x = 1,4 $
```

will define a variate called x of length 2;

```
$assign x = x,3,6,x $
```

will increase the length of x to 6, and will now contain the values $1, 4, 3, 6, 1, 4$. Number sequences are available, and take the form `value1`, `value2,...,value3`. The increment is taken to be the difference between `value1` and `value2`. So, for example,

```
$assign z = 1,4,...,23 $
```

will assign the values $1, 4, 7, 10, 13, 16, 19, 22, 23$ to z, whereas

```
$assign w = 1,...,6,x $
```

will assign the values $1, 2, 3, 4, 5, 6, 1, 4, 3, 6, 1, 4$ to w. Negative increments are permitted.

Factors which have a regular structure can be generated automatically using the `$gfactor` or '*generate factor*' directive. So, for example, we might have a table of counts as follows:

		B					
		1			2		
		C			C		
A		1	2	3	1	2	3
1		10	0	0	2	1	0
2		4	6	2	3	7	1
3		0	3	8	2	4	8
4		3	4	11	1	7	12

We read the values of the table into a variate called count, and then use the $gfactor directive to generate the cross-classifying factors.

```
$assign count = 10,0,0,2,1,0,    4,6,2,3,7,1,
                0,3,8,2,4,8,    3,4,11,1,7,12 $
$gfactor 24 a 4 b 2 c 3 $
$tprint count a,b,c $
```

B	1			2		
C	1	2	3	1	2	3
A						
1	10.000	0.000	0.000	2.000	1.000	0.000
2	4.000	6.000	2.000	3.000	7.000	1.000
3	0.000	3.000	8.000	2.000	4.000	8.000
4	3.000	4.000	11.000	1.000	7.000	12.000

The variate count has length 24, so in the $gfactor directive we first specify a length of 24 for the three factors a, b and c which are to be created. The factor varying least rapidly a is specified first, followed by the desired number of levels of a. This will assign the first six values of a to be 1, the next six to be 2, the following six to be 3, and the final six to be 4. The factor varying next least rapidly is b, which has two levels. The factor b will vary within a, and so the first three values of b will be 1, the next three 2, and then the values of b repeat. Finally, the factor varying most rapidly is c with three levels—this will take the values $1, 2, 3, 1, 2, 3, \ldots$, etc. The $tprint or *tabular print* directive provides a way of examining the data in tabular form—the factors a, b and c can be rearranged into any order.

1.5   Displaying data

We now return to the $print directive, which was introduced in Section 1.2. This directive can also be used to display the contents of variates and factors, and indeed of any structure in GLIM.

```
$print hbefore $
```

```
    51.00   54.00   61.00   54.00   49.00   54.00   46.00   47.00   43.00
    86.00   28.00   45.00   59.00   49.00   56.00   69.00   51.00   74.00
    42.00
```

The data values are printed across the screen, to an accuracy of four significant figures and with nine data values on each line.

The accuracy of four significant figures is the default accuracy; this can be altered by introducing the * symbol followed by a positive number in the range 1–9. For example, *6 will give six significant figures. In addition *-1 specifies that no decimal places are to be printed. This method holds for the duration of the `$print` directive—global accuracy settings can be specified using the `$accuracy` directive.

More than one variate (or factor) may be specified in a `$print` directive, but care must be taken, as the values are printed serially and are concatenated; that is, all the values of the first variate are printed first, followed immediately by the values of the next vector in the list, and so on. A new line is not started for each new variate. It is thus often hard to tell from the output where one vector ends and another begins. We can circumvent this by using separate `$print` statements, as the directive always starts a new-line.

```
$print hbefore $print hafter $print sex $
```

This is rather tedious to type; we can improve the above code by using a colon (:) to repeat the previous directive.

```
$print hbefore : hafter : sex $
```

To display the values of vectors in parallel, the `$look` directive can be used.

```
$look hbefore hafter sex $
```

The values of the three vectors are printed in columns, with each row corresponding to one unit. The unit number is displayed as an additional column on the left. Subsets of the vector values may be displayed in parallel by specifying the lower and upper unit numbers of the range of units immediately following the directive name. For example,

```
$look 5 8 hbefore hafter sex $
```

displays the values of the three variables for cases 5 to 8 only and

```
$look 15 hbefore hafter sex $
```

displays the values of the three variables for case 15.

1.6 Data structures and the workspace

So far, we have seen that GLIM has facilities for dealing with vectors, which are initially declared as variates, and which can be redeclared as factors. We now describe the other data structures available in GLIM—scalars, lists and macros. We will also highlight the difference between user structures and system structures, the latter having pre-assigned names starting with the % symbol.

Scalars are used to store single numeric values, and are declared using the `$number` directive (not the `$scale` directive, which has a special meaning in model fitting). Thus

```
$number nseasons = 4 $
```

declares the scalar nseasons and assigns the value 4 to it. There are also a number of system-defined structures available in GLIM. First, there are 26 *ordinary scalars* with names %a, %b up to %z. These are available for use as scalar storage without the need to declare them. Second, there are 74 system scalars, whose names also begin with a % symbol and which are pre-assigned. We have already seen one of these—the system scalar %pi which contains the value of the mathematical constant π. These system scalars will be introduced and described where necessary—a full list can be obtained through the command $manual scalars $.

Similarly, vectors in GLIM are either user- or system-defined. System vectors are mainly defined after a model fit—for example, there are system vectors %fv (which contains the fitted values from the model) and %rs (which contains the residuals). These are discussed in detail in later chapters—a full list can be obtained through the command $manual vectors $.

Lists in GLIM are used to name ordered list of names. So

```
$list mylist = hbefore,hafter,nseasons $
```

will define a list mylist with three elements—the first two being variates and the last a scalar. They are used primarily in GLIM macros.

Finally, a *macro* is an additional type of GLIM structure which is used for storing text or for storing sequences of GLIM directives. We mention them only briefly in this section as they are not normally used to store data values—Section 1.10 contains a more detailed discussion.

The current contents of the workspace can be discovered through the directive $environment d $, which provides information on all user-defined structures and system vectors existing in the workspace.

GLIM has upper limits on both the number of user identifiers which may be defined and on the amount of workspace which may be used. These limits vary according to the size of the GLIM program which is being run, and the directive $environment u $ may be used to find out these limits. For some large problems, the standard version of GLIM is not large enough, and a larger version is needed—this is achieved through the use of the SIZE parameter when loading GLIM.

The $delete directive saves space by deleting user-defined structures from the workspace. So

```
$delete mylist $
```

will delete the list structure mylist but will leave the structures hbefore, hafter and nseasons untouched. These can be deleted individually if required.

There are two further directives which are useful for managing a GLIM session. The $newjob directive clears the workspace of all defined structures, and will also reset all internal settings back to the GLIM default values. It is used for starting a new piece of work or job while still remaining within the same GLIM session.

The GLIM session is completely ended by using $stop. Output from the GLIM session will be stored in the file glim.log which is in the GLIM working directory. In addition, all commands issued by the user are stored in the journal file glim.jou which is found in the same directory. As these files are overwritten at the start of a new GLIM session, we strongly recommend that they are renamed or moved to another location immediately after finishing a GLIM session if they are to be saved.

1.7   Transformations and data modification

We now describe the directives which are used for general calculation, transformation and modification of variates. The $calculate directive provides facilities in GLIM for assigning, transforming and recoding GLIM variates. However, the simplest use of the $calculate directive (which we abbreviate to $calc) is as a calculator. The GLIM command

```
$calc 5 + 6 $
```

returns the value 11.00; the command

```
$calc 29*9/5 + 32 $
```

(used to convert 29°C to °F) returns 84.20.

The symbols +, -, * and /, when used in arithmetic expressions, are known as arithmetic operators, and represent addition, subtraction, multiplication and division, respectively. Multiplication and division have a higher precedence than addition and subtraction, and parentheses may be used to force evaluation of some expressions before others: $calc (84 - 32)*5/9 + 32 $ forces evaluation of $84 - 32$ first of all.

The results of calculations, instead of being displayed on the screen, may be assigned to a scalar or vector. The equals sign is used to assign the result of a calculation to a GLIM data structure. We first declare a new standard length for vectors by using the $slength directive (Section 1.3)

```
$slen 8 $calc %d = 5 $calc d = 5 $
```

The first $calc statement assigns the single number 5 to the scalar %d. The second $calc statement declares d to be a variate of length 8 (the standard length), and then assigns the number 5 to all elements of d.

Transformations of existing variates can also be carried out. For example

```
$assign x = 1...10 $calc y = 2**(x - 1) $
```

creates a variate y with length 10, with each element of y containing the transformation 2^{x-1} of the corresponding element of x. Assigning the result to a scalar will put the last element of the transformed x into the scalar; thus

```
$calc %e = 2**(x - 1) $print %e $
```

will display 128.0.

The contents of defined vectors may be overwritten at any time by using the $calc directive. Vectors may be transformed ($calc y = (y+1)/2 $) and reassigned ($calc y = x**3 $). All previously defined vectors appearing in the same calculate expression must have the same length. Thus the expression $calc d = x/2 $ produces an error message, as the vector x has length 10, whereas d has length 8.

1.8 Functions and suffixing

A range of functions is provided in GLIM—these can be categorized into structure, mathematical, statistical, logical and other functions. A full list can be found through $manual functions $—help on individual functions is available by issuing the $manual directive followed by the function name, for example, $manual %log $.

1.8.1 *Structure functions*

Structure functions return the characteristics of an identifier, for example, its length (%len), number of levels (%lev) and type (%typ). So

```
$assign x = 1,4...,19 $num lenx $calc lenx - %len(x) $print lenx $
```

will store the value 7 in the scalar lenx.

1.8.2 *Mathematical functions*

GLIM provides a small set of mathematical functions, including the natural log (%log), exponential (%exp), absolute value (%abs), truncation (%tr) and incomplete gamma, digamma and trigamma functions. Thus,

```
$assign x = -1, 0,...,4 $calc lx = %log(x) $
```

will assign the natural log of all units of x to the equivalent units of lx. Where x = 0 or −1 and the natural log is undefined, GLIM continues with the calculation, and assigns the value 0 to all elements where the expression is undefined. A warning message is printed:

```
-- invalid function/operator argument(s)
```

This behaviour of the $calculate directive applies to all other functions and expressions that give an invalid result, such as dividing by zero, or taking the square root of a negative number.

A particularly useful function is the cumulative sum function %cu. The GLIM statement

```
$assign x = 1,2...,6 $calc %y = %cu(x) $calc cux = %cu(x) $
```

will calculate the sum of the values of the vector x and store the result in the scalar %y. If the result is assigned to a vector, rather than a scalar, the vector will contain the partial sums of x.

```
$print cux $
```

```
    1.000    3.000    6.000    10.00    15.00    21.00
```

The function therefore provides one simple method for the calculation of sums and sums of squares of a vector. For example,

```
$calc %m = %cu(x)/%len(x) : %v = %cu((x - %m)**2)/(%len(x) - 1) $
```

will store the mean of the vector x in the scalar %m and its variance in the scalar %v.

The index (%ind) function gives a vector of unit or observation numbers for a variate. For example,

```
$calc index = %ind(hbefore) $
```

1.8.3 *Logical functions and operators*

Logical functions are a particular type of function in GLIM which return only two values, zero (false) and one (true). We give a simple example.

```
$assign x = 13,21,11,36,15,19 $
$calc xgt20 = %gt(x,20) $
$print xgt20 $
```

```
0.000 1.000 0.000 1.000 0.000 0.000
```

The function %gt is used to test whether one item is greater than the other. Equivalently, we could use the logical operator > and the above expression can be written as $calc xgt20 = x>20 $.

There are six logical functions available: equality (%eq or ==), inequality (%ne or ?), greater than or equal to (%eq or >=), greater than (%gt or >), less than or equal to (%le or <=) and less than (%lt or <).

Combinations of the logical functions may be made. Thus, multiplying two logical functions will have the effect of a logical "and" operator.

1.8.4 *Statistical functions*

Statistical functions include the cumulative distribution and inverse cumulative distribution (or probability) functions for a range of distributions. For example, the %nd function gives the deviate of a standard normal for a given probability value, and the function %np gives the cumulative distribution function for a given deviate value. Other distributions provided include the chisquared (%chd and %chp), t (%td and %tp), Poisson, Binomial and beta. Many of the functions have other arguments for distributional parameters and degrees of freedom. Thus, $calc %chd(0.95,1) $ returns 3.84; $calc %np(1.96) $ returns 0.975.

1.8.5 *Random numbers*

Random numbers may be generated by using the %sr function. The most straightforward use of this function is to generate a sequence of uniform random numbers in the range $(0, 1)$:

```
$variate 20 rand rand2 $calc rand = %sr(0) : rand2 = %sr(0) $
```

In this example, both rand and rand2 will contain 20 Uniform $(0,1)$ random numbers. The sequence is predictable—every GLIM session starts with the same seed, and so the same sequence of GLIM commands will produce the same sequence of random numbers. However, repeated use of the %sr function will return different values, so the vectors rand and rand2 will be different. New seed values can be set by using the $sseed directive.

1.8.6 *Suffixes in* calculate *expressions*

All vectors may be suffixed in calculate expressions. So, for example,

```
$assign x = 13,21,11,36,15,19 $calc %a = x(1) $
```

will extract the value of the first element of x and place it in a scalar %a. The suffix can itself be a vector, so

```
$assign i = 1,3,6 $calc sx = x(i) $
```

will create a vector called sx of length 3 containing the 1st, 3rd and 6th element of x. This is one of the methods of extracting certain elements of a vector into another vector—another is the $pick directive, which is described later.

Suffixes can also be used to expand variates. If we want to duplicate each element of x three times, then this could be achieved as follows:

```
$assign i = 1,1,1,2,2,2,3,3,3,4,4,4,5,5,5,6,6,6 $
$calc longx = x(i) $
```

Note that the vector i could be generated by using the $gfactor directive.

1.8.7 *Recoding variates and factors into new factors*

There are two straightforward methods of recoding continuous variables into factors with a fixed number of levels. We use as an example the insult data, and recode the hbefore score into a four category factor hb, defined as follows: less than 50, 50 and less than 65, 65 and less than 75, 75 or greater.

One easy method is to use a single $calculate statement:

```
$calc hb =
     1 + %ge(hbefore,50) + %ge(hbefore,65) + %ge(hbefore,75) $
$factor hb 4 $
```

An alternative is to use the $group directive:

```
$assign cutpts = 50,65,75 $
$group hb = hbefore intervals * cutpts * $
$factor hb 4 $
```

The syntax of the $group directive can be found through the command $manual group $.

When recoding factors into a smaller number of categories, the above methods will not necessarily work, as we may need to recode non-contiguous categories into a new combined category. In the above-mentioned example, we might want to identify groups 1 and 4 of hb as "abnormal" with new code 3, group 2 as "normal" with code 1, and group 3 as "problematic" with code 2. The recoding would then be groups (1, 2, 3, 4) recoded to (3, 1, 2, 3). We can achieve this easily:

```
$assign recode = 3,1,2,3 $calc hbnew = recode(hb) $
$factor hbnew 3 $
```

1.8.8 *Extracting subsets of data*

A common requirement is to extract subsets of data from a larger dataset. Using the insult data as an example, these subsets will usually have particular characteristics, such as 'female only' or those with a before hostility score of 65 or more. We may use the $pick directive for this purpose. For example, to extract a subset of the insult dataset for those values of the 'before' hostility 65 or greater, we use

```
$calc select = (hbefore >= 65) $
$pick hb,ha,sx hbefore,hafter,sex select $
```

A selection vector select is set up with value 1 for the included units, and 0 for the excluded units. The $pick directive then creates three new vectors of appropriate length and type, with names ha, hb and sx, which contain only the included units.

1.9   Graphical facilities

Statistical modelling of data is aided considerably by the graphical display of data. Two directives, $plot and $histogram, provide low-quality character-based scatterplots and histograms but are not covered in this book.

We concentrate instead on the $graph directive, which provides a flexible higher quality alternative for scatter and line plots, and the $gtext directive, which is used to add text to a graph. We return to the insult dataset, and illustrate the directive with a scatterplot of the before and after hostility scores.

The simplest way to produce this plot is $graph hafter hbefore $—the *y*-axis variate is given first, and the *x*-axis variate second. This will produce a scatterplot on the screen. For hard copy output, we use the *postscript* graphics produced by GLIM and which can be displayed and printed using software such as GHOSTSCRIPT, or incorporated into WORD and LATEX documents. The file can be produced at the same time as the screen graphics by using the c = 'post' option in the *option list* to the $graph directive.

Option lists provide greater user control on the actions of some directives. The option name and option setting are separated by an equals sign, and the set of options is enclosed in parentheses. Option names may be abbreviated to one character. We specify the copy or c option, which is set to 'post', and at the same time we label the *x* and *y* axes by using the horizontal and vertical options respectively. The text for the *y* and *x* axes labels is enclosed in quotes. The directive is now

```
$graph (c = 'post' h = 'hostility before' v = 'hostility after')
        hafter hbefore $
```

A scatterplot is produced with axis labelling and using the default plotting symbol of ×.

Postscript output is by default written to a file glimplot.ps which is created in the working directory. Postscript plots are accumulated one by one in this file. However, it is often useful to store each separate graph in a separate file. This can be achieved by adding two directives—$open to create a new graphics file on channel 98—(the postscript channel), and $close to close it. Thus,

```
$open (status = new) 98 = 'insult1.ps' $
$graph (c = 'post' h = 'hostility before' v = 'hostility after' )
        hafter hbefore $
$close 'insult1.ps' $
```

will create a new postscript file called insult1.ps which can be included in reports or other documents.

We can see from Fig. 1.1(a) that there is a strong relationship between the hostility scores before and after the insults. We may want to superimpose a line of no change on the graph. This is most easily produced by plotting hbefore

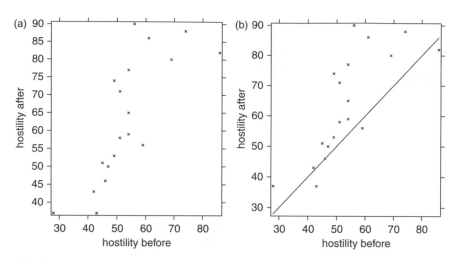

Fig. 1.1. (a) Scatterplot of hafter (*y*-axis) plotted against hbefore (*x*-axis) (b) with superimposed line of no change

against itself, and specifying a second *y*–*x* pair in the graph statement. We do this by having a list of *y*-variates and a list of *x*-variates as the first and second items in the $graph directive, and specifying pen style 1 (the × symbol) for the first pair and pen style 10 (a solid line) for the second pair:

```
$open (status = new) 98 = 'insult2.ps' $
$graph (c = 'post' h = 'hostility before' v = 'hostility after')
        hafter,hbefore   hbefore,hbefore 1,10 $
$close 'insult2.ps' $
```

Figure 1.1(b) shows that most individuals increase their hostility score after the experiment, with only three individuals showing slight decreases. There is also some evidence of increasing spread in the plot as the hbefore score increases. This may be due to different behaviour of the two sexes. However, the plot does not distinguish between male and female subjects; both are plotted with the same symbol. If a factor has been defined with a number of levels, we can choose to identify each of these levels with a different plotting symbol. Here, the factor sex has two levels, and we use the plotting symbol * for females (coded 1) and □ for males (coded 2). These have pen styles 3 and 7 respectively. After the pen styles we include a factor list.

```
$open (status = new) 98 = 'insult3.ps' $
$graph (c = 'post' h = 'hostility before' v = 'hostility after')
        hafter,hbefore   hbefore,hbefore   3,7,10   sex, $
$close 'insult3.ps' $
```

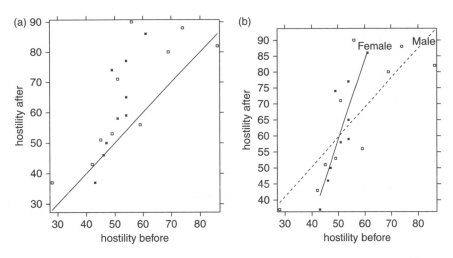

Fig. 1.2. Scatterplot of `hafter` plotted against `hbefore` identifying males (□) and females (*) (a) showing line of no change (b) with separate fitted regression lines for each sex (solid for female, dotted for male)

We now have a list of *y*-variates and a corresponding list of *x*-variates. The first variate in the *y*-list `hafter` is plotted against the first variate in the *x*-list `hbefore`. The first element of the factor list has a variate `sex`, and so we identify the two levels of `sex` with pen styles 3 and 7. Moving on to the second elements of the *y*- and *x*-list, we see that `hbefore` is plotted against `hbefore`, but this time there is no corresponding factor in the factor list, so we plot `hbefore` against `hbefore` with pen style 10. Figure 1.2(b) shows the resulting postscript output.

We have used pen styles without much explanation so far. There are 30 pen styles available; styles 1–16 are predefined. Styles 1–9 are symbol styles with a symbol and colour for each style, whereas styles 10–16 are line styles. The current definition of styles can be found through the `$environment g $` directive (in text form) or the `$environment h $` directive (in graphical form). The output first displays a list of symbol types, line types and colours, and then displays the current line style. Each pen style can be redefined to consist of a combination of a line type, symbol type and colour. So

```
$gstyle 8 line 2 symbol 5 colour 2 $
```

would define pen style 8 to be a dotted red line with circles marking the points. The points are joined up in case order, so the `$sort` directive may be needed to sort the dataset into ascending order of the *x*-variate.

Returning to Fig. 1.2(a) we see that the females appear to have a steeper relationship between before and after scores compared to the males. We can gain some insight by fitting a simple normal linear regression model in GLIM, and plotting the fitted lines. At this stage, we do this without careful investigation of the model—we are interested here in displaying the results of the model in graphical form rather than in statistical model building.

We take the hafter hostility score as the response variate, and fitting hbefore, sex and the interaction between sex and hbefore as explanatory variates. This will give separate regression lines for each sex. After a fit, the fitted values of the model are stored in the vector %fv, and we plot these with the original data. The quality of the graph is improved by plotting the data lines in the ascending order of the x-axis, and so we sort the dataset first (Section 1.11).

```
$open (status = new) 98 = 'insult4.ps' $
$sort hbefore,hafter,sex * sex,hbefore $
$yvar hafter $fit hbefore+sex+sex.hbefore $
$graph (c = 'post' h = 'hostility before' v = 'hostility after')
       hafter,%fv  hbefore,hbefore  3,7,10,11  sex,sex $
```

The $sort directive sorts the three vectors hbefore, hafter and sex into themselves, sorting first by sex and then by hbefore value. As there are 9 female cases and 10 male cases, this means that the largest value of hbefore for females will be hbefore(9) and the largest value for males will be hbefore(19). We then fit the interaction model (Chapter 3) and graph the fitted values. As we want to graph separate fitted lines for each level of sex, we now have two factors in the factor list, and four pen styles in total—two for the observed data and two for the fitted lines.

The resulting plot can be improved by the addition of text annotation on the two lines. We want to add the text 'female' at plotting position (hbefore(9), %fv(9)), and the text 'male' at the plotting position (hbefore(19), %fv(19))—at the end of each of the fitted lines. We proceed as follows:

```
$calc %a = %fv(9) : %b = %fv(19)
    :    %c = hbefore(9) : %d = hbefore(19) $
$ass ytext = %a,%b $ass xtext = %c,%d $number textsize = 1.2 $
$gtext ytext/textsize    xtext    !female', 'male' $
$close 'insult4.ps' $
```

The $gtext directive expects a y and x vector containing plotting positions for each of the text strings. The first element of the y and x vectors ytext and xtext specifies the coordinates for the first text item 'female', and the second element of ytext and xtext gives the coordinates of the second text item 'male'. The scalar textsize specifies the size of the text to be scaled by a factor of 1.2, that is, 20% larger than the default size.

The resulting graph is shown in Fig. 1.2(b). The fitted line for females has a steeper slope than that for males, although no test for statistical significance of the interaction has been carried out.

1.10 Macros and text handling

We now turn our attention to GLIM macros, which may be used to store either text or GLIM code. At any point in a GLIM job, the name of the macro may be used to execute the defined piece of GLIM code, or to substitute a piece of text. The contents of macros are stored as a sequence of characters in the workspace. Macros will therefore compete with vectors for a share of the workspace, and this needs to be taken into account if many long macros are to be defined.

Our first example illustrates the use of macros for storing text strings. For instance, we may want to specify a heading which will appear in any output from the GLIM session. We may specify the text directly in the $print directive

```
$print 'Analysis of White Blood Cell Count Data ....   Run 1 ' $
```

However, if the heading is required to be printed many times in a GLIM job, it is more convenient to use a macro to store the heading as a text string. The directives $macro and $endmacro are used to define a macro, as follows:

```
$macro heading Analysis of White Blood Cell Count Data .... Run 1
$endmac
```

We may now use the equivalent statement

```
$print heading $
```

to produce the desired output.

The $macro directive name is followed by the user-defined macro identifier heading, which in turn is followed by the contents of the macro. The macro definition is terminated by the $endmac directive.

In a similar way, we may define macros to contain text such as option lists or model formulae which may be tedious to repeatedly type in at the terminal. For example, suppose that we wished to produce a series of plots with the x-limits, y-limits and style fixed to constant values. We could define a macro called opt to contain the relevant option list

```
$mac opt (style = 1 xlimits = 0,60 ylimits = 0,60) $endmac
```

We illustrate the use of this macro with the insult data. To produce a plot of hafter against hbefore with the options defined above, we need to substitute the contents of the macro opt in the correct place in the $graph directive. This is achieved by prefixing the substitution symbol # to the macro name:

```
$graph #opt hbefore hafter $
```

The substitution symbol tells GLIM to start reading characters from the macro opt rather than from the current input line. When the end of the macro opt is reached, program control returns to the current input line. The macro `opt` stays defined in the workspace and may be used as many times as required.

The primary function of the macro structure is to store sequences of commands, which may then be executed at appropriate points in a GLIM session. We illustrate with an example to calculate the correlation coefficient r of two variables X and Y with values x_i and y_i, for $i = 1, 2, \ldots, n$.

We use the formula

$$r = \frac{\sum (x_i - \bar{x})(y_i - \bar{y})}{\sqrt{\{\sum (x_i - \bar{x})^2 \sum (y_i - \bar{y})^2\}}}$$

```
$macro correl
   $number len1 len2 mean1 mean2 ss1 ss2 xp corr error $
   $calc len1 = %len(%1) : len2 = %len(%2) $
   $calc error = (len1 /= len2) $
   $fault error 'lengths of input vectors not equal' $
   $calc mean1 = %cu(%1)/len1 : mean2 = %cu(%2)/len2 $
   $calc xp = %cu((%1 - mean1)*(%2 - mean2)) $
   $calc ss1 = %cu((%1 - mean1)**2) $
   $calc ss2 = %cu((%2 - mean2)**2) $
   $calc corr = xp/%sqrt(ss1*ss2) $
   $print 'The correlation coefficient of ' *n %1
         ' and ' *n %2 ' is ' corr $
   $delete len1 len2 mean1 mean2 ss1 ss2 xp corr $
$endmac
```

The macro is defined as before by using the directives `$macro` and `$endmac` to start and finish the macro contents. The macro starts off by first defining nine scalars, which are used by the macro. The items `%1` and `%2` stand for the first and second arguments to the macro which are specified by the user when the macro is executed. The macro is written generally to deal with any two variate arguments of the same length. Lengths of the first and second arguments are then calculated, and a scalar `error` is set to 1 if the lengths are not equal. The `$fault` directive then causes a message to be printed and the macro is stopped if error is non-zero. If the lengths are equal, means and sums of squares terms are then calculated and stored, and the correlation coefficient calculated. Finally the temporary scalars are deleted. The macro is run through the `$use` directive, which is followed by the macro name and the names of any arguments. We wish to calculate the correlation coefficient between `hbefore` and `hafter`.

```
$use correl hbefore hafter $

The correlation coefficient of HBEFORE and HAFTER is    0.7676
```

Many directives are available to enable the efficient use, programming and development of macros. There is a macro editor (`$edmac`), a macro debugger

($debug), switching, looping and user prompts. These are not discussed in detail here.

There are many purpose-written macros available in GLIM. Some of the most useful are to be found in the GLIM macro library, which is provided with GLIM. The GLIM macro library has a subfile structure (that is, a single input file is divided into sections recognized by GLIM), and it is automatically available on input channel %plc. For example, to read in the Box–Cox macros from the macro library, we use

```
$input %plc boxcox $
```

This will read in the subfile boxcox from the macro library. Help on using the macro is displayed on the screen as the macro is read in. A full list of macros in the library and their associated subfiles can either be obtained by $manual library contents $ or by $input %plc inform $—both give the same information.

1.11 Sorting and tabulation

We finish the chapter by a short discussion of the facilities for sorting and tabulation. Sorting is provided through the $sort directive. The simplest use is to sort a vector into itself in ascending order:

```
$assign x = 3,4,2,7,8,1,1 $sort x $
```

However, usually, both the original and the sorted versions of a vector are required, and a destination vector can be specified.

```
$assign x = 3,4,2,7,8,1,1 $sort newx x $
```

The newly-created vector newx will now contain the sorted (ordered) values of x.

The most common use of the $sort directive is to sort sets of variates and factors into another new set according to the values of a particular vector. We may use this to create a new insult dataset ordered according to the hbefore hostility score. This is achieved by:

```
$sort nhbefore,nhafter,nsex   hbefore,hafter,sex   hbefore $
```

Lists of vectors are now specified as the first two items of $sort. The first list is the destination list—the new ordered vectors created by the directive. The second list is the source list—a list of unordered vectors. Finally the third item is a single vector—the key—which determines the order of the first list. In this example, we are creating new ordered vectors nhbefore, nhafter and nsex from hbefore, hafter and sex, sorted according to the ascending order of hbefore.

The $sort directive may also be used to find the maximum, minimum and median values of a vector by extracting the relevant elements of the sorted vector. For example,

```
$num min med max len $calc len = %len(nhbefore) $
$calc min = nhbefore(1) : max = nhbefore(len)
   :   med = nhbefore((len + 1)/2) $
```

would store the minimum value of the vector hbefore in the scalar min, the maximum in max and the median (assuming an odd number of values) in med. However, a more convenient way which avoids the use of calculate expressions is to use the $tabulate or $tab directive, which we now introduce.

```
$tab the hbefore smallest $
$tab the hbefore largest $
$tab the hbefore fifty $
```

The smallest, largest and median values are displayed on the screen.

In this simple use of the $tabulate directive, we are finding summary statistics, and this is achieved by using the phraseword the, the name of the variate and the name of the statistic which is required. Other statistic names available are deviation (the standard deviation), total (the total) and mean (the mean).

The result of any tabulation can be stored rather than displayed on the screen by using a suitable *output-phrase*. The output phraseword into corresponds to (is matched to) the input phraseword the.

```
$tab the hbefore smallest into minhb $
$tab the hbefore largest into maxhb $
```

By default, minhb and maxhb are created as variates of length 1, but scalars can also be specified as output structures if predefined.

We may also use the $tabulate directive to produce a table of counts.

```
$tab for sex $tab for sex using count $
```

```
        SEX    1      2
   TOTAL_WT   9.00  10.00
```

The for input-phrase is used to display a table containing just two cells—we see that there are nine subjects with sex = 1 (female) and ten subjects with sex = 2 (male). How many of each group have a raised initial hostility score? We create a new variate raised which has the value 1 for those subjects with hbefore 65 or more, and 0 otherwise, and use the $tabulate directive. As before, we use the for phraseword to produce the table, but this time we specify astyle of 1 as an option.

```
$calc raised = (hbefore >= 65) $tab (style = 1) for sex, raised $

  -- the table contains empty cell(s)
              +----------------+
   RAISED |   0.000    1.000 |
      SEX |                   |
  +--------+----------------+
  |      1 |   9.000    0.000 |
  |      2 |   7.000    3.000 |
  +--------+----------- ---+
```

A table with four cells is produced, preceded by a warning message, as one of the counts is zero. Three of the males have raised initial hostility score, and none of the females. Note that it is not necessary for raised to be declared as a factor, or even to be coded as a factor. An output variate can be specified: the using output-phrase is matched to the for input-phrase.

```
$tab for sex, raised  using cells$
```

The variate cells is constructed to be of length 4. However, we may be unsure in which order the vector is formed (in fact, the first two values of cell correspond to low and raised scores for females)—and usually we need vectors to cross-classify or label the output counts. We may achieve this by using the by output-phrase to contain a list of new vectors which will contain classification labels for the table. These vectors, which we name tsex and traised, will also be of length 4.

```
$tab for sex, raised  using cells by tsex,traised $
$look cells tsex traised $

        cells    tsex traised
   1    9.000   1.000    0.000
   2    0.000   1.000    1.000
   3    7.000   2.000    0.000
   4    3.000   2.000    1.000
```

We see that the vector tsex contains the values 1 and 2, whereas traised contains the values 0 and 1. As sex is a factor, tsex is automatically declared as a factor. raised, however, is declared as a variate. The types of the vectors specified in the for phrase are transferred to the output vectors specified in the by phrase.

We may use the $tprint directive to print a more informative version of the above table. We specify a style of 1 in the option list.

```
$tprint (style = 1) cells  tsex,traised $
```

Combinations of input phrases may be used to produce tables of summary statistics. For example, to produce a table of means of the variate `hafter` rather than cell counts, we use

```
$tab the hafter mean for sex,raised $
```

Output phrases may additionally be used to store the vector of means, the associated vector of counts and the cross-classifying vectors. All the output vectors will again be of length 4. We may then use $tprint to print both the cell means of hafter and the associated cell counts in one table.

```
$tab the hafter mean for sex,raised into tmean by tsex,traised $
$tprint (style = 1) tmean,cells  tsex,traised $
```

```
                    +----------------+
        TRAISED |   0.000    1.000 |
    TSEX        |                  |
+------------+----------------+
|    1 TMEAN |   61.33     0.00 |
|      CELLS |   9.000    0.000 |
-----------------------------------
|    2 TMEAN |   57.29    83.33 |
|      CELLS |   7.000    3.000 |
+------------+----------------+
```

Thus, the $tabulate directive provides a flexible way of summarizing individual level data into aggregate data suitable for subsequent analysis. It is used extensively in Chapters 4 and 5 to aggregate individual level data into contingency tables and to collapse existing contingency tables.

2
Statistical modelling and inference

2.1 Statistical models

In this book we will be concerned with the statistical analysis of data from experimental and observational studies. In experimental studies, randomization or random assignment of individual experimental units (e.g. human or animal subjects, agricultural plots) to the experimental treatments plays a fundamentally important role.

The experimental units may themselves be sampled randomly according to some sample design from a larger population, in which case conclusions from the experiment can be drawn about the larger population, or they may be all that is available for the experiment (e.g. all patients with a particular disease being assigned to treatments in a hospital clinical trial), in which case conclusions may not be generalizable to a broader population.

In observational studies, observational units are drawn from a population according to some random sample design (which may be the complete examination or *census* of the entire population), and conclusions are to be drawn about the complete population.

In both kinds of study, the random selection or allocation of observations is critical to the analysis and interpretation of the data. Studies which are not based on sample or experimental designs, that is, in which the data collected are accidental, or are subjectively chosen, are particularly hazardous to interpret because of the possibility that inclusion in the study is systematically related to important variables, so that the data are not representative of the population about which conclusions are to be drawn.

The object of statistical modelling is to present a simplified or smoothed representation of the underlying population. This is done by separating systematic features of the data (e.g. sex differences in height or weight) from random variation (natural variability in height or weight within sex). The systematic features are represented by a regression function involving parameters which can be simply related to the structure of the population and to important variables measured on each experimental or observational unit. The random variation unrelated to important variables is represented by a probability distribution depending on a small number of parameters, typically one or two. Interpretation of the data, and conclusions about the population, can then be based on the regression function, with no attempt to interpret random variation.

How do we distinguish between random and systematic variation? This is a familiar problem in hypothesis testing: we suppose first that variation is random, and then test whether the introduction of a systematic effect into the model improves the model, in a well-defined way. If the improvement is substantial, the systematic effect is retained and can be interpreted: if it is only minor, the systematic effect is not retained and is not interpreted. How large an improvement is substantial is partly a subjective matter, but well established rules in hypothesis testing provide a good guide. The importance of hypothesis testing results from a general philosophical approach to scientific inference by statisticians and other scientists, based on the principle of parsimony, known historically as Occam's Razor (from William of Occam or William Ockham, 1280–1349): "entia non sunt multiplicanda praeter necessitatem"—entities should not be needlessly multiplied. Systematic effects should be included in a model only if there is convincing evidence of the need for them: we should not spend time and effort interpreting effects which could just as well be random variation. The principle of parsimony provides an important guide to model simplification: though we may use very complex models for complex sample survey designs, our aim is always to simplify the model as much as possible, while remaining consistent with the observed data.

An important distinction needs to be made between the use of models for *representation* of an existing population, as we have just described, and their use for *prediction* of values of variables for future observations. These uses of models are often confused. The use of models for prediction will be considered in Chapter 3.

It is assumed in subsequent discussions that the data being modelled consist of a random sample of some kind from a population. In some cases, as noted earlier, the data available are a complete enumeration of a population. In such cases the simplification of models using hypothesis testing seems quite artificial, because no sampling is involved, and we already know whether or not the hypothesis is true.

There are two possible approaches to the statistical modelling of complete populations. One is to regard the population as a sample from a random process generating this population, and to model the population as though it were a sample. Such an approach is called *super-population* modelling (Smith, 1976) although no concept of sampling the observed population from a hypothetical superpopulation of populations is actually necessary.

The other approach, which we shall follow, is to regard models simply as smooth approximations to the rough, irregular complete population. The irregular variations about the smooth structure are treated as non-systematic variation which are to be ignored. The degree to which the population can be smoothed is now entirely subjective, since the formal rules of hypothesis testing are not appropriate, but smoothing to the extent appropriate if we had a sample, and not a population, will often be reasonable unless the population size is small.

2.2 Types of variables

The data for which models can be constructed in GLIM consist of values of a *single response variable*, which we will denote by y, and a set of *explanatory* or *predictor* variables $x_1, x_2, \ldots, x_p$, which are assumed to be measured without error and recorded without missing values. These important restrictions are discussed in Chapter 3 and other chapters, and are removed to some extent there. GLIM cannot fit models to multivariate response variables, except for multi-category variables (Chapter 5), and responses linked by a common random effect (Chapter 9). The terms "dependent" and "independent" variables which are often used are confusing (since the "independent" variables are not statistically independent) and will not be used in this book. The response and explanatory variables can take a wide range of forms, as set out in Table 2.1.

Discussions of measurement in the social sciences frequently use Stevens's (1946) classification of nominal, ordinal, interval and ratio scales. Nominal scales correspond to the first two variable types above and ordinal to the third. The distinction between interval and ratio scales (the latter has a fixed zero, the former does not) is not very useful, since it is often not relevant to the kind of analysis which is appropriate.

Quotation marks are used around "continuous" above because all such variables are in practice measured to a finite precision, so they are actually discrete variables with a large number of numerical values. For example, height may be measured to the nearest half-inch or centimetre, survival time to the nearest day, week or month, and Stanford–Binet IQ may be given as an integer though it is defined as the ratio of mental age to chronological age multiplied by 100. Models for "continuous" data usually ignore this measurement precision, but this leads to unnecessary difficulties with statistical theory. Our formulation of models explicitly recognizes this aspect of "continuous" data.

Table 2.1. Types of variables

Variable type	Examples
Categorical, two categories (binary)	Male/female, alive/dead, presence/absence of a disease, inoculated/not inoculated
Categorical, more than two unordered categories	Blood group, cause of death, type of cancer, political party vote, religious affiliation
Categorical with ordered categories	Severity of symptoms or illness, strength of agreement or disagreement, class of University degree
Discrete count	Number of children in a family, number of accidents at an intersection, number of ship collisions in a year
"Continuous"	Height, weight, response time, survival time, Stanford–Binet IQ

2.3 Population models

In this book we use probability models as simplifying approximations to real or conceptual populations of response variables. In the experimental or survey study of the population, we measure or observe a sample of individuals, the value of the quantity or property measured for the i-th individual is denoted by the variable value $y_i, i = 1, \ldots, n$. In the population we have the set of values Y_J^*, with $J = 1, \ldots, N$, where N is the population size. These are not in general all observed, though we may be able to take a census of the whole population in some cases.

Because of the limited precision of any physical measurement, or because of natural discreteness in counting measure, the Y_J^* are generally not all distinct, and we represent them alternatively in terms of the D *distinct* values Y_I which occur with multiplicity N_I, for $I = 1, \ldots, D$, with $\sum_{I=1}^{D} N_I = N$. We adopt the convention that values Y which do not occur in the population are not included in the set of distinct Y_I (though it may sometimes be convenient to include them with multiplicity zero). When the quantity Y is measured on an *ordered* scale, we take the Y_I to be in *increasing order* with I.

The values $P_I = N_I/N$ are the *relative frequencies* or *proportions* of the distinct values Y_I. A graph of the P_I against the Y_I is a *proportion graph*, and one of the N_I against the Y_I, is a *frequency graph*. These graphs differ only in the vertical scale.

More common in introductory courses is the *histogram*, which gives a solid bar plot of the frequencies N_I against the Y_I; these are almost invariably *grouped* into class intervals of Y, either by default or by user specification. The macro ghist in the subfile hist in the GLIM macro library provides a high quality histogram for ungrouped data, with many user-set options. (The macro hist does the same for grouped data; confusingly, the $histogram directive in GLIM gives only a low-quality plot.)

For example, Fig. 2.1 shows the histogram of the counts of birthweights of 648 girls (part of the STATLAB population of Hodges *et al.*, 1975) against birthweight, using the macro defaults. Weight is measured to the nearest 0.1 pound (1 pound = 0.4536 kg), so that a recorded value of 6.4 pounds corresponds to an actual value between 6.35 and 6.45 pounds. The GLIM directives are:

```
$input 'gbirthwt' : %plc hist $
$use ghist bwt $
```

The default grouping gives about 20 class intervals. Smoothing using class intervals results in a loss of information or "resolution". Minimal smoothing results from choosing the interval width equal to the measurement precision of 0.1 pound. Inspection of the data, for example, using

```
$tabulate the bwt largest : the bwt smallest $
```

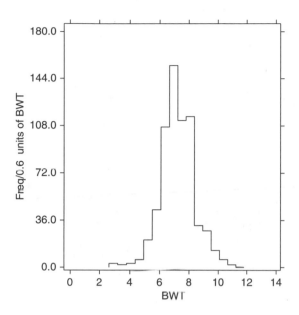

Fig. 2.1. Default birthweight histogram, girls

shows that the range of girl birthweights is from 2.3 to 10.9 pounds. We construct the histogram over a slightly larger range, as an optional argument to the ghist macro, with other unused arguments replaced by asterisks. The styleopt macro gives a border around the graph.

```
$assign bg = 2.0,2.1,...,11.5 $
$macro styleopt s = 1 $endmac
$use ghist bwt bg * * styleopt * $
```

The histogram is now very irregular, though the two histograms suggest a generally symmetrical population model like the normal (Fig. 2.2). The direct modelling of the frequencies to identify a suitable population model was proposed by Lindsey (1974, Lindsey and Mersch, 1992), and a simple example was given in Aitkin (1995). The modelling approach we follow in this book is to construct the *cumulative proportion* graph, in which we plot the cumulative proportions

$$C_I = \sum_{I' \leq I} P_{I'} = \sum_{I' \leq I} N_{I'}/N$$

against the Y_I. We will, in fact, slightly change this definition, and define

$$C_I = \sum_{I' \leq I} N_{I'}/(N+1).$$

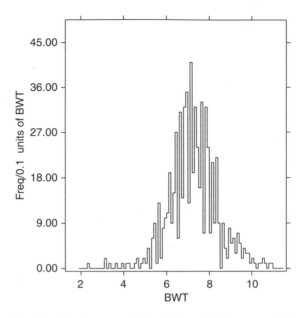

Fig. 2.2. Maximum resolution birthweight histogram, girls

The division by $N + 1$ instead of N is to prevent the last value being 1.0, which causes difficulty in transformations of the probability scale. The difference in the definitions is at most $1/(N + 1)$, which for large N is negligible. (Other definitions are also commonly used, like

$$C_I = \left(\sum_{I' \leq I} N_{I'} - \frac{1}{2} \right) / N.$$

These give very similar results.)

Figure 2.3 shows the cumulative proportions plotted against birthweight. We first use tabulate to save the counts at each birthweight into a variate n.

```
$tabulate for bwt into n by y $
$calc cn = %cu(n) : c = cn/649 $
$graph (s = 1 h = 'birthweight' v = 'cumulative proportion')
      c y 5 $
```

The roughness of the maximum resolution histogram is greatly reduced.

The theoretical or experimental questions we have about the population values Y_I are generally expressed in terms of *parameters*: means or medians, or variances or standard deviations, rather than in terms of the individual proportions P_I themselves. In sample surveys of consumer purchasing, for example, we may be

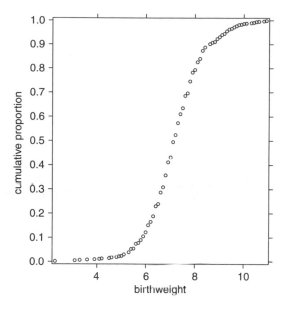

Fig. 2.3. Birthweight cumulative proportion, girls

interested in inference about the average family income, which is

$$\mu = \sum_{I=1}^{D} P_I Y_I$$

for the population.

The role of *statistical models* is to aid this process of inference about parameters like μ by making a *simplifying assumption* about the proportions P_I. Specifically, we assume that the cumulative proportions C_I at Y_I can be represented approximately by a *smooth function* $F(Y_I)$, the *cumulative distribution function* (cdf) of a random variable Y:

$$C_I \doteq F(Y_I).$$

A question of immediate interest is whether this approximating process is *necessary*—can inferences about parameters like μ be made *without* modelling or approximating the C_I? Remarkably, likelihood inferences (defined in Section 2.5) *can* be made without such approximations, through *empirical likelihood* (Owen, 1988, 2001); this is described briefly in Section 2.10.2, together with other non-model-based inferential methods. In this book we concentrate on explicit models for the C_I or P_I.

In general, the approximating cdf F will depend on one or more parameters θ which specify the location and variation of the random variable Y. If the quantity y

could in theory be measured on a continuous scale but is recorded only to a precision of δ, then we can represent the distribution of Y equivalently by its *probability density function*, which we write for a general argument y as $f(y)$:

$$f(y) = \frac{dF(y)}{dy}.$$

The proportions P_I can then be represented approximately by

$$P_I \doteq F(Y_I + \delta/2) - F(Y_I - \delta/2) \doteq \delta f(Y_I).$$

If y is measured on an inherently discrete scale (e.g. the non-negative integers) then the distribution of Y can be represented equivalently by its *probability mass function*

$$f(y) = F(y) - F(y - \delta) = F(y + \delta/2) - F(y - \delta/2),$$

where δ is the discrete measurement scale unit (1 if y is defined on the integers). The proportions P_I are then represented approximately by $P_I \doteq f(Y_I)$. (The second equality above may seem unusual—it is adopted for consistency of notation with the continuous case.)

The choice of the approximating model is based on the closeness of agreement between the population proportions C_I and the approximating model proportions $F(Y_I)$. The appearance of the histogram for the girl birthweights, and the general symmetry of the cumulative proportion graph, suggest a normal distribution model for the C_I. Figure 2.4 shows the normal cdf $F(y)$

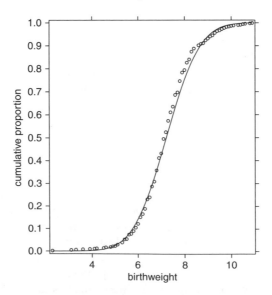

Fig. 2.4. Birthweight cumulative proportion, girls

for the normal distribution with mean 7.22 and standard deviation 1.14, the values for this variable in the population (continuous curve) superimposed on the proportions C_I. The fitted normal cdf is defined by the GLIM directives

```
$calc np = %np((y - 7.22)/1.14) $
$graph (s = 1 h = 'birthweight' v = 'cumulative proportion')
        c,np y 5,10 $
```

Because of the non-linearity and the upper and lower bounds of the cdf, it is difficult to assess the agreement between C_I and $F(Y_I)$. It is therefore convenient to transform the cdf graph by adopting standard probability plotting procedures familiar from residual analysis in normal regression. For a continuous approximating model $f(y)$ we look for scales for F and Y on which the plot is approximately linear. We try to transform the scales so that $g(F)$ is linear in $h(Y)$: $g(F)$ is generally F^{-1}, so that $g(F)$ are the *quantiles* of the approximating distribution F. For a discrete model $f(y)$ such procedures are less satisfactory, but for both continuous and discrete models the goodness-of-fit of the model to the population can be assessed by further modelling.

We try the *inverse normal cdf* scale, or the *equivalent normal deviate* scale, for F, and leave Y unchanged.

Figure 2.5 shows the cumulative proportion graph of girl birthweights using the inverse normal cdf scale for the proportion, together with the straight

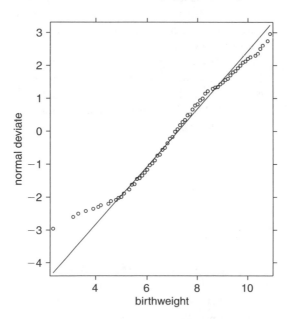

Fig. 2.5. Normal deviate of cumulative proportion, girls

line corresponding to the mean 7.22 and standard deviation 1.14 of this variable. The GLIM directives are:

```
$calc nd = %nd(c) $
    : fn = (y - 7.22)/1.14 $
$graph (s = 1 h = 'birthweight' v = 'normal deviate')
        nd,fn y 5,10 $
```

The plot is generally linear, suggesting a normal distribution model for birth-weight, but there are notable departures from the line for both small and large babies. An acute difficulty in the inspection of such plots is how to decide whether the variation in the plot is too far from a straight line. We are concerned in this section with models for complete populations, but we follow the approach of treating the population as though it were a sample; this may be formalized as a superpopulation model. We will be applying the same approach to samples, and it is especially easy with small samples to overinterpret sampling variation as real model failure. Simulation of samples from normal population models gives some feeling for the behaviour of such plots in small samples, and *simulation envelopes* (Atkinson, 1985) formalize this variation.

In the previous edition of this book, we used the Filliben test (Filliben, 1975) based on the curvature of the plot. This can be used only for normal population models, and it is more useful to have a visual bound on the reasonable sampling variation of the plot. This can also be obtained from the *binomial distribution*, and provides a simultaneous confidence region or band for the population proportions (Owen, 1995). The band construction is discussed briefly in Section 2.7.5.

The band is constructed in the macro binocisim, which requires two argu-ments: the name of a variate holding the cumulative frequency counts at each Y, and the total sample size (as a named vector). The macro computes the upper (pu__) and lower (pl__) endpoints of the 95% simultaneous confidence interval for each Y; these vectors can then be graphed to give the confidence band, together with the fitted normal cdf. The simultaneous band is wider than that based on *pointwise* 95% confidence intervals at each Y.

```
$input 'binocisim' $
$calc %a = %len(n) $
$var %a nn $
$calc nn = 649 $
$use binocisim cn nn $
$graph (s = 1 h = 'birthweight' v = 'cumulative proportion')
      pl__,pu__,np y 9,9,10 $
```

Figure 2.6 repeats Fig. 2.4 with the 95% confidence band instead of the observed proportion, using dot characters for the band, with the fitted normal cdf as a continuous curve. (Note that the vectors nn and cn are not of length 648; we do not need to know their actual length—the number of distinct Y values

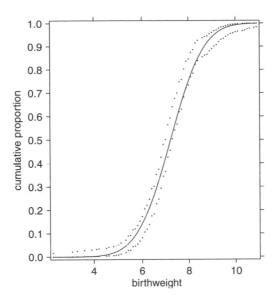

Fig. 2.6. Birthweight cdf and bounds, girls

in the data—as this can be computed using the function %len, the length of the vector. Printing the value of %a provides this value, it is 67.)

The band provides a clear picture of the sampling variability in the cdf to be expected for the given sample size. Systematic crossings of the band by the model cdf indicate failure of the model to adequately represent the data: the cumulative proportions from the fitted model for the population proportions should all fall inside this band for a satisfactory model.

It is difficult to read from the graph whether points fall outside the band. Since we are generally working on a transformed scale (the inverse normal scale in this example), it is convenient to transform the band endpoints in the same way to corresponding band endpoints on the transformed scale. At the extreme values of the cdf and with large N the confidence band may reach the limits 0 and 1; these points cause difficulty when the band is transformed, and are weighted out of the graphs below using the weight option in graph.

```
$calc npl__ = %nd(pl__) $
    : npu__ = %nd(pu__) $
    : wl = %ne(pl__,0) : wu = %ne(pu__,1) $
$graph (s = 1 h = 'birthweight' v = 'normal deviate')
        npl__/wl,npu__/wu,fn y 9,9,10 $
```

Figure 2.7 repeats Fig. 2.5 with the corresponding transformed band. It is now clear that the fluctuations in the cdf are not consistent with a normal model— the apparent departures from linearity for small babies are outside the band

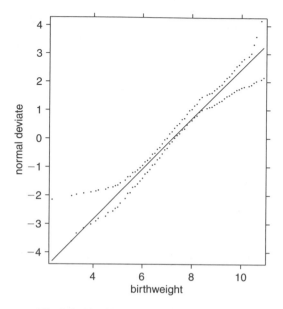

Fig. 2.7. Normal deviate and bounds, girls

range, and those for babies around 8 pounds nearly outside it. The normal popula-tion model is an unsatisfactory fit: there are more large and small babies than are consistent with a normal population model.

What alternative model or models should we consider? A model with heavier "tails" is needed; the central part of the confidence band appears consistent with a normal model, but the tails are not consistent with the same model, though they appear to be consistent with other normal models. In Chapter 7 we discuss *mixtures* of distributions; here we consider the *mixture of normals* distribution as a possible model.

Figure 2.8 shows a plot on the inverse normal cdf scale of the cdf of a three-component normal mixture model, with means 9.13, 7.08 and 3.89 pounds and common standard deviation 0.865 pounds, in corresponding proportions 0.097, 0.886 and 0.017, together with the 95% confidence band. (The mixture distribu-tions described here are fitted by maximum likelihood. See Chapter 7 for a full discussion.)

The mixture cdf follows the population cumulative proportions quite closely, the fitted mixture cdf lying fully inside the band, so the normal mixture distribution provides a good model, with the appealing interpretation of a "normal" sub-population of about 89% with mean 7.1 pounds, a "high birth-weight" sub-population of about 10% with mean 9.1 pounds, and a "low birthweight" sub-population of about 2% with mean 3.9 pounds. This popula-tion representation does not take account of possible explanatory factors like the mother's age and weight, which when modelled might explain the low and

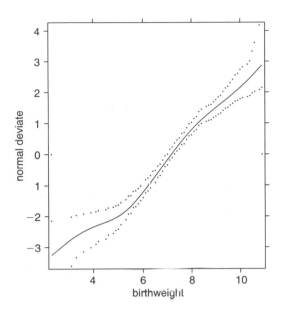

Fig. 2.8. Mixture cdf and bounds, girls

high birthweight groups. Such investigations are the subject of regression analysis or linear modelling; in Chapter 7 we examine these variables.

We now examine the birthweights of the 648 boys in the STATLAB population. We do not repeat the GLIM directives. Figure 2.9 shows the maximum resolution birthweight histogram; the range of boy birthweights is 3.9 to 14.1 pounds. The cumulative proportion graph is shown in Fig. 2.10 with the superimposed normal cdf with mean 7.65 and standard deviation 1.12, the values for birthweight in the boy population. The fitted normal model is shown in Fig. 2.11 on the inverse normal cdf scale with the 95% confidence band. Figure 2.11 shows some departures from the straight line for large babies, with one extreme outlier weighing 14.1 pounds, though all points except the extreme one fall inside the band.

Figure 2.12 shows a three-component normal mixture distribution with means 14.0, 7.7 and 6.9 pounds and common standard deviation 1.08 pounds in corresponding proportions 0.002, 0.949 and 0.049, on the inverse normal cdf scale, together with the 95% band.

The fit is improved for the high birthweight babies, mainly because the extreme outlier is assigned its own component; note that $1/648 = 0.0015$. The fit for low birthweight babies is hardly changed, because the two lower components differ in mean by only 0.69 standard deviations, and so are not really distinguishable: the model is effectively a two-component mixture with one component for the single outlier. (This point is discussed further in Chapter 7.) The normal mixture distribution is little better than the single normal model. In Chapter 7 we discuss how to establish the need for a mixture distribution.

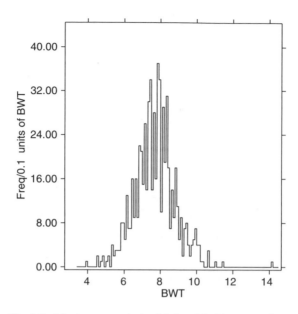

Fig. 2.9. Maximum resolution birthweight histogram, boys

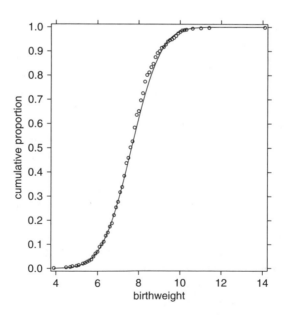

Fig. 2.10. Birthweight cumulative proportion, boys

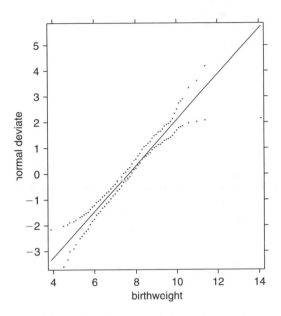

Fig. 2.11. Birthweight cumulative proportion, boys

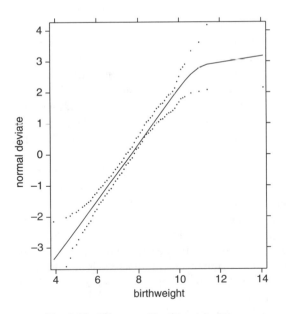

Fig. 2.12. Mixture cdf and bounds, boys

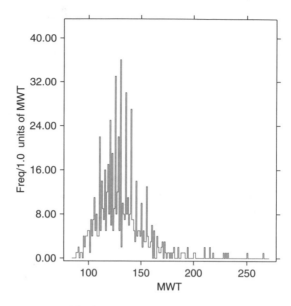

Fig. 2.13. Mothers weight, boys

We now consider models for *skewed* data. We omit the GLIM directives which parallel those above.

Figure 2.13 shows the maximum resolution histogram of weights of the mothers (at the diagnosis of pregnancy) of the 648 boys in the STATLAB population. The histogram is skewed to the right. The cumulative proportion graph is shown in Fig. 2.14. The roughness of the histogram is again greatly reduced, and skewness of the cumulative proportions is visible as asymmetry, with several large values.

In Fig. 2.15 we graph the 95% simultaneous confidence band on the inverse normal cdf scale against weight, together with the straight line corresponding to the normal distribution with mean 130.9 and standard deviation 22.6, the values for this population. The fit is very bad, with much of the fitted model lying outside the confidence band. The consistent *single curvature* of the cdf points strongly to a positively skewed population model. A log transformation of the weight scale before fitting the normal model is a common method for reducing the skew; this is equivalent to using a *lognormal* population model.

In Fig. 2.16 we graph the 95% confidence band on the inverse normal cdf scale against log weight with the fitted (log)normal model. The fit is improved but the lognormal model still falls outside the band at the upper end. The central region of the fitted model is nearly straight but the upper tail is not: the lognormal distribution is not a suitable model. The change in slope of the upper end of the graph suggests a mixture distribution.

In Fig. 2.17 we graph the cdf of a two-component mixture of lognormals on the inverse normal scale, with the 95% confidence band. The fitted distribution

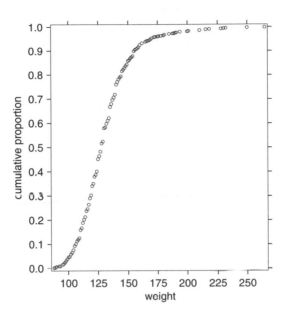

Fig. 2.14. Weight cumulative proportion

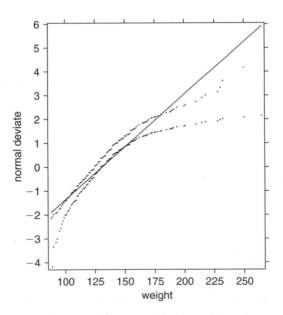

Fig. 2.15. Weight normal deviate and bounds

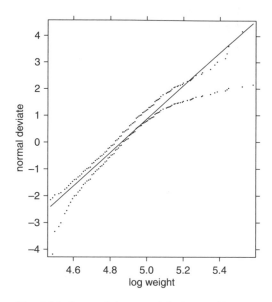

Fig. 2.16. Log weight normal deviate and bounds

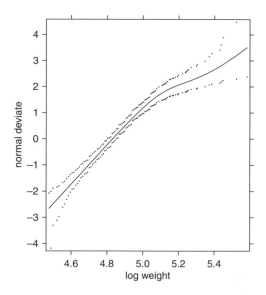

Fig. 2.17. Normal mixture and bounds, log weight

lies near the centre of the band over the whole range, and provides a good model. The two components have means (on the log weight scale) of 5.29 and 4.85, and common standard deviations 0.136, and are in the proportions 0.038 and 0.962. The two components are separated by more than 3.25 standard deviations, and so are nearly distinct. The mothers' weight population appears to be made up of two distinct groups, a "normal" subpopulation (of 96%) with median weight 128 ($e^{4.85}$) pounds, and a "high weight" subpopulation (of 4%) with median weight 198 ($e^{5.29}$) pounds. Each of these groups has a lognormal distribution of weights.

2.4 Random sampling

Almost all investigations in which statistical modelling plays a fundamental role depend on a sampling design of some kind to obtain data from the target popu-lation, either by a formal process of random selection from a population list, or by an act of randomization in the assignment of patients or experimental units to different treatments. The purpose of the random selection or assignment is to ensure representativeness of the population by the sample as far as possible. In developing statistical inference about the population from the observed sample, a critical assumption is that any informative sample design is explicitly represented in the probability model for the data, and hence in the likelihood. Thus, if observations on some sample individuals are missing, the analysis of the observed data by ignoring the missing observations is valid only if the missing data are missing at random (MAR in the notation of Little and Rubin, 1987) and the "missingness process" is ignorable. If not, then the missingness process, that is, the sample design, must be explicitly represented in the likelihood. This is discussed further in Chapter 3.

In the discussion in this book, simple or stratified random sampling will almost always be assumed, except in the discussion of variance component models in Chapter 9, where multi-stage sampling is assumed. In the remainder of this chapter we consider only simple random sampling; stratified sampling is discussed in Section 3.3.

2.5 The likelihood function

We now adopt a consistent notation for the rest of this book. The variable for which a probability model is to be adopted will be denoted by Y, with generic observed values y. These are measured to a finite measurement precision δ, so every possible observed y has a probability modelled by

$$p(y) = \Pr(y - \delta/2 < Y < y + \delta/2)$$
$$= F(y + \delta/2 \mid \theta) - F(y - \delta/2 \mid \theta),$$

where $F(y \mid \theta)$ is the approximating cdf corresponding to the density or mass function $f(y \mid \theta)$ depending on the parameter θ. If Y is inherently discrete (e.g. defined on the non-negative integers) then

$$p(y) = f(y \mid \theta),$$

while if Y is theoretically continuous then if the measurement precision is high (δ small compared with the effective range of Y),

$$p(y) \doteq \delta f(y \mid \theta).$$

If the measurement precision is *not* high, for example, when Y is heavily grouped, the previous cdf definition must be retained.

Observed sample values will be denoted by $y_1, \ldots, y_n$, and when necessary, unobserved random variables corresponding to sample values will be denoted by $Y_1, \ldots, Y_n$. We assume unless otherwise stated that the population size N is large compared to the sample size n. (The consequences of a small population size are discussed later in this chapter.) We may then treat the sample $y_1, \ldots, y_n$ as having been drawn randomly with replacement. The likelihood function is now defined as the probability of obtaining the actual sample values $\mathbf{y} = y_1, \ldots, y_n$, under the approximating model F:

$$
\begin{aligned}
L(\theta \mid \mathbf{y}) &= \Pr(y_1, \ldots, y_n \mid \theta) \\
&= \prod_{i=1}^{n} p(y_i) \\
&= \prod f(y_i \mid \theta), \quad Y \text{ discrete} \\
&= \prod [F(y_i + \delta/2 \mid \theta) - F(y_i - \delta/2 \mid \theta)], \quad Y \text{ continuous}, \quad \delta \text{ large} \\
&\doteq \delta^n \prod f(y_i \mid \theta), \ Y \text{ continuous}, \quad \delta \text{ small}.
\end{aligned}
$$

In definitions of the likelihood function for continuous variables, the constant δ^n (or the equivalent "differential element" $dy_1, \ldots, dy_n$) is usually dropped, and the likelihood function is defined as any function proportional to $\prod f(y_i \mid \theta)$. When we are comparing different continuous probability models for the same data, the constant δ^n always cancels and so can be omitted, but other constants in each model do not cancel in general, and must be retained in the likelihood.

We will generally drop the $\mid \mathbf{y}$ in the likelihood $L(\theta \mid \mathbf{y})$ for θ, but it should be remembered that in this book the likelihood is always defined for *observed* values: the y_i are not regarded as random variables but as observed numerical values. When we want to consider the repeated sampling properties of terms in the likelihood, the algebraic values y_i (or other functions of the data) will be replaced by capital letters Y_i to represent the unobserved random variables.

2.6   Inference for single parameter models

The likelihood function $L(\theta)$ is of fundamental importance in statistical inference, whether from a Neyman–Pearson, Bayes or pure likelihood perspective.
Although there is a lively controversy over the role of models, and of likelihood,
in survey sampling theory, we take the use of and reliance on models as axio-
matic. We make some comments on non-model based methods in the discussion
of empirical likelihood in Section 2.10.2.

We will be principally concerned with models in the *exponential family* and
mixtures of them. The exponential family includes the binomial, Poisson, normal,
gamma and inverse Gaussian distributions; the last of these is more specialized
than the others and we do not discuss it in this book.

We consider the general properties of this family in Section 2.9, but discuss
first the problems of inference in the simplest models. In our approach we follow
closely the "direct" or "pure" likelihood theory of Barnard (1949), Birnbaum(1962)
and Barnard *et al.* (1962), described at book length in Edwards (1972), Clayton
and Hills (1994), Lindsey (1996), Royall (1997) and Sprott (2000). The Likelihood Principle book of Berger and Wolpert (1986) discusses many of the same
issues from a Bayesian viewpoint. We give the very close connections between
likelihood theory and conventional repeated-sampling (Neyman–Pearson) theory
in many models. We give a short discussion of Bayes theory; this provides a unifica-
tion of inferential approaches through the posterior distribution of the likelihood
ratio between models (Aitkin, 1997; Chadwick, 2002; Aitkin *et al.*, 2005), which
gives a simple calibration for nested model comparisons, and for many of the
non-nested and common-family model comparisons needed in later chapters.

In this section we consider single parameter models, and begin with the simplest
normal model.

Normal mean model
A simple random sample $y_1, \ldots, y_n$ of size $n = 25$ is drawn from a population in
which Y has a normal model $N(\mu, \sigma^2)$ with $\sigma^2 = 1$ known, and yields the value
$\bar{y} = 0.4$. We now consider three inferential questions:

(i) Two different substantive hypotheses H_1 and H_2 specify values for μ,
    namely $H_1 : \mu = \mu_1 = 0$ and $H_2 : \mu = \mu_2 = 1$. What evidence does
    the sample provide about hypothesis H_1 relative to hypothesis H_2?
(ii) In the absence of any substantive hypothesis about the value of μ, what
    information does the sample provide about it?
(iii) A substantive null hypothesis H_0 specifies $\mu = \mu_0 = 0$; the alternative is
    the general H_1 that $\mu \neq 0$. What evidence does the sample provide about
    the hypothesis H_0 relative to the hypothesis H_1?

These three questions are simple examples of "hypothesis testing" and
"estimation" problems.

The likelihood function is

$$L(\mu) = \prod_{i=1}^{n} [F(y_i + \delta/2 \mid \mu) - F(y_i - \delta/2 \mid \mu)]$$

$$\doteq \prod_{i=1}^{n} \delta f(y_i \mid \mu)$$

$$= \prod_{i=1}^{n} \delta \cdot \frac{1}{\sqrt{2\pi}\sigma} \exp\left\{-\frac{1}{2\sigma^2}(y_i - \mu)^2\right\}$$

$$= \frac{\delta^n}{(\sqrt{2\pi})^n \sigma^n} \exp\left\{-\frac{1}{2\sigma^2} \left[\sum (y_i - \bar{y})^2 + n(\bar{y} - \mu)^2\right]\right\}.$$

In this expression, δ, σ and $\sqrt{2\pi}$ are known constants (though the value of δ was not specified in the question), and $\bar{y}$ and $\sum (y_i - \bar{y})^2$ are known once the data are observed.

In likelihood theory, we interpret the likelihood $L(\theta)$ directly as the evidence (function), given the model and the data y, for different values of the parameter θ.

2.6.1 *Comparing two simple hypotheses*

To apply the theory to question (i), we assess the relative evidence for the hypothesis H_1 compared with the hypothesis H_2 by computing the *likelihood ratio* LR $= L(\mu_1)/L(\mu_2)$. If this ratio is small, substantially smaller than 1, then the data provide strong evidence against H_1 in favour of H_2. A *large* value of this ratio correspondingly provides strong evidence in *favour* of H_1 and against H_2. A value of 1 for the likelihood ratio means that both hypotheses are equally well supported by the data. An important point in this analysis is that we do not have to specify a "null" hypothesis: the two hypotheses are treated symmetrically—if we change the labels of H_1 and H_2, the conclusions about the parameter values are unaffected.

In computing the likelihood ratio between two hypotheses, we can omit from the likelihood any constants which take the same value under both hypotheses, since these constants cancel in the ratio. So we may as well begin with the *kernel* of the likelihood in which these constants are omitted. Note that it is not necessary to know the measurement precision δ to compute the likelihood ratio (though it *is* necessary to know that the measurement precision is sufficient for the likelihood representation by the normal density, as previously discussed).

For our normal model, the kernel of the likelihood is

$$K(\mu) = \exp\left\{-\frac{1}{2\sigma^2}[n(\bar{y} - \mu)^2]\right\}.$$

Evaluating at $\bar{y} = 0.4, \sigma = 1$, we have LR $= L(0)/L(1) = K(0)/K(1) = 12.18$. An equivalent comparison, of which we will make much use in later chapters, is through the value of the *likelihood ratio test statistic* (LRTS)

$$-2 \log \text{LR} = -2 \log[L(\mu_1)/L(\mu_2)] = \frac{n(\bar{y} - \mu_1)^2}{\sigma^2} - \frac{n(\bar{y} - \mu_2)^2}{\sigma^2}$$

$$= Z_1^2 - Z_2^2,$$

where Z_j ($j = 1, 2$) is the usual Z-test (normal test) statistic for the hypothesis H_j. Here $Z_1 = \sqrt{25} * 0.4 = 2$, $Z_2 = 3$ and $-2 \log[L(\mu_1)/L(\mu_2)] = -5$.

We have sample evidence in favour of H_1 and against H_2, since $12 \gg 1$. This *sample* evidence has to be interpreted in the light of other information about the hypotheses; Bayes theory does this formally through the *prior probabilities* π_1 and $\pi_2 = 1 - \pi_1$ of H_1 and H_2. The *prior odds* $p = \Pr(H_1)/\Pr(H_2) = \pi_1/\pi_2$— the "weight of evidence" in favour of H_1 compared to H_2 before observing the data—is "updated" to the *posterior odds* using Bayes' theorem:

$$\frac{\Pr(H_1 \mid \mathbf{y})}{\Pr(H_2 \mid \mathbf{y})} = \frac{\Pr(\mathbf{y} \mid H_1)\Pr(H_1)}{\Pr(\mathbf{y} \mid H_2)\Pr(H_2)}$$

$$= \frac{L(\mu_1)}{L(\mu_2)} \cdot \frac{\pi_1}{\pi_2},$$

by multiplying the prior odds p by the likelihood ratio (LR) to give the posterior odds $p * \text{LR}$.

The posterior probability of H_1 is $p * \text{LR}/(p * \text{LR} + 1)$. If the prior probabilities of the two hypotheses are equal, the posterior probability of H_1 is $\text{LR}/(\text{LR} + 1)$.

How do we calibrate likelihood ratios as measures of evidence? What value of a likelihood ratio constitutes "persuasive" sample evidence for H_1 compared to H_2? There is no unique answer to this question, just as there is no unique answer to the question of how small a P-value under the null hypothesis H_2, or how large a posterior probability of H_1, constitute "persuasive" evidence in the Neyman–Pearson or Bayes theories. Different observers may be persuaded by different strengths of evidence. It is nevertheless convenient to have some scale reference values, like the 0.05 and 0.01 P-values.

In Bayes theory, a posterior probability of 0.9 for H_1 would generally be regarded as quite strong evidence in favour of H_1; if the hypotheses have equal prior probability, the corresponding likelihood ratio would be 9. So the slightly larger rounded value 10—an *order of magnitude*—for the likelihood ratio would be quite strong evidence, or a value of $-2 \log \text{LR}$ of 4.6, in the absence of prior information favouring one of the hypotheses over the other.

We adopt the scale below, adapted from one suggested by Jeffreys (1961) for Bayes factors, in terms of orders of magnitude.

We regard the sample evidence against H_1 and in favour of H_2, writing LR $= L(\mu_1)/L(\mu_2)$, as

- suggestive if LR $< \frac{1}{3}$ ($-2\log$ LR > 2.20)
- quite strong if LR $< \frac{1}{10}$ ($-2\log$ LR > 4.60)
- strong if LR $< \frac{1}{30}$ ($-2\log$ LR > 6.80)
- very strong if LR $< \frac{1}{100}$ ($-2\log$ LR > 9.21).

If $\frac{1}{3} <$ LR < 3, then we do not have even suggestive sample evidence in support of either model: the evidence is inconclusive.

The Neyman–Pearson theory does not treat the hypotheses symmetrically. We have to specify one of the hypotheses as the "null", and the other as the "alternative". The repeated-sampling distribution of the likelihood ratio for the null to the alternative then has to be found under the null hypothesis, and this hypothesis is rejected if the likelihood ratio is too small, compared with a pre-specified percentage point of its sampling distribution. In practice the tail-area probability—the "P-value" associated with the observed likelihood ratio—is found instead, and referred to standard critical values. A small P-value is interpreted as evidence against the corresponding null hypothesis.

The P-value depends on which hypothesis is specified as the null. Thus, if the null hypothesis is $H_1 : \mu = 0$, the Z_1-test statistic is 2, and its P-value is 0.023, which would lead to rejection of this hypothesis at the usual 5% level. But if the null hypothesis is $H_2 : \mu = 1$, the Z_2-test statistic is -3, and its P-value is 0.0013, which would lead to rejection of *this* hypothesis at any conventional level. Given that both the P-values are small, it is unclear what we should conclude—do the data point to some other value of μ?

The Neyman–Pearson conclusions, therefore, depend on the prior specification of which hypothesis should be regarded as the null: the P-value for a particular specification of the null hypothesis cannot be used as a measure of the strength of evidence against this hypothesis, without also considering the other possible null hypothesis.

We now turn to question (ii).

2.6.2 *Information about a single parameter*

Given a single-parameter likelihood $L(\theta)$, the parameter value $\hat{\theta}$ with the highest likelihood is called the *maximum likelihood estimate* (MLE) of θ, and is a natural reference value for other values of θ. For many (but not all) of the statistical models considered in this book, the likelihood function has a unique maximum interior to the parameter space, and no minima, and in these cases the maximum can be found for a continuous parameter by (in general partial) differentiation of the log likelihood function $\ell(\theta) = \log L(\theta)$. In our example

$$\frac{\partial \ell}{\partial \mu} = \sum (y_i - \mu)/\sigma^2$$

$$\frac{\partial^2 \ell}{\partial \mu^2} = -n/\sigma^2$$

giving the unique maximum at $\hat{\mu} = \bar{y}$. Values of θ near the MLE $\hat{\theta}$ have higher likelihoods than those remote from $\hat{\theta}$, so the *maximum* $L(\hat{\theta})$ of the likelihood is correspondingly a natural reference value for other values of $L(\theta)$. The *relative likelihood* function $R(\theta) = L(\theta)/L(\hat{\theta})$ expresses this directly, giving a measure of "support" or strength of evidence on a 0–1 scale for each value of θ relative to the "best supported" value $\hat{\theta}$. *Likelihood intervals* or *regions* of "well-supported" values of θ are constructed by solving the inequality $R(\theta) > c$ for a suitable c.

The relative likelihood function in our example is easily seen to be

$$R(\mu) = \exp \left\{ -\frac{n(\bar{y} - \mu)^2}{\sigma^2} \right\}.$$

This has the same form as a normal density function, because of the symmetrical appearance of $\bar{y}$ and μ in the likelihood. For this reason this likelihood is called a *normal* likelihood, or equivalently, the log likelihood is *quadratic* (in the parameter μ).

The importance of this name convention follows from the fact that in samples from regular parametric models, with increasing sample size the likelihood in the parameters approaches a normal likelihood. This is easily shown generally, assuming that the likelihood $L(\theta)$ has an internal maximum. We give the result for a single parameter; the general case follows directly. Expanding the log likelihood function about the MLE $\hat{\theta}$ gives:

$$\ell(\theta) = \ell(\hat{\theta}) + (\theta - \hat{\theta})\ell'(\hat{\theta}) + \frac{1}{2!}(\theta - \hat{\theta})^2 \ell''(\hat{\theta}) + \frac{1}{3!}(\theta - \hat{\theta})^3 \ell^{(3)}(\hat{\theta})$$

$$+ \frac{1}{4!}(\theta - \hat{\theta})^4 \ell^{(4)}(\hat{\theta}) + \cdots.$$

Since the log likelihood is a sum of n terms, write $\ell''(\hat{\theta})$ as $n\bar{\ell}''$ and similarly for the higher derivatives. Then $\bar{\ell}''$ and the higher derivatives are $O(1)$ as n increases, and since $\ell'(\hat{\theta}) = 0$,

$$\ell(\theta) = \ell(\hat{\theta}) + \frac{n}{2!}(\theta - \hat{\theta})^2 \bar{\ell}'' + \frac{n}{3!}(\theta - \hat{\theta})^3 \bar{\ell}^{(3)} + \frac{n}{4!}(\theta - \hat{\theta})^4 \bar{\ell}^{(4)} + \cdots.$$

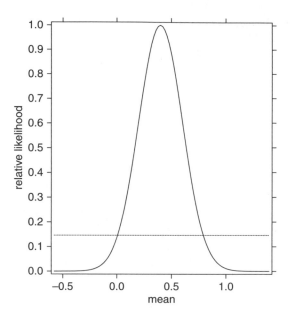

Fig. 2.18. Relative likelihood function

Write $-\bar{\ell}'' = 1/\sigma^2$ and $z = \sqrt{n}(\theta - \hat{\theta})/\sigma$. Then

$$\ell(\theta) = \ell(\hat{\theta}) - \frac{1}{2}z^2 + \sum_{j=3}^{\infty} \frac{n^{1-j/2}}{j!} z^j \bar{\ell}^{(j)} \sigma^j.$$

As n increases the cubic and higher terms in z tend to zero, giving

$$\ell(\theta) \to \ell(\hat{\theta}) - \tfrac{1}{2}z^2,$$

whence

$$L(\theta) \to L(\hat{\theta}) \exp\{-\tfrac{1}{2}z^2\}.$$

The relative likelihood is shown in Fig. 2.18, together with the line $R(\mu) = 0.146$. It is just a rescaling of the likelihood function itself, since only known constants have been omitted. The GLIM directives are:

```
$unit 100 $
$var 100 i mu c rl $
$calc i = %cu(1) $
    : mu = -0.6 + 0.02*i $
    :nrl = %exp(-12.5*(0.4 - mu)**2) $
    :  c = 0.146 $
```

```
$graph (s = 1   h = 'mean'  v = 'relative likelihood')
       nrl,c mu 10,11 $
```

As for the likelihood ratio for the two specified hypotheses, it is not necessary to know the measurement precision δ to compute the relative likelihood.

The likelihood interval for μ based on a relative likelihood of c is easily seen to be $\mu \in \bar{y} \pm \lambda \sigma / \sqrt{n}$ where $\lambda = \sqrt{-2 \log c}$. This will be recognized as the usual Neyman–Pearson confidence interval for μ (but with the random variable $\bar{Y}$ replaced by the observed $\bar{y}$), with the confidence coefficient determined by the choice of c. Thus, if $c = 0.146, \lambda = 1.96$, giving a 95% confidence interval for μ—a *likelihood-based confidence interval*. If $c = 0.0368, \lambda = 2.57$, giving a 99% confidence interval.

It is a feature of likelihood theory that the likelihood interval is determined *only* from the observed likelihood—we have not used the sampling distribution of $\bar{Y}$ to derive it. However, the confidence coefficient of the interval for a particular c depends on the form of repeated sampling assumed for $\bar{Y}$. We discuss this point further below.

The reader may notice that the value of c of 0.146 for a 95% confidence interval is *larger* than the value $\frac{1}{10} = 0.1$ which we proposed above as the value of a likelihood ratio. This would provide quite strong evidence against the "numerator" hypothesis, when we are comparing two simple hypotheses. This reflects one of the differences between the likelihood and Neyman–Pearson theories—in likelihood theory (as in Bayes theory) we do not regard P-values of 0.05 as persuasive evidence against the null hypothesis. We discuss this point further below (Aitkin, 1991, 1997 gave a detailed discussion).

In the Neyman–Pearson theory, in the absence of any specific hypothesis about the parameter θ (here taken to be p-dimensional), the sample information about it is conventionally expressed through the repeated sampling distribution of the MLE, the *score statistic*, or the *likelihood ratio statistic*. We define the log-likelihood function $\ell(\theta)$, score vector $\mathbf{s}(\theta)$, Hessian matrix $H(\theta)$, information matrix $I(\theta)$ and expected information matrix $\mathcal{I}(\theta)$ by

$$\ell(\theta) = \log L(\theta)$$

$$\mathbf{s}(\theta) = \frac{\partial \ell}{\partial \theta}$$

$$H(\theta) = \frac{\partial \mathbf{s}}{\partial \theta'} = \frac{\partial^2 \ell}{\partial \theta \partial \theta'}$$

$$I(\theta) = -H(\theta)$$

$$\mathcal{I}(\theta) = \mathrm{E}[I(\theta)],$$

where the expectation is over the distribution of the random variables $Y_1, \ldots, Y_n$. Then under standard regularity conditions (that θ is not an origin parameter of the support of Y, that $\hat{\theta}$ is an interior point of the parameter space, and that the dimension of θ does not increase with n), the sampling distribution of the MLE $\hat{\theta}$ (regarded as a random variable in repeated sampling) is asymptotically

$$\hat{\theta} \rightarrow N_p(\theta, I^{-1}(\hat{\theta})).$$

Alternatively, the asymptotic distribution of the score statistic is

$$S(\theta) \rightarrow N_p(0, \mathcal{I}(\theta)),$$

while the asymptotic distribution of the likelihood ratio statistic is

$$-2\log\{L(\theta)/L(\hat{\theta})\} \rightarrow \chi_p^2.$$

(Strictly speaking the first two distributions are degenerate asymptotically since the variances tend to zero or infinity, so the MLE and score statistics should be standardized to give non-degenerate limiting distributions, but we give the usual results in the sense of "for large n".)

A "point" estimate of θ is $\hat{\theta}$ (and if we are referring to the sampling distribution, it is of the estimat*or* $\hat{\theta}$), and an "interval" estimate is obtained by including in the interval all those values of θ which would not be rejected by a formal test, using one of the three criteria above.

For the normal model example,

$$L(\mu) = \frac{\delta^n}{(\sqrt{2\pi})^n \sigma^n} \exp\left\{-\frac{1}{2\sigma^2}\left[\sum(y_i - \bar{y})^2 + n(\bar{y} - \mu)^2\right]\right\}$$

$$\ell(\mu) = c - n(\bar{y} - \mu)^2/2\sigma^2$$

$$\hat{\mu} = \bar{y}$$

$$s(\mu) = n(\bar{y} - \mu)/\sigma^2$$

$$H(\mu) = -n/\sigma^2$$

$$I(\mu) = I(\hat{\mu}) = \mathcal{I}(\mu) = \mathcal{I}(\hat{\mu}) = n/\sigma^2.$$

For this example all the asymptotic results are exact and equivalent:

$$\hat{\mu} = \bar{Y} \sim N(\mu, \sigma^2/n)$$

$$S(\mu) = n(\bar{Y} - \mu)/\sigma^2 \sim N(0, n/\sigma^2)$$

$$-2\log\{L(\mu)/L(\bar{Y})\} = n(\bar{Y} - \mu)^2/\sigma^2 \sim \chi_1^2.$$

Thus the point estimate of μ is $\bar{y}$, and the $100(1 - \alpha)\%$ confidence interval for μ in repeated sampling (of fixed size n) is $\bar{Y} \pm \lambda_{1-\alpha/2}\sigma/\sqrt{n}$; this random interval covers the true value of μ with probability $1 - \alpha$. For the observed interval in the given sample $\bar{y} \pm \lambda_{1-\alpha/2}\sigma/\sqrt{n}$ no such statement can be made: it either contains μ or does not.

As we noted earlier, the confidence interval for μ is formally identical to the likelihood interval through the relation between c and the percentage point of the normal distribution. However, their *meanings* are different: the likelihood interval contains those parameter values giving a relatively high likelihood to the observed sample, but the properties of the confidence interval do not refer to the observed sample: they are defined only with respect to the *family* or *reference set* of intervals produced by hypothetical replications of the sample.

Bayes inference about θ requires a *prior distribution* $\pi(\theta)$ for θ, representing the information about θ before the current data are observed. On observing the data y, the prior distribution is updated to the *posterior distribution* $\pi(\theta \mid y)$ using Bayes' theorem:

$$\pi(\theta \mid y) = L(\theta)\pi(\theta) / \int L(\theta)\pi(\theta)\,\mathrm{d}\theta,$$

where the integral is over the parameter space of θ.

There is a vigorous debate amongst Bayesians over the choice of priors, which we will not enter here; where necessary we will generally use *non-informative* or *flat* priors to represent prior information which is uninformative relative to the sample information. Such priors are useful in providing in many cases a *reference* analysis which can be modified by informative priors if needed.

For the normal mean example, a flat prior $\pi(\mu) = 1/2C$ on the interval $-C < \mu < C$, where C is large compared to $\sigma/\sqrt{n}$, gives (as $C \to \infty$) the *normal posterior distribution* $\mu \sim N(\bar{y}, \sigma^2/n)$, and a $100(1 - \alpha)\%$ *highest posterior density* (HPD) or *credible* interval for μ is $\bar{y} \pm \lambda_{1-\alpha/2}\sigma/\sqrt{n}$. This is identical to the $100(1 - \alpha)\%$ confidence interval, but as for the likelihood interval the (posterior) probability statement about μ refers to the observed data on which the distribution of μ is conditioned, and not to hypothetical replications of the sample.

If the prior distribution of μ is *not* flat, the posterior distribution will not have this simple normal form, and the HPD interval will not be the same as the confidence interval or the likelihood interval: this is to be expected from the additional information in the prior.

We now turn to question (iii).

2.6.3 *Comparing a simple null hypothesis and a composite alternative*

This problem is a central one in data analysis and model simplification.

In the Neyman–Pearson theory, the null hypothesis is tested using the previous sampling distribution results for the MLE, score and likelihood ratio statistic.

The previous confidence interval constructions are formalized through the corresponding tests: the *Wald*, *score* and *likelihood ratio* (*LR*) tests. The LRTS for a null hypothesis $H_0 : \theta = \theta_0$ against a general alternative hypothesis $H_1 : \theta \neq \theta_0$ is

$$\text{LRTS} = -2\log\{L(\theta_0)/L(\hat{\theta})\}.$$

As alternatives to the likelihood ratio test, the Wald and score or Lagrange multiplier tests are often used when the likelihood ratio test is difficult or computationally expensive. The Wald test statistic is

$$W = (\hat{\theta} - \theta_0)'I(\hat{\theta})(\hat{\theta} - \theta_0),$$

and the score test statistic is

$$S = \mathbf{s}(\theta_0)'\mathcal{I}^{-1}(\theta_0)\mathbf{s}(\theta_0).$$

Under the null hypothesis H_0, LRTS, W and S are all asymptotically distributed as χ_p^2 under the same regularity conditions (these conditions are in fact weaker for the score test, for which boundary values of $\hat{\theta}$ may not invalidate the asymptotic distribution).

For the simple normal model,

$$s(\mu) = n(\bar{y} - \mu)/\sigma^2$$

$$I(\mu) = n/\sigma^2$$

$$\text{LRTS} = n(\bar{y} - \mu_0)^2/\sigma^2$$

$$W = n(\bar{y} - \mu_0)^2/\sigma^2$$

$$S = \frac{n^2(\bar{y} - \mu_0)^2}{\sigma^4} / \frac{n}{\sigma^2} = W.$$

The three test statistics are identical, and have the exact sampling distribution χ_1^2. For the example, the test statistics have the value 4.0, for which the corresponding *P*-value from the χ_1^2 distribution is 0.0455. This would lead to rejection of H_0 using the common critical *P*-value of 0.05. The three test procedures lead to the same test in this normal model (and the same set of confidence intervals for μ), and do so in large samples in many other models, but in small samples they may give markedly different results (as in the next binomial example). The score and LR tests are invariant to monotone transformation of θ, but the Wald test is not, as we will see shortly.

There is one general likelihood approach corresponding closely to the Neyman–Pearson: the *maximized* likelihood approach. This uses the same LRTS, but interprets it directly as evidence in the same way as the likelihood ratio for simple hypotheses. In our example, the likelihood is maximized at $\mu = \hat{\mu} = \bar{y}$, and the maximized likelihood ratio is therefore

$$L(\mu_0)/L(\bar{y}) = \exp\left\{-\frac{1}{2\sigma^2}[n(\bar{y} - \mu_0)^2)]\right\}$$

which has the value $e^{-2} = 0.135$. Thus the relative likelihood at $\mu = \mu_0$ is quite high (relative to that at $\bar{y}$), and the likelihood ratio 0.135 would not be considered convincing, or even quite strong, evidence against H_0, if $\bar{y}$ were a *specified* alternative value of μ. But $\bar{y}$ has been determined from the data, and any other *specified* alternative $\mu_1 \neq \bar{y}$ will have lower likelihood than $L(\bar{y})$, and so an even greater likelihood ratio $L(\mu_0)/L(\mu_1)$. So the maximized likelihood ratio clearly *overstates* the evidence against H_0—we are acting as though the unknown parameter under H_1 is *known* to be equal to its MLE—and cannot be interpreted directly as a measure of the strength of the sample evidence against H_0 without some further calibration.

This difficulty has limited the use of maximized likelihood ratios as direct measures of evidence. Considerable theoretical attention has been paid (Lindsey, 1996 gave details) to various forms of *adjustment* of the maximized likelihood, to reflect better the sample information about the "nuisance" or unspecified parameters over which we have maximized. These adjusted likelihoods have generally been interpreted in a repeated sampling framework. The number of parameters p in the null hypothesis has to be allowed for as well, in any direct interpretation of the maximized likelihood ratio; Aitkin (1991) and Lindsey (1999) gave discussions of possible approaches to this allowance.

In Bayes theory, the hypothesis comparison problem has a straightforward solution. We need the probability of the data under the alternative hypothesis, and so we integrate the likelihood (the conditional probability of the data, given the parameters) over the marginal (prior) distribution $\pi(\phi)$ of the unspecified parameters ϕ, and compare the specified likelihood under the null hypothesis with the integrated likelihood under the alternative; the ratio is called the *Bayes factor* for the two hypotheses, and is interpreted in the same way as a likelihood ratio for two simple hypotheses; the prior odds on the two hypotheses multiplies the Bayes factor to give the posterior odds as in the comparison of simple hypotheses.

However, this approach encounters serious difficulties, because the integration of the likelihood over the prior is in general very sensitive to the prior specification. For the normal example above, with the flat prior on μ over $(-C, C)$ and likelihood kernel

$$K(\mu) = \exp\left\{-\frac{1}{2\sigma^2}[n(\bar{y} - \mu)^2]\right\}$$

the integrated likelihood kernel is

$$\bar{K} = \int_{-\infty}^{\infty} K(\mu)\pi(\mu)\,d\mu$$

$$= \frac{1}{2C} \int_{-C}^{C} \exp\left\{-\frac{n}{2\sigma^2}(\bar{y}-\mu)^2\right\} d\mu$$

$$= \frac{1}{2C}\frac{\sigma}{\sqrt{n}} \int_{(\sqrt{n}(\bar{y}-C))/\sigma}^{(\sqrt{n}(\bar{y}+C))/\sigma} \exp\left\{-\frac{1}{2}Z^2\right\} dZ$$

$$= \frac{1}{2C}\frac{\sigma}{\sqrt{n}}\sqrt{2\pi}\left[\Phi\left(\frac{\sqrt{n}(\bar{y}+C)}{\sigma}\right) - \Phi\left(\frac{\sqrt{n}(\bar{y}-C)}{\sigma}\right)\right],$$

where Φ is the standard normal cdf. For C large relative to $\sigma/\sqrt{n}$, the term in brackets rapidly approaches 1, so for large C the integrated likelihood kernel is

$$\frac{1}{2C}\frac{\sigma}{\sqrt{n}}\sqrt{2\pi}.$$

This tends to 0 as $C \to \infty$, and it is clear that the integrated likelihood kernel can be made arbitrarily small by choice of C. The Bayes factor is

$$\text{BF} = \frac{K(\mu_1)}{\bar{K}} = 2C \cdot \frac{\sqrt{n}}{\sqrt{2\pi}\sigma} \exp\left\{-\frac{n}{2\sigma^2}(\bar{y}-\mu_0)^2\right\}$$

$$= 2C \cdot \frac{\sqrt{n}}{\sigma}\phi(Z_0)$$

and this can be made arbitrarily large by choice of C. As $C \to \infty$ the Bayes factor $\to \infty$, whatever the (fixed) value of Z_0, and so the evidence in favour of H_0, whatever it is, becomes overwhelming if the prior is sufficiently diffuse!

This result is generally known as Lindley's paradox (Lindley, 1957; Bartlett, 1957), though the "paradox" usually refers to the fact that, as $n \to \infty$ for *fixed* C, BF $\to \infty$ also. This is the same phenomenon: whether the likelihood becomes concentrated with n while the prior stays fixed, or the prior becomes diffuse with C while the likelihood stays fixed, the end result is the same: the data apparently overwhelmingly support the null hypothesis H_0, whatever it is. For the example above we have $Z_0 = 2, n = 25, \sigma = 1$, and the Bayes factor takes the values 0.27, 0.54, 5.4 and 54 for $C = 0.5, 1, 10$ and 100, respectively.

Thus, direct interpretation of Bayes factors is unsatisfactory, and considerable efforts have been devoted to calibrating them. These are reviewed in Aitkin (1991) and Kass and Raftery (1995).

Aitkin (1997) gave an alternative Bayesian solution to this problem that does not suffer from Lindley's paradox, based on earlier results of Dempster (1974, 1997), which allows the general use of maximized likelihood ratios and profile likelihoods, though detailed applications are currently few.

We use the posterior distribution of θ under the alternative hypothesis to find the posterior distribution of the *true likelihood ratio* TLR $= L(\theta_0)/L(\theta)$, evaluated at the true value of θ. If the posterior probability of the event TLR < 0.1 is high, we have strong evidence against the null hypothesis.

For the example, we take the flat prior on μ; the posterior distribution of μ is then $N(\bar{y}, \sigma^2/n)$, which is $N(0.4, 0.04)$. The posterior probability that $\mu > 0$ is $1 - \Phi(-2) = 0.9773 = 1 - P/2$. We work with the posterior distribution of $-2 \log \text{TLR} = Z_0^2 - Z^2$. Then the posterior probability that the true likelihood ratio is less than 0.1, a value which would be quite strong evidence against H_0, is

$$\Pr[\text{TLR} < 0.1 \mid y] = \Pr[-2 \log \text{TLR} > 4.6 \mid y]$$

$$= \Pr[Z^2 < Z_0^2 - 4.6 \mid y]$$

$$= \Pr[Z^2 < -0.6 \mid y]$$

$$= 0$$

since Z^2 must be positive! The true likelihood ratio cannot be as small as 0.1 given the data. In fact, the true likelihood ratio cannot be less than $\exp(-\frac{1}{2}Z_0^2) = 0.135$, since this is the value of the likelihood ratio resulting from maximizing the denominator.

For the much weaker criterion that the true likelihood ratio is less than 1— that is, that the null hypothesis is less well supported than the alternative—the posterior probability is

$$\Pr[\text{TLR} < 1 \mid y] = \Pr[-2 \log \text{TLR} > 0 \mid y]$$

$$= \Pr[Z^2 < Z_0^2 \mid y]$$

$$= \Pr[Z^2 < 4 \mid y]$$

$$= 0.9546 = 1 - P$$

a remarkable result pointed out by Dempster (1974): with normal likelihoods and flat priors, the P-value is equal to *the posterior probability that the true likelihood ratio is greater than 1*, or equivalently, $1 - P$ is the posterior probability that the true likelihood ratio is less than 1.

So the P-value *is* a measure of strength of evidence against the null hypothesis, but it is not a persuasive one, since it refers only to the null hypothesis being better or worse supported than the alternative, and not that the ratio of the likelihoods is small.

We show the value of this approach in the next section.

2.7   Inference with nuisance parameters

The discussion of inference in Section 2.6 was limited to single-parameter models. In practice, almost all inferential problems in statistics involve models with several unknown parameters. Research questions are usually about a subset of these parameters, called the *parameters of interest*, while the remaining parameters—the *nuisance parameters*—are not of interest in themselves, but affect the inference about the parameters of interest.

We begin again with the simplest normal mean model.

Normal mean model

A simple random sample $y_1, \ldots, y_n$ of size $n = 25$ is drawn from a population in which Y has a normal model $N(\mu, \sigma^2)$ with σ^2 now unknown, and yields the values $\bar{y} = 0.4$ and $s^2 = 1.0$, where $s^2 = \sum (y_i - \bar{y})^2 / (n - 1)$. We again consider the three inferential questions above, though we now change their order:

(i)   In the absence of any substantive hypothesis about the value of μ, what information does the sample provide about it?

(ii)   A substantive null hypothesis H_0 specifies $\mu = \mu_0 = 0$; the alternative is the general H_1 that $\mu \neq 0$. What evidence does the sample provide about the hypothesis H_0 relative to the hypothesis H_1?

(iii)   Two different substantive hypotheses H_1 and H_2 specify values for μ, namely $H_1 : \mu = \mu_1 = 0$ and $H_2 : \mu = \mu_2 = 1$. What evidence does the sample provide about hypothesis H_1 relative to hypothesis H_2?

The likelihood function is as before

$$L(\mu, \sigma) = \prod_{i=1}^{n} \left[F(y_i + \delta/2 \mid \mu, \sigma) - F(y_i - \delta/2 \mid \mu, \sigma) \right]$$

$$\doteq \prod_{i=1}^{n} \delta f(y_i \mid \mu, \sigma)$$

$$= \frac{\delta^n}{(\sqrt{2\pi})^n \sigma^n} \exp \left\{ -\frac{1}{2\sigma^2} \left[\sum (y_i - \bar{y})^2 + n(\bar{y} - \mu)^2 \right] \right\}.$$

Now σ is unknown, so the likelihood depends explicitly on the two parameters, though our inferential interest is only in μ. To make an inference about μ, we need to "eliminate" the parameter σ from the likelihood in some way. We will give a general Neyman–Pearson approach based on *maximized* or *profile* likelihoods, calibrated through the sampling distribution of the LRT statistic, while referring in several applications to other non-general approaches based on *marginal* or *conditional* likelihoods. We calibrate the profile likelihood from a Bayes/likelihood theory point of view using the results of Aitkin (1997), as described later; the same approach applies to the general problem with nuisance parameters as to the earlier case of a simple null hypothesis.

2.7.1 *Profile likelihoods*

Consider the general case of a likelihood $L(\theta, \phi)$ depending on a p_1-dimensional parameter of interest θ and a p_2-dimensional nuisance parameter ϕ. If for any outcome data y the likelihood can be written in the form

$$L(\theta, \phi) = L_1(\theta)L_2(\phi),$$

where L_1 is a function of θ only and L_2 a function of ϕ only, and the parameter spaces for θ and ϕ are unrelated, then we call the likelihood $L(\theta, \phi)$ *separable*, and say the parameters θ and ϕ are *orthogonal*. In this case inference about θ may be based on $L_1(\theta)$, since for any specific value ϕ_0 of ϕ,

$$L(\theta_0, \phi_0)/L(\theta_1, \phi_0) = L_1(\theta_0)/L_1(\theta_1).$$

Separable likelihoods are rare. If the likelihood is not separable, we define the *profile* likelihood $P(\theta)$ for θ by

$$P(\theta) = L(\theta, \hat{\phi}(\theta)),$$

where the notation $\hat{\phi}(\theta)$ means the MLE of ϕ as a function of the specified value θ. Thus the profile likelihood is the value of the likelihood as we follow a curved path through the parameter space, defined by $\phi = \hat{\phi}(\theta)$.

We now answer the first two questions above generally. For a general likelihood function $L(\theta, \phi)$, the information provided by the sample about θ is expressed through the profile likelihood $P(\theta)$, interpreted as though it were a likelihood function from a parametric model $f(z \mid \theta)$. The evidence provided by the sample about a null hypothesis $H_0 : \theta = \theta_0$ against a general alternative $H_1 : \theta \neq \theta_0$ is the profile likelihood ratio (profile relative likelihood)

$$P(\theta_0)/P(\hat{\theta}) = L(\theta_0, \hat{\phi}(\theta_0))/L(\hat{\theta}, \hat{\phi}),$$

where $\hat{\phi} = \phi(\hat{\theta})$ is the MLE of ϕ. The profile likelihood ratio is interpreted as though it were a likelihood ratio from the parametric model $f(z \mid \theta)$.

The specification of the same interpretation for a profile likelihood as for a "real" likelihood clearly requires justification, and in some examples this specification fails to provide a satisfactory calibration of the profile likelihood. We discuss this below, but first illustrate with the normal model example.

For fixed μ, the MLE of σ is given by

$$\hat{\sigma}^2(\mu) = \left[\sum (y_i - \bar{y})^2 + n(\bar{y} - \mu)^2 \right]/n.$$

Substituting this value into the likelihood gives the profile likelihood

$$P(\mu) = \frac{\delta^n}{(\sqrt{2\pi})^n \hat{\sigma}(\mu)^n} \exp\left\{-\frac{n}{2}\right\},$$

and the profile likelihood ratio is

$$P(\mu)/P(\hat{\mu}) = [\hat{\sigma}/\hat{\sigma}(\mu)]^n,$$

where

$$\hat{\sigma}^2 = \hat{\sigma}^2(\bar{y}) = \sum(y_i - \bar{y})^2/n = (n-1)s^2/n.$$

Straightforward algebra gives

$$P(\mu)/P(\hat{\mu}) = \left[1 + \frac{n(\bar{y} - \mu)^2}{(n-1)s^2}\right]^{-n/2}$$

$$= \left[1 + \frac{t^2}{(n-1)}\right]^{-n/2},$$

where t is the usual one-sample t-statistic: $t = \sqrt{n}(\bar{y} - \mu)/s$. Remarkably, the profile relative likelihood for μ is identical to the relative likelihood for μ based on just the observation of the t-statistic, which has a t_{n-1} distribution, with density

$$f(t) = \text{const} \cdot \left[1 + \frac{t^2}{(n-1)}\right]^{-n/2},$$

where the normalizing constant disappears in the relative likelihood. Thus, a profile likelihood interval for μ based on a profile relative likelihood of c is identical to a confidence interval for μ based on the t distribution, using the critical value

$$t_c = [(n-1)(c^{-2/n} - 1)]^{1/2}.$$

For our example, with $n = 25$ and the observed $t = 2$, a 14.6% relative likelihood interval for μ uses $c = 0.146$, for which $t_c = 1.998$; the interval is $(0.4 \pm 1.998 \cdot 1.0/\sqrt{25})$, which is $(0.00, 0.80)$. The 95% confidence interval for μ from the t_{24}-distribution uses the critical value 2.064, which is slightly larger; this interval is $(-0.01, 0.81)$. The difference arises because the sampling distributions of the t and normal likelihood ratios, as random variables, are different.

We conclude that the observed t-value of 2 is not very strong evidence against H_0: the P-value just exceeds 0.05.

For smaller df the differences are larger: for $n = 11$ the 14.6% relative likelihood uses $t = 2.047$, while the 95% confidence interval uses $t = 2.230$. For $n = 5$ the values are 2.153 and 2.777 respectively. This illustrates the difficulty

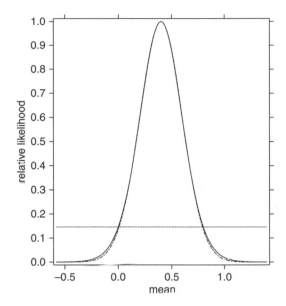

Fig. 2.19. Relative likelihood functions

of direct interpretation of maximized likelihood ratios—for consistent confidence coverage we have to consider the different sampling distributions involved.

Figure 2.19 shows the t likelihood (solid curve) for the example, and the normal likelihood (dashed curve) assuming $\sigma^2 = \hat{\sigma}^2$ for comparison. The t likelihood is slightly more diffuse but very similar in shape. Extending the previous graph of the normal relative likelihood, the GLIM directives are:

```
$var 100 i mu c rl $
$calc i = %cu(1) $
   : mu = -0.6 + 0.02*i $
  :nrl = %exp(-12.5*(0.4 - mu)**2) $
   :  c = 0.146 $
   :  t = 5*(0.4 - mu) $
  : trl = (1 + t**2/24)**(-12.5) $
$graph (s = 1 h = 'mean' v = 'relative likelihoods')
       trl,nrl,c mu 10,12,11 $
```

The profile likelihood ratio is calibrated in the Neyman–Pearson theory through the LRTS, defined by

$$\text{LRTS} = -2\log\{P(\theta)/P(\hat{\theta})\}.$$

The asymptotic repeated-sampling distribution of this statistic (in samples of fixed size n) is $\chi^2_{p_1}$ under the regularity conditions of Section 2.6.2.

In the normal model earlier, this gives

$$n \log[1 + t^2/(n-1)] \to \chi_1^2;$$

for small t compared with n the LHS is approximately $nt^2/(n-1)$. The asymptotic distribution is not needed in this model because of the known exact distribution of t. This normal mean example is one of the few in which the LRTS has an exact small-sample distribution.

The profile likelihood method is readily extended to functions of the model parameters. For example, suppose that a variable Y has a lognormal distribution, with $Z = \log Y \sim N(\mu, \sigma^2)$. We consider an example of family income in Section 2.9. The standard analysis of such data would be to log transform them, and fit the normal model to Z. Suppose, however, we wish to draw an inference about the mean of Y, rather than of Z. We have $E[Y] = \exp(\mu + \sigma^2/2)$, an awkward function. However, it is quite straightforward to compute the profile likelihood in the parameter $\theta = \mu + \sigma^2/2$, and then plot this likelihood against $\exp(\theta)$.

The likelihood in μ and σ is

$$L(\mu, \sigma) = \frac{\delta^n}{(\sqrt{2\pi})^n \sigma^n \prod y_i} \exp\left\{ -\frac{1}{2\sigma^2} \left[\sum (z_i - \bar{z})^2 + n(\bar{z} - \mu)^2 \right] \right\},$$

where $z_i = \log y_i$. We define $T = \sum (z_i - \bar{z})^2$, $\phi = \sigma^2$, $\mu = \theta - \phi/2$, and reparametrize the likelihood in terms of θ and ϕ. The log-likelihood is

$$\ell(\theta, \phi) = c - \frac{n}{2} \log \phi - \frac{1}{2\phi} [T + n(\bar{z} - \theta + \phi/2)^2]$$

$$= c - \frac{n}{2} \log \phi - \frac{1}{2} \left[\frac{T + n(\bar{z} - \theta)^2}{\phi} + n(\bar{z} - \theta) + n\phi/4 \right]$$

$$= c - \frac{n}{2} \log \phi - \frac{n}{2} \left[\frac{\hat{\sigma}^2 + (\bar{z} - \theta)^2}{\phi} + \bar{z} - \theta + \phi/4 \right]$$

$$\frac{\partial \ell}{\partial \phi} = -\frac{n}{2\phi} - \frac{n}{2} \left[-\frac{\hat{\sigma}^2 + (\bar{z} - \theta)^2}{\phi^2} + 1/4 \right].$$

Solving $\partial \ell / \partial \phi = 0$ for ϕ gives

$$\hat{\phi}(\theta) = \hat{\sigma}^2(\theta) = 2 \left[\sqrt{1 + \hat{\sigma}^2 + (\bar{z} - \theta)^2} - 1 \right].$$

Substituting for σ in the likelihood gives the profile likelihood in θ. This is computed in the macro `lognormprof`, and is used in the example in Section 2.9.

2.7.2 Marginal likelihood for the variance

We now consider the same normal model for Y, but suppose σ is the parameter of interest and μ the nuisance parameter. We have immediately $\hat{\mu}(\sigma) = \bar{y}$ for all σ, and so

$$P(\sigma) = \frac{\delta^n}{(\sqrt{2\pi})^n \sigma^n} \exp\left\{-\frac{\sum(y_i - \bar{y})^2}{2\sigma^2}\right\}.$$

The profile likelihood in this case is just the *section* of the likelihood at $\mu = \bar{y}$. The profile likelihood ratio, or profile relative likelihood, is

$$\mathrm{PR}(\sigma) = P(\sigma)/P(\hat{\sigma}) = \left(\frac{\hat{\sigma}}{\sigma}\right)^n \exp\left\{\frac{n}{2} - \frac{\sum(y_i - \bar{y})^2}{2\sigma^2}\right\}$$

$$= \left(\frac{\hat{\sigma}}{\sigma}\right)^n \exp\left\{\frac{n}{2}\left(1 - \frac{\hat{\sigma}^2}{\sigma^2}\right)\right\}.$$

An unsatisfying feature of the profile likelihood is that it is the same as the likelihood which would be obtained if we *knew* that $\mu = \bar{y}$. This is a feature of *all* profile likelihoods, but the fact that $\hat{\mu}(\sigma)$ does not depend on σ in this case accentuates the overstatement of precision in the profile likelihood—surely we know *less* about σ than the profile likelihood suggests.

A re-expression of the likelihood provides an alternative approach. Write

$$L(\mu, \sigma) = L_1(\mu, \sigma \mid \bar{y}) M(\sigma \mid s),$$

where

$$L_1(\mu, \sigma \mid \bar{y}) = \frac{\sqrt{n}}{\sqrt{2\pi}\sigma} \exp\left\{-\frac{n(\bar{y} - \mu)^2}{2\sigma^2}\right\},$$

$$M(\sigma \mid s) = c\left(\frac{s}{\sigma}\right)^{n-1} \exp\left\{-\frac{(n-1)s^2}{2\sigma^2}\right\}$$

and c is a constant not involving μ or σ. Here we denote explicitly the dependence of L_1 and M on the sufficient statistics. L_1 and M are true likelihoods, arising from the normal distribution $N(\mu, \sigma^2/n)$ for $\bar{Y}$ and the independent $\sigma^2 \chi^2_{n-1}$ distribution for $(n-1)s^2$. M depends only on σ, since the distribution of the sum-of-squares statistic $T = (n-1)s^2 = \sum(y_i - \bar{y})^2$ does not depend on μ. The distribution of $\bar{Y}$ however depends on both μ and σ.

If we are willing to ignore the information about σ in L_1, then inference about σ can be based on M, which is called a *marginal* or *restricted* likelihood for σ (Kalbfleisch and Sprott, 1970; Patterson and Thompson, 1971). It is frequently argued, following Kalbfleisch and Sprott, that L_1 does not provide "available" information about σ, since μ is also unknown. This argument is difficult to make precise (though it has been used to define weaker forms of sufficiency by Basu, 1977), and it is simpler to note that the additional information about σ in $\bar{y}$ is worth *at most* one "degree of freedom" in addition to the $n-1$ in s^2, so by using the marginal likelihood we are ignoring a proportion of at most $1/n$ of the information

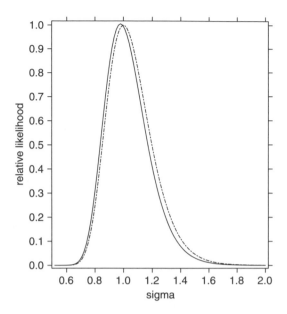

Fig. 2.20. Relative likelihood functions

about σ. As $n \to \infty$ this proportion goes to zero, and the marginal and profile likelihoods are equivalent.

The marginal likelihood is maximized at the MML or REML estimate

$$\tilde{\sigma}^2 = s^2 = \sum (y_i - \bar{y})^2 / (n - 1),$$

and the marginal likelihood ratio, or marginal relative likelihood, is

$$\mathrm{MR}(\sigma) = M(\sigma)/M(\tilde{\sigma}) = \left(\frac{s}{\sigma}\right)^{\nu} \exp\left\{\frac{\nu}{2}\left(1 - \frac{s^2}{\sigma^2}\right)\right\},$$

where $\nu = n - 1$.

The profile and marginal relative likelihoods for σ are shown in Fig. 2.20 for our example. The profile likelihood (solid curve) is offset slightly towards smaller values of σ, and is slightly more concentrated, than the marginal likelihood. As n increases the two likelihoods agree more closely. The GLIM directives are:

```
$calc sig = 0.5 + i*0.015 $
    : prl = (0.98/sig)**25*%exp(12.5*(1 - (0.96/sig**2))) $
    : mrl = (1/sig)**24*%exp(12*(1 - (1/sig**2))) $
$graph (s = 1 h = 'sigma' v = 'relative likelihood')
        prl,mrl sig 10,12 $
```

Profile or marginal likelihood intervals for σ are somewhat tedious to compute by interpolation, Newton methods or grid search, as the relative likelihood functions do not have explicit inverses. Very accurate approximate marginal likelihood intervals for σ may be obtained easily by using the *likelihood normalizing transformation* of σ, which we now describe.

2.7.3 Likelihood normalizing transformations

We find the transformation $\phi = \phi(\theta)$ of θ which makes the likelihood $L(\phi)$ as "normal" as possible. This is achieved (Anscombe, 1964; Sprott, 1973) by making the third derivative of the log likelihood $\ell(\phi)$ zero at the MLE $\hat{\phi}$. From Section 2.6.2, departures from normality at $\hat{\phi}$ are then confined to the fourth and higher derivatives of the log-likelihood in ϕ. Then provided $\hat{\phi}$ is an internal point (not on the boundary of the parameter space), ϕ has an approximately normal likelihood:

$$L(\phi) \doteq L(\hat{\phi}) \exp\{-\tfrac{1}{2}(\phi - \hat{\phi})^2 I_{\hat{\phi}}\},$$

where $I_{\hat{\phi}}$ is the observed information on the ϕ scale, and the likelihood interval for ϕ such that $L(\phi)/L(\hat{\phi}) \geq c$ is given approximately by

$$\phi \in \hat{\phi} \pm \lambda_c \mathrm{SE}(\hat{\phi})$$

where $\mathrm{SE}(\hat{\phi}) = I_{\hat{\phi}}^{-1}$ and $\lambda_c = \sqrt{-2 \log c}$. The corresponding likelihood interval for θ can then be found by reverse-transforming the interval for ϕ. The approximation on the ϕ scale is closer than on the θ scale because the term in $n^{-1/2}$ in the Taylor series expansion has coefficient zero, so the convergence to the normal likelihood is at the rate n^{-1} instead of $n^{-1/2}$.

For the exponential family, the necessary transformation does not depend on the data, only on the form of the variance/mean relationship (Anscombe, 1964). Writing μ for the mean and $V(\mu)$ for the variance, the required transformation is

$$\phi = \int^{\mu} [V(t)]^{-2/3} \, dt.$$

For likelihood inference about $\theta = \sigma^2$ from the marginal likelihood based on T, which has the $\sigma^2 \chi_\nu^2$ distribution (a gamma distribution), we have

$$\mathrm{E}[T] = \mu = \nu\sigma^2, \quad \mathrm{Var}[T] = V(\mu) = 2\nu\sigma^4 = 2\mu^2/\nu.$$

Then the likelihood normalizing transformation is

$$\phi = \int^{\mu} [V(t)]^{-2/3} \, dt = \int^{\mu} t^{-4/3} \, dt = \mu^{-1/3} = \sigma^{-2/3} = \theta^{-1/3},$$

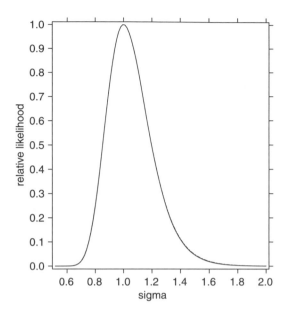

Fig. 2.21. Exact and approximate marginal likelihoods

omitting irrelevant constants. The likelihood for ϕ is very accurately normal even in very small samples, and is expressed in terms of the REML estimate $\tilde{\phi} = s^{-2/3}$ with standard error

$$\mathrm{SE}(\tilde{\phi}) = \frac{\sqrt{2}}{3} \frac{\tilde{\phi}}{\sqrt{\nu}}.$$

So a 14.6% likelihood interval for ϕ is $\tilde{\phi} \pm 1.96\sqrt{2}\tilde{\phi}/(3\sqrt{\nu})$, giving the corresponding interval for σ of

$$s(1 \pm 0.924/\sqrt{\nu})^{-3/2}.$$

This is particularly simple to calculate. Figure 2.21 shows the exact marginal likelihood (solid curve) and the normal approximation (dashed curve) on the σ scale, for our example. The GLIM directives are:

```
$calc %s = 1/(3**%sqrt(12)) $
   : phi = sig**(-2/3) $
   :nmrl = %exp(- 0.5*((phi - 1)/%s)**2) $
$graph (s = 1 h = 'sigma' v = 'relative likelihood')
       mrl,nmrl sig 10,12 $
```

The accuracy of the approximation is remarkable, and is the basis of the very accurate Wilson–Hilferty (1931) normal approximation to the distribution of χ^2. Note that the interval endpoints of s^2/σ^2 for a $100(1-\alpha)\%$ confidence interval are in general $(1 \pm 0.471\lambda_{1-\alpha/2}/\sqrt{\nu})^3$, where λ is the normal percentage point.

The usual Neyman–Pearson approach is to use the χ^2_ν sampling distribution of T/σ^2 to construct equal-tailed confidence intervals for σ^2. It has long been known that these are unsatisfactory (Tate and Klett, 1959): they are not likelihood intervals, are offset relative to the likelihood intervals with the same confidence coverage, and give biased tests for a specific σ.

For our example with $n = 25$ and $s = 1$, the 95% likelihood-based confidence limits for s^2/σ^2 are (0.535, 1.679), while those based on the usual equal-tailed χ^2_{24} distribution are (0.517, 1.640). For small ν the differences are greater: for $\nu = 10$ the corresponding limits are (0.355, 2.156) and (0.325, 2.048), respectively.

It should be noted that for very small ν and α the lower endpoint of the approximate likelihood interval may become negative. The approximation fails in this case, but we rarely ask for very high confidence with very small samples.

2.7.4 *Alternative test procedures*

In the Neyman–Pearson theory, alternatives to the LRT are widely used. The Wald and score tests of Section 2.6 can be readily generalized: let $\ell(\theta, \phi)$ be the log-likelihood, (s_θ, s_ϕ) be the partitioned score vector, and $I_{\theta\theta}, I_{\theta\phi}, I_{\phi\phi}$ be the component submatrices of the observed information matrix, defined for example by

$$I_{\theta\phi} = -\frac{\partial^2\ell}{\partial\theta\partial\phi'},$$

and $\mathcal{I}_{\theta\phi}$ be the corresponding expected information components. Then under the same regularity conditions, the sampling distribution of the MLE $\hat\theta$ is asymptotically

$$\hat\theta \to N_{p_1}(\theta, V_{\hat\theta}),$$

where $V_{\hat\theta} = I^{-1}(\hat\theta \mid \hat\phi) = (I_{\hat\theta\hat\theta} - I_{\hat\theta\hat\phi} I_{\hat\phi\hat\phi}^{-1} I_{\hat\phi\hat\theta})^{-1}$. The corresponding Wald test statistic for a hypothetical $\theta = \theta_0$ is

$$W = (\hat\theta - \theta_0)' V_{\hat\theta}^{-1} (\hat\theta - \theta_0)' = (\hat\theta - \theta_0) I(\hat\theta \mid \hat\phi)(\hat\theta - \theta_0).$$

For the score test, we evaluate the score components and expected information at the MLE of the nuisance parameter under the null hypothesis as $[s(\theta_0), s(\hat\phi(\theta_0))]$ and $\mathcal{I}_{\theta_0\hat\phi(\theta_0)}$. The score test statistic is then

$$S = [s(\theta_0), s(\hat\phi(\theta_0))]' \mathcal{I}_{\theta_0\hat\phi(\theta_0)}^{-1} [s(\theta_0), s(\hat\phi(\theta_0))].$$

Both W and S, like the LRT statistic, are asymptotically distributed as $\chi^2_{p_1}$ under the null hypothesis. However, they may give quite different results in small samples.

The score test has the computational advantage that the information has to be evaluated only under the null hypothesis: the full alternative hypothesis model does not have to be fitted. This is very convenient in many standard models where some additional complexity is introduced by an additional parameter whose ML estimation is difficult. Score tests for overdispersion (Chapter 8) or variance component structure (Chapter 9) are examples.

The Wald test requires the fitting of only the alternative hypothesis model; the likelihood ratio test requires the fitting of both models.

We illustrate the Wald and score tests on the normal model, where the null hypothesis is $H_0 : \mu = \mu_0$, and σ is the nuisance parameter. The likelihood is

$$L(\mu, \sigma) = \frac{\delta^n}{(\sqrt{2\pi})^n \sigma^n} \exp\left\{-\frac{1}{2\sigma^2}\left[\sum(y_i - \bar{y})^2 + n(\bar{y} - \mu)^2\right]\right\}.$$

The score components are

$$\frac{\partial \ell}{\partial \mu} = n(\bar{y} - \mu)/\sigma^2$$

$$\frac{\partial \ell}{\partial \sigma} = -n/\sigma + [T + n(\bar{y} - \mu)^2]/\sigma^3$$

and the elements of the Hessian matrix are

$$\frac{\partial^2 \ell}{\partial \mu^2} = -n/\sigma^2$$

$$\frac{\partial^2 \ell}{\partial \mu \partial \sigma} = -2n(\bar{y} - \mu)/\sigma^3$$

$$\frac{\partial^2 \ell}{\partial \sigma^2} = n/\sigma^2 - 3[T + n(\bar{y} - \mu)^2]/\sigma^4.$$

The observed and expected information are both diagonal. For the Wald test, the MLE of σ^2 is $\hat{\sigma}^2 = T/n$, and hence

$$W = \frac{n(\bar{y} - \mu_0)^2}{\hat{\sigma}^2} = \frac{nt^2}{n-1}.$$

For the score test, the MLE of σ^2 given $\mu = \mu_0$ is

$$\hat{\sigma}^2(\mu_0) = [T + n(\bar{y} - \mu_0)^2]/n,$$

and the σ-component of the score is zero when evaluated at $\mu_0, \hat{\sigma}(\mu_0)$. The expected information for the μ-component of the score, evaluated at $\sigma(\mu_0)$, is $n/\hat{\sigma}^2(\mu_0)$ and so the score test statistic is

$$\left[\frac{n(\bar{y}-\mu_0)}{\hat{\sigma}^2(\mu_0)}\right]^2 \bigg/ \frac{n}{\hat{\sigma}^2(\mu_0)} = \frac{n(\bar{y}-\mu_0)^2}{\hat{\sigma}^2(\mu_0)} = \frac{nt^2}{n-1+t^2}.$$

For the numerical example with $t = 2$ and $n = 25$, the LRTS is 3.85, the Wald test statistic is 4.17 and the score test statistic is 3.57. The score test statistic will be far from the Wald or squared t-statistic if the sample mean $\bar{y}$ is far from the null hypothesis value μ_0. This is a general feature of the score test: since it does not evaluate the nuisance parameter estimate under the alternative hypothesis, the curvature of the likelihood in this parameter at the null hypothesis may be far from that at the MLE, making poor agreement with the LRT.

When the null hypothesis is $\sigma = \sigma_0$ with μ the nuisance parameter, the Wald and score tests can diverge substantially from the LRT in small samples. The reader can verify that the score test statistic is

$$S = n\left(\frac{\hat{\sigma}^2}{\sigma_0^2} - 1\right)^2 \bigg/ 2,$$

while the Wald test statistic is

$$W = 2n\left(\frac{\sigma_0}{\hat{\sigma}} - 1\right)^2$$

for this parametrization, but if σ^2 is taken as the parameter of interest and the null hypothesis is expressed as $H_0 : \sigma^2 = \sigma_0^2$, the Wald test statistic is

$$W^* = n\left(\frac{\sigma_0^2}{\hat{\sigma}^2} - 1\right)^2 \bigg/ 2.$$

While all the test statistics depend only on the ratio $\hat{\sigma}^2/\sigma_0^2$, and all have the same asymptotic χ_1^2 distribution under H_0, their values can be very different. For example, if $n = 10$, $\hat{\sigma} = 1$ and $\sigma_0 = 2$, we have

$$S = 2.813, \quad W = 20, \quad W^* = 5,$$

while the LRT statistic has the value 6.36. The lack of invariance of the Wald test is a serious drawback. The substantial difference between the score and LRT statistics is due to the substantial difference in the curvature $2n/\sigma^2$ evaluated at $\sigma_0 = 2$ (5) and at $\hat{\sigma} = 1$ (20).

Bayes inference

The Bayes analysis of question (i) is again direct. The *joint* prior distribution $\pi(\theta, \phi)$ combines with the likelihood $L(\theta, \phi)$ to give the *joint posterior distribution*

$\pi(\theta, \phi \mid y)$. The *marginal posterior distribution* of θ is then obtained by integrating over ϕ:

$$\pi(\theta \mid y) = \int \pi(\theta, \phi \mid y) \, d\phi$$

$$= \frac{\int L(\theta, \phi)\pi(\theta, \phi) \, d\phi}{\int\int L(\theta, \phi)\pi(\theta, \phi) \, d\phi \, d\theta}.$$

For the normal mean–variance model, if we take the usual flat priors on μ and $\log \sigma$, it is easily shown that the joint posterior distribution of μ and σ can be expressed as the conditional normal distribution $N(\bar{y}, \sigma^2/n)$ of $\mu \mid \sigma$, and the marginal χ^2_{n-1} distribution of T/σ^2. The marginal distribution of μ, integrating over σ, is such that $t = \sqrt{n}(\bar{y} - \mu)/s$ has a t_{n-1} distribution. This choice of priors gives the analogue of the repeated-sampling distributions; other choices of priors do not. The $100(1 - \alpha)\%$ HPD interval for μ is identical to the $100(1 - \alpha)\%$ confidence interval.

Hypothesis testing using Bayes factors suffers the same difficulty as in the simpler case given above. The posterior distribution of the true likelihood ratio was extended to the general case by Aitkin (1997). For a null hypothesis $H_0 : \theta = \theta_0$ with a nuisance parameter ϕ, the posterior distribution $\pi(\theta, \phi \mid y)$ of all the parameters is used to find the posterior distribution of the true likelihood ratio $\text{TLR} = L(\theta_0, \phi)/L(\theta, \phi)$ (evaluated at the true values of the nuisance parameters), or of $-2 \log \text{TLR}$, and hence the posterior probability that the TLR is less than k, for any specified k. This posterior probability can be expressed (in large samples, and with diffuse priors) in terms of the maximized likelihood ratio and hence in terms of the P-value. This allows the simple recalibration of maximized likelihood ratios, or P-values, to provide direct measures of strength of evidence; a particularly striking result (extended from Dempster, 1974) is that, with normal likelihoods and flat priors, the P-value is equal to *the posterior probability that the true likelihood ratio is greater than 1*, or equivalently, $1 - P$ is the posterior probability that the true likelihood ratio is less than 1. So a P-value of 0.05 means that the posterior probability is 0.95 that the true likelihood ratio is less than 1. Since we need a likelihood ratio of $\frac{1}{10}$ or less for quite strong evidence, this P-value does not provide that evidence.

In general, the recalibration requires us to interpret P-values more conservatively: to achieve a posterior probability of $\pi_{k,p}$ that the true likelihood ratio for the null hypothesis to the alternative is less than k, the necessary P-value (for normal likelihoods, and flat priors) is

$$P = 1 - F_\nu(F_\nu^{-1}(\pi_{k,p}) - 2 \log k),$$

where $F_p(x)$ is the cdf of the χ^2_p distribution at x. For reasonably strong evidence against the null hypothesis, we will require $k = 0.1$ and $\pi_{k\nu} = 0.84$,

Table 2.2. P-value required for $\Pr[\text{TLR} < 0.1] = 0.84$

p	1	2	3	4	5	6	7	8	9	10
PV	0.010	0.016	0.021	0.025	0.028	0.031	0.034	0.037	0.039	0.042
p	11	12	13	14	15	16	17	18	19	20
PV	0.044	0.046	0.048	0.049	0.051	0.053	0.054	0.055	0.057	0.058

giving the P-values (PV) shown in Table 2.2 (adapted from Aitkin, 1997, Table 4).

This P-value table is for large samples; small-sample results remain to be developed for specific models. A detailed discussion is given in Chadwick (2002), and some examples in Aitkin *et al.* (2005).

We will use this table as a guideline for model interpretation in subsequent chapters: a single-parameter hypothesis requires a P-value of 0.01 or less (or an equivalent critical Z-value of 2.57, or a relative likelihood of $\exp\{-\frac{1}{2} \cdot 2.57^2\} = 0.037$) for quite strong evidence against the hypothesis, but as the number p of specified parameters in the null hypothesis increases, so does the P-value required, but at a quite slow rate for large p.

For the normal mean–variance model, we have as before

$$\text{TLR} = L(\mu_0, \sigma)/L(\mu, \sigma)$$

$$-2 \log \text{TLR} = \frac{n(\bar{y} - \mu_0)^2}{\sigma^2} - \frac{n(\bar{y} - \mu)^2}{\sigma^2}$$

$$= \frac{n(\bar{y} - \mu_0)^2}{s^2} \cdot \frac{s^2}{\sigma^2} - \frac{n(\bar{y} - \mu)^2}{\sigma^2}$$

$$= t_0^2 \cdot \frac{T}{(n-1)\sigma^2} - Z^2,$$

where $Z \sim N(0, 1)$ conditional on σ, and $T/\sigma^2 \sim \chi_{n-1}^2$ marginally. The posterior distribution of $-2 \log \text{TLR}$ does not have a closed form in this case, though we can evaluate immediately

$$\Pr[\text{TLR} < 1] = \Pr[-2 \log \text{TLR} > 0]$$

$$= \Pr\left[\frac{Z^2}{T/\sigma^2} < t_0^2\right]$$

$$= \Pr[F_{1,n-1} < t_0^2]$$

$$= 1 - P,$$

where P is the P-value of the observed t_0. So as in the simpler cases mentioned earlier, the P-value is again the posterior probability that the TLR is greater than 1. For other values of k, and for other models without closed-form posteriors, *simulation* of the posterior distribution provides a general solution. Further discussion is given in Chadwick (2002) and Aitkin *et al.* (2005).

We finally discuss the third inferential question, assessing the evidence for $H_1 : \mu = \mu_1$ against $H_2 : \mu = \mu_2$. This is the most difficult for the Neyman–Pearson theory, as for the simpler case of two specified null hypotheses, but it has a surprisingly simple answer from the posterior distribution of the TLR. Proceeding as above, we have

$$\text{TLR} = L(\mu_1, \sigma)/L(\mu_2, \sigma)$$

$$-2\log\text{TLR} = \frac{n(\bar{y} - \mu_1)^2}{\sigma^2} - \frac{n(\bar{y} - \mu_2)^2}{\sigma^2}$$

$$= \left[\frac{n(\bar{y} - \mu_1)^2}{s^2} - \frac{n(\bar{y} - \mu_2)^2}{s^2} \right] \cdot \frac{s^2}{\sigma^2}$$

$$= [t_1^2 - t_2^2] \cdot \frac{T}{(n-1)\sigma^2}.$$

Thus $-2\log\text{TLR} \sim [t_1^2 - t_2^2]\chi_{n-1}^2/(n-1)$. Here $[t_1^2 - t_2^2]/(n-1) = -5/24 = -0.208$, and taking $k = 10$

$$\Pr[\text{TLR} > 10] = \Pr[-2\log\text{TLR} < -2\log 10]$$

$$= \Pr\left[\chi_{n-1}^2 < \frac{n-1}{t_1^2 - t_2^2} \cdot (-2\log 10) \right]$$

$$= \Pr[\chi_{24}^2 < (-4.8) \cdot (-4.60) = 22.08]$$

$$= 0.426.$$

The posterior probability that H_1 is quite strongly supported (TLR > 10) relative to $H_2 < 0.5$, but it is certainly better supported.

As n increases $\chi_{n-1}^2/(n-1)$ rapidly approaches $N(1, 2/(n-1))$ which converges to 1 in probability. For large n the disparity or deviance difference between the models is effectively compared with $-2\log k$ as in the case of two simple hypotheses. This result (for large n) applies generally to the comparison of "common-family" models with the same number of parameters which are both nested in a larger family.

We consider now a second example.

2.7.5   *Binomial model*

A random sample of $n = 10$ is drawn from the STATLAB population and the number of "smoking mothers"—mothers smoking at the time of birth of the baby—is found to be $r = 1$. What information do we have about θ, the proportion of smoking mothers in the STATLAB population? There is no specific hypothesis about its value.

The probability model here is a Bernoulli trial model, in which "trials" are drawn of a mother from the population, and θ is the probability of a "success"—the drawing of a smoking mother from the population. The population is large compared to the sample and so sampling with and without replacement are equivalent.

Using the standard device of a dummy or indicator variable y_i which takes the value 1 if the i-th trial results in success and 0 if failure, the probability of the sequence $y_1, \ldots, y_n$ of successes and failures is

$$L(\theta) = \prod_{i=1}^{n} \theta^{y_i}(1 - \theta)^{1-y_i}$$

$$= \theta^r (1 - \theta)^{n-r},$$

where $r = \sum y_i$ is the number of successes, and this does not depend on the particular sequence of zeroes and ones, only on the total number of successes. Thus r is a sufficient statistic for θ. The likelihood function is shown in Fig. 2.22 on a fine grid. It is remarkably skewed.

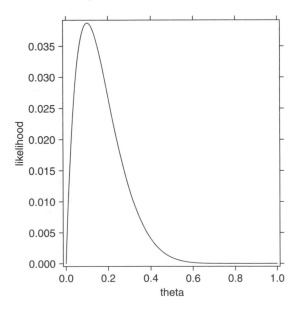

Fig. 2.22. Likelihood for one success in 10 trials

The GLIM directives are:

```
$var 1001 theta bl $
$calc theta = (%cu(1) - 1)/1000 $
      : bl = theta*(1 - theta)**9 $
$graph (s = 1 h = 'theta' v = 'likelihood')
        bl theta 10 $
```

The MLE is $\hat{\theta} = 0.1$, and a likelihood interval for θ is obtained from $R(\theta) < c$. The equation $R(\theta) = c$ cannot be inverted to give θ analytically, so numerical computation is necessary. The simplest approach is to solve it iteratively by the Newton–Raphson method. We do not give details here. For a relative likelihood of 0.146, the likelihood interval endpoints are 0.006 and 0.372 (to 3 dp). The somewhat tedious computation of the interval endpoints can be avoided in single parameter models in the exponential family by using the (approximate) likelihood normalizing transformation, as in the previous discussion. For the binomial or Bernoulli model

$$f(y \mid \theta) = \theta^y (1 - \theta)^{1-y},$$

we have

$$\mathrm{E}[Y] = \mu = \theta, \quad \mathrm{Var}[Y] = V(\mu) = \mu(1 - \mu),$$

and the required transformation is

$$\phi = \int_0^\theta t^{-2/3}(1 - t)^{-2/3} \, \mathrm{d}t,$$

an incomplete Beta integral. This was tabulated by Anscombe (1964) in the form $\frac{1}{3}B_\theta(\frac{1}{3}, \frac{1}{3})$. We adopt a slightly different form in which ϕ has the range (0,1) like θ: define

$$\phi = \int_0^\theta t^{-2/3}(1 - t)^{-2/3} \, \mathrm{d}t / \int_0^1 t^{-2/3}(1 - t)^{-2/3} \, \mathrm{d}t$$

so that ϕ is the cdf at θ of the Beta $(\frac{1}{3}, \frac{1}{3})$ distribution. A graph of ϕ against θ is shown in Fig. 2.23. The transformation is similar in shape to the logit transformation (Chapter 4) but has finite limits 0 and 1, like the arc-sine transformation. The transformation is easily calculated in GLIM by

```
$calc phi = %btp(theta,1/3,1/3) $
$graph(s = 1 v = 'phi' h = 'theta')
      phi theta 10 $
```

The information for $L(\phi)$ is easily obtained by chain differentiation, giving

$$I(\hat{\phi}) = I(\hat{\theta}) / [\phi'(\hat{\theta})]^2$$

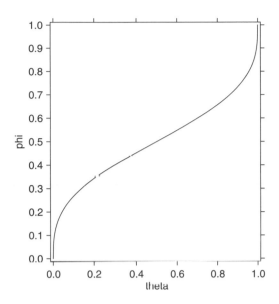

Fig. 2.23. Likelihood normalizing transformation of theta

and so

$$\mathrm{SE}(\hat{\phi}) = \mathrm{SE}(\hat{\theta})\phi'(\hat{\theta}).$$

In our binomial example

$$I(\hat{\theta}) = \frac{n}{\hat{\theta}(1 - \hat{\theta})}$$

$$\phi'(\hat{\theta}) = \hat{\theta}^{-2/3}(1 - \hat{\theta})^{-2/3}/B(\tfrac{1}{3}, \tfrac{1}{3})$$

giving

$$SE(\hat{\phi}) = \hat{\theta}^{-1/6}(1 - \hat{\theta})^{-1/6}/B(\tfrac{1}{3}, \tfrac{1}{3})\sqrt{n}.$$

To evaluate this, we need $B(\tfrac{1}{3}, \tfrac{1}{3}) = \Gamma^2(\tfrac{1}{3})/\Gamma(\tfrac{2}{3}) = 5.300$.

With $\hat{\theta} = 0.1$ and $n = 10$, the value of $\hat{\phi}$ is 0.267 and $\mathrm{SE}(\hat{\phi}) = 0.0891$. The approximate 0.146 likelihood interval for ϕ is then $\hat{\phi} \pm 1.96\mathrm{SE}(\hat{\phi})$, that is (0.092, 0.442), which transforms to $\theta \in (0.004, 0.380)$. This agrees quite well with the exact interval (0.006, 0.372) found earlier, though it is slightly longer. The relative likelihoods at the endpoints are 0.100 and 0.133, not quite equal to 0.146. Figure 2.24 shows the relative likelihood on the ϕ scale (solid line) and the normal approximation (dotted line). The GLIM directives are:

```
$calc %s = 0.1**(-1/6)*0.9**(-1/6)/(5.3*%sqrt(10)) $
    :  nl = 0.1*0.9**9*%exp(-0.5*((phi - 0.267)/%s)**2) $
```

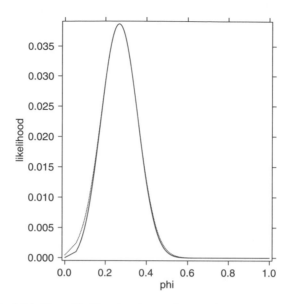

Fig. 2.24. Exact likelihood and normal approximation, phi scale

```
$graph (s = 1 h = 'phi' v = 'likelihood')
       bl,nl phi 10,11 $
```

The visual agreement is close even for this small sample and extreme result except in the tails, where the normal approximation gives intervals which are slightly too long. The normal approximation can be further improved using a t-approximation to the log-likelihood with the degrees of freedom determined by the fourth derivative of the log-likelihood (Sprott, 1980). We do not pursue this as even in small samples the estimated degrees of freedom are usually very large and the normal approximation is recovered.

When $r = 0$ or n the normalizing transformation fails, as the maximum is on the boundary of the parameter space. However, relative likelihood intervals can be easily calculated analytically in these cases: for example, when $r = 0$, the likelihood is

$$L(\theta) = (1 - \theta)^n,$$

and a relative likelihood of c gives the interval for θ of $(0, 1 - c^{1/n})$. If $r = n$ the interval is $(c^{1/n}, 1)$. For $n = 10$ and $c = 0.146$, $c^{1/n} = 0.825$.

The macro binoci uses the normalizing transformation to construct confidence intervals for θ. A version of this macro, binocisim, provides the simultaneous confidence band for the population model cdf discussed in Section 2.3. The macro binoci is used in the same way for a single binomial outcome, except that the standard length of the vectors of interval endpoints has to be set to 1, and a third argument has to be specified, the number of standard errors on the ϕ scale. For the event of $r = 1$ success in $n = 10$ trials, the (approximate) 95% likelihood-based confidence interval is obtained (to 3 dp) by:

```
$slen 1 $
$num r = 1 : n = 10 $
$calc %1 = 1.96 $
$inp 'binoci' $
$use binoci r n %1 $
1.96-sigma confidence interval for p is
0.004, 0.380
```

The confidence interval construction above does not guarantee a minimum 95% coverage over all θ. For this we need the classical interval construction, in which we consider other possible values of r, besides the observed value 1, which might have been generated by the given value $n = 10$. The $100(1 - \alpha)\%$ confidence limits for θ are the solutions of

$$\Pr(R \geq 1 \mid \theta_L) = 1 - (1 - \theta)^{10} = \alpha/2$$
$$\Pr(R \leq 1 \mid \theta_U) = (1 - \theta)^{10} + 10\theta(1 - \theta)^9 = \alpha/2,$$

giving for $\alpha = 0.05$, $\theta_L = 0.0025$, $\theta_U = 0.445$. These values are easily obtained from the cumulative binomial distribution function $\%bip(r,n,p)$ in GLIM. They are substantially wider than the likelihood-based intervals.

The construction of the classical confidence interval from the exact binomial distribution of R is frequently avoided by relying on the large-sample normality of the distribution of $\hat{\Theta}$ or of the score. The log likelihood, score and information are

$$\ell(\theta) = r \log \theta + (n - r) \log(1 - \theta)$$
$$s(\theta) = \frac{r}{\theta} - \frac{n - r}{1 - \theta}$$
$$= \frac{r - n\theta}{\theta(1 - \theta)}$$
$$I(\theta) = \frac{r}{\theta^2} + \frac{n - r}{(1 - \theta)^2}$$
$$I(\hat{\theta}) = \frac{n}{\hat{\theta}(1 - \hat{\theta})}$$
$$\mathcal{I}(\theta) = \frac{n}{\theta(1 - \theta)}.$$

The confidence intervals based on the asymptotic normal distribution of the MLE $\hat{\Theta} \sim N(\theta, \hat{\theta}(1 - \hat{\theta})/n)$ (equivalent to inverting the Wald test) are

$\theta \in \hat{\theta} \pm \lambda_{\alpha/2}\sqrt{\hat{\theta}(1-\hat{\theta})/n}$. In our example with $\hat{\theta} = 0.1$ and $n = 10$, the approximate 95% confidence interval using $\lambda = 1.96$ is $(-0.086, 0.286)$ which includes impossible negative values of θ. Truncating the interval to $(0, 0.286)$ does not help matters because we know θ must be *positive*, since one success has been observed. If $r = 0$ successes are observed, this interval construction collapses with a zero standard error.

Proper intervals may be obtained by not substituting $\hat{\theta}$ for θ in the information (equivalent to inverting the score test rather than the Wald test). The confidence interval is

$$\theta \in \hat{\theta} \pm \lambda_{\alpha/2}\sqrt{\theta(1-\theta)/n}$$

which is equivalent to

$$\theta \in (\theta_L, \theta_U),$$

where $\theta_L < \theta_U$ are the roots of the quadratic equation (writing $\lambda = \lambda_{\alpha/2}$)

$$(\theta - \hat{\theta})^2 = \lambda^2 \theta(1-\theta)/n.$$

The roots are

$$\frac{\hat{\theta} + \lambda^2/2n \pm \lambda\sqrt{\hat{\theta}(1-\hat{\theta})/n + \lambda^2/4n^2}}{1 + \lambda^2/n}.$$

This interval always lies in [0,1]. If $r = 0$ the interval endpoints are 0 and $(\lambda^2/n)/(1 + \lambda^2/n) = 0.278$ for $n = 10$. In our example with $\lambda = 1.96$, $\theta_L = 0.018$, $\theta_U = 0.404$.

Bayes inference is straightforward. The uniform prior on $(0,1)$ gives a Beta posterior density:

$$\pi(\theta \mid y) = \theta^r(1-\theta)^{n-r}/B(r+1, n-r+1),$$

which is the likelihood scaled to integrate to 1. Tail area probabilities or critical values are easily found from the Beta distribution using %btp or %btd, but the HPD interval is more difficult.

For the example with $r = 1$ and $n = 10$, a 95% equal-tailed credible interval for θ from the Beta(2,10) distribution is $(0.023, 0.413)$. This is shifted upwards substantially relative to the likelihood interval.

If $r = 0$, the (one-tailed) 95% credible interval is $(0, 0.238)$.

2.7.6 *Hypergeometric sampling from finite populations*

We noted in Section 2.5 that the construction of the binomial likelihood depended on the assumption that the population size N was large compared to the sample size n, and so sampling with and without replacement were equivalent. For general response variables Y, it is difficult to construct the likelihood for sampling without replacement, since the approximating population model has to change with the sampled population, but for categorical response variables the likelihood construction is straightforward. We illustrate with the binary response model for the smoking mothers example.

Let R be the number of smoking mothers in the population, with $\theta = R/N$. We draw a random sample of size n without replacement, and find r smoking mothers. The likelihood function (in the parameter $R = N\theta$) is given by the hypergeometric sampling model

$$L(R) = \binom{R}{r}\binom{N-R}{n-r} / \binom{N}{n}, \quad R = r, r+1, \ldots, N - (n - r).$$

This can be readily computed as a function of R using %lga, the log gamma function, to evaluate the factorials. Expanding the binomial coefficients, we have

$$L(R) = c \cdot \prod_{i=1}^{r}(R - i + 1) \prod_{j=1}^{n-r}(N - R - j + 1)$$

$$= c' \cdot \prod_{i=1}^{r}\left(\frac{R}{N} - \frac{i-1}{N}\right)\prod_{j=1}^{n-r}\left(\frac{N-R}{N} - \frac{j-1}{N}\right)$$

$$= c'\theta^r(1-\theta)^{n-r}\prod_{i=1}^{r}\left(1 - \frac{i-1}{R}\right)\prod_{j=1}^{n-r}\left(1 - \frac{j-1}{N-R}\right)$$

$$= c'\theta^r(1-\theta)^{n-r}$$

$$\cdot \exp\left[-\frac{r(r-1)}{2R} - \frac{(n-r)(n-r-1)}{2(N-R)} + O\left(\frac{1}{R^2}\right) + O\left(\frac{1}{(N-R)^2}\right)\right].$$

If r and $n - r$ are small compared to R and $N - R$, the exponential term will be close to 1, and the likelihood will be equivalent to the previous binomial likelihood. It is easily shown by considering the ratio $L(R + 1)/L(R)$ that the likelihood is maximized when

$$R = \hat{R} = [(N + 1)r/n + 1] = [N\tilde{\theta} + 1 + \tilde{\theta}],$$

where $\tilde{\theta} = r/n$ is the binomial model MLE and $[x]$ is the integral part of x. For our example

$$\hat{\theta} = \hat{R}/N = 130/1296 = 0.1003,$$

and the binomial and hypergeometric likelihoods are indistinguishable since n is very small compared to N.

2.8   The effect of the sample design on inference

In our second example we have assumed that a simple random sample of fixed size n is taken, and r "successes" are observed. But other sample designs might have led to the same data. For example, individuals might have been sampled randomly until the r-th "success"—smoking mother—was found, this occurring at the n-th observation. The likelihood function would, as a function of θ, be the same as before:

$$L(\theta) = \theta^{r-1}(1 - \theta)^{n-r} \cdot \theta$$

but now r is fixed by the sample design, and the sample size N is the random variable. It is a striking feature of the difference between the Bayes and likelihood theories, and the Neyman–Pearson theory, that this different sample design affects the inference about θ in the latter theory, but not in the former.

For the Bayes and likelihood theories, this is obvious: the likelihood interval depends only on the likelihood from the observed data, so for the same likelihood and the same prior the same Bayes and likelihood intervals will be obtained.

To see the effect on the Neyman–Pearson theory, note that the sufficient statistic N has a negative binomial distribution with parameters r and θ:

$$\Pr(N = n \mid r, \theta) = \binom{n - 1}{r - 1}\theta^r (1 - \theta)^{n-r}$$

for $n = r, r + 1, \ldots$ since the last trial must result in a success, and the preceding $r - 1$ successes can occur anywhere in the preceding $n - 1$ trials. Note that $E[N] = r/\theta$. To construct a Neyman–Pearson 95% confidence interval for θ, we now have to consider other possible values of n, besides the observed value 10, which might have generated the given value $r = 1$. Now the confidence limits for θ are the solutions of

$$\Pr(N \geq 10 \mid \theta_U) = \alpha/2,$$

$$\Pr(N \leq 10 \mid \theta_L) = \alpha/2,$$

which require the cdf of the negative binomial distribution, not available in GLIM. The solutions are for $\alpha = 0.05, \theta_L = 0.0025, \theta_U = 0.336$. The lower endpoint is the same, but the upper endpoint is quite different from the endpoint 0.445 for binomial sampling.

This dependence of the family of intervals on the sampling design used to obtain the data, commonly called the *stopping rule*, is a major complication of conventional confidence interval construction. A striking example of the difficulty of the Neyman–Pearson inference in such cases is the ECMO trial data, discussed in Begg (1990), in which eight different P-values are given for the same 2×2 contingency table. In this example the precise design or stopping rule was not clearly specified, or if specified, was not followed, and so strictly speaking *no* classical confidence interval statement can be made! The different P-values are based on different hypothetical replications of the study. Chapter 4 gives more details.

In the Neyman–Pearson theory we have to be aware of the possibility of different forms of hypothetical replication, each of which may lead to a different statement of precision of the likelihood interval. Defining the *reference set* of hypothetical replications with which our sample is to be compared becomes an important issue in this theory. This is another aspect of the stopping rule issue discussed above: if we do not know exactly how our sample was obtained, how can we know which reference set should be used for its evaluation?

In the next section we discuss the exponential family of distributions.

2.9 The exponential family

The exponential family of distributions is central to data analysis in most scientific fields. Members of this family can represent variation in proportions, counts and time durations as well as the usual "continuous" measurements for which the normal distribution was traditionally used. Regression modelling using the binomial, Poisson and gamma distributions has largely replaced the older approximate methods involving transformations of the response to more nearly approximate a normal distribution.

The probability function $f(y)$ for this family has the general form

$$\log f(y) = [y\psi - b(\psi)]/\phi + c(y, \phi).$$

The parameter ψ is the "natural" or "canonical" parameter of the distribution, and ϕ is a "scale" parameter. The binomial and Poisson distributions do not have a scale parameter (formally we can set it to 1.0), while in the normal distribution ϕ is related to σ^2 and in the gamma distribution to the shape parameter r. We now consider some general properties of the exponential family.

2.9.1 *Mean and variance*
Differentiating with respect to ψ, we have

$$\frac{\partial \log f}{\partial \psi} = [y - b'(\psi)]/\phi$$

$$\frac{\partial^2 \log f}{\partial \psi^2} = -b''(\psi)/\phi.$$

Since $\int f(y)\,dy = 1$, differentiating with respect to ψ gives

$$0 = \int \frac{\partial f}{\partial \psi}\,dy = \int \frac{\partial \log f}{\partial \psi}\cdot f\,dy$$

$$= E\left[\frac{\partial \log f}{\partial \psi}\right]$$

$$= \{E[Y] - b'(\psi)\}/\phi$$

$$0 = \int \frac{\partial^2 f}{\partial \psi^2}\,dy = \int \left\{\frac{\partial^2 \log f}{\partial \psi^2}\cdot f + \left(\frac{\partial \log f}{\partial \psi}\right)^2 \cdot f\right\}\,dy$$

$$= E\left[\frac{\partial^2 \log f}{\partial \psi^2}\right] + E\left[\frac{\partial \log f}{\partial \psi}\right]^2.$$

Thus

$$\mu = E[Y] = b'(\psi)$$

$$\mathrm{Var}[Y] = E[Y - b'(\psi)]^2$$

$$= \phi^2 E\left[\frac{\partial \log f}{\partial \psi}\right]^2$$

$$= -\phi^2 E\left[\frac{\partial^2 \log f}{\partial \psi^2}\right]$$

$$= \phi b''(\psi).$$

The mean involves only ψ while in the variance ϕ appears as a scale multiplier (whence its name).

2.9.2 *Generalized linear models*

A generalized linear model is defined by the following three components:

(1) an exponential family distribution for the response variable Y;
(2) a linear regression function or *linear predictor* η in the explanatory variables $x_1, \ldots, x_p$:

$$\eta = \boldsymbol{\beta}'\boldsymbol{x} = \beta_0 x_0 + \beta_1 x_1 + \cdots + \beta_p x_p,$$

where x_0 is identically 1;
(3) a *parameter transformation* or *link function* $g(\mu)$ which relates the linear predictor η to the mean μ:

$$\eta = g(\mu).$$

We use the standard abbreviation of GLM for generalized linear model, though this causes frequent confusion with the older meaning of GLM—*general* linear model, which is the generalized linear model in the normal distribution.

It is an important restriction of GLMs that the explanatory variables *do not affect the scale parameter*. This restriction is relaxed in Chapter 3 on the normal distribution and in Chapter 6 on the gamma distribution, where we consider *double modelling* of both parameters by regression models.

GLIM provides general methods for fitting GLMs, for a wide range of standard link functions for each distribution, and also for non-standard user-defined link functions. We have to specify the response variable Y to be modelled, the probability ("error") distribution for Y, the link function and the regression model to be fitted.

1. The response variable is specified by the directive $yvar whose argument is the name of the response variable.
2. The error distribution is specified by the directive $error, which has distributional arguments n (normal), b (binomial), p (Poisson), g (gamma) or i (inverse Gaussian).
3. The link function is specified by the $link directive.
4. The regression model is specified as the argument of the $fit directive.

Much more general models (e.g. non-linear in the parameters) can be fitted by generalizations of the computational algorithm used in GLIM. We shall concentrate on the standard models in GLIM, with extensions to other distributions (like the exponential, Weibull and extreme value, and mixtures of the exponential family) which are fitted by extensions of the standard models.

2.9.3 *Maximum likelihood fitting of the GLM*

We consider the GLM with $\eta_i = g(\mu_i) = \boldsymbol{\beta}'\mathbf{x}_i$. The log-likelihood function is

$$\ell(\boldsymbol{\beta}, \phi) = \sum_i [y_i \psi_i - b(\psi_i)]/\phi + \sum_i c(y_i, \phi)$$

and its derivatives, the score function components, are

$$\mathbf{s}(\boldsymbol{\beta}) = \frac{\partial \ell}{\partial \boldsymbol{\beta}} = \sum_i [y_i - b'(\psi_i)] \cdot \frac{\partial \psi_i}{\partial \boldsymbol{\beta}} / \phi$$

$$s(\phi) = \frac{\partial \ell}{\partial \phi} = -\sum_i [y_i \psi_i - b(\psi_i)]/\phi^2 + \sum_i c'(y_i, \phi),$$

where c' is the derivative of c with respect to ϕ. Substitution of

$$\frac{\partial \psi_i}{\partial \boldsymbol{\beta}} = \frac{d\psi_i}{d\mu_i} \cdot \frac{d\mu_i}{d\eta_i} \cdot \frac{\partial \eta_i}{\partial \boldsymbol{\beta}}$$

$$= \frac{1}{b''(\psi_i)} \cdot \frac{1}{g'(\mu_i)} \cdot \mathbf{x}_i$$

and

$$\phi b''(\psi_i) = \mathrm{Var}[Y_i] = V_i, \quad b'(\psi_i) = \mu_i, \quad g'(\mu_i) = g_i'$$

gives

$$\mathbf{s}(\boldsymbol{\beta}) = \frac{\partial \ell}{\partial \boldsymbol{\beta}} = \sum_i (y_i - \mu_i)\mathbf{x}_i / V_i g_i'.$$

For the second derivatives, we have

$$\frac{\partial^2 \ell}{\partial \boldsymbol{\beta} \partial \boldsymbol{\beta}'} = \sum_i [-b''(\psi_i)] \frac{\partial \psi_i}{\partial \boldsymbol{\beta}} \frac{\partial \psi_i}{\partial \boldsymbol{\beta}'} / \phi + \sum_i [y_i - b'(\psi_i)] \frac{\partial^2 \psi_i}{\partial \boldsymbol{\beta} \partial \boldsymbol{\beta}'} / \phi$$

$$\frac{\partial^2 \ell}{\partial \boldsymbol{\beta} \partial \phi} = -\sum_i [y_i - b'(\psi_i)] \cdot \frac{\partial \psi_i}{\partial \boldsymbol{\beta}} / \phi^2 = -\mathbf{s}(\boldsymbol{\beta}) / \phi$$

$$c\frac{\partial^2 \ell}{\partial \phi^2} = 2\sum_i [y_i \psi_i - b(\psi_i)] / \phi^3 + \sum_i c''(y_i, \phi).$$

Straightforward algebra gives

$$\frac{\partial^2 \ell}{\partial \boldsymbol{\beta} \partial \boldsymbol{\beta}'} = -\sum_i \mathbf{x}_i \mathbf{x}_i' / V_i g_i'^2 - \sum_i (y_i - \mu_i)\mathbf{x}_i \mathbf{x}_i' (V_i g_i'' + V_i' g_i') / V_i^2 g_i'^3$$

$$= -X'W^*X,$$

where W^* is a diagonal weight matrix with elements

$$w_i^* = w_i + (y_i - \mu_i)(V_i g_i'' + V_i' g_i') / V_i^2 g_i'^3$$

and

$$w_i = (V_i g_i'^2)^{-1}, \quad V_i' = \frac{dV_i}{d\mu_i}, \quad g_i'' = \frac{d^2 g(\mu_i)}{d\mu_i^2},$$

while X is the $n \times (p + 1)$ design matrix of the explanatory variables $\mathbf{x}$. If the likelihood has an internal maximum over $\boldsymbol{\beta}$ at $\hat{\boldsymbol{\beta}}$, then $\mathbf{s}(\hat{\boldsymbol{\beta}}) = \mathbf{0}$. Also the expected score in $\boldsymbol{\beta}$ is zero, and hence both the observed and expected information matrices have a block-diagonal structure, with the off-diagonal block of cross-derivatives between β and ϕ equal to zero. Further, since ϕ appears in the score equation for β only as a scale constant in V_i, the MLE of β can be obtained independently of ϕ. The MLE of ϕ can then be found by solving the single score equation in ϕ evaluated at the MLE $\hat{\beta}$. This process is familiar from the normal distribution.

We now give the form of the Newton–Raphson (NR) and the Fisher scoring (FS) algorithms for the GLM. We solve iteratively for the zero of the score function $s(\beta)$. Let the estimate at the r-th iteration be β_r. Then the estimate at the $(r+1)$-th iteration of the NR algorithm is given by solving

$$0 = s(\beta_r) + \frac{\partial s(\beta)}{\partial \beta'}(\beta_{r+1} - \beta_r)$$

giving

$$\beta_{r+1} = \beta_r - H^{-1}(\beta_r)s(\beta_r)$$
$$= \beta_r + (X'W_r^*X)^{-1}s(\beta_r),$$

where H is the Hessian matrix and W_r^* is W^* evaluated at the estimate β_r. Now

$$s(\beta) = \sum_i (y_i - \mu_i)g'(\mu_i)w_i\mathbf{x}_i$$
$$= X'W\mathbf{u},$$

where

$$u_i = (y_i - \mu_i)g'(\mu_i).$$

Then each iteration of the NR algorithm can be expressed as

$$\beta_{r+1} = \beta_r + (X'W_r^*X)^{-1}X'W_r\mathbf{u}_r$$
$$= (X'W_r^*X)^{-1}(X'W_r^*X\beta_r + X'W_r\mathbf{u}_r)$$
$$= (X'W_r^*X)^{-1}(X'W_r^*\eta_r + X'W_r\mathbf{u}_r)$$
$$= (X'W_r^*X)^{-1}X'W_r^*\mathbf{z}_r^*,$$

where η_r is the vector of "fitted values" of the linear predictor η at the r-th iteration, and

$$\mathbf{z}^* = \eta + W^{*-1}W\mathbf{u}.$$

Thus, each iteration of the NR algorithm can be expressed as a weighted least squares regression of an "adjusted dependent variate" $\mathbf{z}^*$ on the explanatory variables $\mathbf{x}$ with weight vector (diagonal weight matrix) W^*.

GLIM uses the FS algorithm rather than the NR algorithm. This uses the expected rather than the observed information. Since $E[Y_i - b'(\psi_i)] = 0$ for all i, the second term in the Hessian matrix is identically zero, and the weights w_i and w_i^* are identical. Thus each iteration of the FS algorithm can be expressed as

$$\boldsymbol{\beta}_{r+1} = \boldsymbol{\beta}_r + (X'W_rX)^{-1}X'W_r\mathbf{u}_r$$
$$= (X'W_rX)^{-1}(X'W_rX\boldsymbol{\beta}_r + X'W_r\mathbf{u}_r)$$
$$= (X'W_rX)^{-1}X'W_r\mathbf{z}_r,$$

where

$$\mathbf{z} = \boldsymbol{\eta} + \mathbf{u}.$$

The adjusted dependent variate in this case is much simpler. The FS algorithm is generally more stable than the NR, and converges at about the same rate. In rare cases it may not converge but oscillate in successive iterations (Ridout, 1990).

The NR algorithm gives the correct observed information for $\boldsymbol{\beta}$, after scaling by the MLE of ϕ (used in the weight matrix W^*), though as noted the parameter estimates do not depend on ϕ. The FS algorithm gives the (estimated) expected information matrix, when correspondingly scaled. This scaling is not needed for the Poisson and binomial distributions, where $\phi = 1$. In GLIM the value of ϕ may be set externally by the $scale directive.

If it is *not* set by the user, it is automatically calculated by GLIM for the normal and gamma distributions. Details are given in the relevant chapters.

The observed information represents the curvature of the observed log-likelihood at the MLE. The expected information represents the *average* curvature over outcomes Y with the same explanatory variable values. In models with canonical links (when $\eta = \psi$) these informations are identical, but for other links they are different.

For individual parameter estimates $\hat{\beta}_j$, the standard error is s_j, where s_j^2 is the j-th diagonal element of the inverse of the (observed or expected) information matrix.

2.9.4 *Model comparisons through maximized likelihoods*

The comparison of competing regression models for the data is based on the LRTS.

Consider two candidate models for the data: a "full" model $\eta_f = \boldsymbol{\beta}'\mathbf{x}$ and a "reduced" model $\eta_r = \boldsymbol{\beta}'_r\mathbf{x}_r$, where $\mathbf{x}_r$ is a vector of length $p_1 + 1$, made up of the constant 1 plus p_1 of the explanatory variables in $\mathbf{x}$, and $\boldsymbol{\beta}_r$ is the corresponding vector of regression coefficients. We say that the reduced model is "nested" in the full model. Write $\mathbf{x}_d$ for the vector of explanatory variables deleted from $\mathbf{x}$ in defining $\mathbf{x}_r$, and $\boldsymbol{\beta}_d$ for the corresponding vector of regression coefficients, so that

$$\boldsymbol{\beta}'\mathbf{x} = \boldsymbol{\beta}'_r\mathbf{x}_r + \boldsymbol{\beta}'_d\mathbf{x}_d.$$

Then if $\boldsymbol{\beta}_d = \mathbf{0}$, the full model η_f can be simplified to the reduced model η_r, and so the use of the reduced model is justified (through considerations of parsimony discussed in Section 2.1) if the evidence against $\boldsymbol{\beta}_d = \mathbf{0}$ is not strong. This evidence is provided in the Neyman–Pearson theory by the likelihood ratio test of the hypothesis $\boldsymbol{\beta}_d = \mathbf{0}$, which compares the maximized likelihood functions under the two models. The LRTS is

$$\lambda = -2\left\{\ell(\hat{\boldsymbol{\beta}}_r, \hat{\phi}_r) - \ell(\hat{\boldsymbol{\beta}}, \hat{\phi})\right\},$$

where $\hat{\phi}_r$ is the MLE of ψ in the reduced model. If the parameter estimates under both models are not on the boundary of the parameter space, and if the number of variables p in the full model does not increase with the sample size n, then in large samples the distribution of the LRTS under the null hypothesis is $\chi^2_{p-p_1}$.

We adopt the name *disparity* for the expression $-2\log L(\hat{\boldsymbol{\theta}})$ for a specific model, when the likelihood $L(\theta)$ includes all the constants (apart from the measurement precision term δ^n) from the probability model $f(y \mid \theta)$. As we noted earlier, this allows the comparison of maximized likelihoods across different ("non-nested") models; considerable confusion has arisen in the past from the omission of such constants, making cross-family comparisons very difficult. The macros that we use in this book have been standardized to include all constants in the density or mass function, and so they allow a direct comparison of disparities across different families.

The disparity for a model is closely related to the *deviance* for a model as defined by Nelder and Wedderburn (1972) and as computed in GLIM: the deviance is the disparity for the model relative to the disparity for a "saturated" model with a parameter for each observation. *Differences* in disparities between nested models for the single-parameter Poisson and binomial distributions are thus identical to differences in deviance, and the disparity does not have to be calculated explicitly for nested model comparisons in these distributions. For the two-parameter normal and gamma distributions, mixture distributions, and more generally when different distributions are being compared for the same data, disparities have to be calculated explicitly, retaining all the constants in the density or mass function.

We conclude this chapter with a discussion of likelihood methods without a specific model. This section is not central to the book and may be skipped on first reading.

2.10   Likelihood inference without models

The title of this section may seem paradoxical. If we do not have an explicit model for the population probabilities p_I of a variable Y, how can we write down a likelihood function? Two different approaches provide answers to this question, by constructing likelihoods for population parameters even though the population

Table 2.3. Family incomes for random sample of 40 families (hundreds of dollars)

26	35	38	39	42	46	47	47	47	52	53	55	55	56
58	60	60	60	60	60	65	65	67	67	69	70	71	72
75	77	80	81	85	93	96	104	104	107	119	120		

model itself is not given. In the first case, the object of inference is a *percentile* of the population; in the second, it is the conventional mean of the population.

2.10.1 *Likelihoods for percentiles*

We begin for simplicity with the median. For the population cumulative proportions C_I, the median ψ is defined as the value of y_I such that

$$C_I \leq 0.5 \quad \text{for } y_I \leq \psi$$
$$C_I > 0.5 \quad \text{for } y_I > \psi.$$

Suppose now that we have a simple random sample of n observations from the population, giving values $y_1, \ldots, y_n$ of Y. For example, we give in Table 2.3 the values of family income for a random sample of 40 families from the STATLAB population. For subsequent convenience the 40 values have been ordered from smallest to largest. Given the population median ψ, if the random variable Y is assumed *continuous*, that is, if the measurement precision is very high, then it will be approximately true that

$$\Pr[Y < \psi] = \Pr[Y > \psi] = 0.5$$

and

$$\Pr[Y = \psi] = 0.$$

Then the probability that of the n sample values, r are less than the median and $(n - r)$ greater than the median is the symmetric binomial probability

$$b(r; n, 0.5) = \binom{n}{r} \cdot \frac{1}{2^n}.$$

This provides immediately a likelihood function for the median:

$$L(\psi) = \binom{n}{r} \cdot \frac{1}{2^n}$$

for $y_{(1)} \leq \cdots \leq y_{(r)} \leq \psi \leq y_{(r+1)} \leq \cdots \leq y_{(n)}, r = 0, 1, \ldots, n$. This function is *piecewise constant* between successive distinct ordered observations and can take only the values of the symmetric binomial probabilities $b(r; n, 0.5)$.

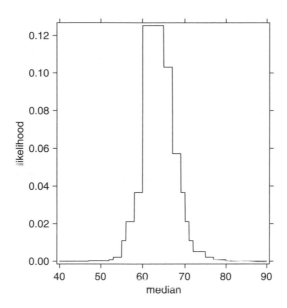

Fig. 2.25. Likelihood for median

Figure 2.25 shows the likelihood for the median for the above sample of 40 family incomes. The GLIM directives are:

```
$unit 40 $
$data inc $
$read
   26  35  38  39  42  46  47  47  47  52  53  55  55  56
   58  60  60  60  60  60  65  65  67  67  69  70  71  72
   75  77  80  81  85  93  96  104  104  107  119  120
$var 1000 percent lik r $
$calc %m = 40 : %d = 0.05 : %n = 1000 : %i = 1 : %p = 0.5 $
$macro prcntlik
   $calc percent(%i) = %m $
           :          %r = %cu(%lt(inc, %m)) $
           : lik(%i) = %bip(%r,40,0.5) - %bip(%r - 1,40,0.5) $
           : %i = %i + 1 : %n = %n - 1 : %m = %m + %d $
$endmac
$while %n prcntlik $
$graph (s = 1 h = 'median' v = 'likelihood')
        lik percent 10 $
```

Note that the likelihood does not quite go to zero as $\psi \rightarrow \pm\infty$, because there is a non-zero, though extremely small, probability that all 40 observations exceed the median, whatever its value. A likelihood interval for the median is

Table 2.4. Binomial probabilities for $n = 40, p = 0.5$

r	11	12	13	14	15	16	17	18	19	20
b	0.0021	0.0051	0.0109	0.0211	0.0366	0.0572	0.0807	0.1031	0.1194	0.1254

r	21	22	23	24	25	26	27	28	29
b	0.1194	0.1031	0.0807	0.0572	0.0366	0.0211	0.0109	0.0051	0.0021

constructed as before, with the unusual feature that only certain values of the relative likelihood are possible, corresponding to the probabilities of the discrete binomial distribution. The probabilities $b = b(r; 40, 0.5)$ are shown in Table 2.4 for $r = 11, \ldots, 29$. A probability (likelihood) ratio of 0.168 occurs for $r = 14$ or 26 relative to $r = 20$, and one of 0.087 for $r = 13$ or 27 relative to $r = 20$. So a relative likelihood of 0.168 is achieved for the interval (y_{14}, y_{26}), that is $(56,70)$, and one of 0.087 for the interval (y_{13}, y_{27}), that is $(55,71)$.

The likelihood approach to interval construction here is quite different from inverting the usual Neyman–Pearson hypothesis testing approach. If we want to test the null hypothesis H_0 that $\psi = \psi_0$ against a general alternative H_1, we would proceed in the Neyman–Pearson approach by rejecting H_0 if the number of observations exceeding ψ_0 were too large or too small (the median test). As for the binomial confidence interval procedures described in Section 2.7.5, we then have to find the values of ψ for which the hypothesis would just be rejected at the 5% level. The resulting interval for ψ will in general be different from the likelihood interval.

The above approach to the median generalizes directly to the case of an arbitrary percentile. Let ψ_p be the $100p$-th percentile of the distribution of Y, so that

$$\Pr[Y \le \psi_p] = p, \quad \Pr[Y > \psi_p] = 1 - p,$$

with $\psi_{0.5} = \psi$, and the variable Y again assumed continuous. Then for the sample $y_1, \ldots, y_n$ ordered as before, the likelihood function for ψ_p is

$$L(\psi_p) = \binom{n}{r} p^r (1 - p)^{n-r}$$

for $y_{(1)} \le \cdots \le y_{(r)} \le \psi_p \le y_{(r+1)} \le \cdots \le y_{(n)}, r = 0, 1, \ldots, n$. The likelihood is again piecewise constant but asymmetric in its values. For p near 0 or 1 the corresponding tail of the likelihood is relatively flat in small samples, reflecting the obvious fact that a small sample can provide little information about extreme percentiles, in the absence of a specific distributional model.

Figure 2.26 shows the likelihood for ψ_p for $p = 0.75$, and Fig. 2.27 that for $p = 0.9$ for the family income sample. The GLIM directives are:

```
$calc %m = 60 : %d = 0.05 : %n = 1000 : %i = 1 : %p = 0.75 $
$while %n prcntlik $
```

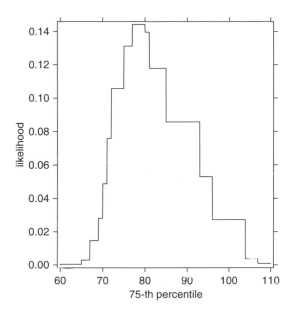

Fig. 2.26. Likelihood for 75-th percentile

```
$graph (s = 1 h = '75-th percentile' v = 'likelihood')
       lik percent 10 $
$calc %m = 80 : %d = 0.05 : %n = 1000 : %i = 1 : %p = 0.9 $
$while %n prcntlik $
$graph (s = 1 h = '90-th percentile' v = 'likelihood')
       lik percent 10 $
```

For $p = 0.9$ the likelihood is constant for all values of $\psi_p > 120$; it is possible, though very unlikely, that *all* the sample observations are less than the 90-th percentile. Thus as noted above precise information about extreme percentiles requires large samples.

2.10.2 *Empirical likelihood*

This term was introduced by Owen (1988, 2001) to describe a likelihood function for a population parameter, a function of the population proportions p_I, when no specific model is assumed for the p_I. The original use of the empirical likelihood was by Lindsey (1974), though his purpose was different, namely to model the multinomial probabilities.

Consider the family income sample above. As in Section 2.3, let Y_I be the distinct values of Y occurring in the population. We now allow Y values with multiplicity zero to be included in this set, as we will see this makes no difference to the conclusions. Without any loss of generality, we may assume that the Y_I

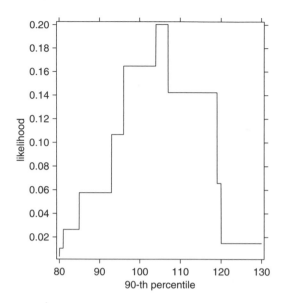

Fig. 2.27. Likelihood for 90-th percentile

form a grid, or mesh, at intervals given by the measurement precision δ. Let D be the number of distinct values of Y, with non-zero or zero probabilities. We want to draw inferences about the *mean* family income $\mu = \sum_I P_I Y_I$ without assuming a model for the P_I. The likelihood function for the P_I is just the multinomial likelihood

$$L(P_1, \ldots, P_D) = \prod_{I=1}^{D} P_I^{n_I}$$

with

$$\sum_{I=1}^{D} P_I = 1.$$

We base inference about μ on the profile likelihood in μ, which is

$$P(\mu) = L(\hat{P}_1(\mu), \ldots, \hat{P}_D(\mu)),$$

where the likelihood is maximized over the P_I subject to the constraints $P_I \geq 0$, $\sum P_I = 1$, and $\sum P_I y_I = \mu$. Note that μ is constrained to lie in the interval $(y_{(1)} < \mu < y_{(D)})$. The maximization is easily accomplished using Lagrange multipliers: let

$$G(P_1, \ldots, P_D) = \log L(P_1, \ldots, P_D) - \phi\left(\sum P_I - 1\right) - n\lambda\left(\sum P_I y_I - \mu\right).$$

Then

$$\frac{\partial G}{\partial P_I} = \frac{n_I}{P_I} - \phi - n\lambda y_I = 0$$

$$\frac{\partial G}{\partial \lambda} = \sum P_I y_I - \mu = 0$$

$$\frac{\partial G}{\partial \phi} = \sum P_I - 1 = 0$$

for a maximum of the constrained likelihood. Multiplying the first equation by P_I and summing over I, we have

$$n = \hat{\phi}(\mu) + n\mu\hat{\lambda}(\mu)$$

giving $\hat{\phi}(\mu) = n(1 - \mu\hat{\lambda}(\mu))$ and

$$\frac{n_I}{\hat{P}_I(\mu)} = n[1 + \hat{\lambda}(\mu)(y_I - \mu)]$$

so that

$$\hat{P}_I(\mu) = \frac{n_I/n}{1 + \hat{\lambda}(\mu)(y_I - \mu)} = \frac{\tilde{P}_I}{1 + \hat{\lambda}(\mu)(y_I - \mu)},$$

where $\tilde{P}_I = n_I/n$. Summing over I again gives $\hat{\lambda}(\mu)$ as the implicit solution of

$$f(\lambda) = 1 - \sum_{I=1}^{D} \frac{\tilde{P}_I}{1 + \lambda(y_I - \mu)} = 0.$$

Note that $\lambda = 0$ always satisfies this equation: it gives the *unrestricted* maximum of the likelihood since the constraint $\sum P_I y_I = \mu$ is not active. Since the $\hat{P}_I(\mu)$ must be non-negative, we require $1 + \hat{\lambda}(\mu)(y_I - \mu) > 0$ for all I, which implies

$$-(y_{(D)} - \mu)^{-1} < \hat{\lambda}(\mu) < (\mu - y_{(1)})^{-1}.$$

Values of Y_I which do not occur in the sample give values of $\tilde{P}_I$ and therefore $\hat{P}_I(\mu)$ of zero, and so the implicit equation for $\hat{\lambda}(\mu)$ involves only the Y_I with $n_I > 0$.

Since this section is not central to the book, we do not give details of the computational method to compute the empirical likelihood. Owen (2001) gave full details.

Figure 2.28 shows the empirical profile relative likelihood for the example. It is shifted considerably to the right compared to that for the median ψ, reflecting the skewness of the income distribution, with the mean income larger than the median. Despite this skewness, the profile likelihood in μ is only slightly skewed.

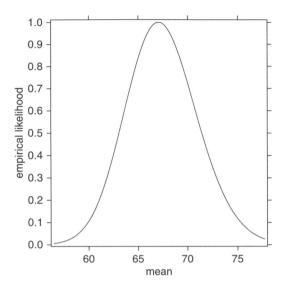

Fig. 2.28. Empirical likelihood

Figure 2.29 repeats the empirical profile relative likelihood together with the profile relative likelihoods for μ, assuming lognormal (dotted curve) and gamma (dashed curve) distributions for income.

The gamma and empirical likelihoods are very close; the lognormal likelihood is less close on the right to the empirical likelihood. Computation of the lognormal and gamma profile likelihoods is discussed in Section 2.7.1 and Chapter 6, respectively.

In Owen's derivation of the empirical likelihood, the asymptotic repeated sampling result $-2\log\{L(\mu)/L(\hat{\mu})\} \sim \chi_1^2$ is used to set confidence intervals for μ. Remarkably, this asymptotic result holds despite the apparent violation of the condition that the number of nuisance parameters P_I does not increase with n, since the number of distinct possible values D certainly increases (though slowly) with n. This is a special case of a very general result by Murphy and van der Vaart (2000) to which we refer in Chapter 8.

Owen and others (see Owen, 2001 for full details) have extended the empirical likelihood approach to other parameters, including variances and regression coefficients. *Joint* empirical likelihood regions in several parameters can be constructed in the same way. However, the small-sample coverage of empirical likelihood intervals and regions may be more affected by the large number of nuisance parameters than profile likelihood intervals and regions in parametric models.

This approach to likelihood construction in the absence of a model provides an important theoretical response to criticisms of model-based approaches, as being based on models which may be incorrect and which may therefore give

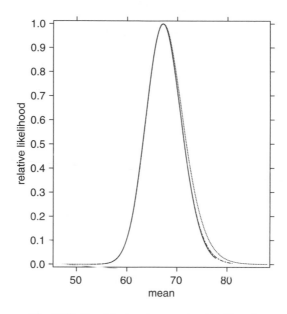

Fig. 2.29. Empirical and parametric likelihoods

misleading inferences. Since no approximating population model is assumed for the P_I, the empirical likelihood provides a completely model-robust likelihood inference about μ. In theory this approach can be extended very widely. In practice, computational complexity has so far limited its generality.

Intervals based on the empirical likelihood have also been regarded as an alternative to bootstrap confidence intervals (Hall and La Scala, 1990), which also do not make any population model assumption but perform physical resampling from the observed sample.

3
Regression and analysis of variance

3.1 An example

We consider a practical example from a psychological study. A sample of twenty-four children was randomly drawn from the population of fifth-grade children attending a state primary school in a Sydney suburb. Each child was assigned to one of the two experimental groups, and given instructions by the experimenter on how to construct, from nine differently coloured blocks, one of the 3×3 square designs in the Block Design subtest of the Wechsler intelligence scale for children (WISC, see Wechsler, 1949). Children in the first group were told to construct the design by starting with a row of three blocks (row group), and those in the second group were told to start with a corner of three blocks (corner group). The total time in seconds to construct four different designs was then measured for each child.

Before the experiment began, the extent of each child's "field dependence" was tested by the embedded figures test (EFT), which measures the extent to which subjects can abstract the essential logical structure of a problem from its context (high scores corresponding to high field dependence and low ability).

The data are given in Table 3.1, and are held in the file solv. The file also contains a group factor taking values 1 and 2 for row and corner groups, respectively.

The experimenter was interested in knowing whether the different instructions produced any change in the average time required to construct the designs, and whether this time was affected by field dependence.

Table 3.1. Block design completion times

Row group						
time	317	464	525	298	491	196
eft	59	33	49	69	65	26
time	268	372	370	739	430	410
eft	29	62	31	139	74	31
Corner group						
time	342	222	219	513	295	285
eft	48	23	9	128	44	49
time	408	543	298	494	317	407
eft	87	43	55	58	113	7

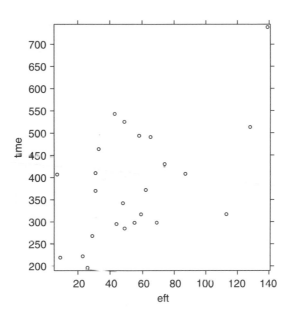

Fig. 3.1. Completion time vs EFT

In practical data analysis, we usually begin with a graph of the data, shown in Fig. 3.1.

```
$input 'solv' $
$graph (s = 1 h = 'eft' v = 'time')
       time eft 5 $
```

There appears to be a general increase in time required as eft increases, with much individual variation. The graph does not distinguish between the two experimental groups: this is achieved with the option for different plotting characters defined by a group factor, shown in Fig. 3.2.

```
$graph (s = 1 h = 'eft' v = 'time')
       time eft 5,7 group $
```

Separate graphs for the two groups can be obtained with the weight vectors g1 and g2 (Figs 3.3 and 3.4). We fix the axis scales to be the same in both graphs.

```
$calc g1 = %eq(group,1) $
$graph (s = 1 h = 'eft' v = 'time' x = 5,140 y = 195,740)
       time/g1 eft 5 $
$calc g2 = %eq(group,2) $
$graph (s = 1 h = 'eft' v = 'time' x = 5,140 y = 195,740)
       time/g2 eft 7 $
```

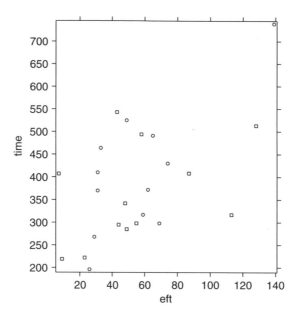

Fig. 3.2. Completion time vs EFT

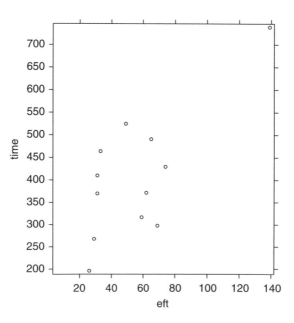

Fig. 3.3. Completion time vs EFT, row group

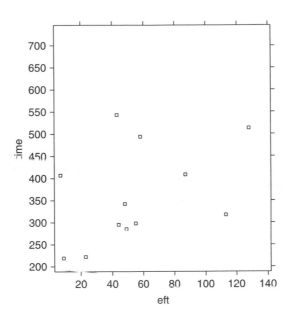

Fig. 3.4. Completion time vs EFT, corner group

It looks as though the corner group takes less time to construct the designs than the row group, but the difference is not marked and may be simply due to sample fluctuations. The assessment of this question is based on comparing a series of statistical models.

The usual approach to the analysis of the data mentioned above is by multiple regression, or analysis of covariance. We formulate the model, expressed in a traditional form,

$$Y_i = \beta_0 + \beta_1 x_{1i} + \beta_2 x_{2i} + \epsilon_i, \quad i = 1, \dots, 24,$$

where Y_i is the time taken by the i-th child, x_{1i} is the eft score of the i-th child, and x_{2i} is a dummy variable, with $x_{2i} = 0$ for the row group children and $x_{2i} = 1$ for the corner group children. The "error" variables ϵ_i are assumed to be independent and normally distributed in repeated sampling with mean zero and common variance σ^2, written $\epsilon_i \sim N(0, \sigma^2)$. Dummy variables for categorical (here binary) explanatory variables are automatically defined by GLIM if the variables are declared to be factors. This can be achieved by the directives

```
$calc group = %gl(2,12) $
$factor group 2 $
```

or more simply by $gfactor group 2 $, since the group sizes are equal. Thus labels 1 and 2 for the group values do not have to be read in as data: they can be calculated instead. The number of levels of the group factor has to be specified. The dummy variable used by GLIM for x_2 in the regression is denoted by GROU(2). For a factor A with k levels ($k \geq 2$), GLIM defines $(k - 1)$ dummy variables A(2),...,A(k) with A(j) = 1 for the j-th level of A, and zero otherwise. These variables cannot be accessed by the user, but it is easy to define such variables explicitly, using the calculate directive. Thus to define a dummy variable for the second level of the factor A, we use $calc A2 = %eq(A,2) $ and similarly for the other levels of A.

In the formal model structure of Chapter 2, the *systematic part* of the model is the regression function or linear model $\beta_0 + \beta_1 x_1 + \beta_2 x_2$ for the mean time taken, and the *random part* is the normal error variation about the regression with variance σ^2. If we write μ_i for the mean time taken for the i-th child, then

$$\mu_i = \beta_0 + \beta_1 x_{1i} + \beta_2 x_{2i}, \quad Y_i = \mu_i + \epsilon_i, \quad \epsilon_i \sim N(0, \sigma^2),$$

which we will call *Model 2*: in all we will consider five models for these data. In the remainder of this discussion we will suppress the specification of the random part of the model as it is the same for all the regression models for μ_i.

Since x_2 is a dummy variable, we may write this regression function equivalently as

$$\mu_i = \beta_0 + \beta_1 x_{1i}, \qquad x_{2i} = 0\text{: Row group,}$$
$$\mu_i = \beta_0 + \beta_2 + \beta_1 x_{1i}, \quad x_{2i} = 1\text{: Corner group.}$$

The two regression lines for the row group and the corner group for the five models discussed are shown in Fig. 3.5.

The regression lines for model 2 are parallel, with slope β_1 representing the increase in mean time taken for a one-point increase in eft score, β_2 representing the difference in mean time taken between the corner and row groups with the same value of eft, and β_0 the (hypothetical) mean time taken for a child with eft score zero in the row group.

If the instructions have no differential effect on completion time, then $\beta_2 = 0$, and model 2 reduces to *model 3*:

$$\mu_i = \beta_0 + \beta_1 x_{1i},$$

identical regressions for the two instruction groups. If the instructions have a different effect, but eft is unrelated to completion time, then $\beta_1 = 0$, and model 2 reduces to *Model 4*:

$$\mu_i = \beta_0 + \beta_2 x_{2i}.$$

It is also possible that neither eft nor instructions has any effect, in which case the null *Model 5* results:

$$\mu_i = \beta_0.$$

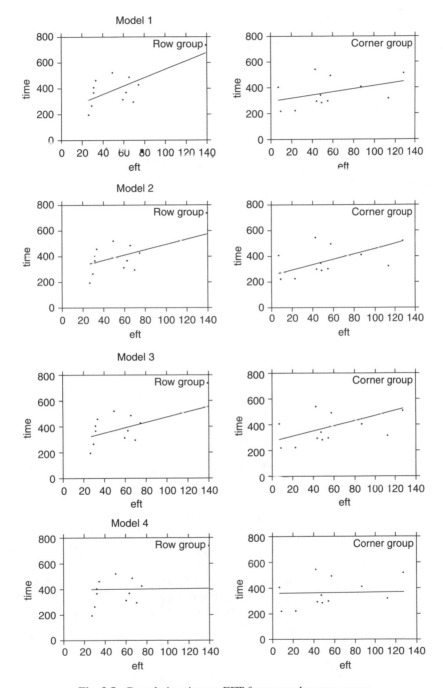

Fig. 3.5. Completion time vs EFT for row and corner groups

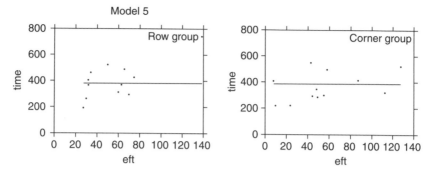

Fig. 3.5. (Continued)

A further possibility is that the regressions of mean time on eft score for the two instruction groups are linear, but not parallel. This can be represented by the *interaction model 1*:

$$\mu_i = \beta_0 + \beta_1 x_{1i} + \beta_2 x_{2i} + \beta_3 x_{1i} x_{2i},$$

which is equivalent to

$$\mu_i = \beta_0 + \beta_1 x_{1i} \qquad\qquad x_{2i} = 0: \text{Row group}$$
$$\mu_i = (\beta_0 + \beta_2) + (\beta_1 + \beta_3) x_{1i} \quad x_{2i} = 1: \text{Corner group.}$$

For model 1, the difference between mean completion time in the two experimental groups, for the same eft score x, is

$$(\beta_0 + \beta_2) + (\beta_1 + \beta_3)x - (\beta_0 + \beta_1 x) = \beta_2 + \beta_3 x,$$

a linear function of x. Thus the effect of different instructions on mean completion time depends on eft score, a much more complicated interpretation than that of model 2, where the difference is a constant β_2 because β_3 is zero.

The interpretation of the experiment depends, therefore, on which of the five (or possibly other) models is the "best" representation of the data, in the sense of "simplest consistent with the data". In Chapter 2 we developed the general theory for this question; we now apply it in this chapter.

It is worth stressing first though that *none* of the models is a true representation of the population. If we could take a complete census of the population of fifth-grade children in the school, and administer the EFT and WISC test to all of them, we would find that the mean completion time for children with each eft score in each experimental group did not lie on a straight-line: the means would be quite irregular about a general trend. Our model ignores the variation, and uses only a linear trend: it thus represents an idealized or smooth version of the population. This is most obvious when we are in fact modelling complete populations: we still smooth the complete population using the model so that the gross systematic features of the population are retained, but the minor irregularities or fine structure are lost, being represented as random variation.

To fit the models in GLIM, we declare:

(1) the *response-variable Y* is time:
    `$yvar time;`
(2) the *probability distribution* for Y is normal $N(\mu, \sigma^2)$ with mean μ and variance σ^2:
    `$error n` (this is the default distribution and can be omitted);
(3) the *link function* is the identity $\eta = g(\mu) = \mu$:
    `$link i` (this is the default link for the normal distribution and can be omitted);
(4) the *linear predictor* is $\eta = \beta_0 + \beta_1 x_1 + \beta_2 x_2$, where x_1 is eft and x_2 group, or one of the other models 1–5:
    `$fit eft + group $disp e $.`

GLIM provides minimal default output: from the `$fit` directive itself only the GLIM deviance (the residual sum of squares for the normal distribution) and its degrees of freedom df ($n - p - 1$, where p is the number of variables in the linear predictor) are printed automatically; the parameter estimates and standard errors are printed using the argument e (for estimates) for the `$display` directive (which can be abbreviated to `$disp`).

For model comparisons by the likelihood ratio test (LRT), it is easily shown (Section 2.9.4—we do not give the derivation) that for a regression model $\beta'\mathbf{x}$ the MLEs of the model parameters are

$$\hat{\boldsymbol{\beta}} = \left(\sum \mathbf{x}_i \mathbf{x}_i'\right)^{-1} \sum \mathbf{x}_i y_i = (X'X)^{-1} X'\mathbf{y}$$

$$\hat{\sigma}^2 = \sum (y_i - \hat{\boldsymbol{\beta}}' \mathbf{x}_i)^2/n = \text{RSS}/n,$$

where X is the "design matrix": $X = (\mathbf{x}_1, \ldots, \mathbf{x}_n)'$, $\mathbf{y} = (y_1, \ldots, y_n)'$ and RSS is the residual sum of squares from the fitted model. Note that GLIM uses the unbiased variance estimate $s^2 = \text{RSS}/(n - p - 1)$ instead of $\hat{\sigma}^2$ to calculate standard errors. The value of the disparity as we define it is

$$-2\ell(\hat{\boldsymbol{\beta}}, \hat{\sigma}) = c + n \log(\text{RSS}),$$

where

$$c = n[1 + \log(2\pi) - \log n],$$

and $\ell(\beta, \sigma)$ is the log-likelihood for the normal model. A formal comparison of two models is expressed in Section 2.9.4 through the LRT of the hypothesis that the variables $\mathbf{x}_d$ omitted from $\mathbf{x}$ in the partition $\mathbf{x}' = (\mathbf{x}_r, \mathbf{x}_d)'$ have zero regression

coefficients and do not contribute to the regression function, so that $\boldsymbol{\beta}_d = \mathbf{0}$. The
likelihood ratio test statistic (LRTS) for this hypothesis is

$$
\begin{aligned}
\lambda &= -2\{\ell(\hat{\boldsymbol{\beta}}_r, \hat{\sigma}_r) - \ell(\hat{\boldsymbol{\beta}}, \hat{\sigma})\} \\
&= n\{\log \text{RSS}_r - \log \text{RSS}_f\} \\
&= n\log(\text{RSS}_r/\text{RSS}_f).
\end{aligned}
$$

(Note that the constant c disappears from the LRTS. For this reason numerical constants which are not functions of the parameters can be omitted in the comparison
of nested models, as noted in Section 2.6 and elsewhere.)

If the hypothesis is true, then the RSS for the reduced model will be not much
larger than that for the full model, and so λ will be small. If the hypothesis is false,
then λ will be large.

For the hypothesis that a single component β_j of $\boldsymbol{\beta}$ is zero, the Wald test compares
$\hat{\beta}_j$ with its estimated asymptotic standard error $\sqrt{\hat{v}_{jj}}$, and treats $W = t_j^2 = \hat{\beta}_j^2/\hat{v}_{jj}$ as
approximately χ_1^2 under the hypothesis. For large samples in which the likelihood
function is normal in the parameters, this test will be closely equivalent to the LRT,
with $W \approx \lambda$. For small samples and for regression models in which the response-
variable is not normally distributed, the likelihood may be far from normal and the
value of W may be substantially different from λ.

Critical values for λ are based on its asymptotic distribution when the null
hypothesis is true, which is $\chi_{p-p_1}^2$ if both X and X_r are of full rank p and p_1
respectively.

For the normal distribution this asymptotic result is not used, since exact results
are available. If we define the *hypothesis sum of squares* HSS by

$$
\text{HSS} = \text{RSS}_r - \text{RSS}_f
$$

then

$$
\begin{aligned}
\lambda &= n\log(1 + \text{HSS}/\text{RSS}_f) \\
&= n\log\left\{1 + \frac{p - p_1}{n - p - 1}\left[\frac{\text{HSS}/(p - p_1)}{\text{RSS}_f/(n - p - 1)}\right]\right\} \\
&= n\log(1 + (p - p_1)F/(n - p - 1)),
\end{aligned}
$$

where

$$
F = \left[\frac{\text{HSS}/(p - p_1)}{\text{RSS}_f/(n - p - 1)}\right]
$$

is the usual F-statistic (with $p - p_1$ and $n - p - 1$ df) for the test of the hypothesis
$\boldsymbol{\beta}_d = \mathbf{0}$ in multiple regression or the analysis of variance. The exact distribution

of λ could therefore be obtained from that of F (or from the beta distribution of RSS_f/RSS_r), but it is simpler to use the F-distribution itself which is well tabulated.

Under the hypothesis that a single component β_j of $\boldsymbol{\beta}$ is zero, the distribution of $t_j = \hat{\beta}_j/\sqrt{\hat{v}_{jj}}$ is exactly t_{n-p-1}. Here the t-test is identical to the F-test, as $t_j^2 = F$, and the distribution of t_{n-p-1}^2 is $F_{1,n-p-1}$.

In practice, the single F-test described above is insufficient to identify parsimonious models, because we need to examine the importance of individual terms in the model. In our example we have five candidate models, so *multiple* tests will be necessary for this identification. These tests are constructed by *partitioning* the HSS into individual sums of squares (SS) for the model terms, by fitting a hierarchical sequence of models of increasing complexity, from the simplest model to the full model. The table of SS for individual terms is called an *analysis of variance* (ANOVA) table. For example, for the sequence of models 5, 4, 2, 1:

```
$fit : + group : + eft : + eft.group $
```

The continuation sign replaces the repeated $fit directive. The use of + eft avoids the complete rewriting of the previous model formula. Terms can be removed from the model in the same way using -.

The RSSs are 367537, 257274, 355522, 245531 and 218076. Listing the models in order and differencing their RSSs gives the ANOVA table (Table 3.2) of SS attributable to the various terms being considered.

The residual in Table 3.2 is simply the RSS from the final model in the sequence and corresponds to the RSS from the full model 1. The usual approach to model comparisons in analysis of variance tables is to calculate mean squares by dividing each SS by its degrees of freedom and then to assess the importance of each effect by testing its mean square against the residual mean square using the F-test with 1 and 20 df. However, this procedure is ambiguous in the analysis of survey and much experimental data because the SS for each effect generally depends on its order of fitting in the model. Thus, if we fit the models in the order 5, 3, 2 and 1, and difference the RSS, we obtain the different ANOVA Table 3.3.

```
$fit : + eft : + group : + eft.group $
```

Table 3.2. ANOVA table

	Model	RSS	df	Source	SS	df
5	1	367537	23			
4	group	355522	22	group	12015	1
2	+ eft	245531	21	eft	109991	1
1	+ eft.group	218076	20	interaction	27455	1
				residual	218076	20
				$s^2 =$	10903.8	

Table 3.3. Second ANOVA table

	Model	RSS	df	Source	SS	df
5	1	367537	23			
3	eft	257274	22	eft	110263	1
2	+ group	245531	21	group	11743	1
1	+ eft.group	218076	20	interaction	27455	1
				residual	218076	20
				$s^2 =$	10903.8	

The analysis of variance tables differ because the variables eft and group are slightly correlated, this correlation being due to the small difference in eft means between the row and corner groups. The difference is only small because of the random allocation of children to groups. Only for *orthogonal* experimental designs is the SS breakdown unique, independent of the fitting order. In observational studies such differences can be large; we discuss this important issue in Section 3.16.

Qualitatively, we can see that the eft mean square is large (more than 10 times the residual mean square in either model) while the group mean square is small (about the same size as the residual mean square). The interaction mean square is the same in both tables, since it is always the last term fitted, and its F-value is 2.52 (27455/10903.8). This value is nowhere near the 5% level for the $F_{1,20}$ distribution, and so we can clearly reduce Model 1 to Model 2. Comparing Model 2 with Models 3 and 4, we see that Model 2 can be reduced to Model 3, since the SS for group is of the same order as the residual mean square. However Model 2 cannot be reduced to Model 4, because the eft variable cannot be omitted as its F-value is more than 10. We conclude that Model 3 is the most appropriate and parsimonious of the models considered to represent the data.

The fitted values of the response-variable are stored in the system vector %fv, and can be plotted as a line with the observed times as follows:

```
$fit eft $
$graph (s = 1 h = 'eft' v = 'time')
        time,%fv eft 5,10 $
```

We obtain the graph in Fig. 3.5 for Model 3 with the two groups plotted on one graph.

The fitted regression is obtained from $disp e $:

$$271.1 + 2.04 \text{ eft,}$$

a one-point increase in eft score being associated with a two-second increase in mean time. Thus field dependence, as measured by the EFT, is associated with increased completion time, but the type of instruction given is not.

Completion times vary randomly about the mean with variance estimated by $s^2 = 257274/22 = 11694$, or standard deviation $s = 108$ seconds, slightly larger than $104 = \sqrt{10903.8}$ from the full model.

To readers used to classical analysis of variance and covariance, this conclusion may seem unsatisfying. Where is the estimate of the treatment effect and its standard error? In simplifying the model, we have set the treatment effect (the coefficient β_2 of group) equal to zero because there is no strong evidence from the data against this value. If we are particularly interested in certain parameters of the model because these represent experimentally controlled variables, then non-parsimonious models containing these variables may be presented as the final summary of the data. The parameter estimates and standard errors for the fitted Model 2 group + eft are

```
293.4 + 2.04 eft - 44.2 group(2)
(48.4) (0.67)      (44.1)
```

The residual standard deviation estimate is $s = 108$ seconds. A 95% confidence interval for β_2 is

$$-44.2 \pm t_{0.025,21}(44.1), \text{ i.e. } (-136, 48).$$

The corner group has an *estimated* mean of 44 seconds below that for the row group (for the same value of eft), but the population value could be any value in the above wide interval, including zero.

3.2   Strategies for model simplification

The simplification of the model in the above sequence—interaction, then main effects—is the most common and useful strategy for model simplification. High-order interactions are usually small, and their retention in a model makes the interpretation of the model much more difficult. Successively removing interactions from the model as far as possible, starting with the most complex, is therefore appealing.

This is, however, not the only possible strategy. Another possible approach is to examine the lower-order interactions or main effects and to simplify them, while retaining the high-order interactions in a simplified form. Section 3.1 provides a good example. The parameter estimates and their standard errors (in parentheses) from the full interaction Model 1 are

```
226.1 + 3.25 eft + 70.5 group(2) - 2.07 eft.group(2)
(63.1) (1.00)      (83.9)          (1.30)
```

As we have already seen, the interaction term can be omitted from the model. However, examination of the regression coefficients for eft in the two instruction groups reveals that in the row group, the slope is large: 3.25, while in the corner group it is small: $3.25 - 2.07 = 1.18$. This suggests that a different simplification of the model is possible: we could set the slope of the eft regression to zero in the corner group, but keep it non-zero in the row group. Since the slope is $(\beta_1 + \beta_3)$ in the corner group, this constraint is achieved by setting $\beta_1 + \beta_3 = 0$, or $\beta_3 = -\beta_1$. The interaction model (model 5) then becomes *Model 6*:

$$\mu = \beta_0 + \beta_1 x_1 + \beta_2 x_2 - \beta_1 x_1 x_2$$
$$= \beta_0 + \beta_1 x_1 (1 - x_2) + \beta_2 x_2.$$

This model specifies a linear regression

$$\mu = \beta_0 + \beta_1 x_1$$

for the row group, and a constant mean

$$\mu = \beta_0 + \beta_2$$

for the corner group. Model 6 is equivalent to Model 1 for the row group and Model 5 for the corner group. It is important to note that β_2 again does not represent a constant difference in means between the groups for children with the same eft score x_1: this difference is now $\beta_2 - \beta_1 x_1$.

The model can be fitted by constructing a new variable $x_1(1 - x_2)$ and fitting it with group. Recall that g2 is an indicator for group 2:

```
$calc eft1 = eft*(1 - g2) $
$fit eft1 + group $disp e $
```

The RSS is 239766, giving an $F_{1,20}$-value of 1.99 for the comparison of the two models, or an equivalent t-statistic of 1.41. The parameter estimates and standard errors are

```
   226.1 + 3.25 eft1 + 135.9 group(2)
   (64.6) (1.02)        (71.6)
```

The slope of the regression on eft1 for the row group is the same as in the interaction model, though its standard error is slightly different since the model is different.

If we omit eft1 from this model we obtain Model 4 which we know is not adequate. The only other possibility is to omit group. Omission of this variable is equivalent to fitting the model

$$\mu = \beta_0 + \beta_1 x_1$$

for the row group, and

$$\mu = \beta_0$$

for the corner group. This implies that the group regression lines intersect at $x_1 = 0$, so that the mean completion time for a corner group child is the same as that for a row group child with an EFT score of zero. Since there are no such children, this model has a strong property that is not verifiable from the observable data. In the absence of compelling experimental reasons for this strong mathematical property, we do not proceed further with this model, but return to model 6.

The interpretation of model 6 is quite different from that of model 3. Mean completion time is affected by field dependence under row instructions, but not under corner instructions. The mean time taken under row instructions is less than that taken under corner instructions for low field dependence, but greater for high field dependence.

These conclusions are quite complicated. The interpretation of model 3 was much simpler: no difference in mean time between instructions, and a common effect of field dependence for both instruction groups. But both models are consistent with the data, and we have no statistical criterion for choosing between them other than parsimony: model 3 has only one variable, model 6 has two.

A problem occurring frequently with models with non-parallel regressions is that of determining the values of the explanatory variable for which a significant difference between means can be asserted. In model 6, the difference between the mean completion time between the corner and row groups, for a given EFT score x, is $\beta_2 - \beta_1 x$. For what values of x can we assert that this is non-zero? The fitted regression lines cross when $x = \hat{\beta}_2 / \hat{\beta}_1 = 41.8$, so around this value the population means are indistinguishable. For any x the variable $\hat{\beta}_2 - \hat{\beta}_1 x$ is normally distributed with mean $\beta_2 - \beta_1 x$ and variance $(v_{22} - 2v_{12}x + v_{11}x^2)$, where $V = (v_{jk})$ is the covariance matrix of the parameter estimates, and is estimated by $\hat{V} = s^2 (X'X)^{-1}$. The hypothesis $\beta_2 - \beta_1 x = 0$ can be rejected by a simultaneous test of level γ valid for all x if

$$t^2(x) = \frac{(\hat{\beta}_2 - \hat{\beta}_1 x)^2}{\hat{v}_{22} - 2\hat{v}_{12}x + \hat{v}_{11}x^2} > 2F_{\gamma, 2, n-p-1}.$$

This procedure was first proposed by Johnson and Neyman (1936) and is often called the Johnson–Neyman "technique"; the simultaneous test formulation was given by Potthoff (1964). The above inequality is equivalent to

$$Q(x) = (\hat{\beta}_1^2 - c\hat{v}_{11})x^2 - 2(\hat{\beta}_1 \hat{\beta}_2 - c\hat{v}_{12})x + (\hat{\beta}_2^2 - c\hat{v}_{22}) > 0,$$

where $c = 2F_{\gamma, 2, n-p-1}$. Thus if $Q(x) = 0$ has real roots $x_L < x_U$, the rejection region is $x > x_U$, $x < x_L$ if $\hat{\beta}_1^2 - c\hat{v}_{11}$ is positive, or $x_L < x < x_U$ if $\hat{\beta}_1^2 - c\hat{v}_{11}$ is negative. If there are no real roots, the rejection region in x is empty.

In the above example, we will take $\gamma = 0.05$ for illustration, so that $c = 2F_{0.05, 2, 21} = 6.94$. The parameter estimates are given earlier in this section: we reproduce them here with the variances and covariances of the parameter estimates, obtained using `$disp v $`:

$$\hat{\beta}_1 = 3.249, \quad \hat{v}_{11} = 1.041,$$
$$\hat{\beta}_2 = 135.9, \quad \hat{v}_{22} = 5120, \quad \hat{v}_{12} = 57.89 = \hat{v}_{21}.$$

The quadratic is

$$Q(x) = 3.3315x^2 - 2(39.7825)x - 17064$$

and the roots of $Q(x) = 0$ are -60.6 and 84.5. The first is far outside the range of the data, but the second is well inside the data. For values of `eft` 85 or greater, the mean completion time under row instructions is significantly greater than that under corner instructions. For values of `eft` less than 85, the mean completion time do not differ significantly.

We should note that only three observations have `eft` > 85, one in the row group with `time` 739, and two in the corner group with `time` 513 and 317. These three observations are thus *influential* in this model; we discuss influential observations in Section 3.4.3.

Given the quite different conclusions from models 3 and 6, what can we confidently say? The answer is that we are tantalizingly short of data. The conclusions from both models are strongly influenced by two of the observations, as we will see in the later sections. With small experimental or observational studies it will frequently happen that several different models are equally well supported by the data. In such cases all the competing models and their implications should be presented, and firm judgements suspended. This example is considered later in the chapter.

3.3   Stratified, weighted and clustered samples

In most sample surveys the sample design is not simple random sampling, but some form of stratified design with known but unequal sampling fractions. Multi-stage *cluster* sampling is often used, and this form of sampling induces correlations between the observations in a cluster, invalidating the assumption of independence used in constructing the likelihood function in Chapter 2. Standard regression methods, as applied above to the data of Section 3.1, are invalid when applied to the observations from clustered samples. Regression models can be adapted for the cluster design, leading to *variance component* or *mixed* models; these are considered in Chapter 9.

Stratified and weighted sample designs which are not clustered can be analysed by the standard methods for generalized linear models, with slight modifications.

For the psychological example of Section 3.1, we stated that the sample was randomly drawn from the population of fifth-grade children in one primary school. In fact, the sample design was stratified: the population was divided into sexes, and twelve children of each sex were randomly drawn from the sex sub-populations.

The twelve children of each sex were then assigned to each experimental group by restricted randomization so that each group had six children of each sex. Thus sex and experimental group are orthogonal in the analysis.

The first six children in each experimental group in Section 3.1 are girls, the last six are boys. Sex is a stratifying factor in the design, and it is modelled as an explanatory variable like experimental group. The sex identification is not given in the data listing, but we can generate it as follows:

```
$factor sex 2 $
$calc sex = %gl(2,6) $
```

Why is it necessary to model the stratifying factor? There are two reasons. First, there may be substantial parameter differences between different strata: this is often the reason for stratification in the first place. Omission of the stratum variable from the model may then substantially bias the estimates of the model parameters, and will certainly increase their standard errors.

Second, the estimates of parameters based on aggregating over the strata assume that the population has the same structure over strata as the sample. Unequal sampling fractions (that is, unequal proportions of the population from different strata included in the sample) combined with large stratum differences may give quite misleading population estimates if the stratum variable is ignored. On the other hand if there are no stratum differences in parameters then unequal sampling fractions make no difference and the model can be collapsed over strata.

One point which often causes confusion is the use of "sample weights" in regression. Survey studies sometimes substantially over-sample small sub-populations or strata to provide sample sizes similar to those from (under-sampled) large sub-populations. A "sample weight" is often provided for each observation in the sample data set to allow the re-aggregation of the final model to provide population predictions. The sample weight is the reciprocal of the probability of inclusion in the sample of an observation from each sub-population. The sample weight will be high for the large sub-populations, and low for the small sub-populations.

These weights can be used formally to define a *weighted* or *pseudo* likelihood: for the sample weight w_i for y_i, the weighted likelihood is

$$WL = \prod_{i=1}^{n} f(y_i \mid \theta)^{w_i}.$$

Then the weighted ML estimates $\tilde{\theta}$ from the score equation satisfy

$$\sum_i w_i \frac{\partial \log f(y_i)}{\partial \theta} = 0.$$

If θ is the population mean and the model for Y is $N(\mu, \sigma^2)$, the weighted ML estimate is $\tilde{\mu} = \sum w_i y_i / \sum w_i$. This correctly weights for the disproportionate sampling.

However, it is an important point that these sample weights should *not* be used as formal weights in a regression analysis: the observations should be equally weighted (i.e. unweighted) in this analysis, and the model should always include the stratifying factor, together with its interactions with the other variables in the model. Model reduction then proceeds as in Chapter 2.

Weights are used only at the *predictive* stage, when population predictions are required taking the sample design into account. If the model reduction has allowed the elimination of the stratum interactions and main effect, then no weighting adjustment is needed, as the strata are homogeneous with respect to the residuals from the regression of the outcome variable on the other explanatory variables. If the stratum interactions or main effect *are* needed in the model, then the predicted values for the final model are aggregated across the strata using the sampling weights.

These weights are inappropriate as formal weights in regression for two reasons: they change the parameter estimates by giving higher weight to the samples from the larger populations, though each observation in fact represents a single individual, not an aggregate (like a mean) of several individuals, and the weights increase the standard errors of the estimated coefficients. Pfefferman (1993) noted "Clearly, the use of (1.1) [weighted regression coefficients] cannot be justified in general based on optimality considerations." (p. 318).

The reason generally given for weighting is that it accounts for non-random sampling, if the sample inclusion probabilities depend on the response Y. In this case the likelihood has to represent explicitly the sample design. This kind of *biased sampling mechanism* is discussed in Section 3.17 on missing data.

Returning to the example, the number of possible models now increases considerably since it is possible that regressions of time on eft are different for each sex. A sequence of models of increasing complexity could be fitted as before. For reasons which will be made clear in the next section, we will consider only the most complex model, in which the regression of time on eft is different for each sex/instruction group:

```
$fit group*sex*eft $disp e $
    deviance =    141739.
 residual df =       16
```

	estimate	s.e.	parameter
1	303.9	126.6	1
2	-84.93	145.4	SEX(2)
3	-120.5	141.8	GROUP(2)
4	1.553	2.404	EFT
5	373.5	177.7	SEX(2).GROUP(2)
6	1.931	2.599	SEX(2).EFT

```
7          1.023        2.611        GROUP(2).EFT
8         -5.512        3.020        SEX(2).GROUP(2).EFT
scale parameter 8859.
```

The six observations for each group and the fitted regressions are shown in Fig. 3.6.

It is immediately striking that two observations (one in the girl/corner group and one in the boy/row group) are remote from the other observations, and without these two observations there would be little evidence of any regression on eft at all. We have already noticed these observations in Model 6.

We now turn to an examination of such features of a model.

3.4   Model criticism

The conclusions drawn from a statistical model depends on the validity of the model. We pointed out in Chapter 2 that models are not exact representations of the population. We require only that they reproduce the main features of the population without major distortion.

A careful examination of the correspondence between data and model should be part of any statistical modelling of data. Examination of the data for failure of the model has been called *model criticism* by Box (1980, 1983), a term which we shall adopt, though the systematic examination of the model through *residuals* has

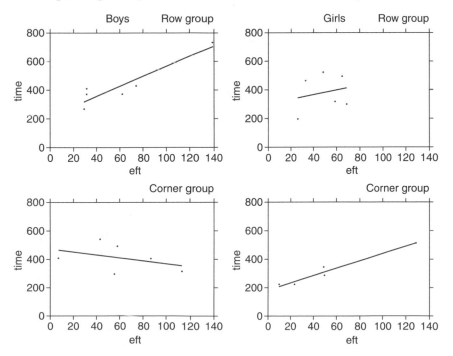

Fig. 3.6. Completion time vs EFT for row and corner groups

a long history. Draper and Smith (1998) gave a discussion of residuals in normal models, and the field is extensively developed (see, e.g. Atkinson, 1982; and the books by Atkinson, 1985; Belsley *et al.*, 1980; Barnett and Lewis, 1994 and Cook and Weisberg, 1982).

There are four important areas in which failures of the model may occur:

(1) mis-specification of the probability distribution for y, leading to an inappropriate likelihood function and inappropriate MLEs and standard errors for the parameters;
(2) mis-specification of the link function;
(3) the occurrence of aberrant observations, distorting either the probability distribution or the parameter estimates from the model;
(4) mis-specification of the systematic part of the model, leading to incorrect interpretations.

Before considering these failures, we quote some standard results for the normal model.

The (raw or "Pearson") residual e_i for the i-th observation from the model is

$$e_i = y_i - \hat{\mu}_i$$
$$= y_i - \mathbf{x}_i'\hat{\boldsymbol{\beta}}.$$

The GLIM directive $disp r $ following the fitting of a normal model displays these residuals; they are available in the system vector %rs following a fit.

In matrix terms

$$\mathbf{e} = \mathbf{y} - X\hat{\boldsymbol{\beta}}$$
$$= (I - X(X'X)^{-1}X')\mathbf{y}$$
$$= (I - H)\mathbf{y},$$

where H is the projection or "hat" matrix

$$H = X(X'X)^{-1}X',$$

so-called since $\hat{\boldsymbol{\mu}} = H\mathbf{y}$: the fitted values are a projection of the original data $\mathbf{y}$ into the space spanned by X. The distribution of $\mathbf{e}$ as a random variable in repeated sampling is

$$\mathbf{E} \sim N_n(\mathbf{0}, \sigma^2(I - H))$$

which is a singular n-dimensional multivariate normal distribution with rank $n - p - 1$, since there are only $n - p - 1$ linearly independent residuals e_i. Denote the diagonal elements of H by h_i; then

$$h_i = \mathbf{x}_i'(X'X)^{-1}\mathbf{x}_i,$$

$$\text{var } E_i = \sigma^2(1 - h_i), \quad \text{cov}(E_i, E_j) = -\sigma^2 h_{ij}, \quad i \neq j,$$

$$\text{var } \hat{\mu}_i = \sigma^2 h_i.$$

The values of h_i are between zero and unity with $\Sigma h_i = p + 1$. A value of zero for h_i means the fitted value of μ_i must be fixed, independently of the y_i, while a value of unity for h_i means that the residual is identically zero, and the model must exactly reproduce the i-th observation y_i. Thus large values of h_i are an indication that the corresponding observations may be influential in determining the position of the fitted model; for this reason the h_i are often called "leverage" values. They are accessible in the system vector %lv, but have to be extracted first (using $extract %lv $).

The residuals have different variances, but can be converted to *standardized residuals* f_i:

$$f_i = e_i / s(1 - h_i)^{1/2}.$$

Most of the assessment of model failure is based on the standardized residuals. Other forms of residual are sometimes useful: the *jack-knife* or *cross-validatory* residuals are

$$j_i = f_i / ((n - p - 1 - f_i^2)/(n - p - 2))^{1/2}.$$

The individual j_i have t_{n-p-2} distributions, though they are not independent; the distribution of the f_i, though it is not normal or t, does not depend on σ. Other statistics, functions of the residuals and/or the influence values h_i, are also used by several authors. GLIM provides *Cook's distances* (Cook, 1977), defined as

$$\text{cd}_i = \frac{h_i}{p(1 - h_i)} f_i^2.$$

These are available after extraction in the system vector %cd.

We now consider the four areas of model failure introduced above.

3.4.1 *Mis-specification of the probability distribution*

Since the probability distribution $f(y)$ is fundamental to the likelihood function and therefore to the parameter estimates and standard errors in the model, an incorrect specification of the distribution can have serious consequences for interpretation of the data.

Examination of the data for failure of this assumption is usually through a *probability plot* (actually a *quantile* or Q–Q plot) of the standardized (or sometimes even the raw) residuals. The residuals f_i are ordered, and the ordered values $f_{(i)}$ plotted against the normal quantiles

$$z_i(a) = \Phi^{-1}\{(i-a)/(n+1-2a)\},$$

where a is a suitably chosen constant ($0 \le a < 1$), and $\Phi(x)$ is the standard normal cumulative distribution function. The usual choice of a is either zero or $\frac{1}{2}$: in the first edition of this book we used the value $a = 0.3175$, since this allowed the use of a test by Filliben (1975) for normality. For discussions of such choices, see Barnett (1975), Filliben (1975) and Draper and Smith (1998). If the probability distribution is correctly specified, the plot should be roughly a straight line. Systematic curvature, or individual observations far from the straight line, indicate failures of the probability distribution specification. The file qplot contains the macro qplot and other macros for calculating the different forms of residuals.

As noted in Chapter 2, an acute difficulty in the inspection of such plots is how to decide whether the variation in the plot is too far from a straight line. In Chapter 2 we used the simultaneous confidence band from the *cdf* of the observed data to check the probability model specification; the same approach may be used for residuals though it is now less theoretically sound since the residuals are not independent and do not have identical normal distributions.

This difficulty can be resolved by a *simulation envelope* constructed by simulating from the fitted model (Atkinson, 1981, 1982, 1985).

We illustrate with the example of Section 3.1. We will first ignore sex and fit the eft*group model; we need to sort the residuals before constructing the simultaneous band. Residual examination is usually done with the full or most complex model, since simplifications of this model may be affected by failure of the normal distribution assumption.

```
$yvar time $
$fit eft*group $disp e $
$sort sortres %rs $
$input 'binocisim' $
$calc nn = 25 : i = %cu(1) : np = %np(sortres/%sqrt(%sc)) $
$use binocisim i nn $
$graph(s = 1 h = 'residual' v = 'cumulative proportion')
        pl_ _,pu_ _,np sortres 9,9,10 $
```

It is immediately clear from Fig. 3.7 that the residual distribution is poorly defined in a sample of 24 (with 20 df): the confidence band is so wide that only severe non-normality could be identified. There is no need to transform to the normal deviate scale.

What should we do if the graph shows a major discrepancy from normality? There are two possibilities, as we discussed in Chapter 2: try a transformation of the variable to produce a closer agreement—usually the log transformation, or less commonly the reciprocal transformation—or use a different type of continuous

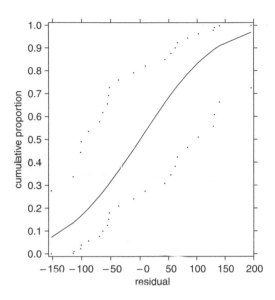

Fig. 3.7. Cumulative proportion, residuals

probability distribution. The first of these possibilities is examined in detail below, and the second is examined in Chapter 6.

The standardized and jack-knife residuals can be explicitly calculated into named variables and used subsequently, provided that %lv, the leverage value, is first extracted. (The raw residuals are available in %rs following a fit *without* extraction.) For example, to obtain the standardized residuals (given here to 3 dp), say sres, we use:

```
$input %plc qplot $
$extract %lv $
$calc sres = #stan $
$print sres $
-1.009   1.342   1.401  -1.537   0.540  -1.199  -0.542  -0.556   0.446
 1.105  -0.372   0.859  -0.113  -1.057  -0.957   0.822  -0.538  -0.695
 0.090   1.968  -0.635   1.290  -1.292   1.118
```

The macro stan in the qplot file contains the GLIM expression for the standardized residuals.

A second and less common form of failure of the probability model assumption is that the nuisance parameters, assumed constant for all observations, are in fact varying randomly, or systematically with the explanatory variables. A simple example of this would be if the variances were different for the two instruction groups in the example of Section 3.1. This could be assessed by plotting the residuals separately for each group against the fitted values. In general, it is necessary to plot the residuals against the fitted values and each explanatory variable: if different variability is evident then the model can be modified appropriately.

Heterogeneity of variance is examined in a factorial design in Section 3.15, and the "double modelling" of both mean *and* variance is discussed in detail in Section 3.17.

3.4.2   *Mis-specification of the link function*

The choice of parameter scale on which the systematic effects are modelled is an important part of statistical modelling. When the response variable is normal it is almost universal practice to work on the scale of the normal mean μ, using the identity link: this has therefore been taken as the default option in GLIM for the link function with a normal error specification.

In non-normal distributions the choice of scale is much more open, as we shall see in later chapters. However, even for normal distribution models, the possibility of working on the log scale particularly should be borne in mind. This possibility is investigated in Section 3.7, where both the lognormal distribution for the response-variable and a log link with the normal distribution are examined.

3.4.3   *The occurrence of aberrant and influential observations*

Good data analysis depends heavily on the correct recording of variable values. Misrecording of data values can result in substantial changes to the fitted model. If an observation happens to lie on the boundary of the space of the explanatory variables, misrecording of its response variable may have a powerful effect on the position of the fitted regression, but may not produce a large residual. The same effect can occur by misrecording an explanatory variable, thus moving the observation out of the cluster of correctly recorded observations. The influence values h_i are particularly useful for diagnosing observations of this type.

Even when no misrecording occurs it is valuable to know whether a small number of observations is substantially influencing the position of the fitted regression. These observations can then be carefully examined and referred back to the original survey or experimental scientists for comment. Observations with large residuals or influence values are not automatically rejected from the model: they always require evaluation.

We illustrate the use of influence values with the example from Section 3.1. For the group*eft model, we can graph the influence values against observation number:

```
$fit group*eft $
$extract %lv $
$calc i = %cu(1) $
$graph %lv i 5 $
$pr %lv $
```

The influence values are given in Table 3.4 to two decimal places.

Table 3.4. Influence values

0.08	0.13	0.09	0.10	0.09	0.16	0.15	0.09	0.14	0.72	0.11	0.14
0.09	0.15	0.22	0.42	0.09	0.09	0.15	0.09	0.08	0.08	0.30	0.23

How should we interpret the influence values? The model specifies a linear regression in each experimental group, with different slopes for each group. For a linear regression model, it is easily shown (see, e.g. Hoaglin and Welsch, 1978) that

$$\mu_l - \beta_0 + \beta_1 x_i$$

gives

$$h_i = \frac{1}{n} + \frac{(x_i - \bar{x})^2}{\sum_{j=1}^{n}(x_j - \bar{x})^2}.$$

Thus h_i takes its minimum value of $1/n$ for observations at the explanatory variable mean, and its maximum value for the observation furthest from the mean: extreme values of x are the most influential. If all x_i are equidistant from the mean, $h_i = 2/n$; if all but one observation have identical values of x_i, these will have $h_i = 1/(n-1)$, and the remaining observation will have $h_i = 1$. In this case the position of the regression is determined completely by the single observation, and we could have no confidence at all in the regression.

For the general regression model $\beta_0 + \boldsymbol{\beta}' \mathbf{x}_i$ with p explanatory variables $\mathbf{x}$, the corresponding result is

$$h_i = \frac{1}{n} + (\mathbf{x}_i - \bar{\mathbf{x}})' S^{-1}(\mathbf{x}_i - \bar{\mathbf{x}}),$$

where $\bar{\mathbf{x}}$ is the vector of means and S is the SSP matrix of the explanatory variables. In general $(p+1)/n$ is the average value of all the h_i and also corresponds to equally influential observations. Hoaglin and Welsch proposed regarding $2(p+1)/n$ as a value of h indicating important influence.

For the model group*eft with $p = 3$ and $n = 24$, $(p+1)/n$ is 0.167. Since the model group*eft corresponds to unrelated simple linear regressions in each instruction group of 12 children, observations near the eft mean in each group will have influence values near $1/12 = 0.0833$.

One value stands out: 0.72 for the 10th observation in the row group. This observation has the largest eft value of 139, and also the largest completion time of 739 seconds. It has a powerful effect on the fitted regression for any subgroup in which it appears. For the eft + group model, its influence value is reduced to 0.35, because the slope of the regression is determined by both row and column groups, and the effect of this observation is diluted by the doubled sample size.

The influence value of 0.42 for the fourth observation in the corner group (in the group*eft model) is also high. This observation has the largest eft value of

128 in the corner group, though its completion time value of 513 seconds is not exceptionally large. Its influence value is reduced to 0.28 in the main effect model.

It is notable that neither of these observations has a large residual: in both cases the fitted regression passes close to the observed value, partly because of the influence of the observation on the position of the fitted line.

In Model 6, only the first of these observations is influential: the 10th observation in the row group has influence 0.72 as before, because the regression on eft is determined only by the row group. The column group observations all have the same influence 0.083, as a constant mean is being fitted for this group.

In the full sex*group*eft model, the two influential observations above (again given to 2 dp) have influence values of more than 0.8.

```
$fit sex*group*eft $
$extract %lv $print %lv $
0.22 0.36 0.17 0.40 0.31 0.55 0.28 0.17 0.27 0.84 0.19 0.27
0.17 0.25 0.37 0.88 0.17 0.17 0.27 0.21 0.17 0.17 0.58 0.60
```

The reasons are clear from Fig. 3.6: the more we subdivide or stratify the observations by explanatory variables, the greater the effect these observations have on the regressions within subgroup. The regressions in the corner group for girls and the row group for boys are being very largely determined by the single largest observation in each case. We did not pursue the simplification of the sex*group*eft model in Section 3.3 because the conclusions from this model depend so heavily on these two observations. This is a frequent problem when samples are extensively cross-classified by many explanatory factors, so that many cells have very small numbers of observations.

To examine the effect of omitting these observations, we set up a weight variate wt, defined to be zero if the influence value $h > 2(p + 1)/n$, and one otherwise:

```
$calc %a = 2*%pl/%nu $
     : wt = %le(%lv,%a) $
$weight wt $
$fit . $disp e $
   deviance =  133907. (change =   -7833.)
  residual df =     14 (change =      -2 ) from 22 observations
```

	estimate	s.e.	parameter
1	303.9	131.5	1
2	-6.606	174.1	SEX(2)
3	-121.0	168.1	GROUP(2)
4	1.553	2.498	EFT
5	295.6	219.4	SEX(2).GROUP(2)
6	0.04795	3.409	SEX(2).EFT
7	1.038	3.715	GROUP(2).EFT
8	-3.644	4.541	SEX(2).GROUP(2).EFT

scale parameter 9565.

The . argument to $fit means "the previous model". The three-way interaction parameter estimate is now smaller than its standard error, and successive elimination of terms shows that all the eft terms can be omitted from the model. The two-way interaction group.sex now becomes quite large (as is its standard error). This interaction can be identified as due to one discrepant cell: girls in the corner group have a mean completion time of 272.6 seconds, while boys in both groups and girls in the row group have similar means, with a common mean completion time of 388.7 seconds.

These conclusions are markedly different from those obtained by retaining the two influential observations: In the latter case the sex*group*eft model can be reduced to the eft model 3 of Section 3.1, with no group difference and a common regression on eft.

What action, if any, should we take? The original data should be checked with the experimenter: are the eft and time values correct for these two observations? If they are not, the correct values can be entered. If they are, then the experiment can be interpreted as before. The difficulty, as previously mentioned, is that because the sample sizes are small several different conflicting interpretations of the data are possible.

It may not be possible to check the correctness of the original data. In this case an analysis of the data might be carried out as above by including and excluding the two highly influential observations, and comparing the results. It is not, however, good statistical practice to set aside routinely or exclude observations with large residuals and/or influence simply because these values are large.

3.4.4 Mis-specification of the systematic part of the model

This form of model failure is different from the others because it is rectified by further modelling, though diagnostic plotting of the standardized residuals against other variables may be useful.

When the explanatory variables are continuous, one form of model failure is the reliance on linear regressions when the relationships are non-linear. Inclusion of quadratic or higher-order polynomial terms in the regression is a common solution; however, this may lead to unreasonable features in the model, for example, the fitted values may reach a maximum and then decrease while the observed values approach an asymptote. Transformations of the variables or of the scale of μ (a different link function) may be preferable.

Interactions between variables may also be necessary. In fitting such terms in complex models a careful eye should be kept on the influence values, because in small samples the possibility of single observations producing apparent interactions is considerable, as in the above example.

The need for non-linear terms or interactions is usually assessed graphically by plotting the raw or standardized residuals against the explanatory variables and the fitted values. Non-random scatter patterns suggest the need for model changes.

We do not discuss this further here as the examples presented in later chapters are analysed in considerable detail.

3.5   The Box–Cox transformation family

In this section we extend the usefulness of the normal distribution by embedding it in a larger family of distributions: the Box–Cox transformation family (Box and Cox, 1964).

The motivation for this general family is the frequent occurrence of skewed data in which a log transformation, or less commonly a reciprocal transformation, of the response-variable produces a nearly normal distribution. This should be carefully distinguished from the use of a log or reciprocal link function: in the first case the observed data have a lognormal or reciprocal normal distribution, in the second they have a normal distribution. We will illustrate both cases with an example in Section 3.7.

The general form of the transformation can be represented by

$$y^{(\lambda)} = (y^\lambda - 1)/\lambda, \quad \lambda \neq 0$$
$$= \log y, \quad \lambda = 0,$$

where λ is the transformation parameter and y the response-variable. Here y must be *positive*. For $\lambda = 1$, $y^{(1)} = y - 1$, and as $\lambda \to 0$, $y^{(\lambda)} \to \log y$, so that $y^{(\lambda)}$ is a continuous function of λ. For $\lambda = -1$, $y^{(-1)} = 1 - y^{-1}$.

Given data $(y_i, \mathbf{x}_i)$ we assume that there is some value of λ for which $Y_i^{(\lambda)}$ has a normal distribution with mean $\boldsymbol{\beta}'\mathbf{x}_i$ and variance σ^2. We want to estimate λ, and decide the appropriate scale of y on which to fit the regression model. The choice of the scale for y, that is, of the value of λ, is determined both by the plausible values of λ from the fitted model and by the interpretability of the scale. Because the transformation is defined by a parameter, we can proceed by ML. For each fixed value of λ, we estimate $\boldsymbol{\beta}$ and σ by ML, and substitute these values in the likelihood. The resulting function of λ only is the *profile likelihood in* λ; its log is the *profile log-likelihood* (these are discussed in Chapter 2). It is used to construct likelihood-based confidence intervals for λ.

The probability density function of $Y^{(\lambda)}$ is normal (μ, σ^2) and hence that of Y is for $\lambda \neq 0$,

$$f(y|\lambda, \mu, \sigma) = \frac{1}{\sigma\sqrt{2\pi}} |\lambda| y^{\lambda - 1} \exp\{-(y^{(\lambda)} - \mu)^2/2\sigma^2\},$$

and for $\lambda = 0$,

$$f(y|0, \mu, \sigma) = \frac{1}{\sigma\sqrt{2\pi}} y^{-1} \exp\{-(\log y - \mu)^2/2\sigma^2\}.$$

We assume that, whatever the value of λ, $\mu = \boldsymbol{\beta}'\mathbf{x}$. For the given observations $(y_i, \mathbf{x}_i)$, the log-likelihood function is

$$\ell(\lambda, \boldsymbol{\beta}, \sigma) = -n \log \sqrt{2\pi} - n \log \sigma + n \log |\lambda| + (\lambda - 1) \sum \log y_i$$
$$- \sum (y_i^{(\lambda)} - \boldsymbol{\beta}'\mathbf{x}_i)^2 / 2\sigma^2,$$

where for $\lambda = 0$, $\log |\lambda|$ is defined to be zero. For fixed λ, the partial derivatives with respect to $\boldsymbol{\beta}$ and σ are then

$$\frac{\partial \ell}{\partial \boldsymbol{\beta}} = \Sigma \mathbf{x}_i (y_i^{(\lambda)} - \mathbf{x}_i'\boldsymbol{\beta}) / \sigma^2$$

$$\frac{\partial \ell}{\partial \sigma} = -n/\sigma + \Sigma (y_i^{(\lambda)} - \mathbf{x}_i'\boldsymbol{\beta})^2 / \sigma^3.$$

Denote the solutions of the equations $\partial \ell / \partial \boldsymbol{\beta} = 0$, $\partial \ell / \partial \sigma = 0$ by $\hat{\boldsymbol{\beta}}(\lambda), \hat{\sigma}(\lambda)$. These are the same as in Section 3.1, with $y_i^{(\lambda)}$ replacing y_i:

$$\hat{\boldsymbol{\beta}}(\lambda) = (X'X)^{-1} X' \mathbf{y}^{(\lambda)}$$

$$\hat{\sigma}^2(\lambda) = \Sigma \{y_i^{(\lambda)} - \mathbf{x}_i' \hat{\boldsymbol{\beta}}(\lambda)\}^2 / n = \mathrm{RSS}(\lambda) / n$$

where $\mathrm{RSS}(\lambda)$ is the residual sum of squares for the given λ.

Substituting into the log-likelihood function, we obtain the profile log-likelihood function

$$p\ell(\lambda) = -n \log \sqrt{2\pi} - \frac{n}{2} \log n - \frac{n}{2} - \frac{1}{2} n \log \mathrm{RSS}(\lambda) + n \log |\lambda|$$
$$+ (\lambda - 1) \sum \log y_i.$$

An (approximate) $100(1-\alpha)\%$ confidence interval for λ consists of those values of λ for which $p\ell(\lambda)$ is within $\frac{1}{2}\chi_{\alpha,1}^2$ units of its maximum: the interval is most simply found by tabulation of $p\ell(\lambda)$ over a grid of values of λ. This interval is used to identify the values of λ giving an interpretable scale.

The file boxcox contains several macros for the construction and plotting of the profile log-likelihood. The macro boxcox constructs a plot of the disparity $-2p\ell(\lambda)$ rather than $p\ell(\lambda)$ itself, so that differences from the minimum value of $-2p\ell(\lambda)$ are compared with $\chi_{\alpha,1}^2$. Illustration of its use is given in the next section.

An important feature of response-variable transformations is that on the transformed scale the model represents variation in the *mean* of the (normal) transformed variable, but on the original scale the variation is in the *median* of the variable.

This is most simply seen for the log transformation. Suppose that $\log Y \sim N(\mu, \sigma^2)$. Then Y has a lognormal distribution, and

$$\text{median}[Y] = \exp(\mu)$$

$$E[Y] = \exp(\mu + \tfrac{1}{2}\sigma^2)$$

$$\text{Var}[Y] = [\exp(\sigma^2) - 1]\exp(2\mu + \sigma^2).$$

Thus the (additive) regression model for the mean of $\log Y$ is a multiplicative model for the median of Y, and also for the mean of Y, though the intercept is changed by $\tfrac{1}{2}\sigma^2$, and the variance of Y is not constant. Though the model is fitted on the log scale, we usually want to interpret the model on the original scale: if fitted values from the model are transformed by `%exp(%fv)` to the original scale, these are fitted values for the median response, not for the mean: for the mean the fitted values are `%exp(%fv + %sc/2)`, where here `%sc` is `%dv/%df`, the unbiased estimate s^2 of σ^2.

For transformations Y^λ with $\lambda \neq 0$, if μ is the mean of Y^λ then

$$\text{median}[Y] = \mu^{1/\lambda}$$

$$E[Y] \approx \mu^{1/\lambda}\{1 + \sigma^2(1 - \lambda)/(2\lambda^2\mu^2)\}$$

$$\text{Var}[Y] \approx \mu^{2/\lambda}\sigma^2/(\lambda^2\mu^2).$$

Here the apparent discontinuity between $\lambda = 0$ and $\lambda \neq 0$ is caused by the use of y^λ rather than $(y^\lambda - 1)/\lambda$.

3.6   Modelling and background information

The most useful models are those which use background information or theory from the field of application. However, theory may be incorrect, or speculative, or there may be no adequate theory and the modelling may be meant to assist the development of an adequate theory.

We illustrate with a set of data on tree volumes taken from the Minitab Handbook, (Ryan *et al.*, 1976) and discussed at length by Atkinson (1982) and other authors. The volume of usable wood v in cubic feet (1 foot = 30.48 cm) is given for each of a sample of 31 black cherry trees, and the height h in feet and the diameter d in inches (1 inch = 2.54 cm) at a height 4.5 feet above the ground. We want to develop a model which will predict the usable wood volume from the easily measured height and diameter.

The data are in the file `trees`, and are listed in Table 3.5.

```
$input 'trees' $
   : %plc qplot normac $
```

Table 3.5. Tree data

d	h	v	d	h	v	d	h	v	d	h	v
8.3	70	10.3	8.6	65	10.3	8.8	63	10.2	10.5	72	16.4
10.7	81	18.8	10.8	83	19.7	11.0	66	15.6	11.0	75	18.2
11.1	80	22.6	11.2	75	19.9	11.3	79	24.2	11.4	76	21.0
11.4	76	21.4	11.7	69	21.3	12.0	75	19.1	12.9	74	22.2
12.9	85	33.8	13.3	86	27.4	13.7	71	25.7	13.8	64	24.9
14.0	78	34.5	14.2	80	31.7	14.5	74	36.3	16.0	72	38.3
16.3	77	42.6	17.3	81	55.4	17.5	82	55.7	17.9	80	58.3
18.0	80	51.5	18.0	80	51.0	20.6	87	77.0			

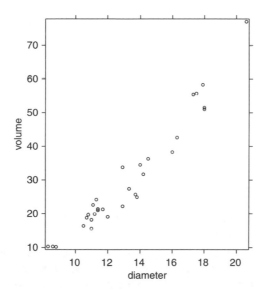

Fig. 3.8. Tree volume vs diameter

We begin by graphing v against d and h (Figs 3.8 and 3.9).

```
$graph (s = 1 h = 'diameter' v = 'volume')
       v d 5 $
     : (s = 1 h = 'height' v = 'volume')
       v h 5 $
```

Clearly accurate prediction of v can be made from d: the variation about a smooth curve is quite small. Height is less useful. Curvature in the v, d graph is evident: the extreme points at both ends are well above the nearly linear ellipse in the centre. A straight-line relationship will not be adequate. We will use the multiple correlation R (or more usually its square R^2) as a measure of the predictability of the response from the explanatory variables: R is just the correlation between the

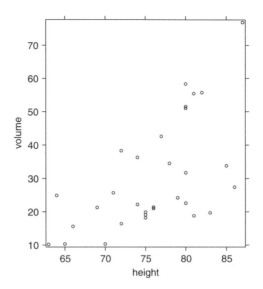

Fig. 3.9. Tree volume vs height

response and the fitted values from the model. Equivalently,

$$R^2 = 1 - \text{RSS}/\text{TSS},$$

where RSS is the residual sum of squares from the given model and TSS is the residual sum of squares from the null model: $\text{TSS} = \sum (y_i - \bar{y})^2$ is usually called the "total sum of squares".

The value of R^2 is not calculated by GLIM, but it can be easily obtained after any fit using the macro rsq in the file normac. (We ignore the "adjusted" R^2 given by the macro; it is based on a multivariate normal model for the explanatory variables which is not relevant to our modelling.)

We first try the model using h and d.

```
$yvar v $
$fit h + d $disp e $
   deviance =   421.92
 residual df =    28

            estimate          s.e.       parameter
       1       -57.99         8.638          1
       2        0.3393        0.1302         H
       3        4.708         0.2643         D
 scale parameter 15.07


$use rsq $
R-squared equals    0.9480
```

For this model R^2 is very high, implying close agreement between observed and fitted volumes. We now examine the residuals and influence values.

```
$extract %lv $
$print %lv $
0.116   0.147   0.177   0.059   0.121   0.156   0.115   0.052   0.092
0.048   0.074   0.048   0.048   0.073   0.038   0.036   0.131   0.144
0.067   0.211   0.036   0.045   0.050   0.111   0.069   0.088   0.096
0.106   0.110   0.110   0.227
```

The 20th and the last observations have influence values of 0.21 and 0.23 respectively, just larger than $6/31 = 0.194$. The last tree is the largest, the 20th has a large diameter for its height.

```
$calc rres = #raw $
$graph (s = 1 h = 'diameter' v = 'residual')
       rres d 5 $
    :  (s = 1 h = 'height' v = 'residual')
       rres h 5 $
```

The graph against h (not shown) looks fairly random but that against d (Fig. 3.10) shows a marked dip in the middle, with large positive residuals at each end and smaller and negative residuals in the centre. This suggests that the model is mis-specified, and the curvature we saw in the v versus d graph suggests that we may

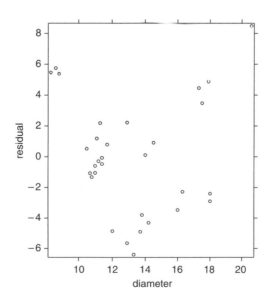

Fig. 3.10. Residual vs diameter

be working on the wrong scale. To investigate this we construct a Box–Cox plot. We first input the file boxcox:

```
$input %plc boxcox $
```

To use the boxcox macro, we first need to specify the untransformed y-variate and the model as the contents of two macros called by boxcox:

```
$macro yvar v $endm
$macro model h + d $endm
$use boxcox $
                --- Model is h + d
                --- Y-Variate is v
Max. value of lambda?
$DIN? 2
Increment?
$DIN? 0.5
Min. value of lambda?
$DIN? -2
```

The boxcox macro prompts the user for the range of values of λ over which the values of the disparity are to be plotted. It is seldom necessary to choose the range wider than -2 to $+2$; initially steps of 0.5 are sufficient. We do not reproduce the plot, but give a table of disparities below; the second decimal place is obtained by specifying higher accuracy before calling boxcox, by using $accuracy 5 $. The Box–Cox macro prints the "deviance", which is the disparity as we have defined it. The reader should note that the table entries differ from those in the first edition because of the inclusion of the constants in the likelihood. (The present entries are less than the former by 18.5.)

```
----------------------------------------
lmb_    -2.0    -1.5    -1.0    -0.5     0.0
dev_   250.08  226.30  201.10  173.80  142.93

lmb_     0.5     1.0     1.5     2.0
dev_   137.84  168.91  201.59  231.22
----------------------------------------
```

The minimum of the disparity occurs near $\lambda = 0.5$. A finer tabulation in steps of 0.1 over the range 0 to 1 is required.

```
$use boxcox $
                --- Model is h + d
                --- Y-Variate is v
Max. value of lambda?
$DIN? 1
Increment?
$DIN? 0.1
```

```
Min. value of lambda?
$DIN? 0
--------------------------------------------------
lmb_    0       0.1    0.2    0.3    0.4    0.5
dev_   142.93 138.22 134.96 133.69 134.69 137.84

lmb_    0.6    0.7    0.8    0.9    1.0
dev_   142.68 148.63 155.20 162.03 168.91
--------------------------------------------------
```

Computing the disparity over a fine mesh, we find the minimum of 133.68 occurs
at $\lambda = 0.31$, and the approximate 95% confidence interval, based on a value of
133.68 ± 3.84, is (0.12, 0.49), which is quite narrow. How should we interpret this
result?

The nature of the problem throws light on this question. Volume is measured in
cubic feet, but height and diameter in feet (or inches). We are attempting to predict
a volumetric measurement from a linear measurement. This suggests that the "side
of the equivalent cube" might be more appropriate as a response, the cube root of
volume. The profile likelihood in λ points very closely to this value. To see why,
we calculate $v^{1/3}$ and graph it against d and h (Figs 3.11 and 3.12):

```
$calc vth = v**(1/3) $
$graph (s = 1 h = 'diameter' v = 'cube root v')
        vth d 5 $
     :  (s = 1 h = 'height' v = 'cube root v')
        vth h 5 $
```

The graph against d is now very closely linear. The curvature has been removed
by the cube root transformation: the extreme points at each end now line up with
the central ellipse. We now refit the model:

```
$yvar vth $
$fit #model $disp e $
   deviance =    0.19209
 residual df =   28

          estimate       s.e.      parameter
     1    -0.08539       0.1843     1
     2     0.01447       0.002777   H
     3     0.1515        0.005639   D
 scale parameter 0.006860

$use rsq $
R-squared equals   0.9777
   :  qplot $
```

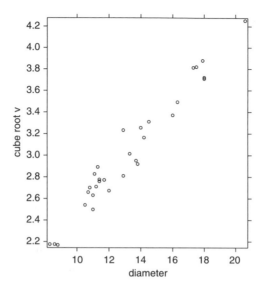

Fig. 3.11. Cube root volume vs diameter

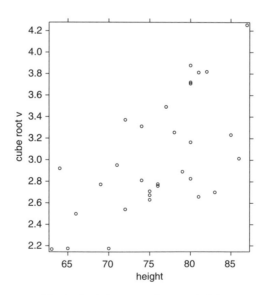

Fig. 3.12. Cube root volume vs height

Here we use the text substitution symbol #: the contents of the macro model are used as the argument of $fit. The value of R^2 is a substantial increase over that for $\lambda = 1$. Residual graphs against d and h (not shown) show a random scatter, though the residuals seem to increase noticeably in magnitude with h, and a quantile plot gives a closely linear fit.

This model can also be fitted using the macro boxfit, which prompts for a specified value of λ:

```
$use boxfit $
--- Model is h + d
--- Y-Variate is v
-- model changed
Value of lambda?   0.3333
```

The value of the disparity is 133.76, almost the same as for $\lambda = 0.31$.

The fitted median values of v from the model are $\hat{\mu}^3$, and the fitted mean values of v are approximately $\hat{\mu}^3(1 + 3s^2/\hat{\mu}^2)$. These can be compared with the observed values, for example using

```
$calc fv = %fv**3$
$print fv :: v $
10.43   10.05   10.07   16.53   19.86   20.85   16.32   18.96   20.89
19.62   21.25   20.61   20.61   19.38   22.38   25.41   29.77   31.99
27.49   25.20   31.69   33.51   32.23   38.64   42.83   50.98   52.85
54.20   54.86   54.86   79.22

10.30   10.30   10.20   16.40   18.80   19.70   15.60   18.20   22.60
19.90   24.20   21.00   21.40   21.30   19.10   22.20   33.80   27.40
25.70   24.90   34.50   31.70   36.30   38.30   42.60   55.40   55.70
58.30   51.50   51.00   77.00
```

Fitted median and mean values both agree closely with the observed values. We have a closely fitting model to which we are led by physical considerations of dimensionality, and the Box–Cox plot.

Other physical considerations, however, lead to a different model. The curvature which we noted in the v against d graph can also be removed by a log transformation (Figs 3.13 and 3.14):

```
$calc lv = %log(v) : ld = %log(d) : lh = %log(h) $
$graph (s = 1 h = 'log diameter' v = 'log volume')
        lv ld 5 $
    :   (s = 1 h = 'log height' v = 'log volume')
        lv lh 5 $
```

This suggests a regression of log v on the logged explanatory variables.

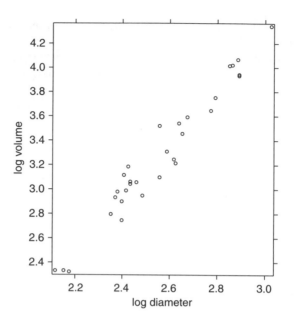

Fig. 3.13. Log volume vs log diameter

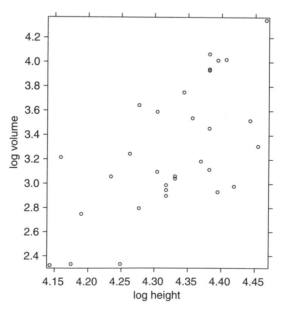

Fig. 3.14. Log volume vs log height

```
$yvar lv $fit lh + ld $disp e $
   deviance =    0.18546
residual df =  28

             estimate          s.e.       parameter
   1          -6.632         0.7998       1
   2           1.117         0.2044       LH
   3           1.983         0.07501      LD
scale parameter 0.006624

$use rsq $
R-squared equals    0.9777
  : qplot $
$extract %lv $print %lv $
0.151   0.167   0.198   0.059   0.121   0.152   0.119   0.051   0.091
0.047   0.072   0.046   0.046   0.073   0.036   0.036   0.116   0.126
0.071   0.243   0.038   0.046   0.054   0.115   0.071   0.086   0.092
0.098   0.101   0.101   0.180
```

The value of R^2 is exactly the same (to four decimal places) as that for the $v^{1/3}$ model. Residual plots (not shown) show little unusual apart from three rather large negative residuals, though the same pattern of residuals increasing with h is visible as with the cube root transformation. The quantile plot of the standardized residuals shows some variation. The influence values are much the same as before, though that for the 20th observation is larger (0.24) and that for the last is smaller (0.18).

Should we be working on the log scale of v? What is the effect of the log transformations of h and d on the Box–Cox transformation of v? We specify a new model for the boxcox macro by overwriting the contents of the macro model.

```
$macro model lh + ld $endm
$use boxcox $
```

Using the grid $-2(0.5)2$ for λ shows a disparity minimum near $\lambda = 0$. A finer tabulation in steps of 0.01 around zero shows a minimum of 131.61 at $\lambda = -0.07$, and a 95% confidence interval of $(-0.24, 0.11)$, which includes zero but excludes $\lambda = 1/3$. Thus if we log transform h and d, the cube root transformation of v is not appropriate, but the log transformation is. The value of the disparity at $\lambda = 0$ of 132.20 is slightly smaller than for $v^{1/3}$, h and d, which was 133.76.

How can we compare these models? It would appear that we can use the LRT, since both models can be expressed in the form

$$v^* \sim N(\beta_0 + \beta_1 h^* + \beta_2 d^*, \sigma^2)$$

with

$$v^* = (v^{\lambda_v} - 1)/\lambda_v$$
$$h^* = (h^{\lambda_h} - 1)/\lambda_h$$
$$d^* = (d^{\lambda_d} - 1)/\lambda_d.$$

The comparison is of the set of values $\lambda_v = 1/3, \lambda_h = \lambda_d = 1$ with the set $\lambda_v = \lambda_h = \lambda_d = 0$, the other parameters being nuisance parameters in each case, the likelihood being maximized over them. The disparity difference is 1.56, but the models are not nested and have the same number of parameters, so this difference cannot be distributed as χ^2 with any number of degrees of freedom. (The χ^2 comparison we gave in the first edition is valid only for comparing each model with the model in which *all* the parameters are fully estimated, but that model is not of interest and has not been fitted.)

Model comparisons of this kind can be treated as discussed in Section 2.7.4 through the posterior distribution of the true likelihood ratio for the two models, evaluated at the true values of the nuisance parameters β and σ. For large samples the disparity difference, or the ratio of maximized likelihoods, is interpreted directly as the weight of evidence for one model relative to the other.

The disparity difference of 1.56 is equivalent to a maximized likelihood ratio of $e^{0.78} = 2.18$; the sample evidence gives only a very weak preference for the log model over the cube root model.

In the fitted model for log v, the coefficient of lh is close to 1, and that of ld close to 2. Again the nature of the problem throws light on this result. The tree may be idealized (modelled) as a regular solid figure, like a cylinder or a cone. The volume v of a cylinder of height h and diameter d is $\pi d^2 h/4$, and that of a cone of height h and base diameter *d* is $\pi d^2 h/12$. In either case

$$\log v = c + \log h + 2 \log d,$$

where c is log $\pi/4$ for the cylinder and log $\pi/12$ for the cone. These volume formulae assume that h and d are measured in the same units. In our example h is in feet but d is in inches. We first convert d to feet:

```
$calc df = d/12 : ldf = %log(df) $
```

We redefine the y-variate and refit the model.

```
$yvar lv
$fit + ldf - ld $disp e $
```

The intercept is now -1.705. Now we fit the model in which the coefficients are fixed at 1 and 2. The model is then

$$\mu = \beta_0 + x_1 + 2x_2$$

in which only β_0 is to be estimated. Here $(x_1 + 2x_2)$ is treated as a single variable whose coefficient is to be fixed at 1.0. This is achieved by declaring this variable to be an *offset* and fitting the null model in which only β_0 is estimated.

```
$calc z = lh + 2*ldf $
$offset z $
$fit $disp e $use rsq $
   deviance =    0.18769
 residual df =   30

            estimate          s.e.      parameter
      1        -1.199        0.01421       1
  scale parameter 0.006256


  R-squared equals    0.9774
```

The value of R^2 is a trivial change from the full model. Does the value of β_0 correspond to a cylinder or a cone or neither?

```
$calc %log(%pi/4) : %log(%pi/12) $
        -0.2416
        -1.340
```

The tree is closer to a cone than a cylinder, but it has a greater volume than the cone, which seems reasonable.

The fitted median values of v from the model are $\exp(\hat{\mu})$, which can be compared with the observed values:

```
$calc fv = %exp(%fv) $
$print fv :: v $
  10.09 10.06 10.21 16.61 19.41 20.26 16.71 18.99 20.63
  19.69 21.11 20.67 20.67 19.77 22.60 25.77 29.60 31.84
  27.89 25.51 32.00 33.76 32.56 38.58 42.82 50.74 52.56
  53.65 54.25 54.25 77.27

  10.30 10.30 10.20 16.40 18.80 19.70 15.60 18.20 22.60
  19.90 24.20 21.00 21.40 21.30 19.10 22.20 33.80 27.40
  25.70 24.90 34.50 31.70 36.30 38.30 42.60 55.40 55.70
  58.30 51.50 51.00 77.00
```

The fitted mean values of v are $\exp(\hat{\mu} + \frac{1}{2}s^2)$, which are only 0.3% greater than the median values.

The two fitted models agree closely with the observed values (a maximum error of 4.59 cubic feet for the $v^{1/3}$ model and 4.66 cubic feet for the `log` v model), with the error standard deviation of 0.079 in the log model corresponding to a percentage error of 8% in the fitted volume. Here the data are insufficient to discriminate between models based on different physical arguments.

3.7   Link functions and transformations

We noted above that even in normal models the choice of the link function relating the parameter of the probability distribution to the linear predictor is not always obvious. The preceding example provides a good illustration. For the first model relating tree volume to diameter and height, a cube root transformation of v linear-ized the graph of v against d. It is not, however, necessary to assume that $v^{1/3}$ has a normal distribution with mean θ equal to the linear predictor $\beta_0 + \beta_1 h + \beta_2 d$, in order to fit such a model. We can assume instead that V is normal, but the link function relating V to the linear predictor is the cube root, that is, if μ is the mean tree volume,

$$\eta = g(\mu) = \mu^{1/3} \quad \text{or} \quad \mu = \eta^3 = (\beta_0 + \beta_1 h + \beta_2 d)^3,$$

while $V \sim N(\mu, \sigma^2)$.

The fitting of this model by ML is equivalent to fitting by non-linear least squares, since the model for μ is now non-linear in the parameters. A simple change in the link function to the Exponent link is all that is needed; the offset also needs to be removed:

```
$yvar v $link e 0.3333 $
$offset $
$fit h + d $disp e $
   deviance =   184.16 at cycle 3
 residual df =    28

            estimate          s.e.        parameter
      1     -0.05098         0.2240        1
      2      0.01428         0.003342      H
      3      0.1503          0.005837      D
 scale parameter 6.577

$use rsq$
 R-squared equals    0.9773
```

The value of R^2 is very slightly less than for the $v^{1/3}$ model. The fitted model is almost the same as in Section 3.6. The RSSs for this model is 184.16, and converting this to the disparity $n(\log RSS + \log(2\pi) + 1 - \log n)$ gives a value of 143.21. Residual plots show the same feature as for the cube root transformation of v, the residuals again increasing in absolute size with h, though not with d.

```
$use qplot $
```

The quantile plot (not shown) looks linear; this model also seems satisfactory.

We can repeat this procedure using the log link instead of the lognormal distribution for v:

```
$link l
$fit lh + ld $disp e $
   deviance =  179.66 at cycle 3
 residual df =   28

             estimate        s.e.       parameter
    1         -6.537        0.9435         1
    2          1.088        0.2421         LH
    3          1.997        0.08207        LD
 scale parameter 6.416
```

```
$use rsq $
 R-squared equals    0.9778
```

The fitted model is again very similar to that in Section 3.6, and the RSS gives a slightly better fit, with a disparity of 142.44. The residual graph against h shows the same feature of rapidly increasing residuals.

Now we have four possible models. Can we choose between response-variable and link transformations? Extending the argument of Section 3.6, we can write the competing models in terms of *four* λ parameters:

$$v^* \sim N(\eta^*, \sigma^2)$$

with

$$\eta^* = (\eta^{\lambda_\ell} - 1)/\lambda_\ell$$

and

$$\eta = \beta_0 + \beta_1 h^* + \beta_2 d^*$$

as in Section 3.6. Here λ_ℓ is the link function parameter. The two models of Section 3.6 had $\lambda_\ell = 1$ with $\lambda_v = 1/3$ and $\lambda_v = 0$; the two models above have $\lambda_\ell = 1/3$ and $\lambda_\ell = 0$ with $\lambda_v = 1$.

Compared with the disparity value of 132.20 for the lognormal model, both the link function transformation models are considerably worse (a disparity difference of 10 is equivalent to a ratio of maximized likelihoods of 0.0067, with the same number of parameters). We therefore conclude that the log or cube root variable transformation models are better, and we prefer the log model because of its "solid body" interpretation. Thus the maximized likelihoods enable us to choose between transforming the distribution and transforming the link function.

A detailed discussion of transformations of the explanatory variables was given by Box and Tidwell (1962). Scallan *et al.* (1984) extended the idea of the Box–Cox transformation to "Box–Coxing the link function" by parametrizing the link function as $\eta = g(\mu) = (\mu^\lambda - 1)/\lambda$ and estimating λ.

3.8   Regression models for prediction

In Chapter 2 we noted two uses of regression models: for simplified descriptions of populations, and for prediction of values of the response variable for new observations. We now consider prediction.

The tree volume example provides a good illustration. The trees were felled and the volume of usable wood measured so that the volume of usable wood could be estimated for stands or forests of similar trees. The real purpose of the model is not for simple description—though this is an essential part of the modelling—but for the prediction of wood volume of similar trees.

This prediction is easily obtained from the fitted model. We are given $Y_i \sim N(\boldsymbol{\beta}'\mathbf{x}_i, \sigma^2)$ independently, $i = 1, \ldots, n$, and a new observation $\mathbf{x}_{n+1}$ on $\mathbf{x}$, and we want to predict the corresponding value Y_{n+1} of Y. We assume that the same distributional model applies for Y_{n+1}. For simplicity of notation we write

$$Y^* = Y_{n+1}, \quad \mathbf{x} = \mathbf{x}_{n+1}.$$

Then

$$Y^* \sim N(\boldsymbol{\beta}'\mathbf{x}, \sigma^2)$$

and

$$\hat{\boldsymbol{\beta}}'\mathbf{x} \sim N(\boldsymbol{\beta}'\mathbf{x}, \sigma^2\mathbf{x}'(X'X)^{-1}\mathbf{x}),$$

$$\text{RSS} \sim \sigma^2\chi^2_{n-p-1}$$

independently of Y^*. It follows immediately that

$$Y^* - \hat{\boldsymbol{\beta}}'\mathbf{x} \sim N(0, \sigma^2(1+h))$$

$$(Y^* - \hat{\boldsymbol{\beta}}'\mathbf{x})/s(1+h)^{1/2} \sim t_{n-p-1},$$

where $s^2 = \text{RSS}/(n-p-1)$ and $h = \mathbf{x}'(X'X)^{-1}\mathbf{x}$ is the variance of the estimated linear predictor $\hat{\boldsymbol{\beta}}'\mathbf{x}$, divided by σ^2.

Thus probability statements may be made about the random variable Y^* from the t-distribution, for example,

$$\Pr(|Y^* - \hat{\boldsymbol{\beta}}'\mathbf{x}| < t_{\alpha/2, n-p-1}s(1+h)^{1/2}) = 1 - \alpha$$

is a $100(1 - \alpha)\%$ *prediction* interval for y. It is not a *confidence* interval since Y^* is a random variable, not a parameter. Confidence interval statements can be made about the mean of Y^*, $\boldsymbol{\beta}'\mathbf{x}$, as this is a parametric function of $\boldsymbol{\beta}$. Since

$$\hat{\boldsymbol{\beta}}'\mathbf{x} \sim N(\boldsymbol{\beta}'\mathbf{x}, \sigma^2 h),$$

a $100(1 - \alpha)\%$ confidence interval for $\boldsymbol{\beta}'\mathbf{x}$ is

$$|\boldsymbol{\beta}'\mathbf{x} - \hat{\boldsymbol{\beta}}'\mathbf{x}| < t_{\alpha/2, n-p-1}sh^{1/2}.$$

This is shorter than that for Y^* as the additional variation in the random variable does not have to be allowed.

The same results for Y^* can be obtained by treating Y^* as a formal unknown parameter in the model for $y_1, \ldots, y_n, Y^*$ and constructing a profile likelihood for Y^* by maximizing the likelihood of $y_1, \ldots, y_n, Y^*$ over $\boldsymbol{\beta}$ and σ for a fixed value of Y^*. This is known as a *predictive likelihood*; see Butler (1986) for details and references. This device is very useful in non-normal models, and is used in subsequent chapters.

Construction of the prediction interval depends on the predicted value $\hat{Y}^* = \hat{\boldsymbol{\beta}}'\mathbf{x}$ and the variance $s^2 h$ of the linear predictor $\hat{\boldsymbol{\beta}}'\mathbf{x}$. These are easily obtained in GLIM as system vectors for the $\mathbf{x}$ values to be predicted following a fit: the vectors %plp, %pfv and %pvl contain the linear predictor, the fitted values and the variance of the linear predictor for the specified values of $\mathbf{x}$.

We illustrate with the tree data. Suppose in addition to the 31 observations we have a new tree with height h = 80 feet and diameter d = 15 inches. What predictive statement can be made about this tree's usable wood volume?

We first fit the required model—we will use the log model for illustration.

```
$yvar lv $
$fit lh + ld $
$calc %a = %log(80) : %b = %log(15) $
$predict lh = %a ld = %b $
prediction for the current model with LH=4.382, LD=2.708
$print %plp %pfv %pvl $
 3.633    3.633   0.0003
$calc %sqrt(%pvl + %sc) $
 0.08340
```

The fitted value is 3.633, the variance is 0.000333 and the residual mean square estimate s^2 is 0.006624. A 95% prediction interval for the tree log-volume is then

$3.633 \pm t_{0.025,\,28} \cdot 0.0834$ which is $(3.462, 3.804)$. The corresponding interval for the predicted tree volume is $(31.9, 44.9)$.

3.9 Model choice and mean square prediction error

We noted in Chapter 2 that regression models are used both for smooth representation and for prediction. The model selection procedure described in Section 3.1 is intended to provide the model which gives the simplest representation of the population consistent with the data. It does not necessarily follow that this model is the best one for prediction.

Since the outcome of the prediction process is a predicted value (and an interval for the true value), it seems reasonable to choose the model which gives the closest prediction. This can be defined in several ways; we will adopt the usual convention that the *squared prediction error*

$$\sum_{i \in N} (y_i - \hat{y}_i)^2$$

is to be minimized, where the minimization is over the set N of future observations. Since this is usually unknown, we are forced to characterize the set of future observations by reference to the calibration set of observations already available: the new observations $\mathbf{x}$ are assumed to be similar to the existing observations $\mathbf{x}_i$. (If they were not, then we should have reservations about extrapolating the fitted regression into regions of the explanatory variable space not represented in the data.)

This similarity is expressed in one of the two ways: the existing $\mathbf{x}_i$ are thought of as fixed, and the new observations $\mathbf{x}$ have a uniform distribution over the $\mathbf{x}_i$; or the existing $\mathbf{x}_i$ are thought of as values of a random variable $\mathbf{X}$ (apart from the constant 1), and the new $\mathbf{x}$ are independent values of this random variable. Minimization of the *expected* or *mean* square prediction error (MSPE) is then taken to be the aim of the model selection procedure, where the expectation is over the uniform distribution of the $\mathbf{x}_i$, or the common multivariate distribution of $\mathbf{X}$. Because of the difficulty of specifying the latter distribution (a multivariate normal distribution is a common, but unrealistic, specification) we consider only the first case.

For the normal model $Y \sim N(\boldsymbol{\beta}'\mathbf{x}, \sigma^2)$, the MSPE is (e.g. Aitkin, 1974)

$$E_f = (n + p + 1)\sigma^2/n.$$

For the subset model $\boldsymbol{\beta}_r'\mathbf{x}_r$, omitting the component $\boldsymbol{\beta}_d'\mathbf{x}_d$ from $\boldsymbol{\beta}'\mathbf{x}$, the MSPE is (using the notation of Section 3.1)

$$E_r = (n + p_\dagger + 1)\sigma^2/n + \boldsymbol{\beta}_d' S_{r.d} \boldsymbol{\beta}_d/n$$

where

$$S_{d.r} = X_d' X_d - X_d' X_r (X_r' X_r)^{-1} X_r' X_d.$$

Thus prediction is improved (MSPE decreases) if

$$E_f - E_r > 0,$$

that is, if

$$p_2 \sigma^2 > \boldsymbol{\beta}_d' S_{d.r} \boldsymbol{\beta}_d.$$

Thus if $\boldsymbol{\beta}_d$ is small, prediction will be improved by omitting $\mathbf{x}_d$. It is possible to test formally whether this inequality is violated for any subset $\mathbf{x}_r$, thus giving a simultaneous test for the non-reduction of the MSPE (Aitkin, 1974) for all possible subsets.

However, this test does not identify the model with the *smallest* MSPE, only those models with MSPEs *not significantly larger* than that of the full model. Replacing parameter values in E_f and E_r by sample estimates from each model allows a comparison of the *estimated* MSPEs, though a formal test is not available.

The above inequality becomes, on substituting sample estimates,

$$\frac{\hat{\boldsymbol{\beta}}_d' S_{d.r} \hat{\boldsymbol{\beta}}_d}{p_2 s^2} < 1$$

which is equivalent to $F_{d.r} < 1$ where $F_{d.r}$ is the F-statistic for the significance of the omitted variables $\mathbf{x}_d$.

The use of this criterion is considered in an example in Section 3.11, but we first examine the use of *cross-validation*.

3.10   Model selection through cross-validation

Considerable experience with the use of regression models for prediction has shown that when applied to new observations or samples, the model predicts much less well than in the calibration sample. This phenomenon is called "shrinkage on cross-validation", the shrinkage referred to being that of R^2. Stone (1974) gave some historical examples. Copas (1983) gave a detailed discussion.

In regression studies with large samples it is possible to divide the sample (randomly) into two halves. The model is fitted on the first half, and used for the prediction of the values of y in the second half. This *cross-validation* of the model on the second independent sample gives a more realistic assessment of its predictive value, and of the value of reduced models. An example with small samples was given by Copas (1983).

With small samples cross-validation can be achieved by omitting each observation in turn from the data, fitting the model (or models) to the remaining observations, predicting the value of y for the omitted observation, and comparing the prediction with the observed value. Let $\hat{y}_{(i)}$ be the predicted value of y_i when the i-th observation is omitted from the data. Then

$$\text{CVE} = \sum_{i=1}^{n} (y_i - \hat{y}_{(i)})^2 / n$$

is the *cross-validation estimate* (CVE) of the MSPE; it is an unbiased estimate of σ^2. The sum $n\text{CVE} = \sum (y_i - \hat{y}_{(i)})^2$ was called prediction error sum of squares (PRESS) by Allen (1971) who gave a more general definition. The computation of PRESS for any normal linear model is easily achieved, without omitting observations, from the standard output of a $fit directive.

Let $X_{(-i)}$ be the matrix of explanatory variables with $\mathbf{x}_i$ deleted and $\mathbf{y}_{(-i)}$ the vector of response values with y_i deleted, and let $\hat{\boldsymbol{\beta}}_{(-i)}$ be the corresponding estimate of $\boldsymbol{\beta}$. Then

$$\hat{y}_{(i)} = \mathbf{x}_i' \hat{\boldsymbol{\beta}}_{(-i)}$$

and

$$X'X = X_{(-i)}'X_{(-i)} + \mathbf{x}_i\mathbf{x}_i'$$
$$X'\mathbf{y} = X_{(-i)}'\mathbf{y}_{(-i)} + \mathbf{x}_i y_i.$$

Then

$$\left(X_{(-i)}'X_{(-i)}\right)^{-1} = (X'X - \mathbf{x}_i\mathbf{x}_i')^{-1}$$
$$= (X'X)^{-1} + (X'X)^{-1}\mathbf{x}_i(1 - \mathbf{x}_i'(X'X)^{-1}\mathbf{x}_i)^{-1}\mathbf{x}_i'(X'X)^{-1}$$
$$X_{(-i)}'\mathbf{y}_{(-i)} = X'\mathbf{y} - \mathbf{x}_i y_i$$
$$\hat{\boldsymbol{\beta}}_{(-i)} = (X_{(-i)}'X_{(-i)})^{-1}X_{(-i)}'\mathbf{y}_{(-i)}$$

and

$$\hat{y}_{(i)} = \mathbf{x}_i'\hat{\boldsymbol{\beta}}_{(-i)} = (\hat{y}_i - h_i y_i)/(1 - h_i)$$
$$e_{(i)} = y_i - \hat{y}_{(i)} = (y_i - \hat{y}_i)/(1 - h_i) = e_i/(1 - h_i),$$

where

$$h_i = \mathbf{x}_i'(X'X)^{-1}\mathbf{x}_i$$

is the influence of the i-th observation in the complete sample. The PRESS criterion is thus

$$\text{PRESS} = \sum_{i=1}^{n} e_{(i)}^2 = \sum_{i=1}^{n} e_i^2/(1 - h_i)^2$$

which is easily calculated using the library macro press available in the file press; this macro also provides the CVE of σ^2, and the cross-validation R_{CV}^2 (the squared correlation between the y_i and $\hat{y}_{(i)}$).

Again, we might choose for prediction the model with the smallest value of PRESS, or more realistically, one of the models with a small value of this criterion.

We examine in the next section the use of these criteria for prediction model choice on a complex example. We conclude this section with a simple illustration on the trees data, using the log-volume model. The coefficients of log height and log diameter have simple physical interpretations: should we use the model 2*1d + 1h for prediction, or the model with estimated coefficients?

```
$input 'trees' $
  : %plc normac press $
$calc lv = %log(v) $
    : ld = %log(d) $
    : lh = %log(h) $
$yvar lv $fit ld + lh $disp e $
$use rsq $use press $
```

For the full model we have RSS $= 0.1855$, $s^2 = 0.006624$, PRESS $= 0.2186$ and CVE $= 0.007050$ which is 6% larger than s^2. The cross-validation R_{CV}^2 is 0.9737, little less than the value 0.9777 for R^2 itself.

```
$calc z = 2*ld + lh $
$offset z $
$fit $disp e $
$use press $
```

For the reduced model RSS $= 0.1877$, $s^2 = 0.006256$, PRESS $= 0.2004$, $R^2 = 0.9774$ and CVE $= 0.006465$ is now only 3% larger than s^2. The cross-validation R_{CV}^2 is 0.9759, *greater* than R_{CV}^2 for the full model. Both s^2 and PRESS decrease in the reduced model, by 6–8% of their values in the full model. This shows that prediction should be based on the reduced model.

3.11   Reduction of complex regression models

The value of modelling and model simplification becomes clear when we are dealing with complex data sets with many possible explanatory variables. A good example is given by Henderson and Velleman (1981), also discussed by Aitkin and Francis (1982). The data are shown below and are in the file car.

```
s c t g disp   hp  cb drat  wt    qmt   mpg   car
-----------------------------------------------------------------
0 6 1 4 160.0 110  4 3.90  2620  16.46 21.0  MAZDA RX-4
0 6 1 4 160.0 110  4 3.90  2875  17.02 21.0  MAZDA RX-4 WAGON
1 4 1 4 108.0  93  1 3.85  2320  18.61 22.8  DATSUN 710
1 4 0 3 258.0 110  1 3.08  3215  19.44 21.4  HORNET 4 DRIVE
0 8 0 3 360.0 175  2 3.15  3440  17.02 18.7  HORNET SPORTABOUT
1 6 0 3 225.0 105  1 2.76  3460  20.22 18.1  VALIANT
0 8 0 3 360.0 245  4 3.21  3570  15.84 14.3  DUSTER 360
1 4 0 4 146.7  62  2 3.69  3190  20.00 24.4  MERCEDES 240D
1 4 0 4 140.8  95  2 3.92  3150  22.90 22.8  MERCEDES 230
1 6 0 4 167.6 123  4 3.92  3440  18.30 19.2  MERCEDES 280
1 6 0 4 167.6 123  4 3.92  3440  18.90 17.8  MERCEDES 280C
0 8 0 3 275.8 180  3 3.07  4070  17.40 16.4  MERCEDES 450SE
0 8 0 3 275.8 180  3 3.07  3730  17.60 17.3  MERCEDES 450SL
0 8 0 3 275.8 180  3 3.07  3780  18.00 15.2  MERCEDES 450SLC
0 8 0 3 472.0 205  4 2.93  5250  17.98 10.4  CADILLAC FLEETWOOD
0 8 0 3 460.0 215  4 3.00  5425  17.82 10.4  LINCOLN CONTINENTAL
0 8 0 3 440.0 230  4 3.23  5345  17.42 14.7  IMPERIAL
1 4 1 4  78.7  66  1 4.08  2200  19.47 32.4  FIAT 128
1 4 1 4  75.7  52  2 4.93  1615  18.52 30.4  HONDA CIVIC
1 4 1 4  71.1  65  1 4.22  1835  19.90 33.9  TOYOTA COROLLA
1 4 0 3 120.1  97  1 3.70  2465  20.01 21.5  TOYOTA CORONA
0 8 0 3 318.0 150  2 2.76  3520  16.87 15.5  DODGE CHALLENGER
0 8 0 3 304.0 150  2 3.15  3435  17.30 15.2  AMC JAVELIN
0 8 0 3 350.0 245  4 3.73  3840  15.41 13.3  CHEVROLET CAMARO Z-28
0 8 0 3 400.0 175  2 3.08  3845  17.05 19.2  PONTIAC FIREBIRD
1 4 1 4  79.0  66  1 4.08  1935  18.90 27.3  FIAT X1-9
0 4 1 5 120.3  91  2 4.43  2140  16.70 26.0  PORSCHE 914-2
1 4 1 5  95.1 113  2 3.77  1513  16.90 30.4  LOTUS EUROPA
0 8 1 5 351.0 264  4 4.22  3170  14.50 15.8  FORD PANTERA L
0 6 1 5 145.0 175  6 3.62  2770  15.50 19.7  FERRARI DINO 1973
0 8 1 5 301.0 335  8 3.54  3570  14.60 15.0  MASERATI BORA
1 4 1 4 121.0 109  2 4.11  2780  18.60 21.4  VOLVO 142E
-----------------------------------------------------------------
```

The data are quarter-mile acceleration time in seconds (qmt) and fuel consumption in miles per (US) gallon (mpg) for 32 cars tested by the US *Motor Trend* magazine in 1974. Nine explanatory variables are given: shape of engine s (straight $= 1$, vee $= 0$), number of cylinders c, transmission type t (automatic $= 0$, manual $= 1$), number of gears g, engine displacement in cubic inches disp, horsepower hp, number of carburettor barrels cb, final drive ratio drat, and weight of the car in pounds wt. In the original analyses referenced above, quarter-mile time is taken as an explanatory variable for mpg, but it is not a basic design variable and is therefore omitted in this analysis of fuel consumption. A separate analysis of qmt as a response variable is given in Section 3.12.

Our object is to obtain a simple model relating mpg to the explanatory variables. The 32 cars are not a random sample of the car population to which this model can be generalized: the *Motor Trend* sample is heavily weighted to European and US high-performance sports and luxury cars.

We begin by graphing mpg against the explanatory variables. This can help identify outlying observations, suggest the important explanatory variables, and give a general feel for the data. We show here only a few of these graphs.

```
$input 'car' : %plc normac $
$graph (s = 1 h = 'displacement' v = 'mpg')
      mpg disp 5 $
```

Curvature in the graph against disp is very noticeable (Fig. 3.15), suggesting that a scale change is required. We try the log scale.

```
$calc lmpg = %log(mpg) : ldis = %log(disp) $
$graph (s = 1 h = 'log displacement' v = 'log mpg')
      lmpg ldis 5 $
```

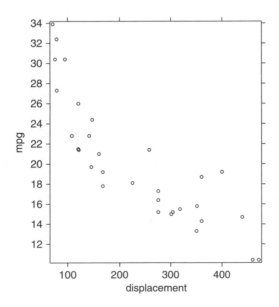

Fig. 3.15. Mpg vs displacement

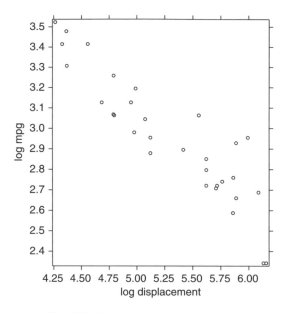

Fig. 3.16. Log mpg vs log displacement

The graph is now much more nearly linear (Fig. 3.16). The graphs against hp and wt also show curvature and these are similarly linearized by a log transformation; we show below both these variables graphed on the log scale (Figs 3.17 and 3.18).

```
$calc lhp = %log(hp) : lwt = %log(wt) $
$graph (s = 1 h = 'log hp' v = 'log mpg')
       lmpg lhp 5 $
    : (s =1 h = 'log wt' v = 'log mpg')
       lmpg lwt 5 $
```

The graphs do not show any marked outliers, and it is clear that hp, wt and disp are important explanatory variables, as one might expect.

We now proceed to fit a model using all the explanatory variables, with hp, wt and disp replaced by their log transformations. Should the model be fitted to mpg or log mpg, or perhaps gpm = 1/mpg—gallons per mile, the European standard for fuel consumption (litres per km) used by Henderson and Velleman (1981)? The Box–Cox family suggests a clear answer.

```
$input %plc boxcox $
$macro yvar mpg $endmac
$macro model s + c + t + g + ldis + lhp + cb + drat + lwt
$endmac
$use boxcox $
```

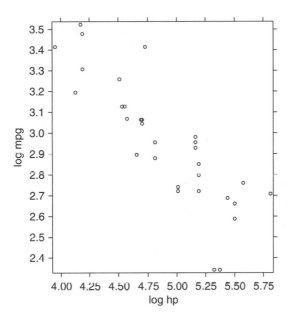

Fig. 3.17. Log mpg vs log hp

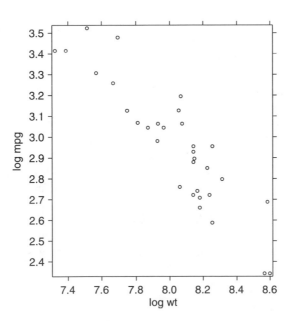

Fig. 3.18. Log mpg vs log wt

Specifying the range -2 to 2 in steps of 0.5 in response to the prompts, we obtain the values of the disparity reproduced below.

```
----------------------------------------------------------------
lam_    -2.0  -1.5  -1.0  -0.5   0.0   0.5   1.0   1.5   2.0
dev_   195.7 181.7 168.5 157.0 149.2 147.6 152.5 162.1 174.0
----------------------------------------------------------------
```

The minimum of 147.3 occurs near $\lambda = 0.4$, and the approximate 95% confidence interval for λ includes $\lambda = 0$ but neither $\lambda = -1$ nor $\lambda = 1$. Thus the log scale for mpg is indicated.

Does the transformation of disp, hp and wt to their logs affect the transformation of mpg? To check, we can repeat the Box–Cox transformation without transforming these variables.

```
$macro model s + c + t + g + disp + hp + cb + drat + wt
$endmac
$use boxcox $
```

The values of the disparity are now:

```
----------------------------------------------------------------
lam_    -2.0  -1.5  -1.0  -0.5   0.0   0.5   1.0   1.5   2.0
dev_   184.8 172.6 162.2 154.9 152.2 154.8 161.7 171.5 182.8
----------------------------------------------------------------
```

Now the minimum occurs almost at $\lambda = 0$, and again neither $\lambda = -1$ nor $\lambda = 1$ is included in the approximate 95% confidence interval. The transformation of the explanatory variables has had very little effect on the estimation of λ, though the disparity is reduced by 3 at $\lambda = 0$ by the log transformations, showing a slightly better fit.

We proceed with an examination of the first model above, using ldis, lhp and lwt. One-point we need to consider first is the status of the Mercedes 240D. The "D" stands for "diesel": this car is the only diesel-engined one in the sample. Since the fuel consumption of diesel engines is generally much less than that of petrol engines of equivalent power, we need to treat this car separately from the others. We do this by defining a dummy variable d which takes the value 1 for this car (the 8-th in the data list) and 0 for all the others, and then fit this variable in the model, in effect defining a two-group structure; the coefficient of d is the difference in mean log mpg between diesel and petrol engines in otherwise identical cars.

```
$calc d = 0 : d(8) = 1 $
$macro model s + c + t + g + ldis + lhp + cb + drat + lwt + d
$endmac
$use boxfit $
```

We specify $\lambda = 0$. We suppress the coefficients, and report only that for d, which is -0.06647, with standard error 0.1564. The t-value for this parameter is only -0.425; there seems to be no real difference in fuel consumption between the diesel and the petrol engines; this is also clear from comparing the very similar Mercedes 230 and 240D: although the 240D has a higher mpg it is much less powerful, and would therefore be expected to be more economical. We therefore omit this dummy variable in the remaining analyses, and treat the diesel- and petrol-engined cars together. We now examine the residuals and influence values from the model.

```
$macro model s + c + t + g + ldis + lhp + cb + drat + lwt
$endmac
$use boxfit $
                    --- Model is s+c+t+g+ldis+lhp+cb+drat+lwt
                    --- Y-Variate is mpg
  -- model changed
  Value of lambda?
$DIN? 0
  -- data list abolished

                    --- Transformed Y-variate is log(mpg)
     deviance =    0.28609
  residual df =   22

             estimate         s.e.       parameter
       1         7.353        1.481          1
       2       -0.02823      0.08830          S
       3      -0.002945      0.03604          C
       4       -0.04464      0.09116          T
       5        0.08111      0.06315          G
       6        -0.1170       0.1543         LDIS
       7        -0.1652       0.1389         LHP
       8       -0.02928      0.02881          CB
       9       -0.02226      0.07097         DRAT
      10        -0.3809       0.1970         LWT
  scale parameter 0.01300

  -2 log l =    149.2
     lambda =   0.
```

The quantile plot for the model residuals is closely linear, apart from two rather large negative residuals. A graph of residuals against observation number shows that these large values are for the Cadillac and Continental, the luxury American cars. Which observations are influential?

```
$extract %lv $
$print %lv $
0.256   0.245   0.232   0.564   0.163   0.331   0.233   0.448   0.299
0.313   0.313   0.185   0.147   0.150   0.186   0.180   0.191   0.230
0.625   0.189   0.375   0.223   0.160   0.326   0.155   0.163   0.618
0.538   0.677   0.417   0.512   0.354
```

Here $p = 9, n = 32$ and $2(p+1)/n = 0.625$. Only two values exceed this, 0.625 for observation 19 and 0.677 for observation 29. Observation 27 might also be considered influential with a value of 0.618. What is unusual about these cars? The Ford Pantera (29) has the second largest hp, but has low gearing for its power—the third highest value of drat. Most of the powerful cars have low values of drat. The Honda Civic (19) has the lowest hp, the highest drat, and the second lowest disp and wt. The Porsche 914 (27) has low disp and hp, the second highest drat and low wt. These three cars lie on the edge of the hp/drat scatter, as can be seen from a graph of hp against drat. The Cadillac and Continental, which had the largest negative residuals, by contrast have very little influence, with values of 0.186 and 0.180.

What happens if we exclude the three observations with high influence? We return to this point below, but for the moment assume that no action is needed. We now examine the fitted model.

At first glance the regression coefficients in the model output look disappointingly non-significant. Only the coefficient of lwt appears even near twice its standard error, and most of the others are around one SE or even less. The regression appears nearly useless for prediction of fuel consumption from the design variables. This impression is however quite false: as we will see below the value of R^2 for this model is very high. There is no doubt of the importance of at least some of the variables, but because of the high correlations among them, many are redundant given the others. This phenomenon occurs quite commonly in regressions with highly correlated explanatory variables. Our job is to determine which are the important variables. We proceed to simplify the full regression model.

A convenient model simplification procedure is *backward elimination*, in which at each step the least important variable is dropped from the current model. We prefer this approach to the common alternative of *forward selection*, in which variables are added to the model, starting from the null model. Forward selection requires a "look ahead" feature in which each variable in turn is added to the model; this increases the number of models examined but does not allow the joint structure of all the explanatory variables to be considered. *Stepwise* procedures combine forward selection and backward elimination but also do not consider the full joint structure of the explanatory variables.

Importance is assessed by the t-ratios $\hat{\beta}_j/\text{s.e.}(\hat{\beta}_j)$; these are not provided as part of the standard output in GLIM, but they may be obtained using the macro tval in the file normac, or by direct calculation from the model output. Elimination of variables continues as long as the t-statistics do not reach large values; how large is "large" was expressed in the first edition through a simultaneous test (Aitkin, 1974): here we follow the likelihood approach in Chapter 2, relating the P-values for the F-tests with different numbers of omitted variables through the table in Section 2.7.4.

To fit the full model we have to declare the y-variate to be lmpg, since the boxfit macro resets the y-variate to that specified in the macro yvar.

```
$yvar lmpg $
$fit #model $disp e $
$use rsq : tval : press $
```

We find $R^2 = 0.8959$, $R^2_{CV} = 0.8081$. Here $R = 0.9468$ is the correlation between the observed lmpg values and the fitted values from the full regression model, which is very high: the full model gives a close reproduction of the actual lmpg values. The estimate of σ^2 from the full model is 0.0130 with $s = 0.114$.

Backward elimination now proceeds from the full model in the following sequence, using at each step

```
$fit - [variable] $disp e $use rsq : tval : press $
```

At each step we list the model RSS, R^2, R^2_{CV} and s^2 values, the CVE of σ^2, the variable with the smallest t-value in magnitude of those remaining in the model, and the cumulative F-statistic for the set of omitted variables with its P-value:

Step	RSS	Rsq	RsqCV	s^2	CVE	Variable to be omitted	Omitted variables t	P-value F	P
0	0.2861	0.8959	0.8081	0.0130	0.0171	c	-0.082	0.007	0.934
1	0.2862	0.8959	0.8298	0.0124	0.0150	drat	-0.320	0.054	0.948
2	0.2875	0.8954	0.8337	0.0120	0.0146	s	-0.340	0.069	0.976
3	0.2888	0.8949	0.8422	0.0116	0.0137	t	-0.461	0.100	0.982
4	0.2913	0.8940	0.8476	0.0112	0.0132	ldis	-0.723	0.155	0.977
5	0.2971	0.8919	0.8529	0.0110	0.0127	cb	-1.070	0.363	0.895
6	0.3097	0.8873	0.8569	0.0111	0.0123	g	1.052	0.460	0.854
7	0.3220	0.8829	0.8549	0.0111	0.0125	lhp	-4.372	2.726	0.028
	0.5342	0.8057	0.7766	0.0178	0.0192				

After step 6 elimination ceases, as the omission of lhp results in a sudden jump in all the statistics in the table at this point. The P-value for the F-test drops from 0.854 for seven omitted variables to 0.028 for eight, well below the value of 0.037

in the table in Section 2.7.4. As in the first edition, the final model found by backward elimination uses lhp and lwt.

The fitted model for lmpg, with standard errors, is

```
8.72 - 0.255 lhp - 0.562 lwt
(0.53) (0.058)      (0.087)
```

while $s = 0.105$; the error standard deviation corresponds to a percentage error of 11% in the fitted mpg.

It is worth noting that both coefficients are substantially larger in this model than in the full model, as these two variables now subsume the effects of the other variables correlated with them. This model could also be used for prediction, if this were appropriate: its values of s^2 and CVE are very slightly greater than the minimum values. It is notable that the cross-validation R^2_{CV} is initially substantially less than R^2, showing shrinkage of R^2 from the full model on cross-validation, but as the model is progressively simplified R^2_{CV} *increases*, though its maximum value of 0.8569 is still well below R^2 for the full model.

We noted above that three observations had substantial influence on the model. If these are deleted by being given weight zero in a weighted fit, we find that backward elimination leads to the same model with very similar estimated coefficients:

```
8.98 - 0.268 lhp - 0.587 lwt
```

with slightly larger standard errors, reflecting the smaller sample size. Thus despite their high influence values these observations are not discrepant from the rest. We retain the preceding model as the final model.

It is tempting to interpret the final model causally: if weight were increased by 9% (an increase of 0.087 in lhp) keeping horsepower constant, then predicted lmpg would decrease by 0.049, giving a 4.8% decrease in mpg. The Mercedes 450 cars provide an example of this: all their design variables are identical apart from wt, which is 4070 (with mpg = 16.4) for the SE, 3780 (mpg = 15.2) for the SLC, and 3730 (mpg = 17.3) for the SL. The mpg (median) predictions from the model are 15.3, 15.9 and 16.0 respectively, with a maximum relative error of 8% for the SL.

Whether such an interpretation is possible or not, it is more realistic to think of the final model as a simple, parsimonious and accurate representation of the fuel consumption results for this set of cars. It may not generalize to other cars, or to the same cars tested in other years, because of the selective sample of cars on which the model has been based. Consequently, this model is inappropriate for prediction to other cars.

It is important to stress, as we noted above, that backward elimination is only one way of proceeding to simplify the model. Generating all possible regressions (2^p models with p explanatory variables) is feasible if p is not too large; since many of these models will not be adequate, the amount of computation can be substantially reduced by careful bookkeeping. We have followed the practice in this book of not searching extensively for additional models; it sometimes happens however that several parsimonious and adequate models exist using quite different subsets of variables. Causal interpretations in such cases are extremely hazardous.

We conclude by noting that the model predicts median mpg by

$$\exp(\hat{\mu}) = 126.0 \, \text{hp}^{-0.255} \, \text{wt}^{-0.562};$$

the prediction of median gpm is just the reciprocal of this.

3.12   Sensitivity of the Box–Cox transformation

Regression diagnostics allow us to identify influential observations which may substantially affect the position of the fitted model. The extension of the model to include the Box–Cox transformation parameter raises the possibility that the estimate of this parameter may also be strongly affected by influential observations. The identification of observations influencing the transformation parameter was discussed by Atkinson (1982); he described the use of an additional "constructed variable" in the model. The need for caution in the use of the Box–Cox transformation is clearly shown in the next example, the modelling of acceleration time qmt in the car data.

```
$newjob $
$input 'car' : %plc boxcox $
$macro yvar qmt $endmac
$macro model s + c + t + g + disp + hp + cb + drat + wt
$endmac
$use boxcox $
```

We first use the explanatory variables disp, hp and wt without transformation. Tabulating over the usual range -2 to $+2$, we find that the disparity decreases monotonically with λ, and is still decreasing rapidly at $\lambda = -2$. Extending the grid, the minimum is found to occur at about $\lambda = -3.5$. (Note that it is possible for numerical instability to occur for large negative values of λ. This problem, however, is easily resolved by scaling, e.g. by dividing y by 10.)

How should we interpret this unusual value? A normal quantile plot on this scale ($use boxfit with argument -3.5) shows a good straight-line fit, with no marked outliers, but the unusual estimate of λ makes us suspect an error in the data. Refitting the model on the log scale with $\lambda = 0$ gives a quantile plot with a large negative residual for observation 9, the Mercedes 230. Background knowledge of the scientific field is helpful, as in the tree example. The Mercedes has the slowest of all acceleration time, and is substantially slower than the diesel Mercedes 240D, which is much less powerful, has higher gearing, and is slightly heavier. The body shapes are identical, so this slow acceleration is very hard to understand. One possible explanation is that the acceleration time of these two cars have been reversed at some point in their publication, or in transcription from *Motor Trend* magazine.

We now examine the effect of excluding the Mercedes 230 from the analysis.

```
$calc w = 1 : w(9) = 0 $
$weight w $
$use boxcox $
```

The Box–Cox macros allow for the weighting-out of observations, so the likelihood calculations and plotting are still correct. The minimum disparity of 25.47 is now very close to $\lambda = -1.5$, while the values at $\lambda = -1, 0$ and 1 are 25.71, 27.50 and 30.73 respectively. Both the reciprocal and log transformations of qmt are appropriate but qmt itself is not. Thus the omission of a single observation has considerably changed the conclusion about the appropriate scale of y. The large outlier is accommodated in a normal model only by an extreme transformation of scale.

The reciprocal transformation corresponds to using average speed over the quarter-mile as the response-variable. The interpretation of the model coefficients (not shown here) is hindered by their small size; if we note that sp = 900/qmt is the average speed in miles per hour (1 mph = 1.6 kph) over the distance, the corresponding estimates and standard errors for the model for sp are just 900 times those for qmt, and are given below:

```
$calc sp = 900/qmt $
$yvar sp $
$fit . $disp e $
44.9 - 3.91 s + 0.163 c - 0.718 t + 1.63 g + 0.162 disp +
(5.0) (1.02)    (0.42)    (1.04)    (0.73)    (0.008)

 0.0304 hp + 0.562 cb + 0.670 drat - 0.00342 wt
(0.011)      (0.38)      (0.78)      (0.00078)
```

The full model gives $s^2 = 1.665$ and an R^2 of 0.9468. The number of gears appears to be important, but we may question whether treating it as a continuous variable is appropriate: this implies that average speed increases uniformly with each extra gear. We can check this by fitting a fully parametrized model for the number of gears. There are only three values for this variable (3, 4 and 5) so 2 df are required to saturate it. This may be done by declaring g to be a factor with five levels (though two are missing), or by defining a quadratic term in g and adding this to the model. These give the same disparity; for simplicity we use the second.

```
$calc g2 = g**2 $
$fit #model + g2 $disp e $
```

The coefficient of g2 is -0.792 with a standard error of 1.20, giving a t-value of only -0.661. The model linear in g is completely satisfactory, and we now reduce it by backward elimination as in Section 3.11, leading to the model containing s, g, hp and wt, with $s^2 = 1.757$ and an R^2 of 0.9305.

```
46.9 - 4.82 s + 1.60 g + 0.0485 hp - 0.00208 wt
(2.5) (0.70)   (0.45)   (0.0062)    (0.00045)
```

Speed increases with the number of gears and horsepower, and decreases with weight and a straight rather than vee engine.

Now we examine the effect of log transforming hp, disp and wt as in Section 3.11.

```
$calc lhp = %log(hp) : lwt = %log(wt): ldis = %log(disp) $
$macro model s + c + t + g + ldis + lhp + cb + drat + lwt
$endmac
$use boxcox $
```

The disparity minimum of 22.64 now occurs at $\lambda = 0.25$, with values at $\lambda = -1, 0$ and 1 of 24.60, 22.72 and 23.31 respectively. All three values are now plausible for λ, though $\lambda = 0$ is better supported than $\lambda = -1$. The effect of log transforming the explanatory variables has been to reduce the disparity at $\lambda = -1, 0$ and 1 by 1.11, 4.78 and 7.42 respectively. Thus if log qmt or qmt itself are to be used, the explanatory variables should be transformed; if 1/qmt is to be used, the transformation of these variables has little effect.

Backward elimination using these logged design variables with log qmt allows the elimination of only t, c and cb, while with qmt the minimal model contains s, g, lhp and lwt:

```
12.9 + 1.67 s - 0.591 g - 2.34 lhp + 2.21 lwt
(3.8) (0.27)   (0.16)    (0.36)      (0.53)
```

with an s^2 of 0.226 and an R^2 of 0.9190, the latter somewhat less than for the sp response. This model has the same interpretation as that for speed, though the signs of the variable coefficients are naturally reversed.

3.13   The use of regression models for calibration

In some applications of regression, we want to estimate a new value x_{n+1} of a single explanatory variable, given data (y_i, x_i) and a new observed response value y_{n+1}. This *calibration* problem arises, for example, when x is a precise but expensive measurement, and y a cheap but imprecise measurement of the same quantity. The calibration sample (y_i, x_i) is chosen to cover the range of values of x of interest to the experimenter.

A likelihood treatment of the problem was given by Minder and Whitney (1975); Brown (1982) gave a more recent discussion and references. Confidence intervals for x_{n+1} are easily obtained. For simplicity of notation we write $Y = y_{n+1}$, $x = x_{n+1}$. Then

$$Y - \hat{\alpha} - \hat{\beta}x \sim N(0, \sigma^2 + v_{11} + 2v_{12}x + v_{22}x^2),$$

where

$$V = (v_{jk}) = \sigma^2 (X'X)^{-1}.$$

Then

$$(Y - \hat{\alpha} - \hat{\beta}x)/(s^2 + \hat{v}_{11} + 2\hat{v}_{12}x + \hat{v}_{22}x^2)^{1/2} \sim t_{n-2}$$

and therefore with probability $1 - \alpha$,

$$|Y - \hat{\alpha} - \hat{\beta}x| < t_{1-\alpha/2, n-2}(s^2 + \hat{v}_{11} + 2\hat{v}_{12}x + \hat{v}_{22}x^2)^{1/2}.$$

This is equivalent to the quadratic inequality, for the observed y,

$$Q(x) = x^2(\hat{\beta}^2 - c\hat{v}_{22}) - 2x(\hat{\beta}(y - \hat{\alpha}) + c\hat{v}_{12}) + (y - \hat{\alpha})^2 - c(s^2 + \hat{v}_{11}) < 0,$$

where

$$c = t^2_{1-\alpha/2, n-2},$$

and $\hat{V}$ is the estimated covariance matrix of the parameter estimates, available from GLIM using $disp v $. The confidence interval for x is defined by the roots of $Q(x) = 0$.

As a numerical illustration, for the data in Section 3.1 on completion time for the Block Design test, we obtained a final model using only EFT (model 3). Suppose we have an additional observation with completion time equal to 400 seconds. What can we say about EFT for this observation?

```
$newjob $
$input 'solv' $
$yvar time $fit eft $
$acc 8 $disp e v $
```

Increased accuracy is required to calculate the roots of the quadratic: four digits are not sufficient. The estimates are

$$\hat{\alpha} = 271.128, \quad \hat{v}_{11} = 1845.44,$$
$$\hat{\beta} = 2.0405, \quad \hat{v}_{12} = -24.490, \quad \hat{v}_{22} = 0.441594,$$
$$y = 400, \quad s^2 = 11694.3,$$

and for a 95% confidence interval, $t_{0.975,22} = 2.074$, so $c = 4.301$. The quadratic in x is

$$Q(x) = 2.264x^2 - 2(157.6)x - 41626.4$$

and the equation $Q(x) = 0$ has roots -82.8 and 222.0. For $Q(x)$ to be negative, x must be between the roots so the 95% interval for x is $(-82.8, 222.0)$. Negative values are impossible from the nature of EFT, which is an unsatisfying feature of the confidence interval construction procedure. (This difficulty can be avoided by constructing a profile likelihood for x over the range of possible values of x. Minder and Whitney constructed a marginal likelihood for x.) The obvious (and ML) estimate for x is $(y - \hat{\alpha})/\hat{\beta}$ which is 63.2; the interval here is extremely wide because of the large residual variation about the regression line. For regressions which are not well determined, it is possible for $Q(x)$ to have no real roots, with $Q(x) < 0$ for all x. Then *all* values of x are consistent with the observed y. This may seem strange, but the special case $\beta = 0$ makes the result obvious: if β is actually zero, then y can provide no information about x, and a value of $\hat{\beta}$ near zero (relative to its standard error) will lead to the same result.

3.14   Measurement error in the explanatory variables

We noted in Chapter 2 that an important assumption of the model was that the explanatory variables are measured without error. In survey studies in the social sciences, this is frequently not the case: many variables contain measurement error, in the sense that repeated measurements of the same individual under the same circumstances give different values of the variable. In psychometric testing it is a standard practice to quote the reliability of a test, which is the correlation of two parallel measurements on the same individual under identical conditions. For example, the quoted reliability for the Block Design Subtest of the WISC is 0.87 for ten-year-old (American) children (Wechsler, 1949).

How should measurement error be allowed for in the model and the analysis? It is currently difficult to allow for measurement error when modelling with GLIM except when replicate or "parallel" measurements on the unreliable explanatory variables are available: we give therefore only a short discussion of the problem. More detailed discussions are given in Schafer (2001) and Aitkin and Rocci (2002).

Consider the simple linear model with one explanatory variable. Let T_i be the true value of the explanatory variable for the i-th observation, assuming it could be measured without error, and suppose that the observed value X_i is related to T_i by

$$X_i = T_i + \xi_i$$

where ξ_i is the measurement error, which we will assume to be normally distributed $N(0, \sigma_m^2)$, independently of the random component of the model for Y.

One possibility, perhaps the most commonly used, is to ignore measurement error. Suppose Y is a student achievement test at the end of an academic year, and X is a measure of initial ability, for example, Stanford–Binet IQ or Verbal Reasoning Quotient. We might argue that the model

$$Y_i|X_i \sim N(\beta_0 + \beta_1 X_i, \sigma^2)$$

is still appropriate despite unreliability in X, because X is the only measure we have of the student's initial achievement, and the regression has to be based on observable data. For a descriptive use of the model, this may be reasonable, but if the observations come from an experimental design (as in the first example of Chapter 3) from which causal conclusions are to be drawn, the ignoring of measurement error may give quite misleading conclusions about experimental variables in the model.

To proceed further we need to make an assumption about the distribution of the T_i, the unobserved true values of the X_i. We will assume that they are normally distributed $N(\mu, \sigma_t^2)$. Then the complete model is

$$Y_i|X_i, T_i \sim N(\beta_0 + \beta_1 T_i, \sigma^2)$$

$$X_i|T_i \sim N(T_i, \sigma_m^2)$$

$$T_i \sim N(\mu, \sigma_t^2).$$

Write

$$\sigma_y^2 = \sigma^2 + \beta_1^2 \sigma_t^2,$$

$$\sigma_x^2 = \sigma_m^2 + \sigma_t^2.$$

Then the joint (marginal) distribution of (Y_i, X_i) is bivariate normal with means $(\beta_0 + \beta_1\mu, \mu)$, variances (σ_y^2, σ_x^2) and covariance $\beta_1\sigma_t^2$.

The conditional distribution of Y_i given the observable X_i is then

$$Y_i|X_i \sim N(\beta_0 + \beta_1(\rho X_i + (1-\rho)\mu), \sigma^2 + \beta_1^2\rho(1-\rho)\sigma_x^2)$$

$$= N(\beta_0^* + \rho\beta_1 X_i, \sigma^2 + \beta_1^2\rho(1-\rho)\sigma_x^2)$$

where $\beta_0^* = \beta_0 + (1-\rho)\beta_1\mu$ and $\rho = \sigma_t^2/\sigma_x^2$.

Thus the slope of the regression of Y_i on X_i is not β_1 but $\rho\beta_1$, and the standard least squares estimate of β_1 systematically underestimates the true parameter value (i.e. is inconsistent). The estimates of any other parameters in a regression function will also be inconsistent. Further, the variance of the residuals, though constant, also depends on β_1, so that MLE of β_1 even if ρ is known is not equivalent to least squares estimation. A consistent estimate of the regression coefficient β_1 can be obtained if ρ is known by dividing the least squares estimate by ρ; the estimate obtained however will not be efficient since the additional information about β_1 in the variance is neglected; further, the other parameters in the model are not correctly estimated.

A particular difficulty with this model is that there are six model parameters but only five sufficient statistics for these parameters, so the model is unidentifiable without knowledge of one parameter. MLE in such models can be achieved in several cases: when reliabilities or error variances are known; when the ratio of the error variance to the residual variance in Y is known; when some ("gold standard") observations have both T and X measured, or when "parallel" measurements of the unreliable explanatory variables are available (from parallel forms or split-half measures).

This discouraging unidentifiability result is however peculiar to the normal true-score/normal error-score model; if either T or $X|T$ is *not* normal, the model *is* identifiable (Reiersøl, 1950). This result allows ML estimation even in models without the additional information specified above, so long as the observed score distribution of X is *not* normal. Discussions of this case are given in Schafer (2001) and Aitkin and Rocci (2002).

Extensive discussions of consistent estimation in general regression models with measurement error in all the explanatory variables can be found in Fuller (1987) and Carroll *et al* (1995).

Quite general models with measurement error in the explanatory variables can be fitted by ML (given parallel measurements) using LISREL (Jöreskog and Sörbom, 1981). A very recent development in STATA, `gllamm` is described in Skrondal and Rabe-Hesketh (2003).

3.15   Factorial designs

We noted in Section 3.1 that the choice of an appropriate model is greatly simplified in orthogonal factorial designs because the analysis of variance table is unique. Orthogonal factorial designs have many other well-known advantages, which we will not discuss here. In many sample surveys with categorical explanatory variables, formally similar cross-classifications result, but the sample sizes in the cells of the table are generally unequal, and sometimes show marked associations with the explanatory variables. In such cases great care is necessary in model simplification and interpretation. We conclude this section with an orthogonal two-factor design discussed by Box and Cox (1964); unbalanced cross-classifications are discussed in Section 3.16.

The file `poison` contains the survival time of rats in units of 10 hours after poisoning and antidote treatment. The design is a completely randomized 3×4 factorial with three replicates, with three types of poisons and four treatments. The data are given in Table 3.6.

Table 3.6. Poison data

type	treat	time			
1	1	0.31	0.45	0.46	0.43
1	2	0.82	1.10	0.88	0.72
1	3	0.43	0.45	0.63	0.76
1	4	0.45	0.71	0.66	0.62
2	1	0.36	0.29	0.40	0.23
2	2	0.92	0.61	0.49	1.24
2	3	0.44	0.35	0.31	0.40
2	4	0.56	1.02	0.71	0.38
3	1	0.22	0.21	0.18	0.23
3	2	0.30	0.37	0.38	0.29
3	3	0.23	0.25	0.24	0.22
3	4	0.30	0.36	0.31	0.33

We input the data.

```
$input 'poison' $
   : %plc boxcox qplot $
$yvar time
$fit : + type : + treat : + type.treat $
$disp e $
   deviance =    3.0051
 residual df =  47

   deviance =    1.9721 (change =    -1.033)
 residual df =  45       (change =   -2    )

   deviance =    1.0509 (change =  -0.9212)
 residual df =  42       (change =   -3    )

   deviance =    0.80072 (change =  -0.2501)
 residual df =  36        (change =  -6    )

           estimate        s.e.       parameter
     1      0.4125       0.07457      1
     2     -0.09250      0.1055       TYPE(2)
     3     -0.2025       0.1055       TYPE(3)
     4      0.4675       0.1055       TREAT(2)
     5      0.1550       0.1055       TREAT(3)
     6      0.1975       0.1055       TREAT(4)
     7      0.02750      0.1491       TYPE(2).TREAT(2)
     8     -0.1000       0.1491       TYPE(2).TREAT(3)
```

```
 9        0.1500        0.1491      TYPE(2).TREAT(4)
10       -0.3425        0.1491      TYPE(3).TREAT(2)
11       -0.1300        0.1491      TYPE(3).TREAT(3)
12       -0.08250       0.1491      TYPE(3).TREAT(4)
scale parameter 0.02224
```

Table 3.7. ANOVA table, time

Source	SS	df	MS	F
type	1.0330	2	0.5165	23.27
treat	0.9212	3	0.3071	16.71
interaction	0.2501	6	0.0417	1.88
residual	0.8007	36	0.0222	

The analysis of variance table is given in Table 3.7.

Since the design is orthogonal, the SS of the main effects are invariant under reordering, and each mean square can be compared directly with the residual mean square by the usual F-test.

The main effects are highly significant, the interaction not significant at the 10% level ($F_{0.10, 6, 36} = 1.945$), though one of the type.treat interaction parameters is rather large.

Inspection of the mean survival time is possible, but not convenient, from the parameter estimates for the interaction model. It is simpler to construct the table of means using $tabulate:

```
$tab the time mean for type,treat $
```

The means are arranged by rows for type and columns for treat.

```
      1      2      3      4

1   0.4125 0.8800 0.5675 0.6100
2   0.3200 0.8150 0.3750 0.6675
3   0.2100 0.3350 0.2350 0.3250
```

Since the design is balanced, the marginal means can be inspected by simple one-way tabulations:

```
$tab the time mean for type $
  TYPE      1       2       3
  MEAN   0.6175  0.5444  0.2763
  : the time mean for treat $
  TREAT     1       2       3       4
   MEAN   0.3142  0.6767  0.3925  0.5342
```

Tabulation of the variances is always useful in cross-classifications. We store them in the vector `var`:

```
$tab the time variance for type,treat into var $
  :  the time variance for type,treat $
         1         2         3         4
1   0.0048250 0.0258667  0.0245583  0.0127333
2   0.0056667 0.1131000  0.0032333  0.0734250
3   0.0004667 0.0021667  0.0001667  0.0007000
```

There are remarkable differences in variance, with

$$s^2_{max}/s^2_{min} = 0.1131/0.000167 = 677.2.$$

This is very large compared with $F_{max,0.01,12,3} = 361$. (An ordinary $F_{3,3}$-test is not appropriate because this value is the largest of all the possible variance ratios; the F_{max} test due to Hartley 1950 must be used. Critical values can be found in the Biometrika Tables by Pearson and Hartley and elsewhere.) Further, the small variances are associated with small means, as is easily seen by comparing the two tabulated outputs. Stronger evidence comes from a χ^2_3 quantile plot of the ordered variances, shown in Fig. 3.19. The plotting points are $(q_i, s^2_{(i)})$ where $s^2_{(i)}$ is the i-th smallest variance, and we take

$$q_i = F^{-1}(i/(n+1))$$

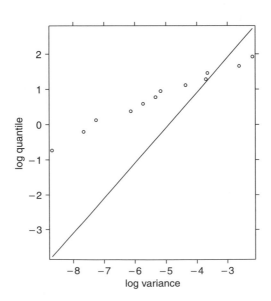

Fig. 3.19. Log quantile vs log variance

as for normal quantile plotting, but F is now the cdf of the χ_3^2 distribution. Also plotted is the line through the origin $q = 3s^2/\tilde{\sigma}^2$, where $\tilde{\sigma}^2$ is the residual mean square from the full model. The points should fall close to the line since each variance should be distributed as $(\sigma^2\chi_3^2)/3$ if the model is correct. We use log scales for both axes; note that the axes are reversed relative to those in the first edition.

```
$var 12 q i$
$sort svar var $
$calc i = %cu(1) $
    :  q = %chd(i/13) $
    : qf = 3*svar/0.0222 $
    : lq = %log(q) $
    :lqf = %log(qf) $
    : lv = %log(svar) $
$graph (s = 1 h = 'log variance' v = 'log quantile')
        lq,lqf lv 5,10 $
```

The assumption of constant variance is clearly violated, and the mean–variance association suggests the scale of the response is wrong. We look for an appropriate scale.

```
$macro yvar time $endm
$macro model type*treat $endm
$use boxcox $
```

The disparity changes rapidly over the range −2 to +2, and a finer tabulation is necessary over the range −1.5 to −0.2, shown below.

```
-----------------------------------------------------------
lmb_   -1.5     -1.4     -1.3     -1.2     -1.1     -1.0     -0.9
dev_  -103.22 -105.28 -107.03 -108.48 -109.60 -110.39 -110.85

lmb_   -0.8     -0.7     -0.6     -0.5     -0.4     -0.3     -0.2
dev_  -110.96 -110.74 -110.17 -109.26 -108.01 -106.44 -104.53
-----------------------------------------------------------
```

The minimum occurs at $\lambda = -0.82$, and the approximate 95% confidence interval is $(-1.29, -0.34)$ which includes $\lambda = -1$ but not $\lambda = 0$ or 1. The reciprocal scale is clearly indicated; the response-variable on this scale can be interpreted as the *rate of dying*.

```
$calc rate = 1/time $
$yvar rate $
$fit   : + type : + treat : + type.treat $
```

Table 3.8. ANOVA table, rate

Source	SS	df	MS	F
type	34.877	2	17.439	72.46
treat	20.414	3	6.805	28.35
interaction	1.571	6	0.262	1.09
residual	8.643	36	0.240	

The analysis of variance table is given in Table 3.8.

The F-values for the main effects are now much larger than they were before, but that for interaction is smaller: the interaction mean square is almost the same size as the residual mean square.

```
$tab the rate mean for type,treat $


        1      2      3      4

1    2.487  1.163  1.863  1.690
2    3.268  1.393  2.714  1.702
3    4.803  3.029  4.265  3.092


    : the rate variance for type,treat $


        1       2       3       4

1   0.24667 0.03980 0.23949 0.13302
2   0.67622 0.30602 0.17432 0.49267
3   0.28051 0.17761 0.05514 0.05956
```

The variance heterogeneity is greatly reduced, with $s^2_{max}/s^2_{min} = 16.99$, much less than $F_{max\,0.05,\,12,\,3} = 124$. More importantly, there is no association between cell means and variances. A χ^2_3 (log) quantile plot of the variances is shown in Fig. 3.20, with the theoretical straight line which now fits closely to the plotted points.

```
$tab the rate variance for type,treat into var $
$sort svar var $
$calc qf = 3*svar/0.240 $
    :lqf = %log(qf) $
   : lv = %log(svar) $
$graph (s = 1 h = 'log variance' v = 'log quantile')
        lq,lqf lv 5,10 $
```

The reciprocal transformation therefore seems satisfactory. (In Section 3.17 however we modify this conclusion.)

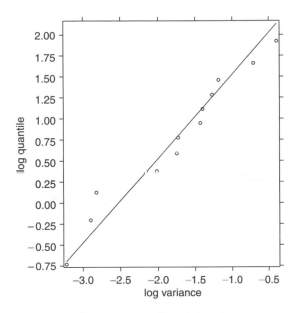

Fig. 3.20. Log quantile vs log variance

The rate means for `type` are 1.801, 2.269 and 3.797, and those for `treat` are 3.519, 1.862, 2.947 and 2.161. Transforming back the means, the fitted median survival times are 0.555, 0.441 and 0.236 for the poison types, and 0.284, 0.537, 0.339 and 0.463 for the treatments.

The effect of the reciprocal transformation on the precision of the analysis in this example is important. The assumption of constant variance on the original scale was invalid, and the pooled variance estimate on that scale was inflated by the large cell variance estimates for the cells with large means. The reciprocal transformation stabilizes the variance estimates, increasing the precision of the analysis.

We saw earlier in this chapter that the estimate of λ in the Box–Cox transformation could be affected by a change in model. What happens if we fit the main effects model instead of the interaction model?

```
$macro model type + treat $endm
$use boxcox $
```

We suppress the output. The disparity minimum now occurs at -0.75, and the 95% confidence interval is $(-1.14, -0.36)$. The model change has had very little effect.

Box and Cox (1964) pointed out that the transformation in this case is trying simultaneously to achieve three ends: normality of the residuals, additivity of the means and homogeneity of variance on the transformed scale. By including the interaction, we require the transformation to attempt to satisfy only two of these requirements. We could relax the requirements further by fitting the interaction model with a different variance term for each cell, thus removing the requirement of homogeneity of variance. Box and Cox discussed this; we do not pursue it here because the Box–Cox macro does not allow non-homogeneous variances. The requirement of homogeneity of variance is a strong one; without it, any value of λ from below -1 to about 2.5 is plausible (Box and Cox, 1964, p. 236).

3.16   Unbalanced cross-classifications

We illustrate the difficulties of the analysis of unbalanced cross-classifications with an example.

3.16.1   *The Bennett hostility data*

Bennett (1974) studied the emotional reactions of husbands to unsuccessful suicide attempts by their wives. He studied 25 husbands of wives who had been admitted to hospital after suicide attempts by taking drug overdoses. The husbands were interviewed in the hospital after the wife was out of danger, and responses to three specific questions were tape recorded, and subsequently analysed for emotional affect using the Gottschalk-Gleser (1969) scales of hostility and affection. The interview length was standardized between husbands.

Bennett used a "control" group of 42 husbands of wives admitted to hospital with acute organic abdominal conditions, to allow for the effect of the hospital admission and environment on affect. The control husbands were interviewed in the same way. Bennett subsequently divided this control group into two, as some of the controls had experienced previous difficulties in the marriage requiring counselling, which might affect their responses in the stressful hospital environment. Controls without previous marriage difficulties are termed "true controls", the others are termed "false controls"; the group factor g is defined by g = 1 overdoses, g = 2 false controls, g = 3 true controls. Two other explanatory factors were recorded: the national origin n of the husband (n = 1 Australian, n = 2 British), and the previous occurrence po of a suicide attempt or acute abdominal hospitalization (po = 1 no previous occurrence, po = 2 a previous occurrence).

Bennett analysed the tape recordings for word frequency on five measures: inward hostility i (guilt), ambivalent hostility a (shown by others to self—paranoia), overt outward hostility o (shown by self to others—anger), covert outward hostility c (shown by others to others) and positivity p (affection). The word counts on these measures are normalized by a square root transformation and can be assumed normally distributed. The values of the five scale variables, and the factor values for g, n and po are in the file hostility. Parts of the data are listed in Table 3.9.

Table 3.9. Hostility data

g	n	po	i	a	o	c	p
1	1	1	1.90	1.26	2.21	1.51	1.69
1	1	1	1.58	1.02	2.11	0.67	2.62
1	1	1	1.80	0.81	2.59	1.86	2.12
1	1	1	0.81	1.65	3.30	2.17	2.30
1	1	1	0.49	0.54	1.65	1.40	0.90
1	1	1	0.85	1.81	2.53	1.89	1.17
1	1	1	0.39	0.29	2.87	1.07	2.15
1	1	1	0.61	1.31	2.94	2.01	1.39
1	1	1	1.15	0.81	3.14	1.39	2.02
1	1	1	0.56	1.47	2.67	1.29	2.16
1	1	1	1.09	1.48	2.47	0.99	2.33
1	1	1	0.41	0.30	1.33	2.07	2.22
.	.	.					
3	2	2	0.48	0.38	0.68	1.56	3.16
3	2	2	0.99	0.30	1.81	0.91	1.39
3	2	2	1.47	1.20	1.83	0.60	3.23
3	2	2	0.83	1.31	1.57	0.70	2.56
3	2	2	1.49	0.36	1.53	0.85	2.31
3	2	2	1.07	0.24	1.37	0.44	2.84
3	2	2	1.15	0.47	1.69	2.10	2.61
3	2	2	0.38	0.38	2.41	0.38	2.89
3	2	2	0.31	0.43	1.61	1.35	1.68

An initial issue is the non-orthogonality of the three-way classification, which can be assessed by tabulation:

```
$input 'hostility' $
$tab for g,n,po $

 N    1                 2
PO    1     2           1      2
 G
 1    12    8           4      1
 2    4     5           3      1
 3    13    6           1      9
```

For the Australian husbands, previous occurrences are nearly proportional over the three groups, with 40% having had a previous occurrence, but for British husbands they are nearly all in the true control group—90% of true control husbands, but only 22% of overdose or false control husbands, had had a previous occurrence.

3.16.2   *ANOVA of the cross-classification*

We analyse the positivity response. Since the cross-classification is non-orthogonal, the SS in the ANOVA tables will depend on the order of fitting the terms in the model. We give three tables constructed by permuting the orders of main effects and two-way interactions so that each term is fitted last in the set in one of the three tables, allowing its (conditional) contribution to be assessed, given the other terms in the model.

```
$yvar p $
$fit : + g : + n : + po : - n : - g :  n : + po : + g $
$fit   + g.n : + g.po : + n.po : - g.po : - g.n $
   :      g + n + po + g.po : + n.po : + g.n $
   :   + g.n.po $
```

The ANOVA tables for main effects and interactions are constructed from the RSSs (Table 3.10).

There is a considerable change in the SS for the main effects depending on their order of fitting. The changes in the interaction SS are much smaller.

The three-way interaction can be omitted, with an $F_{2,55}$ value of 0.169 ($P = 0.845$). From the first column we see that n.po and g.po can be omitted similarly (giving a pooled $F_{5,55}$ value of 0.552, $P = 0.736$), while the F-value for g.n of 2.00 has a P-value of 0.145. This is not significant at conventional levels, but the 2 df may be concealing one larger component. We examine the parameter estimates for the g.n model:

```
$fit g + n + po + g.n $
$disp e $
   deviance =  16.131
residual df =  60
```

	estimate	s.e.	parameter
1	2.053	0.1289	1
2	0.3240	0.2093	G(2)
3	0.4136	0.1665	G(3)
4	-0.1745	0.2608	N(2)
5	-0.4025	0.1407	PO(2)
6	0.6232	0.4056	G(2).N(2)
7	0.6572	0.3470	G(3).N(2)

scale parameter 0.2688

The parameter estimates for the two components of the g.n interaction are not individually large, but they are nearly equal, and we equate them before assessing whether they can be omitted.

Table 3.10. ANOVA tables, positivity

Source	SS	df	MS	Source	SS	df	MS	Source	SS	df	MS
g	4.353	2	2.177	po	0.842	1	0.842	n	1.148	1	1.148
n	0.601	1	0.601	g	5.007	2	2.504	po	1.233	1	1.233
po	1.830	1	1.830	n	0.935	1	0.935	g	4.403	2	2.202
g.n	1.111	2	0.556	n.po	0.030	1	0.030	g.po	0.628	2	0.314
g.po	0.669	2	0.335	g.n	1.161	2	0.581	n.po	0.089	1	0.089
n.po	0.051	1	0.051	g.po	0.640	2	0.320	g.n	1.114	2	0.557
g.n.po	0.094	2	0.047	g.n.po	0.094	2	0.047	g.n.po	0.094	2	0.047
resid	15.316	55	0.278	resid	15.316	55	0.278	resid	15.316	55	0.278

```
$calc g23 = %ne(g,1) $
$fit - g.n + g23 + g23.n $
$disp e $
   deviance =  16.133 (change =  +0.002024)
residual df =  61      (change =  +1       )

           estimate          s.e.       parameter
     1        2.051          0.1266      1
     2        0.3167         0.1905      G(2)
     3        0.4170         0.1604      G(3)
     4       -0.1737         0.2585      N(2)
     5       -0.3986         0.1323      PO(2)
     6        0.000          aliased     G23
     7        0.6451         0.3150      N(2).G23
scale parameter 0.2645
```

The aliased term g23 is included to obtain comparable parameter estimates. The change in deviance or disparity is almost zero, and we try replacing g in the same way by g23; that is, we collapse the group factor into overdoses versus controls:

```
$fit - g $
$disp e $
   deviance =  16.223 (change =  +0.09008)
residual df =  62      (change =  +1       )

           estimate          s.e.       parameter
     1        2.050          0.1259      1
     2       -0.1732         0.2571      N(2)
     3       -0.3960         0.1315      PO(2)
     4        0.3848         0.1498      G23
     5        0.6473         0.3133      N(2).G23
scale parameter 0.2617
```

We now observe that the n(2) parameter estimate is less than its standard error, though n is involved in an interaction with g23. The interpretation of this "main effect" in the presence of interaction is however quite clear. With the GLIM factor coding, n(2) represents the British/Australian difference for the first level of g23, that is, for overdoses. There is no significant mean difference in affection for British and Australian husbands for the overdose group, but there is for the two control groups. So we may set this non-significant difference to zero to simplify the model:

```
$fit - n $
$disp e $
   deviance =  16.341  (change = +0.1187)
residual df =  63      (change = +1     )

            estimate         s.e.      parameter
      1       2.013        0.1121      1
      2      -0.3870       0.1302      PO(2)
      3       0.4192       0.1402      G23
      4       0.4712       0.1719      N(2).G23
scale parameter 0.2594
```

Model simplification could conclude at this point, with a pooled $F_{8,55}$ value of 0.460. However we can smooth the fitted values a little more, and simplify the resulting interpretation, by noting that the three parameter estimates are very close to each other, and can be equated with little increase in disparity.

```
$calc comp = g23 + %eq(n,2)*g23 - %eq(po,2) $
$fit - po - g23 - n.g23 + comp $
$disp e $
   deviance =  16.399  (change = +0.05729)
residual df =  65      (change = +2      )

            estimate         s.e.      parameter
      1       2.035        0.06835     1
      2       0.4266       0.07759     COMP
scale parameter 0.2523
```

We finally tabulate the fitted means, and the cell means, on positivity:

```
$tab the %fv mean for g,n,po $
  : the    p mean for g,n,po $
  N    1                2
 PO    1    2      1    2
  G
  1  2.04  1.61   2.04  1.61
```

```
2   2.46   2.04      2.89   2.46
3   2.46   2.04      2.89   2.46


N     1                2
PO    1       2        1       2
G
1   1.92   1.85      1.90   1.38
2   2.46   1.91      2.84   2.37
3   2.53   1.93      3.20   2.52
```

The fitted mean affection scores are on a four-point scale, with steps of 0.43:

(1) previous occurrence in an overdose: 1.61;
(2) no previous occurrence in an overdose, or previous occurrence in an Australian control: 2.04;
(3) no previous occurrence in an Australian control, or previous occurrence in a British control: 2.46;
(4) no previous occurrence in a British control: 2.89.

It will be seen that this final model is not easily expressed in terms of a classical ANOVA table with SS for each effect in the model. While such a table can be constructed from the successive steps of the model simplification, it is simpler to observe that the ANOVA table construction is only a formalization of regression model simplification, and that the whole process can be expressed as a backward elimination from the full model, which we now describe.

3.16.3   *Regression analysis of the cross-classification*

The process of reducing the full model can be simplified by following the usual backward elimination procedure for a regression model, though factors with multiple degrees of freedom need care. The description of the process takes some space, since new sets of parameter estimates have to be inspected at each step, but it is very important to follow this process, as we show in Section 3.16.4.

The parameter estimates for the full model are:

```
$fit g*n*po $
$disp e $
    deviance =   15.316
 residual df =   55

              estimate         s.e.       parameter
        1       1.999        0.3573       1
        2       1.020        0.6796       G(2)
        3       1.135        0.5096       G(3)
        4       0.4263       0.8274       N(2)
        5      -0.07625      0.2409       PO
```

6	-0.1286	1.292	G(2).N(2)
7	0.3215	1.401	G(3).N(2)
8	-0.4803	0.4282	G(2).PO
9	-0.5271	0.3548	G(3).PO
10	-0.4463	0.6373	N(2).PO
11	0.5294	0.9501	G(2).N(2).PO
12	0.3685	0.8851	G(3).N(2).PO

scale parameter 0.2785

Inspection of the parameter estimates suggests that the three-way interactions are negligible, there is a large g(3) effect, and no other important effects. However it is very important not to misinterpret these estimates; as with the general multiple regression model, the *t*-values for these model coefficients represent the importance of each variable if it *alone* is omitted from the model, all other variables being retained. But as we have seen in backward elimination procedures, the importance of each variable in general depends on which other variables are in the model. This issue is discussed extensively in Aitkin (1978) in terms of *effect coding*; the form of dummy variable coding used in the statistical package determines the parameter estimates and standard errors for these estimates. Whatever this coding, to determine the importance of the variables requires a backward elimination in the non-orthogonal cross-classification, just as it does in the general multiple regression model.

Thus in inspecting the model estimates above, the only relevant estimates are those for the three-way interaction, since the estimates for the other parameters will change when these interaction terms are omitted. These terms are omitted *first* from the model because of their complexity in interpretation; as in Section 3.1 we generally simplify models according to the hierarchy of main effects, first-order interactions, second-order interactions and so on.

Omitting the three-way interaction, we obtain new estimates and standard errors:

```
$fit - g.n.po $
$disp e $
   deviance =   15.411 (change =  +0.09433)
residual df =   57     (change =  +2      )
```

	estimate	s.e.	parameter
1	1.939	0.1473	1
2	0.4896	0.2834	G(2)
3	0.5865	0.2021	G(3)
4	-0.08552	0.2768	N(2)
5	-0.1172	0.2260	PO(2)

6	0.5458	0.4207	G(2).N(2)
7	0.8206	0.4158	G(3).N(2)
8	-0.3780	0.3753	G(2).PO(2)
9	-0.4714	0.3201	G(3).PO(2)
10	-0.1596	0.3691	N(2).PO(2)

scale parameter 0.2704

The two omitted terms give an F-value of 0.169 with 2 df. In addition to the large g(3) term, we now see a large g(3).n(2) interaction which was not present in the full interaction model.

The n.po interaction is small and can be omitted; the g.po interaction terms are also small. We try omitting both of them:

```
$fit   g.po   n.po $
$disp e $
   deviance =   16.131  (change =   +0.7200)
residual df =   60      (change =   +3      )
```

	estimate	s.e.	parameter
1	2.053	0.1289	1
2	0.3240	0.2093	G(2)
3	0.4136	0.1665	G(3)
4	-0.1745	0.2608	N(2)
5	-0.4025	0.1407	PO(2)
6	0.6232	0.4056	G(2).N(2)
7	0.6572	0.3470	G(3).N(2)

scale parameter 0.2688

The F-value for the 3 df omitted is less than 1 (0.2400/0.2688). We now proceed as above for the further reduction of the model.

3.16.4   *Statistical package treatments of cross-classifications*

The backward elimination procedure, and the multiple ANOVA tables resulting from the non-orthogonality of the cross-classification, complicate the analysis of such classifications. There have been many attempts to simplify the analysis by constructing a "composite" ANOVA table with one unique entry for each effect. This is done in many packages by reporting the SS for each effect, "adjusted for" (after fitting) all other effects of the same order in the hierarchy (or less commonly, adjusted for all other effects of all orders, discussed below). In the Bennett data, this leads to a composite "ANOVA" table, which can be obtained from the full decompositions above with each effect fitted last, as shown in Table 3.11.

Inspection of the table suggests that there are important main effects of g and po, but no important interactions and no significant main effect of n. Tabulation

Table 3.11. Composite "ANOVA" table, positivity

Source	SS	df	MS	F	P-value
g	4.403	2	2.202	7.92	0.0009
n	0.935	1	0.935	3.36	0.072
po	1.830	1	1.830	6.58	0.013
(Total)	7.168				
g.n	1.114	2	0.557	2.00	0.145
g.po	0.640	2	0.320	1.15	0.324
n.po	0.051	1	0.051	0.18	0.673
(Total)	1.805				
g.n.po	0.094	2	0.047	0.17	0.844
res	15.316	55	0.278		

of the p means for g and po gives

```
PO      1       2    | mean
G
1      1.92    1.79  | 1.87
2      2.63    1.98  | 2.32
3      2.58    2.28  | 2.43
       ----    ----
mean   2.30    2.08
```

This analysis misses the g.n interaction between nationality and the overdoses and controls (which requires careful examination of the parameter estimates to identify), but the main difficulty in the use of such composite "ANOVA" tables is that they do not represent a breakdown of the "hypothesis" sum of squares for all the full model effects into additive components from a hierarchical sequence of added or omitted terms. This is most clearly seen if we add the SS for the three main effects in the composite table: the sum is 7.168, while the sum of the additive components in the three breakdowns is 6.784, whichever sequence is followed, since this represents the SS difference between the null model and the model with all three main effects.

Thus the composite table overstates the importance of the main effects, as it does (slightly) for the two-way interactions. In other examples the composite ANOVA table can drastically *understate* the importance of effects, when these are positively correlated: the SS in the composite table can add to much less than the real total in the hierarchical sequence: there may be a "missing sum of squares" in the composite analysis. Aitkin (1978) gave examples.

If the composite "ANOVA" table is produced by adjusting each effect for *all* other effects in the model, an even more serious problem occurs, in which the quoted sum of squares for each effect depends critically on the form of dummy variable (or contrast) coding used for the effects. Changes in the form of coding (for example from $(0,1)$ to $(1,0)$) can completely change the composite table: such tables cannot be used to identify important terms in the model.

We emphasize that to obtain correct conclusions from unbalanced cross-classifications, it is *essential* to proceed by elimination of effects from the full interaction model, in a sequence of model fits.

This conclusion contradicts a large part of the literature on unbalanced cross-classifications, which attempts to produce a single composite ANOVA table (by the method above, or other methods) without explicit fitting of reduced models. Such tables are unfortunately produced by the default option in many statist-ical packages, and great care is needed in using these packages for unbalanced cross-classifications. The user should choose the option for "full hierarchical partitioning" to avoid this problem. Whatever package is used, the backward elimination regression procedure from the full interaction model described above will identify the appropriate model (or models—there may be more than one with highly correlated factors); it is a serious and unfortunate consequence of non-orthogonality that no *single* model analysis can guarantee this.

Our conclusions may be thought controversial; the reader should consult Nelder (1977) and Aitkin (1978) for very detailed statements and examples (with contributed discussions).

3.17   Missing data

Throughout this chapter we have assumed that complete data on response and explanatory variables are available for every sample member. The reality of survey data collection, however, is that missing values always occur, through non-response, misrecording or accident. The data matrix of individuals by variables generally has a scatter of missing values, and this considerably complicates model fitting by ML. In some cases the entire set of variables may be missing because the respondent could not be contacted—in the language of sample surveys, this is a missing "unit-level" observation, rather than a missing "item-level" observation on an individual variable.

The simplest case, which we consider first, is when the missing values are "missing completely at random" (MCAR—this and other terms were introduced by Rubin, 1976). We may imagine a completely random "failure process", in which the value of any variable for each individual has a constant probability (possibly different for each variable) of failing, that is, of being missing, and failures occur independently. More realistically, the missingness probability for a variable may depend on other observed variables, but not on the missing variable—the value of the variable is "missing at random" (MAR).

In both these cases fully efficient MLEs of the model parameters may be obtained from the observed data using the EM algorithm (Little and Rubin, 1987). We do not have to model explicitly the missingness process, since the likelihood factors into independent components for the response model and the missingness model. However it is necessary to assume an explicit probability model for the variables for which values are missing. Little and Schluchter (1985) gave computational details for a normal regression model with normal or categorical explanatory variables. The EM algorithm does not provide standard errors, and computation of these from the information matrix using the result of Louis (1982) is particularly complicated.

These cases of values missing at random may not hold. The probability of a variable value being missing may depend on the value that *would have been* observed. This case of "non-random missingness" or "missing non-randomly" (MNR) is particularly difficult to deal with as the model for missingness is in general unidentifiable, since we are missing the data which would allow it to be estimated. In this case the most that can be done is to construct a model (usually logistic) relating the missingness probability to the explanatory variables, and assume a value for the parameter governing the dependence of the missingness on the value of the missing variable; this model is then identifiable, and a sensitivity analysis can be carried out to assess the sensitivity of parameter estimates and tests to the magnitude of the non-random dependence. Copas (1997) gave a detailed discussion.

As a simple example, consider a survey of monthly income of individuals classified by age, sex, educational level and industrial sector of employment. The object of analysis is to model average income by a regression on the other explanatory variables, but the value of income only is missing for some individuals. If these values are missing at random, then ML estimation of the regression model parameters is equivalent to omitting the individuals with missing values and analysing only the complete cases.

Suppose now that the probability of an income value being missing increases with age of the respondent. This can be established by a logistic regression (Section 4.2) of the missing data indicator on the explanatory variables. Then higher ages are under-represented in the complete cases, and if we intend to report the overall mean monthly income for the population, the sample means will have to be weighted by the inverse of the sample inclusion probabilities to allow the under-representation of older people. (See Section 3.3 for a discussion of weighting.)

The fitted regression of earnings on the explanatory variables is, however, not affected by the selection occurring in the sample, because the fitted model is conditional on the observed values of the explanatory variables. It is only marginal, or unconditional inferences which are affected. The variances of the regression coefficients are however increased by the restriction of range.

Suppose finally that the probability of an income value being missing increases with decreasing income. Then low incomes are under-represented in the sample, and the regression ignoring the missingness process will be systematically biased. In some "biased" sample designs the same result may be achieved by design: we may deliberately over-sample high income respondents to ensure a sufficient sample with high incomes. Model fitting then requires either weighting by the inverse of the sample selection probabilities, or complex modelling of the sample distribution of the observed responses incorporating the design. Lawless *et al.* (1999) gave a recent comprehensive discussion.

When the response-variable is fully observed but the explanatory variables are missing, *imputation* methods are widely used to "fill in" the missing explanatory variable values, with the aim of providing a "clean data set" for a standard analysis. *Single imputation* methods, in which a single value is provided, do not give MLEs of the model parameters from the standard analysis of the "completed" data set even under MCAR assumptions, whatever form of imputation is used, and standard errors are always underestimated.

Multiple imputation methods (Rubin, 1987; Schafer, 1997) provide M multiple imputed values simulated from the conditional distribution of the variables with missing data given the observed variables, for each case with any missing data. These M multiple imputed values are used to construct M completed data sets, to each of which the standard analysis is applied; the M sets of estimates and their covariance matrices are then combined to give a single estimate and covariance matrix. Fay (1996) gave an alternative method of combining the M data sets, by weighting each set with weight $1/M$ in a single analysis.

Multiple imputation is implemented in the S-plus functions of Schafer (1996), available at http://www.stat.psu.edu/jls/, and in the package SOLAS. Multiple imputation methods give a close approximation to ML estimates and information-based standard errors without the difficult computation of the latter; their performance has been discussed in several review papers (see, e.g. Fay, 1996).

We conclude with a discussion of approximate methods available in statistical packages.

3.18   Approximate methods for missing data

Three approximate methods are in common use. The first omits all observations with missing data on any variables, and uses only complete records ("complete case" analysis). There are two serious disadvantages of this method. The first is the obvious one that the number of observations may be substantially reduced. In extreme cases there may be no complete records. The second is that serious bias in the model parameter estimates may result. When observation are missing on both response and explanatory variables, the parameter estimates from the complete cases are no longer consistent.

The second approximate method estimates the mean and variance of each variable from all the complete observations on that variable, and the covariance between each pair of variables from the complete observations on those pairs of variables ("available case" analysis). This does not avoid bias in the parameter estimates, and may result in a covariance matrix for the explanatory variables which is not positive definite.

The third approximate method fills in the missing values with single imputed values, with the object of creating a completed data set which can be used for standard analyses. As noted above, single imputation by any method does not provide ML estimates, and always underestimates the standard errors of the estimates.

3.19   Modelling of variance heterogeneity

We have already seen in the poison example that variances may not be constant. There are several ways of allowing for variance heterogeneity.

The first is by adapting the variance expression of the usual regression estimator to allow for a general variance pattern. For the general regression model, suppose that

$$Y_i|\mathbf{x}_i \sim N(\mu_i, \sigma_i^2)$$

with

$$\mu_i = \boldsymbol{\beta}'\mathbf{x}_i,$$

where the variances σ_i^2 are different but unspecified.

Then the MLE $\hat{\boldsymbol{\beta}}$ of $\boldsymbol{\beta}$ under homogeneity has the repeated sampling distribution $\hat{\boldsymbol{\beta}} \sim N(\boldsymbol{\beta}, V)$ where

$$V = (X'X)^{-1}X'DX(X'X)^{-1}$$

and D is the diagonal matrix of variances σ_i^2. The variances are unknown, but can be estimated from the squared residuals from the regression as $\widetilde{\sigma}^2_i = e_i^2$, leading to the *sandwich* or *robust* variance estimate

$$\tilde{V} = (X'X)^{-1}X'\tilde{D}X(X'X)^{-1},$$

due to Huber (1967) and White (1982). This estimate is provided in many statistical packages (though not in GLIM) in addition to the usual estimate assuming variance homogeneity.

Without knowledge of V we cannot improve on the least squares estimate. It might appear that we can use the estimated variances as inverse weights, and define the *weighted* least squares estimate

$$\tilde{\boldsymbol{\beta}} = (X'\tilde{D}^{-1}X)^{-1}X'\tilde{D}^{-1}\mathbf{y}$$

with variance

$$(X'\tilde{D}^{-1}X)^{-1},$$

or perhaps an iterated version of this, but it is easily seen that the estimate of $\boldsymbol{\beta}$ is inconsistent, because as $n \rightarrow \infty$ the variance estimates $\widetilde{\sigma}^2_i$ do not converge to the true variances σ_i^2—the number of these variances, each with (at most) 1 df, goes to infinity as well. To improve on the least squares estimate we need to *model* the variances by a parsimonious structure of small dimension relative to the sample size.

We now extend the standard model by explicit modelling of the variance heterogeneity in terms of explanatory variables. Such models are called *dispersion models*, and are discussed in great generality in Jorgensen (1997). In Chapter 8 we allow for *random* variances, leading to the *t-distribution* for the observed response Y.

We generalize the regression model to the form:

$$Y_i | \mathbf{z}_i \sim N(\mu_i, \sigma_i^2)$$

with

$$\log \sigma_i^2 = \boldsymbol{\lambda}' \mathbf{z}_i,$$

a *log-linear model* for the variances. This ensures positivity of variances in the fitted model. The variables $\mathbf{z}_i$ may be a subset or superset of $\mathbf{x}_i$, or may be unrelated. We first consider the special case of *known* means μ_i, and write $e_i = y_i - \mu_i$. The likelihood is

$$L(\boldsymbol{\lambda}) = \prod_i \frac{1}{\sqrt{2\pi}\sigma_i} \exp\left\{-\frac{e_i^2}{2\sigma_i^2}\right\}.$$

Recall that $E_i^2/\sigma_i^2 \sim \chi_1^2$, and since the χ_ν^2 random variable is twice a gamma variable with degrees of freedom $\nu/2$, the likelihood $L(\boldsymbol{\lambda})$ is proportional to that of a set of n scaled gamma variables, with scale factors $2\sigma_i^2$ and degrees of freedom $1/2$. The log-linear model for σ_i^2 can then be fitted in GLIM directly by specifying the gamma error model for the squared residuals e_i^2, and a GLIM scale parameter of $1/\nu = 2$; the model fitted to the mean σ_i^2 is the log-linear model specified above.

Now consider the general case where μ_i is unknown, but specified by another linear model $\mu_i = \boldsymbol{\beta}' \mathbf{x}_i$. This *double generalized linear model* can be fitted by successive relaxation, by alternately fitting the mean and variance models (Smyth, 1986, 1989; Aitkin, 1987).

Given observations $(y_i, \mathbf{x}_i, \mathbf{z}_i)$, $i = 1, \ldots, n$ the likelihood is

$$L(\boldsymbol{\beta}, \boldsymbol{\lambda}) = \prod_i \frac{1}{\sqrt{2\pi}\sigma_i} \exp\left\{-\frac{e_i^2}{2\sigma_i^2}\right\}$$

where now

$$e_i = y_i - \boldsymbol{\beta}' \mathbf{x}_i.$$

The log-likelihood is

$$\ell(\boldsymbol{\beta},\boldsymbol{\lambda}) = -\frac{1}{2}\left[\sum \log \sigma_i^2 + \sum (y_i - \boldsymbol{\beta}'\mathbf{x}_i)^2/\sigma_i^2\right],$$

and the first and second derivatives with respect to the parameters are

$$\frac{\partial \ell}{\partial \boldsymbol{\beta}} = \sum \mathbf{x}_i (y_i - \boldsymbol{\beta}'\mathbf{x}_i)/\sigma_i^2$$

$$\frac{\partial^2 \ell}{\partial \boldsymbol{\beta}\partial \boldsymbol{\beta}'} = -\sum \mathbf{x}_i \mathbf{x}_i'/\sigma_i^2$$

$$\frac{\partial \ell}{\partial \boldsymbol{\lambda}} = \frac{1}{2}\left[-\sum \mathbf{z}_i + \sum (y_i - \boldsymbol{\beta}'\mathbf{x}_i)^2 \mathbf{z}_i/\sigma_i^2\right]$$

$$= \frac{1}{2}\sum (e_i^2/\sigma_i^2 - 1)\mathbf{z}_i$$

$$\frac{\partial^2 \ell}{\partial \boldsymbol{\lambda}\partial \boldsymbol{\lambda}'} = -\frac{1}{2}\sum (e_i^2/\sigma_i^2)\mathbf{z}_i \mathbf{z}_i'$$

$$\frac{\partial^2 \ell}{\partial \boldsymbol{\beta}\partial \boldsymbol{\lambda}'} = -\sum (e_i/\sigma_i^2)\mathbf{x}_i \mathbf{z}_i'.$$

Taking expectations, the expected information matrix is block-diagonal, with blocks

$$\mathcal{I}_{\boldsymbol{\beta}} = X'W_{11}X$$

$$\mathcal{I}_{\boldsymbol{\lambda}} = \tfrac{1}{2}Z'Z,$$

where

$$X' = [\mathbf{x}_1,\ldots,\mathbf{x}_n]$$

$$Z' = [\mathbf{z}_1,\ldots,\mathbf{z}_n]$$

$$W_{11} = \mathrm{diag}(1/\sigma_i^2),$$

since $E[E_i] = 0$ and $E[E_i^2] = \sigma_i^2$. Thus a Fisher scoring algorithm for the simultaneous MLE of $\boldsymbol{\beta}$ and $\boldsymbol{\lambda}$ reduces to two separate algorithms for $\boldsymbol{\beta}$ and $\boldsymbol{\lambda}$. Since however W_{11} depends on $\boldsymbol{\lambda}$ and e_i depends on $\boldsymbol{\beta}$, it is simplest to formulate the scoring algorithm as a successive relaxation algorithm.

For given σ_i^2, $\hat{\boldsymbol{\beta}}$ is a weighted least squares estimate with weights $1/\sigma_i^2$, and for given $\boldsymbol{\beta}$, $\hat{\boldsymbol{\lambda}}$ is the MLE from a gamma model with response-variable e_i^2 and GLIM scale parameter 2.

The algorithm conveniently begins with an initial unweighted normal regression of y on $\mathbf{x}$, taking $\sigma_i^2 \equiv \sigma^2$. The squared residuals from the least squares fit are defined as a new response-variable with a gamma distribution with scale parameter 2. The linear predictor $\lambda'\mathbf{x}$ is then fitted using a log-link function, and the disparity calculated for the initial estimate of (β, λ). A weighted normal regression of y on $\mathbf{x}$ is now fitted, with scale parameter 1 and weights given by the reciprocals of the fitted values from the gamma model. This process continues until the disparity converges. At this point the standard errors (based on the expected information) from both models are correct. However the log-likelihood in the two sets of parameters may be very skewed, and so the standard errors are not a reliable indicator of variable importance. In addition, the loss of degrees of freedom in the variance model due to the estimation of the mean parameters may be serious, requiring a *marginal* or *restricted* likelihood maximization for the variance model. See Smyth and Verbyla (1999) for a discussion.

The above analysis assumes that the parameters β and λ are functionally independent, which will usually be the case.

This approach to variance heterogeneity modelling provides a routine analysis of the famous *Behrens–Fisher problem*, of how to test the hypothesis of equal means in a two-group problem when the variances are different. This is the simplest case of variance heterogeneity: we fit two models, the "alternative hypothesis" model has the group factor in both mean and variance models, the "null hypothesis" model has a null mean model and the same group variance model. A profile likelihood for the mean difference can be computed by defining this difference as an offset and fitting a null mean model. This test procedure however does not provide a fixed size α test except in large samples (as $n \rightarrow \infty$): the test size depends on the true variance ratio.

The double modelling procedure is implemented in the macro `varmodel` (Aitkin, 1987); its use is illustrated below with two examples, the Box–Cox poison data (Section 3.15) and the Minitab tree data (Section 3.6).

3.19.1 *Poison example*

We adopt the reciprocal scale for time, as in the previous analysis, but we now allow for general variance heterogeneity, though we found little evidence for this previously. With cross-classifications, we follow the general procedure of fitting a saturated model for the mean, and using this to find a suitable model for the dispersion, beginning with the saturated model and using backward elimination. When a final model for dispersion has been found, we simplify the mean model in a similar way. The mean and variance models are specified in two macros, `modm` and `modv`. Both mean and variance models allow for offsets (`ofsm`, `ofsv`); if these are not to be used they must be set equal to zero explicitly.

```
$newjob $
$input 'poison' $
```

```
    :    'varmodel' $
$calc rate = 1/time $
$macro modm type*treat $endm
$macro modv type*treat $endm
$calc ofsm = ofsv = 0 $
$use varmodel rate $
Mean model
            estimate         s.e.      parameter
      1        2.487        0.2151     1
      2        0.7816       0.4160     TYPE(2)
      3        2.316        0.3144     TYPE(3)
      4       -1.323        0.2318     TREAT(2)
      5       -0.6242       0.3019     TREAT(3)
      6       -0.7972       0.2668     TREAT(4)
      7       -0.5517       0.4877     TYPE(2).TREAT(2)
      8        0.06961      0.5006     TYPE(2).TREAT(3)
      9       -0.7697       0.5388     TYPE(2).TREAT(4)
     10       -0.4503       0.3736     TYPE(3).TREAT(2)
     11        0.08646      0.3925     TYPE(3).TREAT(3)
     12       -0.9137       0.3674     TYPE(3).TREAT(4)
scale parameter 1.000


Variance model,log scale
            estimate         s.e.      parameter
      1       -1.687        0.7071     1
      2        1.008        1.000      TYPE(2)
      3        0.1286       1.000      TYPE(3)
      4       -1.824        1.000      TREAT(2)
      5       -0.02955      1.000      TREAT(3)
      6       -0.6176       1.000      TREAT(4)
      7        1.031        1.414      TYPE(2).TREAT(2)
      8       -1.326        1.414      TYPE(2).TREAT(3)
      9        0.3009       1.414      TYPE(2).TREAT(4)
     10        1.367        1.414      TYPE(3).TREAT(2)
     11       -1.597        1.414      TYPE(3).TREAT(3)
     12       -0.9320       1.414      TYPE(3).TREAT(4)
scale parameter 2.000

Iteration number   2 , -2 log Lmax is    39.14
```

The disparity calculated here does not correspond to that in the Box–Cox macro as it is missing the term $\prod 1/y_i^2$ in the likelihood. The comparable value from the constant (null) variance model is 53.92, so the saturated variance model gives a disparity reduction of 14.78 on 11 df. This is not large, but we now examine the contribution of individual model terms to it by backward elimination.

None of the interaction parameters is large, and so the interaction can be omitted:

```
$macro modv type + treat $endm
$use varmodel $
Mean model
            estimate          s.e.        parameter
    1          2.487        0.2348        1
    2         0.7816        0.4325        TYPE(2)
    3          2.316        0.3146        TYPE(3)
    4         -1.323        0.2822        TREAT(2)
    5        -0.6242        0.2810        TREAT(3)
    6        -0.7972        0.2848        TREAT(4)
    7        -0.5517        0.5196        TYPE(2).TREAT(2)
    8        0.06961        0.5174        TYPE(2).TREAT(3)
    9        -0.7697        0.5244        TYPE(2).TREAT(4)
   10        -0.4503        0.3780        TYPE(3).TREAT(2)
   11        0.08646        0.3764        TYPE(3).TREAT(3)
   12        -0.9137        0.3815        TYPE(3).TREAT(4)
scale parameter 1.000

Variance model, log scale
            estimate          s.e.        parameter
    1         -1.511        0.5000        1
    2         0.8718        0.5000        TYPE(2)
    3        -0.2297        0.5000        TYPE(3)
    4        -0.8126        0.5774        TREAT(2)
    5        -0.8408        0.5774        TREAT(3)
    6        -0.7546        0.5774        TREAT(4)
scale parameter 2.000

Iteration number 2 , -2 log Lmax is    45.05
```

The disparity increases by 5.91 on 6 df. The parameter estimates in the saturated mean model are unchanged by changes in the variance model, but their standard errors are changed. For further reductions in the variance model we suppress the mean model output.

```
$macro modv type $endm
$use varmodel $
Variance model, log scale
```

```
          estimate          s.e.      parameter
    1      -2.091          0.3536     1
    2       0.9173         0.5000     TYPE(2)
    3      -0.1401         0.5000     TYPE(3)
scale parameter 2.000
```

```
Iteration number 2 , -2 log Lmax is    48.28
```

The omission of the treatment main effect gives a disparity increase of 3.23 on 3 df.

```
$macro modv 1 $endm
$use varmodel $
Variance model, log scale
          estimate          s.e.      parameter
    1      -1.714          0.2041     1
scale parameter 2.000
```

```
Iteration number 2 , -2 log Lmax is    53.92
```

Omitting the poison type main effect gives a disparity increase of 5.64 on 2 df, and this is almost all concentrated in the contrast of type 2 with the other two types.

We define a dummy variable for type 2:

```
$calc type2 = %eq(type,2) $
$macro modv type2 $endm
$use varmodel $
Variance model, log scale
          estimate          s.e.      parameter
    1      -2.159          0.2500     1
    2       0.9850         0.4330     TYPE2
scale parameter 2.000
```

```
Iteration number 2 , -2 log Lmax is    48.36
```

Replacing `type` by `type2` increases the disparity by only 0.08 on 1 df, the contrast of type 2 with the other two increasing it by 5.56 on 1 df. The final model on the rate scale thus specifies a common variance for types 1 and 3 but a larger variance for type 2.

As noted above, the parameter estimates for the saturated mean model do not depend on the dispersion model fitted, but their standard errors do. The largest interaction of -0.9137 is for `type(3).treat(4)`, and its standard error is 0.367 in the saturated dispersion model, 0.340 in the final dispersion model with a type 2 term, and 0.490 in the common dispersion model. The standard errors are smaller in the first two models because the variance estimate is based on the smaller type 3 variance. (When the null variance model is fitted, the variance estimate is the MLE of σ^2, not the unbiased estimate, so the standard errors are smaller than those reported by GLIM in a standard analysis.)

We now reduce the mean model.

```
$macro modm type+treat $endm
$use varmodel $
Mean model
            estimate        s.e.        parameter
      1        2.676        0.1170      1
      2        0.4686       0.1739      TYPE(2)
      3        1.996        0.1398      TYPE(3)
      4       -1.612        0.1776      TREAT(2)
      5       -0.5758       0.1776      TREAT(3)
      6       -1.314        0.1776      TREAT(4)
  scale parameter 1.000


Variance model, log scale
            estimate        s.e.        parameter
      1       -1.856        0.2500      1
      2        0.7391       0.4330      TYPE2
  scale parameter 2.000


  Iteration number 3 , -2 log Lmax is      58.97
```

Fitting the main effect model for the mean with the type 2 dispersion model gives a disparity increase of 10.61 on 6 df. We add a parameter for the type 3, treatment 4 cell using a dummy variable:

```
$calc d34 = %eq(type,3)*%eq(treat,4) $
$macro modm type + treat + d34 $endm
$use varmodel rate$
Mean model
            estimate        s.e.        parameter
      1        2.603        0.1383      1
      2        0.4686       0.1749      TYPE(2)
      3        2.154        0.1440      TYPE(3)
      4       -1.598        0.1653      TREAT(2)
      5       -0.5769       0.1653      TREAT(3)
      6       -1.033        0.2018      TREAT(4)
      7       -0.6323       0.2730      D34
  scale parameter 1.000
```

```
Variance model, log scale
           estimate          s.e.       parameter
     1       -2.050         0.2500       1
     2        1.031         0.4330       TYPE2
scale parameter 2.000


 Iteration number 4 , -2 log Lmax is     54.31
```

The single interaction parameter reduces the disparity by 4.66 on 1 df, the remaining five interaction parameters accounting for a disparity of 5.95. The importance of this interaction term is not apparent in the common dispersion model, where the disparity change for d34 is 3.48, and that for the remaining five interaction parameters is 4.54.

Tables 3.12 and 3.13 show the cell means and variances (SS divided by 4) for rate, and fitted values from the final model. The tables are arranged so that mean values decrease monotonically down and across the table. The d34 dummy variable reproduces the observed mean in the cell marked with an asterisk.

Table 3.12. Cell means for rate, and [fitted values]

Type	Treat			
	1	3	4	2
3	4.80	4.27	3.09	3.03
	[4.76]	[4.18]	[3.09]*	[3.16]
2	3.27	2.71	1.70	1.39
	[3.07]	[2.50]	[2.04]	[1.47]
1	2.49	1.86	1.69	1.16
	[2.60]	[2.03]	[1.57]	[1.01]

Table 3.13. Cell variances for rate, and [fitted values]

Type	Treat			
	1	3	4	2
3	0.210	0.041	0.045	0.133
	[0.129]	[0.129]	[0.129]	[0.129]
2	0.507	0.131	0.370	0.230
	[0.361]	[0.361]	[0.361]	[0.361]
1	0.185	0.180	0.100	0.030
	[0.129]	[0.129]	[0.129]	[0.129]

3.19.2   *Tree example*

We consider the "solid of revolution" model with log transformations of v, d and h. With continuous explanatory variables there is no obvious "saturated model" for the variance; we adopt here a second-degree response surface in `log h` and `log d` while maintaining the linear model in the mean; Cook and Weisberg (1983) noted some evidence of variance heterogeneity.

```
$input 'trees' $
   , 'varmodel' ¢
$calc lv = %log(v) : lh = %log(h) : ld = %log(d) $
   :  lh2 = lh**2  : ld2 = ld**2 : lhld = lh*ld $
   : ofsm = ofsv = 0 $
$macro modm lh + ld $endm
$macro modv lh + ld + lh2 + lhld + ld2 $endm
$use varmodel lv $
Mean model
              estimate        s.e.        parameter
        1      -6.407        0.2237        1
        2       1.092        0.05288       LH
        3       1.937        0.01453       LD
   scale parameter 1.000


Variance model, log scale
              estimate        s.e.        parameter
        1      -751.8        631.0         1
        2       438.7        307.4         LH
        3      -178.8         83.03        LD
        4       -77.33        38.54        LH2
        5        94.77        26.89        LHLD
        6       -44.91         8.547       LD2
   scale parameter 2.000


   Iteration number 8 , -2 log Lmax is    -92.60
```

As noted above, the standard errors for the terms in the response surface model for the variance are very unreliable indicators of importance in this small sample and heavily parametrized model—the likelihood in these parameters is very skewed. Backward elimination of terms from the variance model leaves a quadratic in `ld` shown in Table 3.14, with the estimates and (standard errors) from the response surface variance model and the constant variance model; the mean model is consistent with the model `log h + 2log d`, and estimates for this model are also shown in the table, obtained by appropriately defining the offset `ofsm` as in Section 3.6.

Table 3.14. Mean and variance models for log v, tree data

Mean			Variance						
1	*lh*	*ld*	1	*lh*	*ld*	*lh*^{2}	*lhld*	*ld*^{2}	dis
−6.41	1.09	1.94	−752	439	−179	−77	94	−45.0	−92.60
(0.22)	(0.05)	(0.01)	(631)	(307)	(83)	(39)	(27)	(8.5)	
−6.39	1.08	1.96	−180	0	134	0	0	−25.5	−88.31
(0.25)	(0.07)	(0.02)	(28)		(22)			(4.2)	
−6.15	1	2	−156	0	115	0	0	−22.0	−85.10
(0.01)			(28)		(22)			(4.2)	
−6.17	1	2	−5.11	0	0	0	0	0	−70.30
(0.01)			(0.25)						

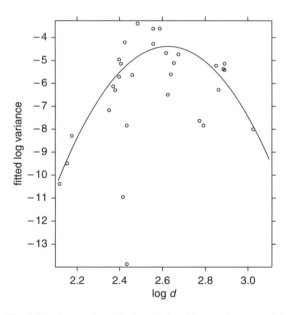

Fig. 3.21. Squared residuals and fitted log variance model

It is difficult to understand the meaning of the quadratic variance function from the estimates; Fig. 3.21 shows the squared residuals and fitted variance function on the log scale, and Fig. 3.22 shows them on the actual variance scale.

There is a striking increase in variability for intermediate values of tree diameter, with small variances for large and small trees. This unexpected result leads us to examine closely the influence of individual observations on the variance model, by extracting them following the variance model fit and graphing them against ld in Fig. 3.23.

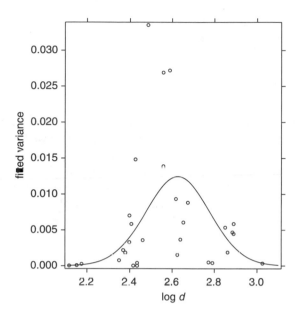

Fig. 3.22. Squared residuals and fitted variance model

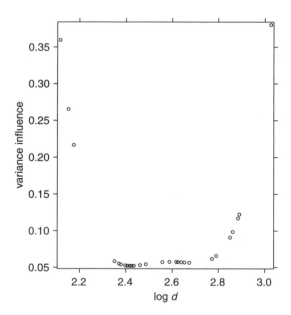

Fig. 3.23. Tree data: influence values from quadratic variance model

```
$macro modm 1h + 1d $endm
$macro modv 1d + 1d2 $endm
$use varmodel $
$extract %lv $
$graph (s = 1 h = 'log d' v = 'variance influence')
        %lv 1d 5 $
```

The four extreme observations, both large and small, appear to have a very powerful effect on the fitted variance model. However if these are removed from the data set the quadratic term in the variance model still gives a disparity reduction of 4.16. It happens that the mean model prediction of tree volume is very accurate at the extremes of the range of tree diameter; it would require more data over a larger range of diameters to determine whether this effect is real or accidental.

4
Binary response data

4.1 Binary responses

Much social science data consist of *categorical* variables. Familiar examples are religion, nationality, residence (urban/rural), type of dwelling, level of education and social class. The categories may be unordered (religion, nationality) or ordered (degree of disablement, attitude to a social question).

The simplest categorical variable is one with just two categories. Simple Yes/No or Agree/Disagree responses to questionnaire items are examples, while in medical science Present/Absent, Susceptible/Not susceptible, Exposed/Not exposed are common examples. In Chapter 3, we considered the use of binary variables, coded 0 or 1, as explanatory variables in a regression model. Here, however, we are interested in the relationship between a binary response variable and other explanatory variables.

We begin with a simple example from Racine *et al.* (1986), of a small study of the effects of $m = 4$ increasing doses of a drug on the mortality of mice through acute toxicity on inhalation of the drug (Table 4.1).

At the i-th dose level d_i, $i = 1, \ldots, m$, (expressed in mg/ml), $n_i = 5$ mice are administered the drug dose, and r_i of them die. The small number of animals reflects the ethical need to minimize animal sacrifice; the dose levels are set to cover the full range of response probability, from 0 to 1. We want to model the relationship between the probability p_i of death and the dose d_i. We graph the *sample proportions* dying against dose in Fig. 4.1; because of the substantial skew in the dose values we use the log scale for dose.

```
$unit 4
$data d r $
$read
422 0   744 1   948 3   2069 5 $
$calc x = %log(d) : n = 5 :   op = r/n $
$graph (s = 1 h = 'log dose' v = 'p') op x 5 $
```

Table 4.1. Toxicity of a drug in mice

i	1	2	3	4
d_i	422	744	948	2069
r_i	0	1	3	5
n_i	5	5	5	5

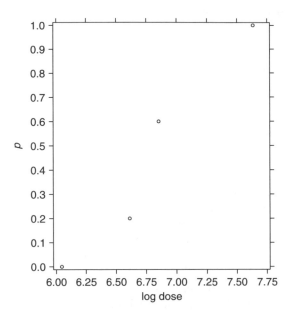

Fig. 4.1. Observed death proportions

For binary responses in general, we define the response variable

$$Y = 1 \text{ "success" (occurrence of the response)}$$
$$= 0 \text{ "failure" (non-occurrence)}$$

and let p be the success probability for a randomly chosen individual at a given explanatory variable value x. Then Y has a *Bernoulli* distribution

$$\Pr[Y = y] = p^y(1-p)^{1-y}, \quad y = 0 \text{ or } 1,$$

where p is related to x through a suitable regression function and link function. The random variable Y has mean p and variance $p(1-p)$.

If n individuals are observed at the same value of x, the number of individuals Y with a success at this value of x has a *binomial* distribution

$$\Pr[Y = r] = \binom{n}{r} p^r(1-p)^{n-r}, \quad r = 0, 1, \dots, n$$

with mean np and variance $np(1-p)$.

We now consider in detail the use of different link functions for probability parameters.

4.2 Transformations and link functions

The relation between the response variable Y and d will be through a regression model for p in which $x = \log d$ is used as the explanatory variable in the linear predictor. How is p to be related to the linear predictor?

In Chapter 3 the regression model was for the mean μ of the normal distribution, although as illustrated in Section 3.7 other functions of the mean could be used. Since the mean of the Bernoulli distribution is p, this suggests a linear regression model

$$p = \beta_0 + \beta_1 x$$

for the probability of death. Such linear models for proportions are sometimes used, but they suffer from obvious difficulties. If $\beta_1 > 0$, then for x sufficiently large, p will exceed 1, while for x sufficiently small, p may be negative. For the mice data, it is clear that a linear model cannot be appropriate since at the extreme dose levels the response observed proportions are already 0 and 1.

In practical data analysis these natural bounds for p are sometimes not exceeded, so that linear models may give sensible answers.

Transformations of the success probability are the standard solution to the problem of the finite range for p. Let $H(\theta)$ be a strictly increasing function of θ, where $-\infty < \theta < \infty$, such that $H(-\infty) = 0, H(\infty) = 1$. This condition is satisfied if H is a cumulative distribution function (*cdf*) for any continuous random variable defined on $(-\infty, \infty)$. Noting that for each value of p there is a value of θ with $p = H(\theta)$, we can define a transformation of p from $(0, 1)$ to $(-\infty, \infty)$ by

$$\theta = H^{-1}(p) = g(p).$$

A linear model for this transformed parameter

$$\theta = \beta_0 + \beta_1 x$$

now has the property that any finite value of the linear predictor will give a value of p in the range $(0, 1)$.

Three such transformations are provided as standard options in GLIM, where they are defined by the link function $g(p)$, introduced in Section 2.9.2. They are the *logit* transformation or link

$$\theta = \operatorname{logit} p = \log\{p/(1 - p)\}, \quad H(\theta) = \{e^\theta/(1 + e^\theta)\},$$

for which $H(\theta)$ is the *cdf* for the logistic distribution, the *probit* transformation or link

$$\theta = \operatorname{probit} p = \Phi^{-1}(p), \quad H(\theta) = \Phi(\theta),$$

for which $H(\theta)$ is the *cdf* for the standard normal distribution, and the *complementary log–log* (CLL) transformation or link

$$\theta = \log\{-\log(1-p)\}, \quad H(\theta) = 1 - \exp(-e^{\theta}),$$

for which $H(\theta)$ is the *cdf* for the extreme value distribution (see Chapter 6).

The probit and logit links are similar, and can be made almost identical by a scale change to equate the variances of the underlying normal (1) and logistic ($\pi^2/3$) distributions. They generally give very similar fitted models, and are symmetric in p and $1-p$. The complementary log–log link is not symmetric, but for small p it is very close to the logit link, since in this case $\log(1-p) \approx -p$, so that

$$\log(-\log(1-p)) \approx \log p, \quad \log\{p/(1-p)\} \approx \log p + p.$$

A fourth link function, the log–log link, can also be used though it is not given as a separate link option in GLIM. It is defined by

$$\theta = -\log(-\log p), \quad H(\theta) = \exp(-e^{-\theta}),$$

for which $H(\theta)$ is the *cdf* for the reversed extreme value distribution (Section 6.13): if z has the extreme value distribution with *cdf*

$$H(z) = 1 - \exp(-e^{z}),$$

then $w = -z$ has the reversed extreme value distribution with

$$H(w) = \exp(-e^{-w}).$$

For p near 1, the log–log link is very close to the logit link.

As $p \to 0$ or 1, all these link functions approach $\pm\infty$. The logistic transformation is also called the *log odds*, $p/(1-p)$ being the *odds in favour* of $y = 1$.

To fit models to binary data in GLIM, we have to specify that the probability distribution is binomial ($error b): the Bernoulli distribution is not provided as a separate distribution, since it is the special case of the binomial when the number of trials n is equal to 1. Since the binomial distribution depends on both n and p, the binomial denominator, that is the number of trials n_i on which the i-th observed number of successes r_i is based, has to be specified as a second argument to the $error directive, using a variate name following the b argument. The yvar is specified as the number of successes r_i. In the Bernoulli case, we have only one observation at each value of x_i and the denominator must be calculated explicitly to be 1. The link function is specified to be either logit (g), probit (p) or complementary log–log (c), and then the model is specified in the $fit directive.

We analyse the dose–response data using various link functions. For each link function we compute the fitted probabilities from the model over a fine grid of log dose values covering the observed range using $predict to obtain a smooth fitted model.

```
$yvar r $error b n $link i $fit : + x $disp e $
scaled deviance =  15.791 at cycle 3
    residual df =   3


** fitted mean out of range for error distribution, at
[it : +x $] on level 2  from channel 1

At cycle 1, the fitted mean for unit 1 is -0.1058, which is
less than or equal to zero when the error is B.   This
indicates an inappropriate model.
```

An error message is produced—the linear model gives an out-of-range fitted probability at the lowest dose level. The null model deviance is 15.79, which does not depend on the choice of link function. We switch to the links which give in-range probabilities.

```
$link g $fit x $disp e $
scaled deviance =  0.046694 at cycle 6
    residual df =  2


          estimate        s.e.       parameter
    1       -53.90        34.34        1
    2        7.930         5.081       X
scale parameter 1.000
$calc op = %yv/%bd $
$assign gridx = 6,6.01,...,7.65 $
$predict x - gridx $
$calc fpg = %pfv $
$link p $fit x $disp e $
-- model changed
scaled deviance =  0.0037626 at cycle 7
    residual df =  2


          estimate        s.e.       parameter
    1       -31.13        22.02        1
    2        4.579         3.264       X
scale parameter 1.000
$predict x = gridx $
$calc fpp = %pfv $
$link c $fit x $disp e $
-- model changed
```

```
scaled deviance =   0.066666 at cycle 7
    residual df =   2

          estimate          s.e.      parameter
     1      -43.45          28.68        1
     2        6.330          4.223       X
scale parameter 1.000
$predict x = gridx $
$calc fpc = %pfv $
$graph (s = 1 h = 'log dose' v = 'p')
       op,fpg,fpp,fpc
       x,gridx,gridx,gridx
       5,10,11,12 $
```

The three link functions all fit very closely, with "scaled" deviances of almost zero—the differences among them are negligible. This is essentially because the parameters are determined by the two intermediate observations, as the extreme 0 and 1 proportions do not clearly locate the model parameters.

The GLIM scaled deviance is the likelihood ratio test statistic for the model relative to the "saturated" model with a parameter for each dose level. In other words, the scaled deviance from the linear model is the likelihood ratio test statistic (LRTS) for *linearity* of the log dose–response relationship, relative to a *cubic* log dose model, the "saturated" polynomial model which exactly interpolates the data. (It is also the LRTS for the log dose model relative to a four-group model with arbitrary death probabilities at each dose level, but this is not of interest in the experimental context.)

The fitted probabilities closely interpolate the observed proportions (Fig. 4.2), as is evident from the very small deviances. But the graph shows a discrepancy between the complementary log–log (CLL) link and the others, at the upper end of the dose range. Since there are no data in this region, the fit of the CLL link is as close as the others. We work with the logit link.

The standard error of the slope coefficient is quite large, with a Wald test statistic of $z^2 = (7.93/5.08)^2 = 1.56^2 = 2.44$, so it appears that the regression of death probability on log dose is not significant. However, the LRTS for the same hypothesis is the deviance difference 15.74, which is very large compared to χ_1^2. In this case the Wald test is not a reliable tool for assessing the importance of explanatory variables—the likelihood ratio test shows clearly that dose is important. We discuss this important point further below.

4.2.1   *Profile likelihoods for functions of parameters*

In non-linear models with small samples, the likelihood function is often far from normality, and we cannot therefore rely on large-sample normality of parameter estimates, particularly for non-linear functions of the parameters.

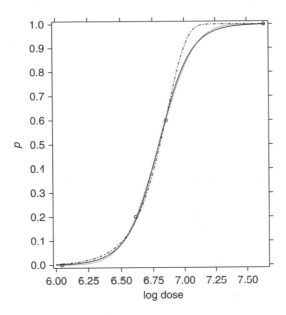

Fig. 4.2. Observed and fitted death probabilities

Profile likelihoods (Section 2.7.1) for the parameters allow for the non-linearity and provide correctly-located regions and intervals for the parameters, though we are still dependent on asymptotic theory for the confidence coverage of these regions or intervals. We consider here several such functions for the very small drug example above.

(i) The slope parameter
This is easily computed using the macro logodds. The macro fixes the value of β_1 over a grid, with minimum grid value set by %a, grid interval by %i, and maximum value by %z, and maximizes the likelihood for the null model with β_0, with $\beta_1 x$ held as an offset. The profile likelihood is held in the vector lik_ and the values of β_1 in the vector lth_. The profile deviance is also held in the vector dv_; these can be printed to obtain an asymptotic confidence interval for β_1.

```
$input 'logodds' $
$link g $
$calc %a = 0 : %i = 0.25 : %z = 25 $
$use logodds x $
```

The profile likelihood (Fig. 4.3) is substantially skewed. The asymptotic 95% confidence interval for β_1 can be found by linear interpolation in the deviance for

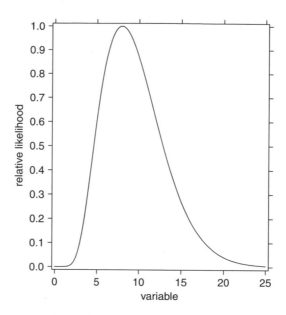

Fig. 4.3. Profile likelihood for slope

the value 3.84.

```
$look lik_ dv_ lth_ $

             LIK_             DV_            LTH_
  ...
      10    0.0205872357    3.8830841       2.2500
      11    0.0379245020    3.2721579       2.5000
  ...
      86    0.0238225702    3.7371218      21.2500
      87    0.0213560686    3.8464193      21.5000
  ...
```

We find the interval (2.27, 21.49), far away from zero. The '*t*' interval based on $\hat{\beta}_1 \pm 1.96 \cdot \mathrm{SE}(\hat{\beta}_1)$ is $(-2.03, 17.89)$, misleadingly including zero.

(ii) The death probability at a given dose x_0
This is easily computed using the macro rprob. We define the parameter of interest to be

$$\theta = \beta_0 + \beta_1 x_0$$

and express β_0 in terms of θ and β_1 in the model:

$$\text{logit}\, p = \theta + \beta_1(x - x_0).$$

The macro fits the model over a grid of θ with the explanatory variable x centred by x_0, an offset of θ and no intercept, and finally converts θ to p by the anti-logit transformation

$$p = \frac{e^{\theta}}{1 + e^{\theta}}.$$

We illustrate with the dose $d = 1100$, corresponding to log dose $x = 7.00$. The grid interval parameters are set as for logodds, though they now refer to the probability scale rather than the β_1 scale. The centring value is set in the scalar %x. The values of p are held in the vector p_.

```
$input 'rprob' $
$calc %a = 0.1 : %i = 0.01 : %z = 0.99 : %x = 7 $
$use rprob x $
```

The profile likelihood is very severely skewed, not surprisingly since the MLE of the death probability at log dose 7 is 0.83 (Fig. 4.4).

```
$look lik_ dv_ p_ $
```

	LIK_	DV_	P_
...			
30	0.1337903	4.0229630	0.3900
31	0.1487319	3.8112197	0.4000
...			
89	0.3842184	1.9130884	0.9800
90	0.2163246	3.0619502	0.9900

The 95% asymptotic confidence interval for p at dose 1100 is, to two dp, (0.40, 1.00).

(iii) The dose needed for a given response probability p_0
An important use of this model in drug trials is to establish the dose at which a desired response occurs. Two common doses of interest are the *medial lethal dose* or *median effective dose*, the LD50 or ED50, and the *90% effective dose*, the ED90. These dose levels are non-linear functions of the model parameters. For the given probability p_0, write x_0 for the corresponding log dose. Then

$$\theta_0 = \text{logit}\, p_0 = \beta_0 + \beta_1 x_0$$

and

$$\text{logit}\, p = \theta_0 + \beta_1(x - x_0).$$

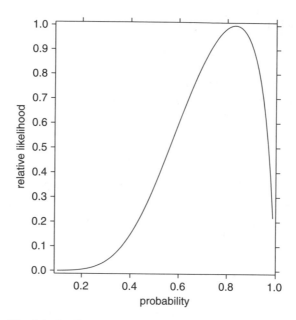

Fig. 4.4. Profile likelihood for death probability at $x = 7$

This is formally similar to the previous case, but now x_0 is the parameter of interest and θ is the nuisance parameter. The macro fits the model over a grid of x_0 with the explanatory variable centred by x_0, a fixed offset of θ_0 and no intercept.

We illustrate with both the ED50 and the ED90. These are computed with the macro ed. The grid interval parameters are set as for logodds, though they now refer to the log dose scale rather than the β_1 scale. The probability value is set in the scalar %p. The explanatory variable (log dose) values are held in the vector ld_.

```
$input 'ed' $
$calc %a = 6.35: %i = 0.01 : %z = 7.5: %p = 0.5 $
$use ed x $

$look lik_ dv_ ld_ $
           LIK_          DV_          LD_
...
    24    0.132210    4.0467238     6.580
    25    0.150183    3.7917955     6.590
...
    77    0.147616    3.8262885     7.110
    78    0.136189    3.9874215     7.120
```

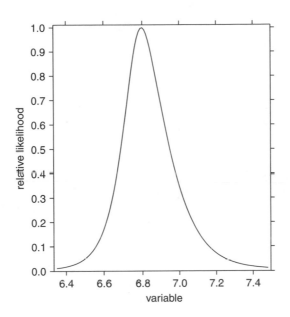

Fig. 4.5. Profile likelihood for ED50

```
$calc %a = 6.7 : %i = 0.02 : %z = 9.5: %p = 0.9 $
$use ed x $

$look lik_ dv_ ld_ $
          LIK_          DV_          LD_
...
    7   0.0869792    4.8841715    6.820
    8   0.1531634    3.7525003    6.840
...
   62   0.1503589    3.7894609    7.920
   63   0.1427163    3.8937924    7.940
```

The profile likelihood for the ED50 (Fig. 4.5) is nearly symmetrical, while for the ED90 (Fig. 4.6) is severely skewed. Asymptotic 95% confidence intervals for the log ED50 and log ED90 are (6.588, 7.111) and (6.828, 7.930), respectively; the corresponding dose intervals are (726, 1225) and (923, 2779).

4.3   Model criticism

In Section 3.4 we discussed model criticism for the normal distribution model. For discrete data the possibilities are more limited, but we follow the same sequence.

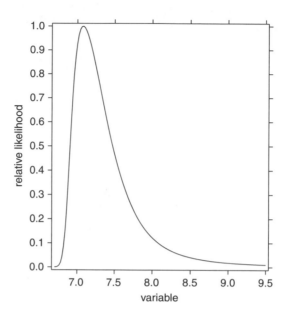

Fig. 4.6. Profile likelihood for ED90

4.3.1 *Mis-specification of the probability distribution*

The $disp r $ directive in GLIM when used with binomial data prints the Pearson residuals

$$e_i = (y_i - n_i\hat{p}_i)/[n_i\hat{p}_i(1 - \hat{p}_i)]^{1/2}.$$

These are approximately standardized variables with mean 0 and variance approximately 1; the variance is approximate because of the estimation of β in the linear predictor. They are not, however, even approximately normally distributed unless the binomial sample sizes n_i are large, so a normal quantile plot is not appropriate and may be misleading. Large values of the Pearson (or other) residuals indicate failure of the model to fit at the corresponding points. The sum of squares of the Pearson residuals is the Pearson goodness-of-fit X^2, which is available in the system scalar %X2.

"Deviance residuals" have been defined by McCullagh and Nelder (1989); these are the signed square roots of the individual components of the deviance: their sum of squares is the deviance. Residual analysis in this chapter will be based on the Pearson residuals.

Mis-specification of the probability distribution resulting in *overdispersion* may arise from the omission of important variables in the model. This important issue is discussed at length in Chapter 8.

4.3.2   *Mis-specification of the link function*

GLIM implements three standard links for the binomial distribution, but other more general families of links have been proposed and used for this distribution, by generalizing the log transformation of the odds to more general transformation families.

Stukel (1988) gave an extensive discussion of several families. We do not discuss these as we do not make use of them in this book.

4.3.3   *The occurrence of aberrant and influential observations*

With binomial data in which aggregation of individual responses has occurred across common explanatory variable values, with all $n_i > 1$, aberrant observations tend to be damped by the aggregation, whether they occur in the response or in the explanatory variables. With individual Bernoulli observations, however, the effect of single aberrant observations can be marked. We discuss this and other issues peculiar to binary data in Section 4.4.

The simple results of Chapter 3 on regression diagnostics can be generalized in various ways (Pregibon, 1981) at the expense of additional computing. We will use the diagonal elements of the hat matrix from the iteratively reweighted least squares algorithm. At the final (r-th) iteration GLIM computes the parameter estimates $\hat{\beta}$ using

$$X'W^{(r)}X\hat{\beta} = X'W^{(r)}\mathbf{z}^{(r)},$$

where $W^{(r)}$ is the diagonal matrix of iterative weights and $\mathbf{z}^{(r)}$ the adjusted dependent variate at the final iteration. The hat matrix is then

$$H = (W^{(r)})^{1/2}X(X'W^{(r)}X)^{-1}X'(W^{(r)})^{1/2}.$$

The diagonal elements of H are obtained in GLIM as the product of the variance of the linear predictor and the iterative weight variate, since

$$\mathrm{var}(X\hat{\beta}) = X\mathrm{var}(\hat{\beta})X'$$
$$= X(X'W^{(r)}X)^{-1}X'.$$

They are stored in the system vector %lv. We illustrate their use in Section 4.4.

4.4   Binary data with continuous covariates

We give an example from Finney (1947), see also Pregibon (1981). Table 4.2 gives the data obtained in a carefully controlled study of the effect of the rate rate and volume vol of air inspired by human subjects on the occurrence (coded 1) or non-occurrence (coded 0) of a transient vaso-constriction response resp in the skin of the fingers. These data are in the file vaso, in subject order. Three subjects were involved in the study: the first contributed 9 observations at different values

Table 4.2. Vaso-constriction response

S1			S2			S3					
vol	rate	r	vol	rate	r	vol	rate	r	vol	rate	r
3.7	0.825	1	0.9	0.45	0	0.85	1.415	1	1.9	0.95	1
3.5	1.09	1	0.8	0.57	0	1.7	1.06	0	1.6	0.4	0
1.25	2.5	1	0.55	2.75	0	1.8	1.8	1	2.7	0.75	1
0.75	1.5	1	0.6	3.0	0	0.4	2.0	0	2.35	0.3	0
0.8	3.2	1	1.4	2.33	1	0.95	1.36	0	1.1	1.83	0
0.7	3.5	1	0.75	3.75	1	1.35	1.35	0	1.1	2.2	1
0.6	0.75	0	2.3	1.64	1	1.5	1.36	0	1.2	2.0	1
1.1	1.7	0	3.2	1.6	1	1.6	1.78	1	0.8	3.33	1
0.9	0.75	0				0.6	1.5	0	0.95	1.9	0
						1.8	1.5	1	0.75	1.9	0
						0.95	1.9	0	1.3	1.625	1

of rate and vol, the second 8, and the third 22 observations. They are identified by S1, S2 and S3 in Table 4.2. The response is abbreviated to r in the table.

There was an error in the data in the first edition: the original value of rate of 0.03 for the 32nd observation has been corrected to 0.3. Figure 1 in Finney graphs the correct data but Finney's data Table 1 has the error. This observation had very high leverage, discussed below.

The experiment was designed to ensure as far as possible that successive observations obtained on each subject were independent, and the initial observations for each subject were discarded to allow for a training period in which they became familiar with the measurement procedure.

Correlation between successive observations on the same subject in such studies is always a possibility. We consider this issue in Chapter 9, but for the moment assume that the responses among and within individuals are independent, and that there are no differences in response level among individuals.

We fit the two-variable logit model.

```
$input 'vaso' $
$calc n = 1 $
$yvar resp $error b n $link g
$fit rate + vol $disp e $
 scaled deviance =  30.457 at cycle 5
     residual df =  36

          estimate        s.e.      parameter
      1     -9.187        3.101      1
      2      2.593        0.9051     RATE
      3      3.660        1.332      VOL
 scale parameter 1.000
```

How is the scaled deviance to be interpreted in this model? The saturated model, with a parameter for every observation, will fit the data exactly and have a deviance of zero, so if we were to use the large-sample theory of the LRT, we would treat the deviance as approximately χ^2_{36} if the fitted model is an adequate representation of the data. For binary data, however, large-sample theory does not apply to the distribution of the residual deviance from the fitted model, because the saturated model is equivalent to m separate models, one for each single observation: we are fitting m models to m samples of size 1.

For binary data, the residuals are

$$e_i = (y_i - \hat{p}_i)/[\hat{p}_i(1 - \hat{p}_i)]^{1/2}$$

and since y_i is equal to 0 or 1, the e_i are equal to a_i or $-1/a_i$ for each i, where

$$a_i = -\{\hat{p}_i/(1 - \hat{p}_i)\}^{1/2}.$$

Large residuals will still be evidence of model discrepancy from the data. We print them.

```
$print %rs $
  0.0389   0.0397   0.3924     3.583     0.3609     0.2937 -0.0802
 -0.6864  -0.1389  -0.0942    -0.0916   -0.9785    -1.483    0.3718
  0.1938   0.1751   0.0355     3.331    -0.8978     0.3554 -0.2813
 -0.3357  -0.6890  -0.9186     0.5260   -0.2121     0.5243 -0.6760
  0.8908  -0.3178   0.2670    -1.101    -0.8124     0.7619  0.8223
  0.3049  -0.6760  -0.4688     1.113
$print %x2 $
  37.78
```

There are two large residuals: 3.583 for the fourth observation and 3.331 for the 18th. These two observations are "successes" at points where the fitted probability of success is very low: 0.072 for observation 4 and 0.083 for observation 18. They contribute a total of 23.93 to the Pearson X^2 of 37.78 (in large samples with binomial data the deviance and the Pearson X^2 agree closely, but they may differ substantially in small samples or with binary data).

A graph shows these points clearly. We graph in Fig. 4.7 successes and failures using plotting characters + and o, respectively, corresponding to the levels of a factor equal to resp + 1:

```
$calc rf = resp + 1 $factor rf 2 $
$graph (s = 1 h = 'vol' v = 'rate' x = 0,4 y = 0,4)
        rate vol 5,6 rf $
```

The two "success" observations with large residuals appear on the bottom left hand side of the cluster of "success" observations.

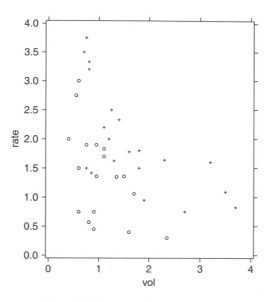

Fig. 4.7. Vaso-constriction response

What now? As with residual examination in normal regression models, we do not automatically remove these observations. They would be checked for correctness with the experimenter; if they are correct, then it is the model which needs modification, or else we accept the poor fit at these points as random variation.

One possible modification is the incorporation of a subject effect, to allow for differences in response level. The data come from three subjects who are not identified in the model. If subjects vary in their base level vaso-constriction response, then the fitted model is misspecified and a better fit may be obtained by the inclusion of the subject effect.

```
$factor s 3 $
$calc i = %cu(1) $
    : s = 1 + %gt(i,9) + %gt(i,17) $
$fit + s $disp e $
  scaled deviance =   26.845 (change =    -3.612) at cycle 5
       residual df =   34      (change =    -2    )

              estimate          s.e.        parameter
       1        -10.42          4.009        1
       2         3.800          1.512        RATE
       3         4.610          1.638        VOL
       4        -4.305          2.631        S(2)
       5        -1.896          1.783        S(3)
  scale parameter 1.000
```

```
$print %rs $
  0.0075   0.0072   0.0886    1.876    0.0661   0.0471  -0.0907
 -1.747   -0.1812  -0.0119   -0.0119  -0.4200  -0.7579   0.7455
  0.2246   0.3475   0.0471    4.518   -0.7998   0.2434  -0.2384
 -0.2511  -0.6193  -0.8919    0.4009  -0.1462   0.4304  -0.7005
  0.9718  -0.1812   0.2248   -0.8443  -0.8665   0.5714   0.6635
  0.1333  -0.7005  -0.4418    1.074
$pr %x2 $
  36.49
```

The deviance is now 26.85, a reduction of 3.61 for 2 df, and the Pearson X^2 is 36.49, a reduction of 1.29. There is no strong evidence of a subject effect. The residual at the fourth observation has decreased to 1.876, but that at the eighteenth has increased to 4.518. We cannot attribute the poor fit at these points to the absence of a subject effect, and we omit this term from subsequent analyses.

Could the form of the model be inappropriate? The analyses in Finney (1947) and Pregibon (1981) used log rate and log vol rather than rate and vol, though there were no strong scientific reasons given for choosing between them.

One data-based reason is the considerable skewness of both rate and vol, evident from Fig. 4.7. Such skewness is substantially reduced by a log transformation. Since these variables are explanatory, and not responses, this reason is not by itself compelling. Another reason is that the curves of constant logit, and therefore constant p, are straight lines cutting the rate and vol axes in our analysis above, but are hyperbolae which do not cut the axes if the log scales are used. We construct the straight lines—the level curves—for response probability values 0.05, 0.1, 0.3, 0.5, 0.7, 0.9 and 0.95, in Fig. 4.8 using the following GLIM code. The corresponding logit values are -2.94, -2.02, -0.85, 0, 0.85, 2.02 and 2.94.

```
$var 100 v r5 r10 r30 r50 r70 r90 r95 $
$assign v = 0.04,0.08,...,4 $
$calc   r5 = (9.19 - 2.94 - 3.66*v)/2.593 $
     : r10 = (9.19 - 2.20 - 3.66*v)/2.593 $
     : r30 = (9.19 - 0.85 - 3.66*v)/2.593 $
     : r50 = (9.19 + 0    - 3.66*v)/2.593 $
     : r70 = (9.19 + 0.85 - 3.66*v)/2.593 $
     : r90 = (9.19 + 2.20 - 3.66*v)/2.593 $
     : r95 = (9.19 + 2.94 - 3.66*v)/2.593 $
$graph (s = 1 h = 'vol' v = 'rate' y = 0,4)
        rate,r5,r10,r30,r50,r70,r90,r95
        vol,v,v,v,v,v,v,v
        5,6,11,13,12,10,12,13,11 rf,,,,,,, $
$gtext 0.6 1.4 '10' : 0.7 2.0 '50' : 0.8 2.5 '90'$
```

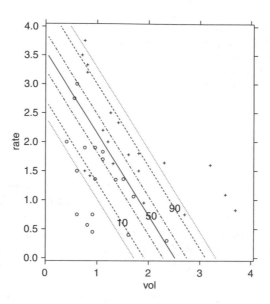

Fig. 4.8. Vaso-constriction response and linear model level curves

The concluding commas in the graph directive are dummies, as the success/fail factor is needed only for the observed response data. The three lines for 10%, 50% and 90% response rate are labelled accordingly.

Now we transform the variable scales.

```
$calc lv = %log(vol) : lr = %log(rate) $
$fit lv + lr $disp e$
 scaled deviance =  29.264 at cycle 5
      residual df =   36


             estimate        s.e.        parameter
       1       -2.924        1.284       1
       2        4.631        1.783       LR
       3        5.220        1.852       LV
   scale parameter 1.000
$print %rs $
 0.2214   0.1343   0.2887   3.575    0.5226   0.6017  -0.0314
-1.016   -0.0904  -0.0277  -0.0352  -0.5066  -0.7778   0.2528
 0.4283   0.1560   0.0698   2.951   -1.060    0.2385  -0.1055
-0.4132  -1.016   -1.361    0.3328  -0.1562   0.3638  -0.8963
 0.9096  -0.0947   0.6284  -0.1327  -1.205    0.5419   0.5385
 0.4765  -0.8963  -0.4836   0.7067
$print %x2 $
  36.49
```

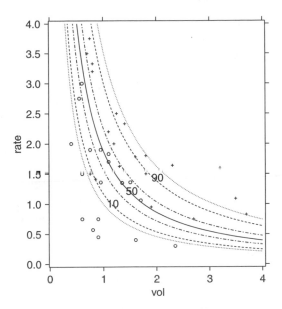

Fig. 4.9. Vaso-constriction response and log model level curves

The deviance is now 29.26, 1.20 less than for the first model. The Pearson X^2 is reduced from 37.78 to 36.49, and the residuals at the fourth and eighteenth observations are reduced to 3.575 and 2.951. The formal LRT cannot discriminate between models with logged and unlogged explanatory variables, as Finney (1947) noted, but as discussed in Chapter 2 the maximized LR between the models is a good indicator of relative support as the models have the same number of parameters. The LR of $\exp(1.20/2) = 1.82$ gives a very slight preference for the logged model. We graph the level curves in Fig. 4.9.

```
$var 100 llv lr5 lr10 lr30 lr50 lr70 lr90 lr95 $
$calc llv = %log(v) $
:  lr5  = (2.924 - 2.94 - 5.22*llv)/4.63 :  ulr5  = %exp(lr5)  $
:  lr10 = (2.924 - 2.20 - 5.22*llv)/4.63 :  ulr10 = %exp(lr10) $
:  lr30 = (2.924 - 0.85 - 5.22*llv)/4.63 :  ulr30 = %exp(lr30) $
:  lr50 = (2.924 - 0    - 5.22*llv)/4.63 :  ulr50 = %exp(lr50) $
:  lr70 = (2.924 + 0.85 - 5.22*llv)/4.63 :  ulr70 = %exp(lr70) $
:  lr90 = (2.924 + 2.20 - 5.22*llv)/4.63 :  ulr90 = %exp(lr90) $
:  lr95 = (2.924 + 2.94 - 5.22*llv)/4.63 :  ulr95 = %exp(lr95) $
$graph (s = 1 y = 0,4 h = 'vol' v = 'rate')
        rate,ulr5,ulr10,ulr30,ulr50,ulr70,ulr90,ulr95
        vol,v,v,v,v,v,v,v
        5,6,11,13,12,10,12,13,11 rf,,,,,,, $
$gtext 0.9 1.1 '10' : 1.2 1.4 '50' : 1.4 1.8 '90' $
```

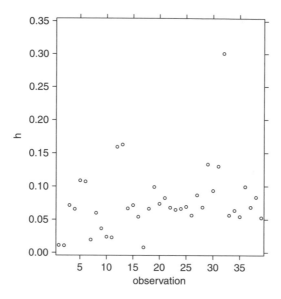

Fig. 4.10. Influence values, linear model

The level curves are quite different in the two models, but the models fit almost equally well, and agree on the two unusual responses.

We now examine the influence of observations in the two models. We noted earlier that the data for this example have been changed from the first edition: the value 0.03 of rate in Finney's (1947) table for the 32nd observation was incorrect, and has been changed to 0.3. This changes the influence of all the observations from that reported in the first edition. We first fit the linear model (Fig. 4.10).

```
$fit rate + vol $
 scaled deviance =  30.457 at cycle 5
     residual df =  36
$extract %lv $calc h = %lv : i = %ind(h) $
$print h $
0.0111  0.0106  0.0714  0.0658  0.1087  0.1071  0.0194
0.0599  0.0364  0.0238  0.0228  0.1593  0.1628  0.0668
0.0717  0.0543  0.0079  0.0663  0.0991  0.0738  0.0825
0.0680  0.0648  0.0663  0.0693  0.0563  0.0868  0.0684
0.1333  0.0934  0.1301  0.3004  0.0563  0.0636  0.0543
0.0993  0.0684  0.0835  0.0527
$graph (s = 1 h = 'observation' v = 'h' y = 0,0.35) h i 5 $
```

The value of $2(p + 1)/n$ is 0.154. The 12th and 13th values exceed this marginally, and the 32nd considerably: it is 0.300. This point has the smallest value of rate and a large vol. Omitting it has a noticeable effect on the fitted model:

```
$calc w1 = 1 : w1(32) = 0 $
$weight w1 $
 -- model changed
$fit . $disp e $
scaled deviance =  28.318 (change =    -2.139) at cycle 5
    residual df =  35       (change =    -1    ) from 38 observations

            estimate        s.e.      parameter
     1       -9.567        2.267      1
     2        2.468        0.8912     RATE
     3        4.279        1.576      VOL
  scale parameter 1.000
```

The deviance changes by 2.14 and the vol parameter estimate changes by half a standard error, that for rate changing much less (Fig. 4.10).

Omission of the two observations with large residuals, on the other hand, has a dramatic effect.

```
$calc w2 = 1 : w2(4) = w2(18) = 0 $
$weight w2 $
 -- change to data values affects model
$fit . $disp e $
 scaled deviance =  13.223 (change =    -15.09) at cycle 7
     residual df =  34       (change =    -1    ) from 37 observations

            estimate        s.e.      parameter
     1       -29.97        13.13      1
     2        7.881        3.531      RATE
     3       12.00         5.307      VOL
  scale parameter 1.000
```

The deviance changes by more than 17 relative to the full model, and the parameter estimates have become much larger. The large estimates for the parameters show that a rapid change from 0 to 1 in the response probability occurs over a small interval of each variable: the successes and failures can be almost perfectly discriminated from the model. We illustrate in Fig. 4.11 with the level curves from this model.

```
$calc   r5 = (29.97 - 2.94 - 12*v)/7.881 $
     : r10 = (29.97 - 2.20 - 12*v)/7.881 $
     : r30 = (29.97 - 0.85 - 12*v)/7.881 $
     : r50 = (29.97 - 0    - 12*v)/7.881 $
     : r70 = (29.97 + 0.85 - 12*v)/7.881 $
     : r90 = (29.97 + 2.94 - 12*v)/7.881 $
```

```
$graph (s = 1 h = 'vol' v = 'rate' y = 0,4)
        rate/w,r5,r10,r30,r50,r70,r90,r95
        vol,v,v,v,v,v,v,v
        5,6,11,13,12,10,12,13,11 rf,,,,,,, $
```

We repeat the analysis for the logged model.

```
$weight $
$fit lr + lv $disp e $
 scaled deviance =  29.264 at cycle 5
     residual df =  36
```

```
              estimate          s.e.        parameter
        1       -2.924          1.284          1
        2        5.220          1.852          LV
        3        4.631          1.783          LR
 scale parameter 1.000
$extract %lv $calc h = %lv : i = %ind(h) $
$print h $
0.0932   0.0426   0.0585   0.0820   0.1107   0.1461   0.0067
0.0554   0.0301   0.0061   0.0083   0.1487   0.1632   0.0528
0.1253   0.0393   0.0165   0.0897   0.1251   0.0515   0.0355
0.0952   0.0732   0.0706   0.0580   0.0513   0.0661   0.0646
0.1601   0.0430   0.2403   0.0873   0.0512   0.0593   0.0548
0.1116   0.0646   0.0982   0.0530
```

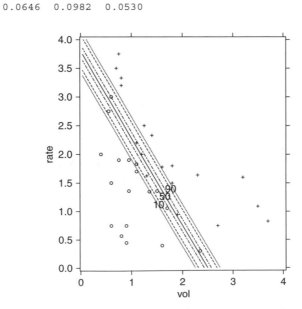

Fig. 4.11. Vaso-constriction response and linear model level curves

The 13th values and 29th values now exceed $2(p + 1)/n$ marginally, and the 31st more substantially: it is 0.24. The 32nd observation does not have high leverage because of the log transformations.

```
$weight w1 $
-- model changed
$fit . $disp e $
 scaled deviance =  29.227 (change =   -0.03660) at cycle 5
 residual df =  35       (change =   -1  ) from 38 observations

            estimate        s.e.      parameter
      1       -2.875        1.318      1
      2        5.179        1.860      LV
      3        4.562        1.834      LR
scale parameter 1.000
```

Excluding the 31st observation gives a small deviance change and small changes in the parameter estimates, by less than 4% of a standard error.

```
$weight w2 $
-- model changed
$fit . $disp e $
 scaled deviance =   7.3610 (change =    -21.87) at cycle 9
 residual df =  34        (change =     -1 ) from 37 observations

            estimate        s.e.      parameter
      1       -24.58       13.93       1
      2        39.55       23.09       LV
      3        31.93       17.64       LR
scale parameter 1.000
```

Excluding observations 4 and 18 again has a dramatic effect, with a deviance change of 21.9 and very large parameter estimates for log rate and log vol, respectively. We graph the level curves for this model in Fig. 4.12.

```
$calc lr5 = (24.58 - 2.94 - 39.55*lv)/31.93 : ulr5 = %exp(lr5)$
: lr10 = (24.58 - 2.20 - 39.55*lv)/31.93 : ulr10 = %exp(lr10) $
: lr30 = (24.58 - 0.85 - 39.55*lv)/31.93 : ulr30 = %exp(lr30) $
: lr50 = (24.58 - 0    - 39.55*lv)/31.93 : ulr50 = %exp(lr50) $
: lr70 = (24.58 + 0.85 - 39.55*lv)/31.93 : ulr70 = %exp(lr70) $
: lr90 = (24.58 + 2.20 - 39.55*lv)/31.93 : ulr90 = %exp(lr90) $
: lr95 = (24.58 + 2.94 - 39.55*lv)/31.93 : ulr95 = %exp(lr95) $
$graph (s = 1 h = 'vol' v = 'rate' y = 0,4)
        rate,ulr5,ulr10,ulr30,ulr50,ulr70,ulr90,ulr95
        vol,v,v,v,v,v,v,v
        5,6,11,13,12,10,12,13,11 rf,,,,,,,, $
```

The simpler model

$$\frac{\hat{p}}{1 - \hat{p}} = \exp(\hat{\theta}) = 0.05(\texttt{vol*rate})^5$$

can therefore be fitted by extending the offset, and fitting the model with no
estimated parameters:

```
$calc ofs = %log(0.05) + ofs $
-- change to data values affects model
$fit - 1 $
 scaled deviance =  29.589 (change =  +0.05147) at cycle 2
     residual df =  39     (change =  +1      )
```

The df is now 39, equal to the number of observations.
   The fitted model and observed data are now easily graphed in one dimension.
First we sort the fitted probabilities into increasing order with lvr to give a smooth
graph (Fig. 4.13).

```
$sort sresp,sfv,slvr,srf resp,%fv,lvr,rf lvr $
$factor srf 2 $
$graph (s = 1 h = 'lvr' v = 'p' x = -1,2) sresp,sfv slvr,slvr
        5,6,10 srf, $
```

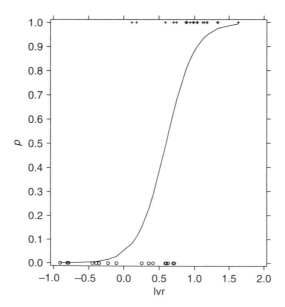

Fig. 4.13. Vaso-constriction response and fitted lvr model

The two observations 4 and 18 with large residuals are those at the top of the graph at the left-hand end of the row of '+'s. It can be seen that when they are removed from the fit, the slope of the regression will be much steeper as there is very little overlap in lvr between the 0 and 1 responses.

Note that the value of 5 for the power of vol*rate is chosen only for simplicity. No strong substantive interpretation should be drawn from it, in the absence of a physical model for an integer power.

4.5   Contingency table construction from binary data

In larger samples with binary data and continuous explanatory variables it is frequently possible to construct a contingency table by grouping the explanatory variables into ordered categories. This allows more detailed model criticism than with the original binary data, at the expense of some loss of precision in parameter estimation. We illustrate with an example given by Wrigley (1976) of data from a 1974 study of chronic bronchitis in Cardiff conducted by Jones (1975). The data consist of observations on three variables for each of 212 men in a sample of Cardiff enumeration districts. The data are too extensive to be listed here; they are in the file bronchitis (Fig. 4.14).

The variables are cig, the number of cigarettes smoked per day, poll, the smoke level in the locality of the respondent's home (obtained by interpolation

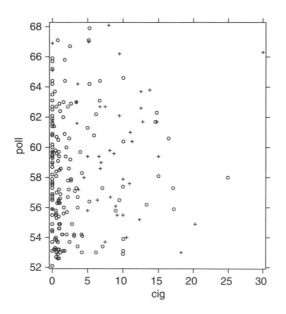

Fig. 4.14. Bronchitis response

from 13 air pollution monitoring stations in the city), and r, an indicator variable taking the value 1 if the respondent suffered from chronic bronchitis, and 0 if he did not. The presence or absence of chronic bronchitis was determined using a special questionnaire devised by the Medical Research Council; detection of chronic bronchitis using this questionnaire had been found to be almost completely consistent with the clinical diagnosis. (The text in Wrigley refers to cig as the total consumption of cigarettes ever smoked in hundreds, but the published data are either total consumption expressed in units of ten thousand, or are current rates of smoking in cigarettes per day. We have assumed the latter.)

Our aim is to develop and fit a statistical model relating r to cig and poll. We begin by graphing the data using + and o as plotting symbols as in the previous example.

```
$input 'bronchitis' $
$calc %s = %cu(r) $print %s $
 46.00
$calc rf = r + 1 $factor rf 2 $
$graph (s = 1 h = 'cig' v = 'poll') poll cig 5,6 rf $
```

There are 46 men with bronchitis, a sample proportion of 0.217. Seven of these are non-smokers; the remainder are more common at higher levels of cigarettes and pollution. Two very heavy smokers have different bronchitis outcomes.

We proceed to fit the logit model. The binomial denominator $n = 1$ is already defined in the file.

```
$yvar r $error b n $link g
$fit cig + poll $disp e $
 scaled deviance =  174.21 at cycle 4
     residual df =  209

            estimate         s.e.      parameter
       1      -10.08         2.950       1
       2       0.2117        0.03812     CIG
       3       0.1318        0.04894     POLL
 scale parameter 1.000
$print %x2 $
    192.6
```

Level curves for the fitted probability of bronchitis are shown in Fig. 4.15.

The deviance of 174.21 with 209 df does not necessarily indicate a good fit. The Pearson X^2 is 192.6. To display the residuals would take many screens of output; instead we display just the large residuals (greater than 2 in absolute value) using the w option for $disp, and setting %re = 1 for the large residuals and zero

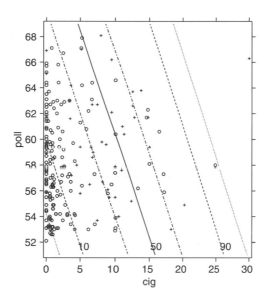

Fig. 4.15. Bronchitis response and level curves

for the rest:

```
$calc e2 = %rs**2 : %re = %gt(e2,4) $
$disp w $
     unit    observed    out of    fitted    residual
        7           1         1     0.181       2.125
       48           1         1     0.125       2.641
       59           1         1     0.124       2.658
       87           1         1     0.098       3.033
      122           1         1     0.062       3.895
      132           1         1     0.184       2.107
      142           1         1     0.089       3.197
      147           0         1     0.944      -4.113
      149           1         1     0.146       2.417
      171           1         1     0.194       2.036
      195           1         1     0.158       2.310
```

There are 11 residuals greater than 2 in absolute value. Where in the variable space are these observations?

```
$graph (s = 1 h = 'cig' v = 'poll') poll cig 5,6 rf $
```

Of the 10 positive residuals, 6 occur for bronchitis sufferers at `cig = 0` and the others for low values of `cig` and `poll`; the negative residual occurs for

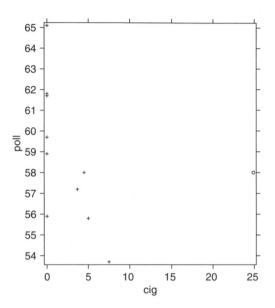

Fig. 4.16. Bronchitis response

a non-occurrence for the heavy smoker, observation 147 with `cig` = 24.9 and `poll` = 58. The model fits badly for non- and light smokers (Fig. 4.16).

How should we amend the model? The 'additive' model assumes that the regression of logit p on `poll` is the same for smokers and non-smokers: is this reasonable? We might hypothesize that the irritating effects of air pollution might be greatly increased by heavy smoking, producing an interaction. Non-smokers might have a response to `poll` quite different from smokers.

This possibility is investigated by fitting models for smokers and non-smokers separately:

```
$calc wns = %eq(cig,0) $weight wns $
-- model changed
$fit poll $disp e $
scaled deviance =   39.382 at cycle 4
    residual df =   52      from 54 observations

          estimate          s.e.        parameter
    1       -12.83          7.633        1
    2        0.1809         0.1247       POLL
scale parameter 1.000
```

The `poll` slope is not very different from that for the two groups combined.

```
$calc ws = 1 - wns $weight ws $
-- model changed
```

```
$fit cig + poll $disp e $
 scaled deviance =  134.00 at cycle 4
     residual df =  155     from 158 observations

            estimate        s.e.      parameter
     1       -9.399         3.234      1
     2        0.2288        0.04522    CIG
     3        0.1172        0.05405    POLL
 scale parameter 1.000
```

The total deviance for the two models is 173.38: compared with the deviance of 174.21 for the original model, almost nothing has been gained by a separate regression on poll for non-smokers. The poor fit for non-smokers is thus not due to a simple interaction.

We now extend the model to a second-degree response surface:

```
$calc p2 = poll**2 : cp = cig*poll : c2 = cig**2 $
$weight $fit + p2 + cp + c2 $disp e $
 -- model changed
scaled deviance =  163.72 (change =   +29.71) at cycle 4
     residual df =  206     (change =   +51    )

            estimate        s.e.      parameter
     1       -36.96         43.42      1
     2        0.4620        0.4969     CIG
     3        1.006         1.446      POLL
     4       -0.007215      0.01203    P2
     5       -0.0003891     0.008617   CP
     6       -0.01286       0.004003   C2
 scale parameter 1.000
```

The deviance relative to the full linear model decreases by 10.49 on 3 df, almost all of which is due to the c2 term.

```
$fit - cp - p2 $disp e$
 scaled deviance =  164.08 (change =   +0.3626) at cycle 4
     residual df =  208     (change =   +2       )

            estimate        s.e.      parameter
     1       -11.04         3.059      1
     2        0.4410        0.08292    CIG
     3        0.1392        0.05034    POLL
     4       -0.01303       0.003763   C2
 scale parameter 1.000
```

Reduction of the model by omitting cp and p2 gives a deviance increase of only 0.36. The rate of increase of bronchitis logit falls off as cig increases, and becomes zero at cig = 0.4410/2(0.0130) = 17.0. This is a surprising result. We again examine the residuals.

```
$calc e2 = %rs**2 : %re = %gt(e2,4) $
$disp w $
     unit    observed    out of    fitted    residual
        2           1         1     0.152       2.366
        7           1         1     0.122       2.682
       48           1         1     0.081       3.375
       59           1         1     0.080       3.398
       87           1         1     0.062       3.906
      122           1         1     0.037       5.088
      142           1         1     0.055       4.129
      149           1         1     0.165       2.247
      195           1         1     0.200       2.002
```

Nine of the residuals remain large, and all are for bronchitis sufferers who are non-smokers or low on smoking and/or pollution. The response surface extension has not corrected the poor fit for non-smokers.

This complexity makes us suspect some systematic failure of the model to represent the data. To investigate the appropriateness of the model, we group both cig and poll into classes and construct a three-way contingency table (actually a two-way cross-classification of r and n). The number of bronchitis sufferers in each cell of the table is then modelled by a binomial distribution.

```
$calc fcig = 1 + %gt(cig,0) + %gt(cig,1) + %gt(cig,3) +
                 %gt(cig,5) + %gt(cig,8) $
    : fpol = 1 + %gt(poll,55) + %gt(poll,57.5) + %gt(poll,60) +
                 %gt(poll,62.5) + %gt(poll,65) $
```

Alternatively, fcig and fpol can be constructed using the $assign and $group directives.

The variables fcig and fpol take the values 1 to 6: we have defined 6 categories for each of the two continuous variables cig and poll. Non-smokers are the first category of fcig. The category boundaries for cig are chosen by eye inspection to give roughly equal numbers in each marginal category; alternatively, $hist may be used to estimate approximate percentiles. The category boundaries for poll are equally spaced with steps of 2.5.

We form two new variates tr and tn to contain the grouped numbers of successes and sample sizes for the 36 cells.

```
$tab the r total for fcig,fpol into tr $
 -- the table contains empty cell(s)
```

```
    : the n total for fcig,fpol into tn $
 -- the table contains empty cell(s)
$print tr :: tn $
     0.          1.000    2.000    2.000    0.000    2.000
     0.          0.       0.       0.       0.       0.
     0.          0.       0.       0.       0.       0.
     0.          2.000    2.000    1.000    2.000    0.
     1.000       1.000    3.000    0.       2.000    2.000
     3.000       5.000    5.000    5.000    3.000    2.000


     6.000      11.00    14.00    11.00    7.000    5.000
    17.00       11.00     8.000    3.000    3.000    1.000
     9.000       8.000    8.000    4.000    4.000    2.000
     6.000       6.000    4.000    2.000    3.000    0.
     3.000       3.000    3.000    2.000    5.000    4.000
     6.000      11.00     7.000    9.000    4.000    2.000
```

The tabulation directive uses the convention that the last factor named varies most rapidly and the first least. Factor levels for cigarettes and pollution for the table variates tr and tn can now be defined:

```
$var 36 tcig tpol $
$calc tcig = %gl(6,6) : tpol = %gl (6,1) $
```

The variates tcig and tpol each take values 1 to 6, but are of length 36. Cell sample sizes are spread fairly evenly from 2 to 11, with two larger values (14 and 17). One cell has no observations.

```
$unit 36$
 -- standard length re-initialised
$factor tcig 6 tpol 6 $
$yvar tr $error b tn $fit tcig + tpol $
 -- number of units changed
 -- binomial denominator unset
 -- model changed
scaled deviance =   14.878 at cycle 10
    residual df =   24      from 35 observations

$disp e $
          estimate          s.e.        parameter
     1      -3.236         0.7533        1
     2     -10.61         57.73          TCIG(2)
     3     -10.58         56.14          TCIG(3)
     4       1.687         0.6680        TCIG(4)
     5       1.761         0.6480        TCIG(5)
     6       2.662         0.5734        TCIG(6)
```

```
      7          0.6253         0.7300         TPOL(2)
      8          1.781          0.7560         TPOL(3)
      9          0.9983         0.7766         TPOL(4)
     10          1.251          0.8010         TPOL(5)
     11          2.353          0.9525         TPOL(6)
scale parameter 1.000


$print %rs $
 -0.4857   0.2948  -0.4437   0.9595  -0.9809   0.5280  -0.0041
 -0.0045  -0.0068  -0.0028  -0.0032  -0.0032  -0.0030  -0.0039
 -0.0069  -0.0033  -0.0037  -0.0046  -1.129    0.2662  -0.2334
  0.3940   0.8419   0.        0.6545   0.1276   1.486   -1.114
 -0.1999  -0.9068   0.7122  -0.3875  -0.3497  -0.3010   0.3671
  0.5809
```

The deviance of 14.88 on 24 df shows a good fit for binomial data, and the largest residual is 1.486 for the 27th cell.

```
$fit - tpol $fit + tpol - tcig $
scaled deviance = 24.966 (change = +10.09) at cycle 9
 residual df   = 29        (change =   +5  ) from 35 observations


scaled deviance = 87.127 (change = +62.16) at cycle 4
 residual df   = 29        (change =    0  ) from 35 observations
```

Both main effects are necessary. Cigarette consumption is much more important than smoke level.

Inspection of the parameter estimates for the main effects model shows that the estimates for tpol generally increase with level, though the value for level 3 is considerably higher than for levels 2 or 4. This suggests a linear trend in logit with increasing tpol. We construct a new variate linp which takes the values of the factor tpol.

```
$calc linp = tpol $
$fit tcig + linp $disp e $
 scaled deviance =   19.473 at cycle 10
     residual df =   28       from 35 observations


            estimate        s.e.        parameter
       1     -3.048        0.6737       1
       2    -10.71        59.58         TCIG(2)
       3    -10.65        58.23         TCIG(3)
       4      1.516        0.6456       TCIG(4)
       5      1.625        0.6213       TCIG(5)
```

```
    6        2.468      0.5447       TCIG(6)
    7        0.3210     0.1403       LINP
 scale parameter 1.000
$print %rs $
-0.6265   0.0988   0.3855   0.3337  -1.286    0.8021  -0.0050
-0.0047  -0.0047  -0.0034  -0.0040  -0.0027  -0.0037  -0.0041
-0.0049  -0.0040  -0.0047  -0.0039  -1.337    0.2279   0.5770
 0.1760   0.5148   0.       0.3364   0.0720    2.180   -1.319
-0.6524  -0.5079   0.3178  -0.4048   0.6448  -0.7239   0.0637
 0.7215
```

The deviance change is 4.59 on 4 df, and the largest residual is now 2.180 for the 27th cell. Inspection of the parameter estimates for tcig shows a very surprising pattern: for levels 2 and 3 the estimates are -10.7, corresponding to a fitted probability of almost zero, and we see from the tabled values of Lr that there are no bronchitis sufferers at any air pollution level among smokers of not more than 3 cigarettes per day, though there are sufferers among non smokers! It is for this reason that the response surface model failed to fit adequately: it was trying to reproduce both a zero proportion of bronchitis sufferers amongst light smokers and a non-zero proportion amongst non-smokers, as well as a high proportion amongst heavy smokers.

Further inspection of the parameter estimates suggests that the classification of fcig is unnecessarily detailed: the fourth and fifth levels, with estimates 1.516 and 1.625, can be collapsed, as can the second and third levels.

```
$calc mcig = %eq(tcig,1) + 2*(%eq(tcig,2) + %eq(tcig,3)) +
            3*(%eq(tcig,4) + %eq(tcig,5)) + 4*%eq(tcig,6) $
$factor mcig 4 $
$fit - tcig + mcig $disp e $
 scaled deviance =  19.498 (change = +0.02529) at cycle 10
     residual df =  30      (change = +2     ) from 35 observations

          estimate       s.e.      parameter
    1       -3.069       0.6620     1
    2        0.3265      0.1361     LINP
    3       -10.68       41.62      MCIG(2)
    4        1.574       0.5325     MCIG(3)
    5        2.473       0.5444     MCIG(4)
 scale parameter 1.000
```

The interpretation of the model is that each step up the air pollution scale increases the logit by 0.327, that is, multiplies the odds on chronic bronchitis by 1.39. Categories 3 and 4 of mcig, smokers of 3–8 and more than 8 cigarettes a day have their odds on bronchitis multiplied by 4.83 and 11.86, respectively, compared with non-smokers. Light smokers (1–3 a day) have no observed cases of bronchitis.

We finally return to the original data and refit the model.

```
$unit 212 $
 -- standard length re-initialised
$calc c = 1 + %gt(cig,0) + %gt(cig,3) + %gt(cig,8) $
$factor c 4 $
$yvar r $error b n $link g $
 -- number of units changed
 -- binomial denominator unset
 -- model changed
$fit c + poll $disp e $
 scaled deviance =   143.65 at cycle 9
     residual df =   207

           estimate          s.e.       parameter
      1     -9.367          3.282        1
      2     -8.394          13.28        C(2)
      3      1.542          0.5296       C(3)
      4      2.466          0.5422       C(4)
      5      0.1241         0.05366      POLL
  scale parameter 1.000
```

Note that mcig cannot be used in the $fit directive since it is of length 36. The equal intervals for the linp scale correspond to 2.5 units of poll. Multiplying the poll coefficient and standard error by 2.5 gives 0.3103 and 0.1341, which agree very closely with the linp values of 0.3265 and 0.1361. The estimate of −8.380 for c(2) tries to reproduce on the logit scale the observed proportion of zero bronchitis sufferers. The large standard error for this parameter estimate occurs because the likelihood in this parameter is nearly flat at the estimated value and hence the second derivative of the log likelihood with respect to this parameter evaluated at the estimate is nearly zero. The constant bronchitis rate for smokers of more than eight cigarettes a day corresponds to the property of the earlier quadratic model, of a decreasing rate of linear increase with increasing cig.

Five of the previous eleven observations, (48, 59, 87, 122, 142) now have residuals larger than 2 in absolute value: these all correspond to non-smokers with bronchitis.

We do not show level curves for this model as they are not informative—for $cig = 0$ they are points, for cig between 0 and 3 they do not appear on the graph at all, and for cig greater than 3 they are lines parallel to the cig axis, but only the 30% and 50% lines appear.

It can be verified that the addition of a quadratic term in poll adds nothing to the model, and that using log poll gives almost the same deviance as poll. Further, the probit, complementary log–log and log–log links all give almost the same deviances.

The paradoxical results for non-smokers and light smokers require explanation. Do non-smokers also include those who have given up smoking? Are the non-smokers much older than the smokers? Without more details of the original survey, these questions remain open.

This example shows the value of categorizing continuous explanatory variables when the response variable is binary. In many cases, as in the last two examples in this chapter, such categorization is routinely carried out before the data are analysed. Even without categorization, however, binary responses can become binomial if the number of explanatory variable categories is limited. The next example illustrates both this feature and the use of binary models for prediction.

4.6 The prediction of binary outcomes

The data in the file ghq were published by Silvapulle (1981), and come from a psychiatric study of the relation between psychiatric diagnosis (as case or non-case) and the value of the score on a 12-item General Health Questionnaire (GHQ), for 120 patients attending a general practitioner's surgery. Each patient was administered the GHQ, resulting in a score between 0 and 12, and was subsequently given a full psychiatric examination by a psychiatrist who did not know the patient's GHQ score. The patient was classified by the psychiatrist as either a "case", requiring psychiatric treatment, or a "non-case". The question of interest was whether the GHQ score, which could be obtained from the patient without the need for trained psychiatric staff, could indicate the need for psychiatric treatment. Specifically, given the value of GHQ score for a patient, what can be said about the probability that the patient is a psychiatric case? Sex of the patient is an additional variable. The data are shown in Table 4.3.

Both men and women patients are heavily concentrated at the low end of the GHQ scale, where the overwhelming majority are classified as non-cases. The small number of cases are spread over medium and high values of GHQ.

The ghq data file gives the number of cases c and non-cases nc at each ghq score, and the total number $n = c + nc$, classified by the factor sex. We first graph the proportion of cases against ghq for each sex separately:

```
$input 'ghq' $
$calc p = c/n   : male = %eq(sex,1) $
$calc %re = male $
$graph (s = 1 h = 'ghq' v = 'p') p ghq 5 $
```

Table 4.3. GHQ score for cases and non-cases

GHQ	Men		Women	
	Cases	Non-cases	Cases	Non-cases
0	0	18	2	42
1	0	8	2	14
2	1	1	4	5
3	0	0	3	1
4	1	0	2	1
5	3	0	3	0
6	0	0	1	0
7	2	0	1	0
8	0	0	3	0
9	0	0	1	0
10	1	0	0	0

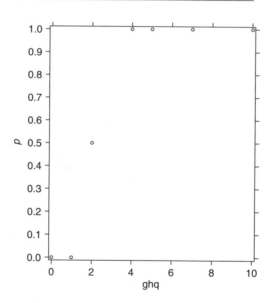

Fig. 4.17. Case response proportion, males

Males (Fig. 4.17) show a rapid change from non-cases to cases between GHQ scores of 1 and 4.

```
$calc female = %eq(sex,2) $
$calc %re = female $
$graph (s = 1 h = 'ghq' v = 'p') p ghq 5 $
```

Females (Fig. 4.18) also show a rapid change, though there is a small proportion of cases even for GHQ zero.

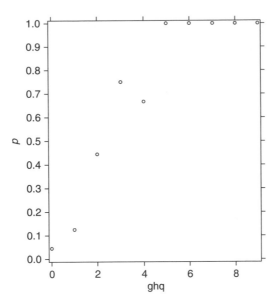

Fig. 4.18. Case response proportion, females

Thus, the GHQ score is a good indicator of being a psychiatric case. To make this statement more precise we fit a model:

```
$yvar c $error b n $link g $fit sex + ghq $disp e $
 scaled deviance =    4.9419 at cycle 5
     residual df =   14

          estimate         s.e.      parameter
    1       -4.072        0.9754     1
    2        0.7938       0.9284     SEX(2)
    3        1.133        0.2900     GHQ
scale parameter 1.000
```

The standard error of the sex parameter is large compared to the parameter estimate, suggesting that the sex effect is not needed in the model. This is confirmed by a formal test based on the deviance difference:

```
$fit - sex $disp e $
 scaled deviance =    5.7440 (change =   +0.8021) at cycle 5
     residual df =   15       (change =   +1     )

          estimate         s.e.      parameter
    1       -3.454        0.5623     1
    2        1.440        0.2972     GHQ
scale parameter 1.000
```

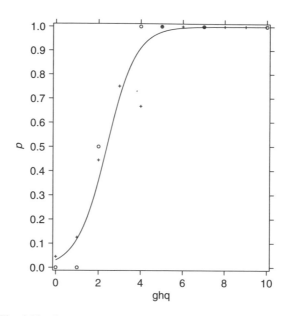

Fig. 4.19. Case response proportions and fitted probabilities

The change in deviance is quite small compared to χ_1^2. Thus, as far as the relation between "caseness" and GHQ is concerned, we can ignore the sex classification and use a single model for both sexes.

Our interest is in the probability of being a case, not in the logit of this probability, so we transform the fitted values back to the probability scale, and graph in Fig. 4.19 the observed and fitted proportions, pooled over sex:

```
$calc fp = %fv/n $
$del %re $
$graph (s = 1 h = 'ghq' v = 'p') p,fp ghq 5,6,10 sex, $
```

The analysis looks complete at this point, but if we examine the data more carefully a strange feature appears. Suppose we check the assumption of no interaction between sex and ghq by including their interaction in the model:

```
$fit sex + ghq + sex.ghq $disp e $
 scaled deviance =    1.5449 at cycle 10
     residual df =   13
 -- (no convergence yet)
 -- unit(s) held at limit

           estimate           s.e.       parameter
      1       -17.38          33.07        1
      2        14.38          33.08        SEX(2)
```

```
    3         8.691          16.58        GHQ
    4        -7.449          16.58        SEX(2).GHQ
scale parameter 1.000
```

Very large positive or negative estimates now appear for all the parameters, with even larger standard errors, and the GLIM scoring algorithm has not converged. Why is this happening?

The fitting of the full interaction model is equivalent to fitting separate ghq models for each sex, with different slopes and intercepts. This can also be achieved by fitting each sex separately using a weight variable.

```
$weight female $fit ghq $disp e $
-- model changed
scaled deviance =   1.5422 at cycle 5
    residual df =   8       from 10 observations

           estimate        s.e.       parameter
    1        -2.999        0.5672      1
    2         1.242        0.2971      GHQ
scale parameter 1.000
$weight male $fit ghq $disp e $
-- model changed
scaled deviance =   0.0026916 at cycle 10
    residual df =   5       from 7 observations
-- (no convergence yet)
-- unit(s) held at limit

           estimate        s.e.       parameter
    1       -17.38         33.07       1
    2         8.691        16.58       GHQ
scale parameter 1.000
```

For males, convergence does not occur in 10 cycles. We increase the number of cycles to 30 and refit the model.

```
$cycle 30 $fit . $disp e $
  scaled deviance = 0.000049332 (change = -0.002642) at cycle 14
    residual df = 5                 (change =  0        ) from 7
                                                          observations

           estimate        s.e.       parameter
    1       -25.38         244.3       1
    2        12.69         122.1       GHQ
scale parameter 1.000
```

At convergence after 14 cycles, the deviance is less than 10^{-4}, and the model fits "exactly". The intercept and slope values are such that at GHQ $= 2$, the fitted logit from the model is zero, corresponding to the observed proportion of 0.5. Any large values of β_0 and β_1 with $\beta_1 > 0$ such that $\beta_0 + 2\beta_1 = 0$ will give an exact fit to the data, in the sense that the fitted values for ghq $= 0$ or 1 and for ghq > 2 can be made arbitrarily close to the observed values. Thus the GLIM estimates in this example are determined solely by the convergence criterion in GLIM: if this is made stricter then $\hat{\beta}_0$ and $\hat{\beta}_1$ will be larger, but still with $\hat{\beta}_0 + 2\hat{\beta}_1 = 0$. The likelihood function approaches its maximum of 1 as $\beta_0 \to -\infty, \beta_1 \to +\infty$ with $\beta_0 + 2\beta_1 = 0$. Thus the maximum likelihood estimates of β_0 and β_1 "do not exist", that is they are not finite. This causes GLIM no difficulty, however, since the convergence criterion is satisfied for finite values of these parameters.

Since the evidence for interaction is not very strong—the deviance reduction on adding the interaction is 3.40 on 1 df—the simpler main effects model can be retained, and the sex effect can be omitted from this model, as we have already seen.

The psychiatric value of the model involving only GHQ is in prediction: for a new patient (assumed similar to those in the study) with a value of GHQ of 2 (say), what can we say about the probability that this patient is a psychiatric case?

The fitted logit from the model at ghq $= 2$ is $-3.454 + 2(1.440) = -0.574$ and the corresponding fitted probability is 0.360. For prediction purposes, however, this value is misleadingly precise. As in Section 4.2, we construct the profile relative likelihoods for the case probability, for the values 0–6 of GHQ. We do not show the separate graphs here, only the combined graph.

```
$weight $
$fit ghq $disp e $
$input 'rprob' $
$calc %a = 0.001 : %i = 0.001 : %z = 0.998 $
$calc %x = 0   $use rprob ghq $calc lik0 = lik_ $
$calc %x = 1   $use rprob ghq $calc lik1 = lik_ $
$calc %x = 2   $use rprob ghq $calc lik2 = lik_ $
$calc %x = 3   $use rprob ghq $calc lik3 = lik_ $
$calc %x = 4   $use rprob ghq $calc lik4 = lik_ $
$calc %x = 5   $use rprob ghq $calc lik5 = lik_ $
$calc %x = 6   $use rprob ghq $calc lik6 = lik_ $
$graph (s = 1 h = 'p' v = 'relative likelihood')
         lik0,lik1,lik2,lik3,lik4,lik5,lik6 p_
         10,10,10,10,10,10,10
$gtext .95 .05 '0' : .95 .14 '1' : .95 .40 '2' $
     : .95 .75 '3' : .95 .94 '4' $
```

The profile likelihoods for GHQ $= 0$ and 1 are quite concentrated, while those for GHQ $= 3$ and 4 are quite diffuse. For GHQ $= 5$ and above the likelihoods become very concentrated close to 1. Asymptotic confidence intervals for the case probability p are obtained as in Section 4.2 by interpolating in the deviance dv_. We give the 95% intervals to 3 dp in Table 4.4, without calculation details. A GHQ score of zero means that the patient is very unlikely to be a case (only 2 out of 60 patients with this score were cases) and a score of 5 or more means that the patient is very likely to be a case (all of the 15 patients with a score of 5 or more were cases). For values from 1 to 4, however, the GHQ score is not a definite indicator of "caseness". Note that for a score of 4, the fitted probability is 0.909, strongly suggesting that such a patient would be a case: however the confidence interval includes values of p near 0.7, and we note that 1 out of 4 patients with this score was not a case (Fig. 4.20).

Table 4.4. Predicted case probability and 95% confidence interval

ghq	Lower limit	Predicted value	Upper limit
0	0.009	0.031	0.077
1	0.055	0.118	0.208
2	0.212	0.360	0.552
3	0.467	0.704	0.891
4	0.701	0.909	0.983
5	0.857	0.977	0.999
6	0.935	0.994	0.999

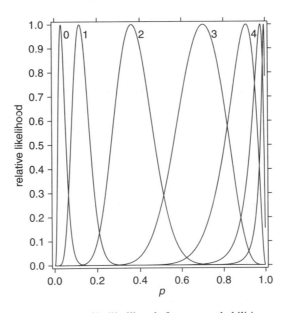

Fig. 4.20. Profile likelihoods for case probabilities

4.7 Profile and conditional likelihoods in 2 × 2 tables

The 2×2 contingency table arising from a randomized two-group design is an important special case of the binomial generalized linear model, and its analysis is still a matter of some controversy (see, e.g. Yates, 1984). Throughout the book we have used the likelihood ratio test, or equivalently the profile likelihood, to draw conclusions about a model parameter. The likelihood is maximized over the values of the other (nuisance) parameters in the model, for a fixed value of the parameter of interest. As discussed in Chapter 2, in small samples both the justification for following this procedure and the exact coverage of the confidence interval are not altogether satisfactory, and other methods of eliminating nuisance parameters are of value.

Consider the 2×2 table in which a binary response is classified by a two-level group factor, with r_i successes in n_i trials with success probability p_i ($i = 1, 2$). The likelihood is

$$L(p_1, p_2) = \Pr[R_1 = r_1, R_2 = r_2]$$

$$= \binom{n_1}{r_1} p_1^{r_1} (1 - p_1)^{n_1 - r_1} \cdot \binom{n_2}{r_2} p_2^{r_2} (1 - p_2)^{n_2 - r_2}$$

$$= \binom{n_1}{r_1}\binom{n_2}{r_2} [p_1/(1 - p_1)]^{r_1} [p_2/(1 - p_2)]^{r_2} (1 - p_1)^{n_1} (1 - p_2)^{n_2}.$$

We now define $r = r_1 + r_2$ and substitute for r_2 in the likelihood:

$$L(p_1, p_2) = \binom{n_1}{r_1}\binom{n_2}{r - r_1} \{[p_1/(1 - p_1)]/[p_2/(1 - p_2)]\}^{r_1}$$

$$\cdot (1 - p_1)^{n_1} (1 - p_2)^{n_2} [p_2/(1 - p_2)]^{r}$$

$$= \binom{n_1}{r_1}\binom{n_2}{r - r_1} \theta^{r_1} (1 - p_1)^{n_1} p_2^{r} (1 - p_2)^{n_2 - r},$$

where $\theta = \exp(\beta_1)$ is the odds ratio from the logistic model

$$\operatorname{logit} p_i = \beta_0 + \beta_1 x_i$$

and x_i is a $(0, 1)$ dummy variable taking the value 1 for the first level of the factor.

Now consider the distribution of the random variable $R = R_1 + R_2$, the total number of successes in both groups. We have

$$\Pr[R = r = r_1 + r_2] = \sum_{u=u_1}^{u=u_2} \Pr[R_1 = u] \Pr[R_2 = r - u],$$

where the limits of the sum are determined by the marginal totals. Since both $0 \leq u \leq n_1$ and $0 \leq r - u \leq n_2$, we have $u_1 = \max(0, r - n_2), u_2 = \min(n_1, r)$.

Thus

$$\Pr[R = r] = \sum_{u=u_1}^{u=u_2} \binom{n_1}{u}\binom{n_2}{r-u}\theta^u(1-p_1)^{n_1}p_2^r(1-p_2)^{n_2-r}$$

and

$$\Pr[R = r_1, R_2 = r_2 \mid R = r] = \Pr[R = r_1, R_2 = r_2]/\Pr[R = r]$$

$$= \binom{n_1}{r_1}\binom{n_2}{r_2}\theta^{r_1} \bigg/ \sum_{u=u_1}^{u=u_2} \binom{n_1}{u}\binom{n_2}{r-u}\theta^u.$$

This conditional distribution depends only on θ—the nuisance parameter β_0 in the logistic model has been eliminated by the conditioning. Given the values of r_1 and r_2, this (non-central hypergeometric) conditional distribution defines the *conditional likelihood*

$$\mathrm{CL}(\theta \mid r) = \binom{n_1}{r_1}\binom{n_2}{r_2}\theta^{r_1} \bigg/ \sum_{u=u_1}^{u=u_2} \binom{n_1}{u}\binom{n_2}{r-u}\theta^u.$$

This distribution was introduced by Fisher (1935), and the conditional likelihood was developed by Fisher as the appropriate analysis of the 2×2 table from this experimental design. Fisher argued that the marginal number of successes R was an *ancillary* statistic which provides no information about θ but does provide information about its precision, and is therefore an appropriate statistic on which to condition. Since the marginal distribution of R does depend on θ, R is not an exact ancillary and so there is some loss of information about θ in the conditioning, but Plackett (1977) showed that the information about θ in R is small compared to the information in R_1 unless the sample sizes n_1 and n_2 are very small.

The hypergeometric probabilities are easily calculated provided n_1 and r are not too large, since $\mathrm{CL}(\theta)$ is a polynomial in θ of degree u_2. The conditional likelihood in β_1 can then be obtained by log transforming the scale of θ.

The usual application of the conditional likelihood is to Fisher's "exact" test of the hypothesis $\theta = 1$, which is equivalent to $p_1 = p_2$. The value $\mathrm{CL}(1)$ is the (Fisher) "exact probability" of the observed table, that is, the hypergeometric probability of the observed table given fixed margins and $\theta = 1$. In the usual application of the "exact" test this probability is assessed by comparison with the probabilities of "more extreme" tables, and a p-value assigned to the observed table which is the cumulative probability of the observed table and all more extreme ones. Opinions differ over whether this probability should be doubled (corresponding to a two-tailed test) or the probabilities of extreme tables in the other direction should be cumulated as well (Yates, 1984 gives a discussion).

These procedures do not lend themselves easily to interval construction, but the conditional likelihood can be used directly for this purpose.

We illustrate with a small numerical extreme example. In samples of $n_1 = 11$ and $n_2 = 1$, $r_1 = 11$ and $r_2 = 0$ successes are observed. The likelihood is

$$L(p_1, p_2) = p_1^{11}(1 - p_2).$$

Assuming that it is appropriate to condition on the marginal total $r = 11$, the conditional likelihood is

$$\mathrm{CL}(\theta) = \frac{\theta^{11}}{11\theta^{10} + \theta^{11}} = \frac{\theta}{11 + \theta}.$$

This has a maximum of 1 as $\theta \to \infty$. We now compare the conditional likelihood with the profile likelihood. For the parameter β_1 in the logistic model, the profile likelihood is obtained using the macro logodds as in Section 4.2.1. Since the full likelihood is maximized at 1 for $\hat{p}_1 = 1, \hat{p}_2 = 0$, the profile likelihood in β_1 will also have a maximum of 1 as $\beta_1 \to \infty$. We graph the two likelihoods over a grid $(-1(0.1)10)$ of $\beta_1 = \log \theta$. The values of $\log \theta$ are held in the vector lth_ after the use of logodds (Fig. 4.21).

```
$newjob $
$unit 2
$data r n group $
$read 11 11 1   0 1 2 $
$factor group 2 $
```

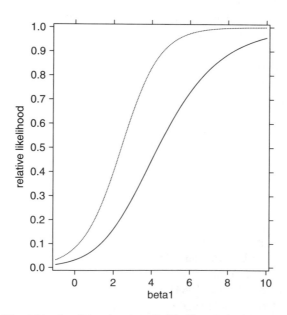

Fig. 4.21. Conditional and profile likelihoods for logodds

```
$yvar r $error b n $link g
$inp 'logodds' $
$calc %a = -1 : %i = 0.1 : %z = 10 $
$use logodds group $
$var 111 cl $
$calc cl = %exp(1th_)/(11 + %exp(1th_)) : cdv = -2*%log(cl) $
$graph (s = 1 h = 'betal' v = 'relative likelihood' y = 0,1)
        lik_,cl lth_ 10,11 $


$look lth_ dv_ cdv $
...
   10   -0.10    7.06802    5.1538897
   11    0.00    6.88400    4.9698133
   12    0.10    6.70139    4.7872653
...
   17    0.60    5.81288    3.9023433
   18    0.70    5.64084    3.7320132
...
   29    1.80    3.91282    2.0722589
   30    1.90    3.77309    1.9455347
...
```

Since both deviances go to zero as $\beta_1 \to \infty$, the deviance at $\beta_1 = 0$ is the likelihood ratio test statistic for $\beta_1 = 0$ against a general alternative. The conditional LRTS is 4.97, the (unconditional) LRTS is 6.88. Both would lead, assuming the asymptotic distributions can be validly applied, to rejection of the hypothesis of equal probabilities, though at different levels: 0.026 for the CLRT and 0.009 for the LRT. The asymptotic 95% confidence intervals for $\beta_1 = \log \theta$ are quite different: $(0.64, \infty)$ for the CLRTS and $(1.84, \infty)$ for the LRTS.

The conclusion from Fisher's "exact" test is expressed differently. The hypergeometric probabilities are evaluated under the null hypothesis $\theta = 1$ for the possible tables with the same margins as the observed table. For our example, there are only two possible tables with the same margins: the observed one and the table with $r_1 = 10, r_2 = 1$. The probabilities of these tables are $\theta^{11}/(11\theta^{10} + \theta^{11})$ and $11\theta^{10}/(11\theta^{10} + \theta^{11})$ respectively. At $\theta = 1$, these probabilities are 1/12 and 11/12 respectively. In this case there are no more "extreme" tables than the observed one, since we already observe all the successes in one group and all the failures in the other. Thus the "exact" probability of the observed table is $1/12 = 0.0833$. This probability would not lead to rejection of the null hypothesis at conventional levels.

The "exact" test assesses the probability of the observed event under the null hypothesis relative to those of other more extreme events under the same hypothesis. This is philosophically different from the conditional likelihood ratio test which assesses the probability of the observed event under the null hypothesis relative to that of the same event under other hypotheses.

In this example with very small sample sizes, the conclusions about the log-odds ratio from the two analyses are not equivalent. In larger samples the conditional and profile likelihoods become equivalent, and so the computational difficulties of the conditional likelihood (the evaluation of high degree polynomials in θ) can be avoided in larger samples by computing the profile likelihood.

The example above may seem contrived. In fact it is a real data set, taken from the extra-corporeal membrane oxygenation (ECMO) study of Bartlett *et al.* (1985). In this study ECMO was compared with the then-standard treatment for neonatal respiratory failure, which had a very high mortality. To reduce mortality in the study for those assigned to the standard treatment, the treatment assignment was a form of "play-the-winner" rule (Zelen, 1969) in which randomization to the new treatment group has a probability which increases with its success relative to the standard treatment on patients already treated. Under this form of assignment the conditioning argument used to derive the conditional likelihood is invalid, and a different conditional reference set is required for p-value calculations. In the actual study, the randomization was stopped after 10 patients had been treated, and the two additional patients included in the study were assigned non-randomly to the new treatment, but the stopping rule for termination of the trial is unknown. Formally this invalidates all p-value calculations since these require a complete statement of the sample design.

However the likelihood is invariant to the sample design, so likelihood and Bayes analyses can be applied to the data without a knowledge of the stopping rule. Further discussion of this study, with many interesting comments on philosophical issues (see Royall's comments in particular) can be found in Begg (1990).

4.8   Three-dimensional contingency tables with a binary response

We now consider two examples of three-dimensional contingency tables with a binary response as one dimension.

4.8.1   *Prenatal care and infant mortality*

The data (Bishop *et al.*, 1975, p. 41) come from a study of two prenatal clinics in a large US city. Mothers attended one of the prenatal clinics for varying periods before the birth of the baby. The length of time the mother attended the clinic has been categorized for analysis into less than one month or more than one month.

The binary response variable of interest is the infant mortality outcome: infant died within one month, or survived at least one month. There are two explanatory variables: amount of prenatal care, and clinic (A or B); the data are given in Table 4.5. An additional column has been added giving the mortality rates for each

Table 4.5. Survival experience of child

Mother's attendance	Died within one month	Survived at least one month	Total number	Mortality rate (%)
Clinic A				
<1 month	3	176	179	1.68
>1 month	4	293	297	1.35
Clinic B				
<1 month	17	197	214	7.94
>1 month	2	23	25	8.00

Table 4.6. Aggregate survival experience of child

Mother's attendance	Died within one month	Survived at least one month	Total number	Mortality rate (%)
Both clinics				
<1 month	20	373	393	5.09
>1 month	6	316	322	1.86

group. Clinic B has a considerably higher mortality rate than Clinic A, but the duration of mother's attendance seems to be unrelated to mortality.

If the data for the two clinics are combined, a quite different picture appears (Table 4.6). Now mothers attending for less than one month seem to experience a higher mortality rate for their infants.

This example clearly illustrates two of the hazards of inference from observational data: the danger of aggregating data over important variables, and the inappropriate assumption of causality—that duration of prenatal care causes or is responsible for the change in infant mortality. Such a conclusion is insupportable in a non-randomized study: it could be supported only if mothers were randomly assigned to the two classes of prenatal care. In fact, there is a systematic biasing factor visible from the complete data: of the mothers attending Clinic B, only 10% per cent attended for more than one month, while this proportion was 62% for mothers attending Clinic A. Since Clinic B has a much higher mortality rate, when the clinic identification is suppressed the variation in mortality rate appears to be associated with duration of prenatal care. This is a classic example of *Simpson's paradox* (Simpson, 1951).

It is equally invalid, however, to conclude from the full data that the difference in mortality rates is caused by different quality of prenatal advice and treatment in the two clinics. There is no randomization of mothers to clinics: each clinic serves the mothers in a local area of the city. The different mortality rates probably reflect different infant mortality rates in the subpopulations of the city served by each clinic. Without randomized assignment in such studies it is impossible to draw strong causal conclusions.

The analysis of the data through statistical models can, however, avoid the invalid conclusions produced by aggregating the data for both clinics. We proceed to model the data.

```
$unit 4
$data death total $
$read
3  179  4  297  17  214  2  25
$factor clinic 2 attend 2 $
$calc clinic = %gl(2,2) : attend = %gl(2,1) $
$yvar death $error b total $link g
$fit : + attend : clinic : + attend : + clinic.attend $
```

The deviances are 17.83, 12.22, 0.08, 0.04, 0.00. Successive differencing of the deviances leads to two analysis of deviance tables (Table 4.7). There is a striking interchange of deviance between Clinic and Attendance when their fitting order is reversed, reflecting the substantial correlation between the variables noted above. It is obvious from the table that the Clinic term is essential in the model, and that once it is included, the Attendance term is unnecessary. We refit the clinic model and calculate the fitted probabilities from the model:

```
$fit clinic $disp e $
$calc fp = %fv/total $
$print fp $
```

The fitted death rates are 0.0147 for clinic A and 0.0795 for clinic B, which are just the death rates for each clinic pooled over the attendance classification.

Thus statistical modelling of the three-way table has shown that the table can be collapsed over Attendance to give a simple representation, but it cannot be collapsed over Clinic without serious distortion.

Table 4.7. Analysis of deviance tables

Source	dev	df	Source	dev	df
attend	5.61	1	clinic	17.75	1
clinic	12.18	1	attend	0.04	1
interaction	0.04	1	interaction	0.04	1

Table 4.8. Blood pressure and cholesterol level in CHD

chol	<127	127–146	147–166	>166
		bp		
<200	2/119	3/124	3/50	4/26
200–219	3/88	2/100	0/43	3/23
220–259	8/127	11/220	6/74	6/49
>259	7/74	12/111	11/57	11/44

4.8.2   Coronary heart disease

The file chd gives the number r of men diagnosed as having coronary heart disease (CHD) in an American study of 1329 men (the data are presented and analysed in Ku and Kullback, 1974). The serum cholesterol level chol and blood pressure bp in mm mercury were recorded for each man, and are reported in one of four categories, giving a 4 × 4 cross-classified table in each cell of which the number r of men with CHD and the total number n of men examined are given as r/n. The data are reproduced in Table 4.8.

The explanatory variables were originally continuous, but have been categorized into ordered categories. We treat them first as unordered factors. bp and chol are defined on the file.

How is the proportion of men suffering from CHD related to blood pressure and serum cholesterol levels? Before fitting any models, we find the proportion p suffering from CHD, and graph it against bp separately for each value of chol (Fig. 4.22) by using different graph characters, joined by lines. For this purpose, we define new graph styles which use both lines and characters:

```
$input 'chd' $
$calc p = r/n $
$gstyle 1 1 1 s 5 $
    :   2 1 2 s 6 $
    :   3 1 3 s 7 $
    :   4 1 4 s 9 $
$graph (s = 1 h = 'blood pressure' v = 'p') p bp 1,2,3,4 chol $
$gtext 0.04 2.5 '1' : 0.01 2.5 '2' : 0.07 2.5 '3' : 0.15 2.5 '4' $
```

Apart from the zero proportion of cases in one cell, the pattern of increase is fairly consistent at each level of the other factor.

Now we try fitting models on the logit scale:

```
$yvar r $error b n $link g
$fit : bp : chol : chol + bp $
 scaled deviance =  58.726 at cycle 4
     residual df =  15
```

```
scaled deviance =  35.163 at cycle 4
    residual df =  12

scaled deviance =  26.805 at cycle 4
    residual df =  12

scaled deviance =  8.0762 at cycle 4
    residual df =  9
```

The two analysis of deviance tables are given in Table 4.9. chol is the more important, but both main effects are necessary in the model. The interaction is not explicitly fitted, but we know that the interaction model is saturated and has a deviance of zero. The value of 8.076, near the mean (or median) of χ_9^2, shows that the main effects model with 7 parameters provides a good fit to the data.

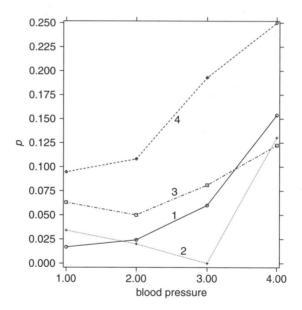

Fig. 4.22. CHD probability

Table 4.9. Analysis of deviance tables

Source	Dev	df	Source	Dev	df
bp	23.56	3	chol	31.92	3
chol	27.09	3	bp	18.73	3
interaction	8.08	9	interaction	8.08	9

To check that the model is fitting well at each sample point, we examine the (Pearson) residuals.

```
$print %rs $
 -0.8351 -0.2977   0.3331    1.076    0.5917 -0.2249  -1.352
  0.9708  0.6036   0.0541   -0.0949 -0.5755 -0.3047   0.2260
  0.5212 -0.4636
```

For the main effects model the largest residual is -1.35, showing a good fit.

Now we examine the parameter estimates to interpret the model:

```
$disp e $
            estimate         s.e.        parameter
     1        -3.482        0.3486       1
     2       -0.2080        0.4664       CHOL(2)
     3        0.5622        0.3508       CHOL(3)
     4         1.344        0.3430       CHOL(4)
     5      -0.04146        0.3037       BP(2)
     6        0.5324        0.3324       BP(3)
     7         1.200        0.3269       BP(4)


  scale parameter 1.000
```

The parameter estimates for both factors show a consistent increase as the factor level increases from 1 to 4, apart from level 2 which for each factor is very little different from level 1. This suggests that we equate levels 1 and 2 and then try smoothing the proportions further by using the factor level as though it were a continuous variable. We then fit a regression model using one parameter for each score by redefining the factors to be variates:

```
$calc ch = chol - %ge(chol,2) : b = bp - %ge(bp,2) $
$factor ch 3 bp 3 $
$fit ch + b $disp e $print %rs $
  scaled deviance =    8.2971 at cycle 3
      residual df =    11

            estimate         s.e.        parameter
     1        -3.592        0.2460       1
     2        0.6478        0.2952       CH(2)
     3         1.431        0.2861       CH(3)
     4        0.5554        0.2809       B(2)
     5         1.223        0.2741       B(3)
  scale parameter 1.000

-0.6749 -0.1795   0.4808    1.245    0.4237 -0.4209  -1.436
```

```
 0.7694   0.6719   0.0001 -0.0908 -0.5710 -0.2447    0.1683
 0.5236 -0.4606
```

The deviance increases by 0.21 on 2 df, and the parameter estimates show nearly equal steps. Now we redefine ch and b as variables, and refit:

```
$var ch b $
$fit ch + b $disp e $print %rs $
 -- change to number of parameters in model
 scaled deviance =   8.4192 at cycle 3
     residual df =   13

            estimate          s.e.        parameter
      1      -4.962          0.3876        1
      2       0.7221         0.1426        CH
      3       0.6069         0.1348        B
 scale parameter 1.000

  -0.6156 -0.1090   0.4634   1.344    0.4944 -0.3627 -1.444
    0.8537   0.5806 -0.1070 -0.2897 -0.6257 -0.1752   0.2584
    0.4520 -0.3540
```

The deviance increases by 0.12 on 2 df and there are no large residuals. The parameter estimates of 0.61 for b and 0.72 for ch (with standard errors about 0.14) suggest that these could be equated: if $\beta_1 = \beta_2 = \beta$, then the model

$$\theta = \beta_0 + \beta_1 x_1 + \beta_2 x_2$$

reduces to

$$\theta = \beta_0 + \beta(x_1 + x_2),$$

a simple linear regression on the total score $x_1 + x_2$.

```
$calc score = b + ch $
$fit score $disp e $print %rs $
 scaled deviance =   8.7422 at cycle 3
     residual df =   14

            estimate          s.e.        parameter
      1      -4.924          0.3783        1
      2       0.6615         0.09360       SCORE
 scale parameter 1.000

  -0.6628 -0.1653   0.3156   1.069    0.4380 -0.4091 -1.508
    0.6185   0.6573 -0.0170 -0.3526 -0.8146   0.0477   0.5487
    0.5536 -0.4133
```

The deviance has increased very little, and the model fits well at all points, with a largest residual of −1.51 for unit 7, where the observed proportion is zero. The regression coefficient of 0.66 is halfway between those for b and ch.

How do we interpret the model? We have effectively a five-point scale for score (2–6), and an increase of 1 point on this scale gives a predicted increase of 0.66 in the log–odds of having CHD, that is, it multiplies the odds in favour of heart disease by $e^{0.66} = 1.93$, so the odds approximately double for each point up the scale. If the regression coefficient were log 2 = 0.6931, the odds would exactly double. We try smoothing further by fixing the slope β at 0.6931, and estimating only the intercept β_0. This is achieved by defining an offset of log 2 × score:

```
$calc ofs = %log(2)*score $
$offset ofs $
$fit $disp e $
 -- model changed
 scaled deviance =     8.8565 at cycle 3
     residual df =   15


           estimate        s.e.      parameter
      1       -5.047      0.1110         1
 scale parameter 1.000
```

Now we are fitting just the intercept as the score slope is specified. The deviance increases by 0.11 on 1 df and the intercept is −5.047. The fitted logits, odds and probabilities at each point on the scale are given in Table 4.10.

The odds values are very nearly 1/40, 1/20, 1/10, 1/5, 2/5. Can we smooth the model any further? If the intercept had been −5.075 instead of −5.047, we would have had exactly the rounded-off odds values above. Since the standard error of $\hat{\beta}_0$ is 0.11, we might as well do the extra smoothing to easily present interpreted values. We finally fit a model with no estimated parameters.

```
$calc ofs2 = -5.075 + ofs $
$offset ofs2 $
$fit - 1 $
 -- model changed
 scaled deviance =     8.9217 (change =   +0.06520) at cycle 2
     residual df =   16       (change =   +1       )
```

Table 4.10. Fitted logits, odds and probabilities

Score	2	3	4	5	6
Logit	−3.660	−2.967	−2.276	−1.581	−0.888
Odds	0.0257	0.0515	0.1029	0.2058	0.4116
Probability	0.0251	0.0489	0.0933	0.1707	0.2916

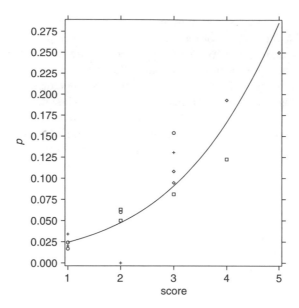

Fig. 4.23. CHD probability against score

In presenting the data finally we can simplify by defining the score on the range 1–5 instead of 2–6, by subtracting 1. The observed and (fitted) probabilities are graphed in Fig. 4.23 against the `score`, using the previous graph symbols for the levels of `chol`.

```
$graph (s = 1 h = 'score' v = 'p') p score 1,2,3,4 chol $
```

We do not quote standard errors for these fitted proportions (or fitted logits) because the purpose of presenting the smoothed fitted values is to give a simple interpretation to the data. Each step up the score scale doubles the odds on CHD, starting from odds of 1/40 at score 1. This gives a very simple and understandable relation between risk and the explanatory variables. A more precise statement might have been possible if the original values of blood pressure and serum cholesterol had been retained: the second example in this chapter shows how the model would then have been used. The categorization of the explanatory variables leads to a *simpler* interpretation, though we note that as far as the relation between CHD and the categorized variables is concerned, the lowest two categories for each variable do not need to be distinguished.

The appearance of the graph suggests that a *linear* probability model might fit well. We investigate this by changing the link and removing the offset:

```
$link i $offset
$fit score $disp e $
  scaled deviance =  10.494 at cycle 4
```

```
residual df =   14

          estimate           s.e.        parameter
   1      -0.02244         0.01132         1
   2       0.04248         0.006522        SCORE
scale parameter 1.000
```

Compared with the logit model, the deviance has increased by 1.75 (for the logit model with two estimated parameters). This is not a large change, so the linear model is a possible one here.

4.9   Multidimensional contingency tables with a binary response

The file byssinosis gives the number of workers in a survey of the US cotton industry suffering (yes) and not suffering (no) from the lung disease byssinosis, together with the values of five cross-classifying categorical explanatory variables: the race, sex and smoking habit smok of the worker, the length of employment emp in three categories, and the dustiness dust of the workplace in three categories. The data were presented and discussed in Higgins and Koch (1977).

The question of primary scientific interest was the relationship between the incidence of byssinosis and the dustiness of the workplace, but smoking habit and length of employment, and hence exposure to dust, are obviously relevant and need to be allowed for.

We first display the completely cross-classified table of proportions suffering from byssinosis.

```
$input 'byssinosis' $
$calc n = yes + no : p = yes/n $
 -- invalid function/operator argument(s)
```

A diagnostic message is printed here: the value of n in some cells is zero, and division by zero results in zero. We construct and print a table of proportions.

```
$accuracy 2 $
$tprint p smok,sex,emp,race,dust $
          RACE    1                          2
          DUST    1        2        3        1        2        3
  SMOK SEX EMP
    1    1    1   0.0750   0.0000   0.0077   0.1524   0.0000   0.0122
              2   0.2759   0.0196   0.0053   0.2105   0.0000   0.0000
              3   0.2870   0.0070   0.0237   0.2439   0.0000   0.0000

         2    1   0.0000   0.0106   0.0164   0.0833   0.0136   0.0114
              2   0.0000   0.0294   0.0208   0.0000   0.0000   0.0000
              3   0.0000   0.0319   0.0168   0.0000   0.0000   0.0000
```

```
    2    1    1   0.0000   0.0000   0.0000      0.0741   0.0208   0.0081
              2   0.2000   0.0588   0.0000      0.1000   0.0000   0.0000
              3   0.0962   0.0000   0.0162      0.1667   0.0000   0.0000

         2    1   0.0000   0.0182   0.0117      0.0400   0.0207   0.0131
              2   0.0000   0.0000   0.0110      0.0000   0.0000   0.0000
              3   0.0000   0.0158   0.0058      0.0000   0.0000   0.0000
```

The table has a large number of zero values, but there is a noticeably high byssinosis
rate at dust level 1. The accuracy setting gives two significant digits in the smallest
value of p. Before modelling the table, we examine marginal tabulations:

```
$accuracy 3 $
$tab the p mean for dust with n $
  DUST       1        2        3
  MEAN   0.1570   0.0138   0.0122
  :   the p mean for race with n $
  RACE       1        2
  MEAN   0.0262   0.0384
  :   the p mean for sex  with n $
   SEX       1        2
  MEAN   0.0439   0.0148
  :   the p mean for smok with n $
  SMOK       1        2
  MEAN   0.0392   0.0179
  :   the p mean for emp  with n $
   EMP       1        2        3
  MEAN   0.0231   0.0365   0.0384
  :   the p mean            with n $
             0.0304
```

The highest dust level has an outstandingly high byssinosis rate. There appear to be
clear racial, sex and smoking differences in byssinosis rate. Length of employment
differences are less clear. The overall byssinosis rate for all workers is 3%.

As we saw in Section 4.8.1, there is a serious hazard in collapsing contingency
tables over important explanatory variables: marginal tabulations may be misleading if explanatory variables are strongly associated. Such associations may
be expected in survey data, and occur in this example, as is easily seen if we
cross-tabulate by dust and race:

```
$tab the p mean for race,dust with n $
  DUST       1        2        3
  RACE
       1   0.1835   0.0140   0.0129
       2   0.1393   0.0135   0.0104
```

Now we see that the higher marginal byssinosis rate for non-whites is spurious: at each dust level the sample rate is in fact lower for non-whites than for whites! The higher marginal rate results from the much higher proportion of non-whites than of whites working in the high dust condition, as can be verified by a further tabulation:

```
$tab the n total for race,dust $
  DUST    1      2      3
  RACE
    1    267.   855.   2394.
    2    402.   445.   1056.
```

This is another example of Simpson's paradox

We now proceed to model the full table. We fit a sequence of logit models, with the main effects and interactions added in an arbitrary (but hierarchical) order. We suppress the output and present only the changes in deviance in the analysis of deviance table.

```
$accuracy
$yvar yes $error b n $link q $
$fit : + dust : + emp : + smok : + sex : + race $
     : + dust.emp : + dust.smok : + dust.sex : + dust.race $
                 : + emp.smok : + emp.sex : + emp.race $
                             : + smok.sex : + smok.race $
                                         : + sex.race $
     : + dust.emp.smok : + dust.emp.sex  : + dust.emp.race $
     : + dust.smok.sex : + dust.smok.race : + dust.sex.race $
     : + emp.smok.sex : + emp.smok.race : + emp.sex.race $
                                         : + smok.sex.race $
```

The models with dust.emp.race and more interactions do not converge in 10 iterations, because they are attempting to reproduce the zero proportions with byssinois in the table. The standard errors for all these parameters, and for some of their marginal terms, become very large as a consequence and it is difficult to identify the important variables from the estimates from the full three-way interaction model. We use the analysis of deviance Table 4.11 to reduce the model.

The immediately striking feature of this is the very large deviance for dust. However, this is fitted first in the model, and it may be substantially less important when fitted after the other main effects. Interactions are much smaller, though some may be important. Note that dust.emp.sex has only 3 df because one interaction parameter is not estimable.

The residual deviance for the estimable four- and five-way interactions is very small: the observed value of 2.71 is at the lower 0.06 percentage point of χ^2_{14}.

Table 4.11. Analysis of deviance table, byssinosis

Source	Deviance	df
dust	252.11	2
emp	15.10	2
smok	11.43	1
sex	0.30	1
race	0.32	1
dust.emp	4.51	4
dust.smok	4.75	2
dust.sex	4.83	2
dust.race	0.25	2
emp.smok	1.94	2
emp.sex	2.04	2
emp.race	7.33	2
smok.sex	0.26	1
smok.race	1.26	1
sex.race	0.45	1
dust.emp.smok	0.91	4
dust.emp.sex	5.32	3*
dust.emp.race	1.52	4
dust.smok.sex	1.47	2
dust.smok.race	0.07	2
dust.sex.race	1.31	2
emp.smok.sex	0.37	2
emp.smok.race	1.22	2
emp.sex.race	0.00	2
smok.sex.race	0.77	1
Estimable four- and five-way interactions	2.71	14

This illustrates the failure of the asymptotic χ^2 distribution for the residual deviance when the table becomes very sparse, as the model more nearly reproduces the observed data. The failure of the residual deviance to follow the χ^2 distribution invalidates the usual test procedure for model selection based on the LRT, since this compares the size of the residual deviance from the current model to the χ^2 percentage point.

However, the same principle of model reduction by backward elimination can be applied to the three-way interaction model instead of the saturated model. We carefully examine the individual components of terms with multiple degrees of freedom, and continue elimination until a point is reached where the removal of any remaining term in the model produces a large change in deviance, so that no further simplification is possible without major distortion.

Inspection of the analysis of deviance table shows that all the three-way interactions are small and can be omitted, with the possible exception of dust.emp.sex. We omit all the three-way interactions except this one, and examine its contribution. Typing of large numbers of interactions in specifying the model can be avoided by using the dot product of the main effects model with itself; GLIM interprets terms like dust.dust as main effects.

```
$fit (dust + emp + smok + sex + race).
     (dust + emp + smok + sex + race) + dust.emp.sex $
 scaled deviance =  10.69 at cycle 8
      residual df =  35    from 65 observations
```

The deviance is 10.69, compared to 15.67 for the model with all two-way interactions. The deviance for the dust.emp.sex interaction is thus 4.98 when it is fitted as the first of the three-way interactions, and this interaction can be omitted as well. We list the estimates only for the interactions.

```
$fit - dust.emp.sex $disp e $
 scaled deviance =  15.667 (change =   ¡4.973) at cycle 5
      residual df =  38     (change =   +3    ) from 65 observations
```

	estimate	s.e.	parameter
...			
9	-0.5733	1.346	DUST(2).EMP(2)
10	-0.4642	1.141	DUST(2).EMP(3)
11	-1.936	1.025	DUST(3).EMP(2)
12	-0.9466	0.7984	DUST(3).EMP(3)
13	1.393	0.7265	DUST(2).SMOK(2)
14	0.7709	0.5646	DUST(3).SMOK(2)
15	0.2158	0.6934	EMP(2).SMOK(2)
16	-0.06715	0.5677	EMP(3).SMOK(2)
17	1.990	0.9661	DUST(2).SEX(2)
18	1.236	0.8108	DUST(3).SEX(2)
19	-0.6776	1.052	EMP(2).SEX(2)
20	-1.205	0.7924	EMP(3).SEX(2)
21	-0.2696	0.5719	SMOK(2).SEX(2)
22	-0.6817	1.143	DUST(2).RACE(2)
23	-1.077	0.8147	DUST(3).RACE(2)
24	-1.857	0.8024	EMP(2).RACE(2)
25	-1.504	0.7114	EMP(3).RACE(2)
26	0.6363	0.5503	SMOK(2).RACE(2)
27	-0.5545	0.8260	SEX(2).RACE(2)

scale parameter 1.000

Inspection of the table suggests that emp.smok, smok.sex, smok.race and sex.race can all be omitted, but emp.race is large, and some other terms may need to be retained. We refit the model omitting the first four interactions and examine the parameter estimates to see which other terms can be omitted.

```
$fit - emp.smok - smok.(sex + race) - sex.race $disp e $
  scaled deviance =  18.704 (change =   +3.037) at cycle 5
      residual df =  43      (change =   +5    ) from 65 observations


            estimate          s.e.        parameter
 ...
       9     -0.7349          1.298        DUST(2).EMP(2)
      10     -0.6501          1.090        DUST(2).EMP(3)
      11     -2.019           1.009        DUST(3).EMP(2)
      12     -1.015           0.7768       DUST(3).EMP(3)
      13      1.057           0.5684       DUST(2).SMOK(2)
      14      0.4390          0.4466       DUST(3).SMOK(2)
      15      2.117           0.9449       DUST(2).SEX(2)
      16      1.430           0.7688       DUST(3).SEX(2)
      17     -0.3540          0.9242       EMP(2).SEX(2)
      18     -0.9906          0.6140       EMP(3).SEX(2)
      19     -0.8853          0.9753       DUST(2).RACE(2)
      20     -1.222           0.7273       DUST(3).RACE(2)
      21     -1.788           0.7858       EMP(2).RACE(2)
      22     -1.396           0.6914       EMP(3).RACE(2)
  scale parameter 1.000
```

The emp.sex parameter estimates are now the least important, and we omit them:

```
$fit - emp.sex $disp e $
  scaled deviance =  21.519 (change =   +2.814) at cycle 5
      residual df =  45      (change =   +2    ) from 65 observations


            estimate          s.e.        parameter
 ...
       9     -1.020           1.092        DUST(2).EMP(2)
      10     -1.383           0.9810       DUST(2).EMP(3)
      11     -2.217           0.8962       DUST(3).EMP(2)
      12     -1.440           0.7193       DUST(3).EMP(3)
      13      1.006           0.5668       DUST(2).SMOK(2)
      14      0.3868          0.4462       DUST(3).SMOK(2)
      15      1.706           0.8552       DUST(2).SEX(2)
      16      1.032           0.7005       DUST(3).SEX(2)
```

```
    17        -0.8428      0.9741      DUST(2).RACE(2)
    18        -1.151       0.7278      DUST(3).RACE(2)
    19        -1.785       0.7843      EMP(2).RACE(2)
    20        -1.389       0.6930      EMP(3).RACE(2)
scale parameter 1.000
```

The dust.race terms are now the least important, and we omit them:

```
$fit - dust.race $disp e $
  scaled deviance =  24.511 (change =   +2.992) at cycle 4
      residual df =  47     (change =   +2   ) from 65 observations

            estimate        s.e.      parameter
  ...
     9        -0.3495      0.7883      DUST(2).EMP(2)
    10        -0.6710      0.5874      DUST(2).EMP(3)
    11        -1.371       0.6801      DUST(3).EMP(2)
    12        -0.5744      0.4241      DUST(3).EMP(3)
    13         0.9957      0.5660      DUST(2).SMOK(2)
    14         0.3782      0.4461      DUST(3).SMOK(2)
    15         1.685       0.8536      DUST(2).SEX(2)
    16         1.021       0.7001      DUST(3).SEX(2)
    17        -1.146       0.6041      EMP(2).RACE(2)
    18        -0.7706      0.4896      EMP(3).RACE(2)
scale parameter 1.000
```

The dust.smok terms are now the least important and we omit them:

```
$fit - dust.smok $disp e $
  scaled deviance =  27.721 (change =   +3.211) at cycle 4
      residual df =  49     (change =   +2   ) from 65 observations

            estimate        s.e.      parameter
  ...
     9        -0.3618      0.7877      DUST(2).EMP(2)
    10        -0.5624      0.5850      DUST(2).EMP(3)
    11        -1.385       0.6787      DUST(3).EMP(2)
    12        -0.5636      0.4234      DUST(3).EMP(3)
    13         1.909       0.8408      DUST(2).SEX(2)
    14         1.082       0.6928      DUST(3).SEX(2)
    15        -1.140       0.6020      EMP(2).RACE(2)
    16        -0.7691      0.4883      EMP(3).RACE(2)
scale parameter 1.000
```

Inspection of the parameter estimates suggests that dust.emp can be omitted, except perhaps for dust(3).emp(2):

```
$calc d3e2 = %eq(dust,3)*%eq(emp,2) $
$fit - dust.emp + d3e2 : - d3e2 $disp e $
 scaled deviance =  29.883 (change =   +2.162) at cycle 4
    residual df =  52     (change =   +3    ) from 65 observations
 scaled deviance =  33.241 (change =   +3.358) at cycle 4
    residual df =  53     (change =   +1    ) from 65 observations
          estimate         s.e.      parameter
  ...
        9       2.045      0.8316     DUST(2).SEX(2)
       10       1.289      0.6825     DUST(3).SEX(2)
       11      -0.7427     0.5575     EMP(2).RACE(2)
       12      -0.5947     0.4738     EMP(3).RACE(2)
 scale parameter 1.000
```

The deviance change on omitting dust(3).emp(2) is only 3.36. The emp.race interactions are now quite small and can be omitted:

```
$fit - emp.race $disp e $
 scaled deviance =  35.721 (change =   +2.480) at cycle 4
    residual df =  55     (change =   +2    ) from 65 observations

          estimate         s.e.      parameter
        1      -1.847      0.2339     1
        2      -3.243      0.5140     DUST(2)
        3      -2.849      0.2482     DUST(3)
        4       0.5030     0.2615     EMP(2)
        5       0.6979     0.2157     EMP(3)
        6      -0.6582     0.1945     SMOK(2)
        7      -1.001      0.6106     SEX(2)
        8       0.1127     0.2065     RACE(2)
        9       2.002      0.8311     DUST(2).SEX(2)
       10       1.261      0.6823     DUST(3).SEX(2)
 scale parameter 1.000
```

The dust.sex interaction is large and is retained, as are the marginal main effects of dust and sex. The race variable can clearly be omitted, but the smok and emp main effects are large and must be retained:

```
$fit - race $disp e $
 scaled deviance =  36.019 (change =   +0.2978) at cycle 4
    residual df =  56     (change =   +1    ) from 65 observations
```

	estimate	s.e.	parameter
1	-1.753	0.1564	1
2	-3.277	0.5101	DUST(2)
3	-2.878	0.2422	DUST(3)
4	0.4640	0.2512	EMP(2)
5	0.6367	0.1836	EMP(3)
6	-0.6578	0.1945	SMOK(2)
7	-0.9990	0.6106	SEX(2)
8	2.006	0.8309	DUST(2).SEX(2)
9	1.258	0.6823	DUST(3).SEX(2)

scale parameter 1.000

The two interaction terms differ by 0.75, or about one standard error, which suggests that they can be equated; the same is true for the main effect parameters. Equating dust(2) and dust(3) is equivalent to collapsing these two categories into one.

```
$factor d2 2 $
$calc d2 = 2 - %eq(dust,1) $
$fit   dust.sex + d2 + d2.sex $disp e $
  scaled deviance =   37.443 (change =    +1.425) at cycle 4
     residual df =   57      (change =    +1    ) from 65 observations
```

	estimate	s.e.	parameter
1	-1.755	0.1565	1
2	-2.873	0.3166	DUST(2)
3	-2.973	0.2362	DUST(3)
4	0.4636	0.2511	EMP(2)
5	0.6409	0.1836	EMP(3)
6	-0.6578	0.1945	SMOK(2)
7	-0.9977	0.6106	SEX(2)
8	0.000	aliased	D2(2)
9	1.447	0.6626	SEX(2).D2(2)

scale parameter 1.000

The aliased factor d2 is included in the model to allow a direct comparison between the parameter estimates for d2.sex and dust.sex.

The deviance increase is very small, and the parameter estimates for dust(2) and dust(3) are now almost equal. We remove the three-level dust factor and retain the two-level d2 factor.

```
$fit - dust $disp e $
  scaled deviance =   37.564 (change =    +0.1203) at cycle 4
     residual df =   58      (change =    +1    ) from 65 observations
```

```
          estimate          s.e.      parameter
    1      -1.754           0.1565     1
    2       0.4617          0.2510     EMP(2)
    3       0.6406          0.1836     EMP(3)
    4      -0.6585          0.1945     SMOK(2)
    5      -0.9980          0.6106     SEX(2)
    6      -2.951           0.2269     D2(2)
    7       1.459           0.6616     SEX(2).D2(2)
scale parameter 1.000
```

A similar recoding of emp can be used to equate levels 2 and 3:

```
$factor e2 2 $
$calc e2 = 2 - %eq(emp,1) $
$fit - emp + e2 $disp e r $
 scaled deviance =  38.105 (change =  +0.5416) at cycle 4
     residual df =  59       (change =  +1     ) from 65 observations
```

```
          estimate          s.e.      parameter
    1      -1.758           0.1565     1
    2      -0.6485          0.1940     SMOK(2)
    3      -0.9931          0.6105     SEX(2)
    4      -2.949           0.2268     D2(2)
    5       0.5923          0.1724     E2(2)
    6       1.453           0.6614     SEX(2).D2(2)
scale parameter 1.000
```

Only one residual (not shown here) exceeds 2.0 in magnitude: 2.27 at the 38th cell, where an observed proportion of $1/17 = 0.059$ with byssinosis is modelled by a fitted proportion of $0.144/17 = 0.008$. Thus the fitted model seems satisfactory.

Before proceeding to interpret this model, a word of warning is in order. We equated levels 2 and 3 of emp because the difference between their parameter estimates was not large compared to the individual standard errors. Close examination of these parameter estimates suggests two other possible interpretations, however:

(1) since emp(2) does not differ significantly from zero, levels 1 and 2 of emp could instead be equated;
(2) since the values 0, 0.462 and 0.641 are roughly linear, we could replace the factor emp by a variate taking the same values.

We illustrate the latter model:

```
$calc line = emp $
$fit - e2 + line $disp e $
```

```
scaled deviance =   37.930  (change =   -0.1749) at cycle 4
    residual df =   59      (change =   0      ) from 65 observations

            estimate        s.e.      parameter
   1         -2.047        0.2182     1
   2         -0.6651       0.1942     SMOK(2)
   3         -1.021        0.6092     SEX(2)
   4         -2.950        0.2269     D2(2)
   5          0.3186       0.09102    LINE
   6          1.478        0.6608     SEX(2).D2(2)
scale parameter 1.000
```

The deviance is 37.93, slightly smaller than the value 38.11 when levels 2 and 3 of emp are equated. We cannot choose between these models: the data do not provide enough evidence to discriminate between the model in which risk is the same for the two longer-term exposure categories and that in which risk increases steadily with exposure. To make such a discrimination the actual number of years worked in the industry for each individual would be necessary, or at least a finer

Table 4.12. Fitted and observed byssinosis proportions, and (sample sizes)

		dust 2/3		dust 1	
		Women	Men	Women	Men
Non-smokers	emp 1	0.007	0.005	0.032	0.084
		0.015	0.006	0.034	0.062
		(676)	(340)	(29)	(97)
	emp 2	0.010	0.007	—	0.112
		0.008	0.012	—	0.150
		(129)	(82)	(0)	(20)
	emp 3	0.014	0.009	0.059	0.147
		0.009	0.012	0.000	0.114
		(537)	(248)	(2)	(70)
Smokers	emp 1	0.014	0.009	0.060	0.151
		0.013	0.007	0.069	0.137
		(687)	(667)	(29)	(204)
	emp 2	0.020	0.013	—	0.196
		0.022	0.007	—	0.239
		(137)	(277)	(0)	(67)
	emp 3	0.027	0.017	0.108	0.251
		0.022	0.019	0.000	0.275
		(275)	(695)	(2)	(149)

classification of emp. The parameter estimates for variables other than emp are very little affected by this ambiguity; we base our interpretation on the last line model.

The odds on having byssinosis for smokers are greater than for non-smokers by a factor of $e^{0.665} = 1.94$, or approximately 2, and increase steadily (in this model) with increasing duration of employment in the industry. Interpretation of other features of the model is simplified if we tabulate the observed and fitted proportions by the factors smok, emp, sex and d2. race can be omitted. We suppress the individual tables.

```
$calc fp = %fv/n $
$tab the fp mean for smok,emp,d2,sex with n $
  :   the  p mean for smok,emp,d2,sex with n $
```

The sample sizes can also be tabulated in the same way.

```
$tab the n total for smok,emp,d2,sex $
```

The fitted and observed proportions and sample sizes are shown in Table 4.12 to 3 dp (in vertical order fitted, observed, n) so that the fitted proportion with byssinosis increases across and down the table. Two cells have no observations. It is now easily seen that there is a striking increase in the incidence of byssinosis for men in the dustiest working conditions. Another interesting feature is that smokers with less than 10 years' employment in the industry have the same incidence rates as non-smokers with more than 20 years' employment. The incidence of byssinosis is very high—one in four—amongst smoking men with more than 20 years' employment who are working in the dustiest conditions. Amongst smoking women under the same conditions the byssinois rate is 11%.

5
Multinomial and Poisson response data

5.1 The Poisson distribution

The file `faults` gives the number n of faults in 32 rolls of material of length ℓ metres; the data come from Bissell (1972), and are reproduced in Table 5.1.

The number of faults is a non-negative integer, and is naturally modelled by the *Poisson* distribution, the standard model for count data.

The number of faults N in a roll of material is modelled by

$$\Pr(N = n|\mu) = e^{-\mu}\mu^n/n!, \quad \mu > 0, \quad n = 0, 1, 2, \ldots,$$

where μ is the mean number of faults, and is related to the linear predictor $\eta = \boldsymbol{\beta}'\mathbf{x}$ through

$$g(\mu) = \eta,$$

where the link function g is generally taken to be the (default) log, guaranteeing positive fitted values:

$$\log \mu = \eta = \boldsymbol{\beta}'\mathbf{x}$$

though the identity link is sometimes used. The fitting of the Poisson model in GLIM is achieved using `$error p` and the appropriate link and model specifications.

For the fault data, the logical model would have the mean number of faults proportional to the length of the roll, so that $\mu = c\ell$ for some constant c. Using the log link gives $\eta = \log \mu = c' + \log \ell$; under the proportionality model the coefficient of $\log \ell$ should be 1.

We fit the regression on $\log \ell$ with log link:

Table 5.1. Fault data

ℓ	551	651	832	375	715	868	271	630	491	372	645
n	6	4	17	9	14	8	5	7	7	7	6
ℓ	441	895	458	642	492	543	842	905	542	522	122
n	8	28	4	10	4	8	9	23	9	6	1
ℓ	657	170	738	371	735	749	495	716	952	417	
n	9	4	9	14	17	10	7	3	9	2	

```
$input 'faults' $
$calc ll = %log(l) $
$yvar n $err p
$fit : + ll $disp e $
scaled deviance =  103.71 at cycle 4
    residual df =   31


scaled deviance =  64.537 (change =   -39.18) at cycle 4
    residual df =   30      (change =   -1   )

          estimate          s.e.      parameter
    1       -4.173          1.135      1
    2        0.9969         0.1759     LL
scale parameter 1.000
```

The coefficient of $\log \ell$ is almost unity, and the proportionality model is well-supported. We fit it explicitly by constraining the coefficient to be 1, using an offset:

```
$offset ll $
$fit 1 $disp e $
-- model changed
scaled deviance =  64.537 at cycle 4
    residual df =   31

          estimate          s.e.      parameter
    1       -4.193          0.05934    1
scale parameter 1.000
```

The deviance change is zero to 3 dp. The fitted log-linear model is shown as the solid straight line with the data on the original scales of mean and length in Fig. 5.1.

There is no sign of non-linearity in the plot, but the deviance from the fitted model (i.e. the LRT statistic for the linear model relative to the saturated model) is very large (64.54) compared with its degrees of freedom (31), and examination of the residuals shows several which are very large. We have clear evidence of *overdispersion*: the most plausible explanation is that there are other factors varying over the data which are affecting the chance of a fault. Since these factors are not included in the model, they inflate the variability of the data beyond the Poisson variance, and appear in the large residuals. We can illustrate this further by graphing the fitted Poisson variance, in the form of 2SD bounds above and below the mean, also shown in Fig. 5.1 by dashed curves:

```
$calc sd = %sqrt(%fv) $
    : lb = %fv - 2*sd $
    : ub = %fv + 2*sd $
```

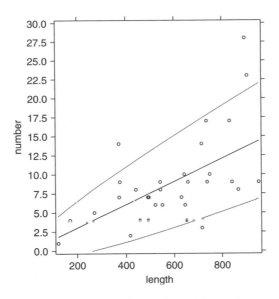

Fig. 5.1. Observed and fitted values with 2SD bounds

```
$sort sn,sfv,slb,sub,sl n,%fv,lb,ub,l l $
$graph (s = 1 h = 'length' v = 'number' y = 0,30)
        sn,sfv,slb,sub sl 5,10,11,11 $
```

Note that the observations need to be sorted by length, otherwise adjacent points in the original data set are connected, instead of points adjacent in roll length.

Four points fall outside the 2SD limits, two of them far outside. Although the log-linear model appears appropriate for the relationship between length of roll and number of faults, the Poisson distribution understates the extent of variability in the data. The precision of the regression coefficient estimates is overstated from the Poisson model, and the estimates themselves may be biased. We deal with these problems in detail in Chapter 8, and postpone further discussion of this example until then.

5.2   Cross-classified counts

The file claims gives the number of policyholders n of an insurance company who were "exposed to risk", and the number c of car insurance claims made in the third quarter of 1973 by these policyholders, arranged as a contingency table cross-classified by three four-level factors: dist, the district in which the policyholder lived (1: rural, 2: small towns, 3: large towns, 4: major cities), car, the engine capacity of the car (1: <1 litre, 2: 1–1.5 litres, 3: 1.5–2 litres, 4: >2 litres), and age, the age of the policyholder (1: <25, 2: 25–29, 3: 30–35, 4: >35). The data are shown in Table 5.2, adapted from Baxter *et al.* (1980).

Table 5.2. Claims data

		age							
		1		2		3		4	
dist	car	n	c	n	c	n	c	n	c
1	1	197	38	264	35	246	20	1680	156
	2	284	63	536	84	696	89	3582	400
	3	133	19	286	52	355	74	1640	233
	4	24	4	71	18	99	19	452	77
2	1	85	22	139	19	151	22	931	87
	2	149	25	313	51	419	49	2443	290
	3	66	14	175	46	221	39	1110	143
	4	9	4	48	15	72	12	322	53
3	1	35	5	73	11	89	10	648	67
	2	53	10	155	24	240	37	1635	187
	3	24	8	78	19	121	24	692	101
	4	7	3	29	2	43	8	245	37
4	1	20	2	33	5	40	4	316	36
	2	31	7	81	10	122	22	724	102
	3	18	5	39	7	68	16	344	63
	4	3	0	16	6	25	8	114	33

We wish to model the relation between the frequency of claims and the explanatory variables. At first sight the appropriate probability model appears to be binomial: we might reasonably suppose that each policyholder has a constant probability p of making a claim, so that the number C would be binomial $b(n, p)$, with p modelled as a logistic function of the explanatory variables. However, it is possible (though unlikely) for any policyholder to submit several claims during the quarter, so the binomial distribution may not be appropriate: we examine the use of the binomial distribution at the end of this section.

In the i-th cell of the contingency table, there are n_i policyholders. We model the number of claims C_{ij} made by the j-th policyholder in the i-th cell as a Poisson distribution with mean μ_i, and for each cell, the C_{ij} are assumed independent. The total number of claims

$$C_i = \sum_{j=1}^{n_i} C_{ij}$$

submitted by all the policyholders in the i-th cell is the sum of n_i independent Poisson variables each with mean μ_i, and has a Poisson distribution with mean $\theta_i = n_i \mu_i$.

Table 5.3. Analysis of deviance

Source	Deviance	df
dist	12.73	3
car	87.24	3
age	84.87	3
car.age	10.51	9
dist.car	7.38	9
dist.age	6.24	9
dist.car.age	27.29	27

If the c_{ij} were observed, the model $\log \mu_i = \boldsymbol{\beta}' \mathbf{x}_i$ could be fitted directly. Since only c_i is observable, we can fit a model only to θ_i. We have

$$\log \theta_i = \log n_i + \log \mu_i$$
$$= \log n_i + \boldsymbol{\beta}' \mathbf{x}_i.$$

Thus to specify the model correctly we must include the term $\log n_i$ as an explanatory variable with a coefficient of 1, that is, $\log n_i$ must be taken as an offset for the model. We suppress the output.

```
$input 'claims' $
$yvar c $err p
$calc ln = %log(n) $
$offset ln $
$fit : + dist : + car : + age $
     : + car.age : + dist.car : + dist.age $
```

Here car.age is fitted before the other interactions as the one most likely to be important. We construct an analysis of deviance table in Table 5.3 for the factors in the order fitted.

The observed values of c are generally large, so the standard asymptotic χ^2 distribution theory for deviance differences should hold adequately. The three-way interaction term is not explicitly fitted as its deviance is just the residual deviance from the model with all two-way interactions. All the interactions are close to their degrees of freedom, suggesting that the main effects model with a residual deviance of 51.42 and 54 df provides a good fit. This model was used by Baxter *et al.* (1980). We check the fit at each sample point.

```
$fit dist + car + age $
$print %rs $
 1.087 -0.046 -1.541 -0.228  1.228 -0.018 -0.485  0.097 -2.279
-0.612  1.774  0.217 -1.080  0.327 -0.207  0.220  2.101 -0.014
 1.008 -0.353 -0.753  0.080 -1.167  0.683 -0.553  1.754  0.084
```

```
-1.311   0.849   1.011  -0.744  -0.256  -0.364   0.271  -0.183   0.415
-0.145  -0.257   0.591  -0.133   0.828   0.738   0.577   0.026   0.649
-1.908  -0.331  -0.815  -1.033  -0.243  -0.744  -0.288  -0.164  -1.515
 0.273   0.031  -0.193  -0.882   0.371   0.269  -1.038   0.575   0.651
 1.853
```

The Pearson residuals for the Poisson distribution are

$$e_i = (y_i - \hat{\theta}_i)/\sqrt{\hat{\theta}_i}.$$

These are approximately standardized variables with mean 0 and variance approximately 1; the variance is approximate because of the estimation of β in the linear predictor.

There are two residuals exceeding 2 in absolute value: -2.279 at observation 9 and 2.101 at observation 17. Since the observed numbers of claims c_i are fairly large, the residuals e_i can be treated as approximately standard normal variables, and a normal quantile plot provides a check on their assumed distribution.

The fit to a straight line (not shown) is very close, so the assumption of a Poisson model is well supported.

In fitting the model, we have specified that ln has a fixed coefficient of 1. This is implicit in the structure of the model, but it may not be in accordance with the data. To check this, we fit ln explicitly, removing the offset.

```
$offset $
$fit + ln $disp e $
-- model changed
   scaled deviance =   49.450 (change =    -1.970) at cycle 3
       residual df =   53      (change =    -1    )

                estimate        s.e.      parameter
        1        -2.778        0.689         1
....
       11         1.202        0.144        LN
scale parameter 1.000
```

The deviance decreases by only 1.97 on 1 df, and the parameter estimate for ln is within 1.4 standard errors of the offset value 1. The value of 1 is in accordance with the data. Now we examine the other parameter estimates.

```
$offset ln $
$fit - ln $d e $
 -- model changed
   scaled deviance =   51.420 (change =    +1.970) at cycle 3
       residual df =   54      (change =    +1    )
```

	estimate	s.e.	parameter
1	-1.822	0.077	1
2	0.026	0.043	DIST(2)
3	0.039	0.051	DIST(3)
4	0.234	0.062	DIST(4)
5	0.161	0.051	CAR(2)
6	0.393	0.055	CAR(3)
7	0.563	0.072	CAR(4)
8	-0.191	0.083	AGE(2)
9	-0.345	0.082	AGE(3)
10	-0.537	0.070	AGE(4)

scale parameter 1.000

The parameters for DIST(2) and DIST(3) are smaller than their standard errors, but that for DIST(4) is nearly four times its standard error. It appears that the first three districts can be amalgamated, only district four—London and other major cities—being kept separate.

```
$calc d4 = %eq(dist,4) $fit - dist + d4 $disp e $
  scaled deviance =  52.135 (change =   +0.715) at cycle 3
      residual df =  56       (change =   +2     )
```

	estimate	s.e.	parameter
1	-1.810	0.075	1
2	0.162	0.051	CAR(2)
3	0.394	0.055	CAR(3)
4	0.566	0.073	CAR(4)
5	-0.189	0.083	AGE(2)
6	-0.342	0.082	AGE(3)
7	-0.533	0.070	AGE(4)
8	0.219	0.058	D4

scale parameter 1.000

The car estimates and the age estimates show similar patterns of nearly linear increase or decrease with factor level, suggesting that each factor can be replaced by the corresponding variable.

```
$calc lcar = car $fit - car + lcar $disp e $
  scaled deviance =  52.993 (change =   +0.8579) at cycle 3
      residual df =  58       (change =   +2     )
```

	estimate	s.e.	parameter
1	-2.026	0.080	1
2	-0.190	0.083	AGE(2)

```
       3      -0.344        0.082       AGE(3)
       4      -0.535        0.070       AGE(4)
       5       0.219        0.059       D4
       6       0.198        0.021       LCAR
scale parameter 1.000


$calc lage = age $fit - age + lage $disp e $

scaled deviance =   53.111 (change =   +0.1185) at cycle 3
    residual df =   60      (change =   +2      )

            estimate          s.e.        parameter
       1     -1.853          0.080        1
       2      0.219          0.059        D4
       3      0.198          0.021        LCAR
       4     -0.177          0.019        LAGE
scale parameter 1.000
```

The parameter estimates for lcar and lage differ in magnitude by one stand-ard error: it appears that we can simplify the model further by equating their magnitudes.

```
$calc score = lcar - lage $fit - lcar - lage + score $disp e $
  scaled deviance =   53.695 (change =   +0.5835) at cycle 3
      residual df =   61      (change =   +1      )

            estimate          s.e.        parameter
       1     -1.795          0.024        1
       2      0.220          0.059        D4
       3      0.186          0.014        SCORE
scale parameter 1.000
```

The coefficient of d4 is close to that for score: being in district 4 is nearly equivalent to one point on the score scale. We try making them equivalent.

```
$calc s2 = d4 + score $fit - d4 - score + s2 $disp e $
  scaled deviance =   54.016 (change =   +0.3217) at cycle 3
      residual df =   62      (change =   +1      )

            estimate          s.e.        parameter
       1     -1.789          0.022        1
       2      0.188          0.014        S2
scale parameter 1.000
```

Thus each point up the score scale adds 0.188 to the fitted mean log number of claims, so the fitted mean is multiplied by $\exp(0.188) = 1.207$—the mean number of claims increases by 20%. The score scale runs from -3 to $+3$, with an extra point for district 4; the fitted mean number of claims for an individual with scale score -3 is $\exp(-1.789 - 3 \times 0.188) = 0.095$.

We present finally the observed mean numbers of claims per individual, c_i/n_i, by car and age, for districts 1–3, and for district 4 separately, together with the fitted means. These are arranged in a table with the age classification reversed, so that score increases from the top left-hand to the bottom right-hand corner. Within each cell of the table, the order of districts is

```
1   2,
3
```

with district 3 below district 1. We label the cells with the value of score $+4$, which runs from 1 to 8.

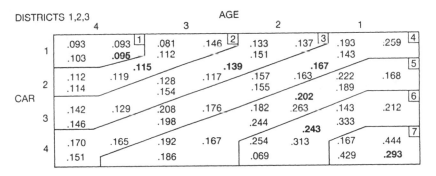

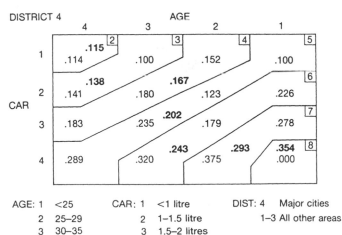

AGE: 1	<25	CAR: 1	<1 litre	DIST: 4	Major cities
2	25–29	2	1–1.5 litre		1–3 All other areas
3	30–35	3	1.5–2 litres		
4	>35	4	>2 litres		

The fitted values accurately reproduce the observed means where the numbers exposed are large. The simplification of the table, as in the example in Chapter 4 on coronary heart disease, is very great: only two numbers are necessary to describe the whole 64-cell table.

It can be verified that a very similar main effects model results, with a deviance of 63.09 on 54 df, if c is assumed to be binomial $b(n, p)$ with a logit link for p (Baxter *et al.*, 1980). This is because the mean claim frequency is low, even in the highest score category, and so the probability of more than one claim in the quarter, which is $\{1 - (\mu + 1)e^{-\mu}\}$, is very small – 0.049 for the largest mean value of 0.353.

5.3   Multicategory responses

The file miners gives the numbers of coalminers classified by radiological examination into one of three categories of pneumoconiosis (n—normal, m—mild pneumoconiosis, s—severe pneumoconiosis), and by period p spent working at the coalface (interval midpoint). Period p is declared as a factor with eight levels, the levels corresponding to midpoints of class intervals: 5.8, 15.0, 21.5, 27.5, 33.5, 39.5, 46.0 and 51.5 years worked at the coalface. The data are discussed in Ashford (1959) and in McCullagh and Nelder (1989, p. 179), and are given in Table 5.4 and in the file miners.

We first graph the proportions of miners in each category against years worked (Fig. 5.2).

```
$input 'miners' $
$calc t = n + m + s : np = n/t : mp = m/t : sp = s/t $
$graph (s = 1 h = 'years' v = 'proportion')
        np,mp,sp years 5,2,8 $
```

With many points in the graph, it becomes difficult to see the trends with years. We add connecting lines to the points to clarify the trends, by defining new styles with both points and lines (Fig. 5.3), and add symbols.

```
$gstyle 21 line 1 sym 5 col 1 :
        22 line 2 sym 2 col 1 :
        23 line 3 sym 8 col 1 $
$graph (s = 1 h = 'years' v = 'proportion')
        np,mp,sp years 21,22,23 $
$gtext 0.52 42 'n' : 0.29 42 's' : 0.21 42 'm' $
```

The proportions free of pneumoconiosis decline steadily with years, and those for the other two categories increase; there is a noticeable increase in the proportion with severe pneumoconiosis in the 7th and 8th periods.

Table 5.4. Pneumoconiosis level

Years	n	m	s
5.8	98	0	0
15.0	51	2	1
21.5	34	6	3
27.5	35	5	8
33.5	32	10	9
39.5	23	7	8
46.0	12	6	10
51.5	4	2	5

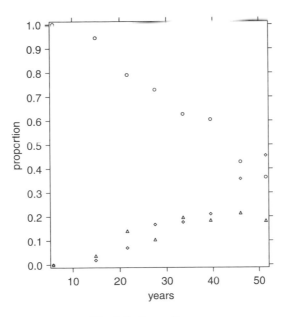

Fig. 5.2. Proportions

How should multiple response proportions be modelled? One simple way would be to convert the three categories to two by collapsing two adjacent categories into one. For example, the second and third categories could be collapsed, and we could model the logit of the probability of no pneumoconiosis.

```
$yvar n $err b t $link g
$fit : + p $disp e $
scaled deviance = 97.564 at cycle 4
          d.f. = 7
```

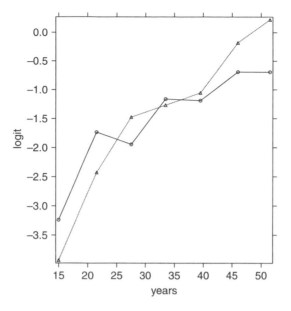

Fig. 5.6. Empirical logits

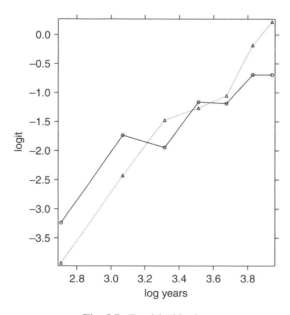

Fig. 5.7. Empirical logits

We want now to fit simultaneously the multinomial logit models

$$\theta_{2i} = \beta_{20} + \beta_{21} \log (\text{years}_i),$$
$$\theta_{3i} = \beta_{30} + \beta_{31} \log (\text{years}_i), \quad i = 1, \ldots, 8,$$

assess the goodness of fit, and convert the fitted models back to fitted multinomial probabilities. This cannot be done directly, since GLIM does not include the multinomial distribution in its family of standard distributions. The model is fitted indirectly by using the relation between the multinomial and Poisson distributions, which we now describe.

5.5   The Poisson-multinomial relation

We begin with the simple example of Cox (1958). Two radioactive sources are being compared. We count the number of particles n_1 and n_2 emitted from the sources in the same fixed time. The question of interest is—are the sources composed of the same radioactive material, or more generally, are the sources emitting particles at the same rate?

We assume the Poisson model for radioactive disintegrations, so the numbers N_1 and N_2 of particles emitted in a fixed time will be Poisson with means μ_1 and μ_2 for the two sources. Define the parameter of interest to be $\theta = \mu_2/\mu_1$, and define a *nuisance parameter* $\phi = \mu_1 + \mu_2$. If the sources are the same, then $\mu_1 = \mu_2$, or $\theta = 1$.

The likelihood in μ_1, μ_2 from observed counts n_1 and n_2 is

$$L(\mu_1, \mu_2) = e^{-\mu_1 - \mu_2} \mu_1^{n_1} \mu_2^{n_2} / (n_1! n_2!).$$

Transforming to the parameters θ and ϕ, we have

$$L(\theta, \phi) = e^{-\phi} \left(\frac{\phi}{1 + \theta} \right)^{n_1} \left(\frac{\theta\phi}{1 + \theta} \right)^{n_2} \bigg/ (n_1! n_2!).$$

Write $n = n_1 + n_2$; then the likelihood can be re-written in the form

$$L(\theta, \phi) = \left[e^{-\phi} \phi^n / n! \right] \cdot \left[\frac{n!}{n_1! n_2!} \left(\frac{1}{1 + \theta} \right)^{n_1} \left(\frac{\theta}{1 + \theta} \right)^{n_2} \right].$$

This striking result shows that the likelihood factors into two components, corresponding to independent random variables: the total count N is Poisson (ϕ), and conditional on $N = n$, N_2 independently has the binomial distribution $b(n, p)$ with $p = \theta/(1 + \theta)$. Further, if the Poisson distributions themselves have log-linear models

$$\log \mu_j = \boldsymbol{\beta}_j' \mathbf{x}, \quad j = 1, 2$$

```
$yvar death $error b total $link g
$fit $disp e $
$fit clinic $disp e $
$fit + attend : + attend.clinic $
```

The deviances and parameter estimates are reproduced below for the models.

```
-------------------------------------------------
Model          Deviance   Estimate (s.e.)
-------------------------------------------------


1                17.828    -3.277
                           (0.200)
CLINIC            0.082    -4.205 + 1.756 CLIN(2)
                           (0.381) (0.450)
ATTEND+CLINIC     0.043

ATTEND*CLINIC     0.000


-------------------------------------------------
```

To use the Poisson analysis, we define a response factor of length 8 and a count vector which contains the survive and death counts, and block the explanatory variables by duplicating their values in factors of length 8. This is achieved very easily using the $assign directive.

```
$var 8 count resp pclin patt $
$assign pclin = clinic,clinic $
     :    patt = attend,attend $
     :   count = survive,death $
$factor pclin 2 patt 2 resp 2 $
$gfactor resp 2 $
```

To fit the Poisson model we redefine the standard length of vectors to be 8:

```
$slen 8
$yvar count $err p $link l
$fit resp + pclin*patt $disp e $
 scaled deviance =  17.828 at cycle 4
     residual df =   3
```

```
             estimate        s.e.     parameter
      1        5.150        0.075     1
      2       -3.277        0.200     RESP(2)
      3        0.179        0.101     PCLIN(2)
      4        0.506        0.095     PATT(2)
      5       -2.653        0.232     PCLIN(2).PATT(2)
```

```
    scale parameter 1.000

$fit + resp.pclin $disp e $
  scaled deviance =  0.082289 (change =   -17.75) at cycle 3
      residual df =  2            (change =   -1   )

               estimate        s.e.        parameter
        1        5.173        0.075        1
        2       -4.205        0.381        RESP(2)
        3        0.111        0.103        PCLIN(2)
        4        0.506        0.095        PATT(2)
        5        1.756        0.450        RESP(2).PCLIN(2)
        6       -2.653        0.232        PCLIN(2).PATT(2)
  scale parameter 1.000

$fit + resp.patt $
  scaled deviance =  0.043256 (change =  -0.03903) at cycle 3
      residual df =  1            (change = -1        )

$fit + resp.pclin.patt $
  scaled deviance =  4.907e-14 (change =  -0.04326) at cycle 2
      residual df =        0    (change = -1        )
```

The coefficient and standard error for resp in the Poisson model are identical
to the intercept estimate and standard error in the null logit model. The nuisance parameters for pclin and patt serve only to reproduce the marginal
totals over the resp factor in each clinic/attendance cell. If these factors
are omitted, the Poisson model does not correspond to the multinomial logit
model because the marginal totals are being treated as random variables rather
than fixed, as assumed in the logit model. It can be verified that the fitted counts added over the two response categories do reproduce the observed
margins:

```
$tab the %fv total for pclin,patt $

    PATT      1         2
    PCLIN
       1  179.00   297.00
       2  214.00    25.00

    :  the %yv total for pclin,patt $
```

```
   4      -0.02572        0.01977      RESP(2).YR
   5       0.000          aliased      RESP(3).YR

   eliminated term: PERIOD
```

The first model is equivalent to the null multinomial model, and the second is equivalent to the multinomial model using yr. The parameter estimates appear confusing because the coefficient of the *last* category of resp has been set to zero instead of the first, as we might have expected. This is because the main effect of yr has not been included in the model. The standard parametrization can be recovered by adding yr to the model:

```
$fit + yr $disp e r $
scaled deviance =   13.928 (change =   0.) at cycle 4
      residual df =   12       (change =   0 )

              estimate         s.e.      parameter
      1        -4.292         0.5213      RESP(2)
      2        -5.060         0.5964      RESP(3)
      3         0.000         aliased     YR
      4         0.08357       0.01528     RESP(2).YR
      5         0.1093        0.01647     RESP(3).YR

      eliminated term:  PERIOD

      unit    observed     fitted    residual
        1         98       94.761      0.333
        2         51       49.972      0.145
        3         34       37.423     -0.560
        4         35       37.964     -0.481
        5         32       34.652     -0.451
        6         23       20.575      0.535
        7         12       10.741      0.384
        8          4        2.912      0.638
        9          0        2.105     -1.451
       10          2        2.395     -0.255
       11          6        3.087      1.658
       12          5        5.171     -0.075
       13         10        7.792      0.791
       14          7        7.639     -0.231
       15          6        6.864     -0.330
       16          2        2.947     -0.552
       17          0        1.134     -1.065
       18          1        1.634     -0.496
```

19	3	2.490	0.324
20	8	4.865	1.421
21	9	8.556	0.152
22	8	9.786	-0.571
23	10	10.395	-0.122
24	5	5.141	-0.062

Note that yr is aliased, because it can be expressed as a linear function of the dummy variables for the levels of period. The fit of the model is consequently unchanged.

The deviance changes by 87.71 for the 2 df, and the residual deviance of 13.93 with 12 df looks reasonable. The largest residual is 1.658 at the 11th observation, but the residuals for the first and third categories show the same pattern of alternating signs as in the binomial logit analysis. We try the log year model.

```
$fit resp + lyr + resp.lyr $disp e r$
scaled deviance =    5.3474 at cycle 3
     residual df =   12
```

	estimate	s.e.	parameter
1	-8.936	1.579	RESP(2)
2	-11.98	1.996	RESP(3)
3	0.000	aliased	LYR
4	2.165	0.4570	RESP(2).LYR
5	3.067	0.5640	RESP(3).LYR

```
eliminated term: PERIOD
```

unit	observed	fitted	residual
1	98	97.290	0.072
2	51	50.380	0.087
3	34	36.502	-0.414
4	35	35.930	-0.155
5	32	32.608	-0.107
6	23	20.272	0.606
7	12	12.078	-0.023
8	4	3.940	0.030
9	0	0.576	-0.759
10	2	2.334	-0.219
11	6	3.687	1.205
12	5	6.184	-0.476
13	10	8.605	0.476
14	7	7.643	-0.233
15	6	6.333	-0.132

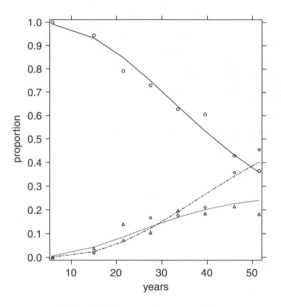

Fig. 5.8. Separate slope model

16	2	2.638	-0.393
17	0	0.135	-0.367
18	1	1.286	-0.252
19	3	2.811	0.113
20	8	5.886	0.871
21	9	9.787	-0.251
22	8	10.085	-0.657
23	10	9.588	0.133
24	5	4.422	0.275

The deviance drops to 5.35, and the slopes are 2.165 (s.e. 0.457) for the second category and 3.067 (s.e. 0.564) for the third. The alternating sign pattern in the residuals is much weaker, and the model provides a good fit. We construct and graph (in Fig. 5.8) the observed and fitted proportions from the model.

```
$tab the %fv total for period into t $
$calc f2p = %fv/t(period) $
  :  op = %yv/t(period) $
$graph (s = 1 h = 'years' v = 'proportion')
       op,f2p yr 5,2,8,10,11,12 resp,resp $
```

The severe proportion increases considerably for large values of years.

Since the estimated slopes differ by about twice either standard error, it seems unlikely that the true slopes could be equal. We investigate this question using option s of $display, which displays the standard errors of differences between parameter estimates.

```
$disp s $
standard errors of parameter estimate differences
    1        0.000
    2        2.372        0.000
    4        2.033        2.111        0.000
    5        1.751        2.558        0.6678        0.000
             1            2            4            5
```

The standard error of the difference between the slopes is 0.668; the observed difference of 0.902 is only 1.35 standard errors, so the slopes can be set equal. The estimated slopes have a large negative covariance, so the variance of the difference is substantially larger than the individual variances.

We now fit the model using a single dummy variable to estimate the common slope.

```
$calc r23 = %gt(resp,1) $
 : r23lyr = r23*lyr $
$fit resp + r23lyr $disp e $
scaled deviance =    7.2080 at cycle 3
     residual df =   13

          estimate        s.e.        parameter
    1        -10.38        1.342        RESP(2)
    2        -10.23        1.341        RESP(3)
    3         2.576        0.3857       R23LYR

    eliminated term: PERIOD
```

The deviance increases by 1.86 on 1 df. The compromise slope is 2.576 and the largest residual is 1.678 at the 11th observation.

We finally construct and graph (in Fig. 5.9) the fitted proportions from this model.

```
$tab the %fv total for period into t $
$calc flp = %fv/t(period) $
   : op = %yv/t(period) $
$graph (s = 1 h = 'years' v = 'proportion')
        op,flp yr 5,2,8,10,11,12 resp,resp $
```

In this case the fitted probabilities of the normal category are the same as those from the binomial logit model, but this happens only because the slopes for the mild and severe categories have been equated. In the resp.lyr interaction model, the fitted probabilities from the binomial and multinomial models are different.

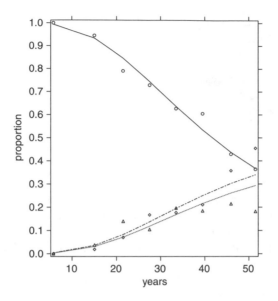

Fig. 5.9. Common slope model

In the common slope model the fitted probabilities for the mild and severe categories are very close to each other, but they are not close to the observed proportions at the upper end, where the sample sizes are small. We do not have enough sample information to discriminate the two models at the upper end of the years range.

5.7   Ordered response categories

The three-category response considered in Sections 5.3 and 5.6 had a natural order to the categories: normal, mild, severe. This is a common feature of categorical responses, and such responses often have many categories: between five and nine is quite common. The number of parameters in the multinomial models then becomes very large, and it is natural to look for simpler models with fewer parameters. Such simpler models make various assumptions about the relations among the category probabilities. We now consider a number of possibilities, in all of which only one regression coefficient vector $\boldsymbol{\beta}$ is used, rather than $(r-1)$ such vectors.

5.7.1   *Common slopes for the regressions*

The general model, in the notation of Section 5.5, is

$$\theta_1 = 0$$

$$\theta_j = \log(p_j/p_1) = \boldsymbol{\beta}'_j \mathbf{x}, \quad j = 2, \dots, r.$$

In the miners example of Section 5.6 we found that the slope estimates $\hat{\beta}_{21}$ and $\hat{\beta}_{31}$ could be equated. If the slopes are equal across categories for *all* the explanatory variables, excluding the intercepts, we can write

$$\theta_1 = 0$$
$$q_j = \gamma_j + \boldsymbol{\beta}'\mathbf{x}, \quad j = 2, \ldots, r.$$

Then

$$p_1 = 1/(1 + \delta e^{\boldsymbol{\beta}'\mathbf{x}})$$
$$p_j = \frac{e^{\gamma_j + \boldsymbol{\beta}'\mathbf{x}}}{1 + \delta e^{\boldsymbol{\beta}'\mathbf{x}}}, \quad j = 2, \ldots, r,$$

where

$$\delta = \sum_{j=2}^{r} \exp(\gamma_j).$$

Since δ can be expressed as e^{α}, it can be absorbed as an additional intercept term into $\boldsymbol{\beta}'\mathbf{x}$. The model for p_1 is then an ordinary binomial logit model for category 1 relative to the other pooled categories. This explains why the fitted logit model for the normal proportion in Section 5.3 gave the same fitted probabilities as those for the multinomial logit model in Section 5.6 in which the regression coefficients β_{21} and β_{31} were equated. The models for p_2 and p_3 are however not binomial logit models since the exponents in the numerator and denominator differ.

If the intercepts are also equated, then

$$\theta_1 = 0$$
$$\theta_j = \boldsymbol{\beta}'\mathbf{x}, \quad j = 2, \ldots, r$$

and

$$p_1 = 1/\{1 + (r-1)e^{\boldsymbol{\beta}'\mathbf{x}}\}$$
$$p_j = e^{\boldsymbol{\beta}'\mathbf{x}}/\{1 + (r-1)e^{\boldsymbol{\beta}'\mathbf{x}}\}, \quad j = 1, \ldots, r.$$

Identical models are now being fitted for all but the first category: the second and third categories are indistinguishable and are effectively being collapsed.

We continue the analysis of the miners example of Section 5.6, where we have already fitted a common slope for the response categories in the regression on lyr. Equal intercepts are fitted by replacing the resp factor in this model by r23:

```
$fit + r23 - resp $disp e $
scaled deviance =    7.6474 (change =  +0.4394) at cycle 3
    residual df =   14        (change =  +1    )
```

	estimate	s.e.	parameter
1	-10.30	1.337	R23
2	2.576	0.3859	LYR

```
$calc fpeq = %fv/t(period) $
$graph (s = 1 h = 'years' v = 'proportion' x = 0,60)
       op,fpeq yr 5,2,8,10,11,12 resp,resp$
```

The deviance increases by only 0.44 on 1 df, not surprising since the estimates for RESP(2) and RESP(3) in the previous model were very close. Note that this deviance does not correspond to the deviance of 3.13 with 6 df for the logit model in Section 5.3 because the variation between the second and third categories is suppressed in that model. The difference of 4.52 with 8 df is the variation due to constraining these two categories to have the same response probabilities. The fitted proportions from this model are shown in Fig. 5.10.

5.7.2   *Linear trend over response categories*

Suppose the regression coefficients β_j, rather than being equal, increase in equal steps with the category index (apart from the intercepts):

$$\theta_1 = 0,$$
$$\theta_j = \gamma_j + (j-1)\boldsymbol{\beta}'x, \quad j = 2,\dots,r.$$

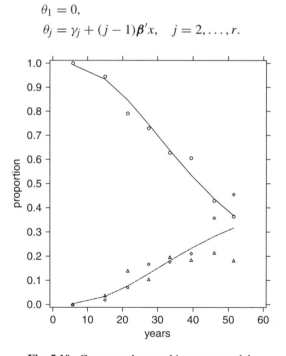

Fig. 5.10. Common slope and intercept model

Thus the regression coefficients β_j have a *linear trend* over the given ordering of the response categories. In using the Poisson representation to fit this model, we replace the interaction between response factor and explanatory variables by its linear component.

We now define the linear component of `resp`, and its interaction with `lyr`:

```
$calc lres = resp - 1 $
```

If the slopes 2.165 and 3.067 were estimates of β and 2β, then β would be roughly $5.232/3 = 1.744$.

```
$fit resp + lyr + lres + lyr.lres $disp e $
scaled deviance -    7.0493 at cycle 4
   residual df =  13
```

	estimate	s.e.	parameter
1	-7.429	0.8771	RESP(2)
2	-13.36	1.832	RESP(3)
3	0.000	aliased	LYR
4	0.000	aliased	LRES
5	1.726	0.2584	LYR.LRES

The deviance increases by 1.70 on 1 df, and the estimate of β is 1.726.

Thus the model with $\beta_{31} = 2\beta_{21}$ fits just as well as the previous model with $\beta_{31} = \beta_{21}$. As is often the case, different smoothed structures fit the data equally well. In the common slopes model the probabilities of mild and severe pneumoconiosis increase at the same rate relative to the normal group; in the linear trend model, the severe probability increases much more rapidly. We cannot distinguish between these two interpretations.

If the *intercepts* γ_j also increase in equal steps, (Haberman, 1974; Goodman, 1979), we can write

$$\theta_j = (j-1)\boldsymbol{\beta}'\mathbf{x} = b(j-1),$$

where

$$b = \boldsymbol{\beta}'\mathbf{x},$$

and

$$p_j = p_1 e^{b(j-1)}, \quad j = 2,\ldots,r$$

which is equivalent to

$$p_j = (1-\theta)\theta^{j-1}/(1-\theta^r), \quad j = 1,\ldots,r$$

with

$$\theta = e^b.$$

Thus the category probabilities $p_1, \ldots, p_r$, follow a *truncated geometric distribution* with parameter $\theta = e^{\beta' x}$.

The *practical* interpretation of this distribution in general is less interesting than the model simplification resulting: the regression coefficients are simply proportional.

Is the linear trend in the intercept model reasonable for the `miners` data? The intercept for category 3 is very close to twice that for category 2: it looks as though the model will fit well.

```
$fit - resp + lres $disp e $
scaled deviance =  24.142 (change =   +17.09) at cycle 4
    residual df =  14       (change =   +1   )


           estimate        s.e.       parameter
      1      0.000      aliased       LYR
      2     -8.085      0.9529        LRES
      3      2.087      0.2709        LYR.LRES
scale parameter 1.000
    eliminated term: PERIOD
```

Surprisingly, the deviance increases by 17.09. Why does this happen? The answer becomes clear if we examine the correlation between the RESP(2) and RESP(3) estimates.

```
$fit + resp - lres $disp c $
scaled deviance =   7.0493 (change =   -17.09) at cycle 4
    residual df =  13       (change =   -1   )

correlations between parameter estimates
    2    1.0000
    3    0.9785   1.0000
    4   -0.9786  -0.9951   1.0000
           2        3        4
```

The correlation is extremely high—0.9785. Thus the (asymptotic) variance of $\hat{\gamma}_3 - 2\hat{\gamma}_2$ is $(1.832^2 - 4 \times 0.9785 \times 1.832 \times 0.877 + 4 \times 0.877^2)$, that is, 0.1442, so the standard error is 0.380, and $(\hat{\gamma}_3 - 2\hat{\gamma}_2)/\text{s.e.}(\hat{\gamma}_3 - 2\hat{\gamma}_2) = 3.95$.

Thus the truncated geometric distribution is untenable for the proportions of normal, mild and severe pneumoconiosis, though the model with a linear trend in the slopes fits well.

5.7.3 *Proportional slopes*

A weaker constraint than the linear trend in Section 5.7.2 is proportional slopes. Suppose

$$\theta_1 = 0$$
$$\theta_j = \gamma_j + \delta_j \boldsymbol{\beta}' \mathbf{x}, \quad j = 2, \ldots, r;$$

then the regression coefficients $\boldsymbol{\beta}$ are proportional across categories (except for the intercepts). This model is called the *stereotype model* by Anderson (1984); it is not easily fitted in GLIM. If there is only one explanatory variable, the model imposes no constraint on the full multinomial logit model.

5.7.4 *The continuation ratio model*

A somewhat different model expresses the multinomial model as a succession of binomial logit models, each of which can be fitted independently. The reduction in the number of parameters comes about through connections between the logit models.

Consider the three-category multinomial model, with response counts n_1, n_2 and n_3:

$$\Pr(N_1 = n_1, N_2 = n_2, N_3 = n_3) = \frac{n!}{n_1! n_2! n_3!} p_1^{n_1} p_2^{n_2} p_3^{n_3}.$$

We define $n_{23} = n_2 + n_3$, and write the above probability function as

$$\frac{n!}{n_1! n_{23}!} p_1^{n_1} (1 - p_1)^{n_{23}} \cdot \frac{n_{23}!}{n_2! n_3!} \left(\frac{p_2}{1 - p_1} \right)^{n_2} \left(\frac{p_3}{1 - p_1} \right)^{n_3}.$$

The first term is the marginal binomial distribution of N_1, and the second is the conditional binomial distribution of N_2, given that the first response did not occur. In our example p_1 is the probability that the response is "normal", while $p_2/(1-p_1)$ is the conditional probability that the response is "mild" rather than "severe", given that it is not "normal". Now consider binomial logit models for p_1 and $p_2/(1 - p_1)$. We immediately see that the model for p_1 is exactly that considered at the beginning of Section 5.3. This model also arises, as we saw in Section 5.7.1, from the multinomial logit model if the regression coefficients β_2 and β_3 are equal, including the intercepts. However the binomial logit model parameters do not correspond in general to those of the multinomial logit model, since

$$\operatorname{logit} p_1 = \log[(p_1/(p_2 + p_3)] = -\log(e^{\boldsymbol{\beta}_2' \mathbf{x}} + e^{\boldsymbol{\beta}_3' \mathbf{x}}),$$

and only if $\boldsymbol{\beta}_2' \mathbf{x} = \boldsymbol{\beta}_3' \mathbf{x} + c$, where c is a constant, is this equivalent to a linear regression on $\mathbf{x}$.

However

$$\operatorname{logit} p_2/(1 - p_1) = \log(p_2/p_3) = (\boldsymbol{\beta}_2 - \boldsymbol{\beta}_3)'\mathbf{x}.$$

So the conditional category 2/category 3 comparison is both an ordinary logit model and equivalent to (part of) the multinomial logit model.

To fit the two logit models, we require a derived data structure, in order to relate the two models. We construct data vectors d1 and d2 of length 16, where the first 8 observations are the "normal" and "abnormal" counts, and the second 8 are the "mild" and "severe" counts, with the years variable duplicated, and a factor defined over the two sets of responses:

```
$var 16 d1 d2 lyear lresp $
$calc d1(i) = n(i) :        d1(i + 8) = m(i) $
   :    d2(i) = m(i) + s(i): d2(i + 8) = s(i) $
   :        tot = d1 + d2 $
$assign lyear = years,years $
$calc lly = %log(lyear) $
$factor lresp 2 $calc lresp = %gl(2,8) $
$slen 16 $
$yvar d1 $err b tot $link g
$fit lresp $
scaled deviance = 101.64 at cycle 4
    residual df = 13 from 15 observations

: + lly $disp e $
scaled deviance = 7.6268 (change = -94.01) at cycle 3
    residual df = 12      (change = -1   ) from 15 observations

            estimate     s.e.     parameter
       1       8.734     1.128     1
       2      -0.6825    0.2750    LRESP(2)
       3      -2.321     0.3266    LLY

$fit + lresp.lly $disp e $
scaled deviance = 4.8784 (change = -2.748) at cycle 3
    residual df = 11      (change = -1   ) from 15 observations

            estimate     s.e.     parameter
       1       9.609     1.339     1
       2      -5.745     3.003     LRESP(2)
       3      -2.576     0.3863    LLY
       4       1.440     0.8515    LRESP(2).LLY
```

The full interaction model gives unrelated logistic regressions on log years for the normal category relative to abnormal (9.609 − 2.576 lly), the same as for our original logit analysis with the mild/severe categories combined, and for the mild category relative to severe (7.033 − 1.136 lly). The deviance for this model is 4.88 with 11 df, one df less than for the multinomial logit model; this is because for the mild/severe data the first observation is 0/0 and is automatically omitted from the analysis.

The fit is similar to the multinomial logit, but slightly closer. The interaction term is not needed; omitting it gives a deviance of 7.63, similar to the common slope multinomial logit model deviance of 7.21. The common slope of −2.321 is similar, but not identical, in magnitude (but of opposite sign) to the common slope of 2.576 in the multinomial logit model. For this model the odds on abnormality versus normality increase at the same rate with years as do the odds on severe versus mild pneumoconiosis: the odds ratio "continues" with increasing severity of the response.

We now convert the three models back to the equivalent multinomial model. We first tabulate the fitted probabilities (to 3 dp) from the three models.

```
$calc fp = %fv/%bd $
$tab the fp mean for lresp,lyear $

LYEAR   5.8    15      21.5   27.5   33.5   39.5   46      51.5
LRESP
     1  0.991  0.920   0.834  0.739  0.642  0.550  0.462   0.398
     2  0.000  0.854   0.717  0.589  0.475  0.382  0.302   0.250
```

The first row of the table gives the fitted probabilities for response n in the multinomial model. The second row gives the fitted conditional probabilities for response m, conditional on m+s. To obtain the fitted multinomial probabilities for responses m and s, we multiply the n entries by the corresponding probabilities $(1 - p_n)$ to give the multinomial probabilities for category m; those for category s are obtained by difference from 1. The observed and fitted proportions are shown in Fig. 5.11.

```
$var 8 fn fm fs $
$calc fn = fp(i) $
   : fm = fp(i + 8)*(1 - fn) $
   : fs = 1 - fn - fm $
$pr 'normal' fn : ' mild' fm :'severe' fs $

normal  0.991  0.920  0.834  0.739  0.642  0.550  0.462  0.398
  mild  0.     0.068  0.119  0.154  0.170  0.172  0.163  0.151
severe  0.009  0.012  0.047  0.107  0.188  0.278  0.376  0.452
```

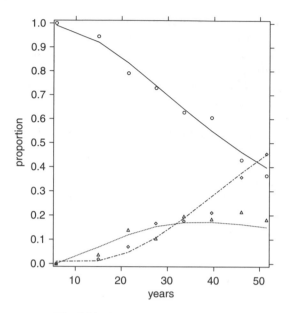

Fig. 5.11. Continuation-ratio model

```
$graph (s = 1 h = 'years' v = 'proportion')
       np,mp,sp,fn,fm,fs years 5,2,8,10,11,12 $
```

The fitted model for the "mild" and "severe" categories is quite different from the common slope multinomial logit model though it has nearly the same deviance, and is similar to the separate slopes model though more extreme, with the "mild" category probability *decreasing* beyond 33 years and the "severe" category increasing rapidly. The sparseness of data for these categories at high values of years allows different models to fit the data equally well.

Note that the unrelated models can be fitted without "blocking" the data into vectors of length 16, but the no-interaction continuation-ratio model cannot be fitted without this device.

5.7.5 *Other models*

Cumulative probabilities may be modelled directly using the same link functions as for the binomial. Define the cumulative probability P_j for the j-th category by

$$P_j = \sum_{s=1}^{j} p_s, \quad j = 1, \ldots, r - 1,$$
$$P_r = 1.$$

Then

$$g(P_j) = \boldsymbol{\beta}_j'\mathbf{x}, \quad j \neq r$$

is a model for the cumulative probability, with g the logit, probit or CLL link function. This model has as many parameters as the multinomial, so is of interest only if restricted. A useful restriction is

$$g(P_j) = \theta_j + \boldsymbol{\beta}'\mathbf{x},$$

where θ_j is an intercept parameter, since then $\boldsymbol{\beta}$ is invariant under the collapsing together of adjacent categories. If g is the logit link, this model is called the *proportional odds model*. If g is the CLL link, the model is called the *proportional hazards model*.

This class of models can be expressed in terms of a *latent variable representation*. If

$$g(P_j) = \theta_j + \boldsymbol{\beta}'\mathbf{x}$$

for some link function g, then

$$P_j = H(\theta_j + \boldsymbol{\beta}'\mathbf{x}),$$

where H is the cumulative distribution function inverse to the link function g (see Section 4.2). Then the category probabilities p_j can be represented as

$$p_j = H(\theta_j + \boldsymbol{\beta}'\mathbf{x}) - H(\theta_{j-1} + \boldsymbol{\beta}'\mathbf{x}).$$

Suppose Z is an unobserved continuous random variable with a location/scale parameter distribution with cumulative distribution function $H\{(z - \mu)/\sigma\}$, and let $\phi_1, \ldots, \phi_{r-1}$ be fixed cut-point values of Z. We observe only the interval (ϕ_{j-1}, ϕ_j) in which Z lies, with probability

$$p_j = H((\phi_j - \mu)/\sigma) - H((\phi_{j-1} - \mu)/\sigma), \quad j = 1, \ldots, r$$

with $\phi_0 = -\infty$, $\phi_r = +\infty$. This is identical to the above model if

$$\phi_j/\sigma = \theta_j$$

and

$$-\mu/\sigma = \boldsymbol{\beta}'\mathbf{x}.$$

Thus an ordered cumulative probability model always has an interpretation in terms of modelling an underlying latent variable, though no such interpretation is necessary to use the model.

These models are discussed by Bock (1975, Chapter 8), McCullagh (1980) and Anderson (1984). The cumulative distribution function H is usually taken to be normal or logistic. Programs for fitting some of these models include MULTIQUAL (Bock and Yates, 1973) and PLUM (McCullagh, 1980); it is possible to fit the grouped normal (latent variable) model in GLIM using a composite link function (Thompson and Baker, 1981).

5.8   An example

A survey of student opinion on the Vietnam War was taken at the University of North Carolina, Chapel Hill in May 1967 and published in the student newspaper. Students were asked to fill in "ballot papers", available in the Student Council building, stating which policy out of A, B, C or D they supported. Responses were cross-classified by sex and by undergraduate year or graduate status. The policies were:

A: The US should defeat the power of North Vietnam by
     widespread bombing of its industries, ports and
     harbours and by land invasion.

B: The US should follow the present policy in Vietnam.

C: The US should de-escalate its military activity,
     stop bombing North Vietnam, and intensify its efforts
     to begin negotiation.

D: The US should withdraw its military forces from Vietnam
     immediately.

Much more detail of this survey and a complementary correspondence analysis of the data are given in Aitkin (1996b).

The data are stored in the file vietnam, and are reproduced in Table 5.5. What can we conclude about students' attitude to the war? A first question concerns the response rates: what proportion of students actually responded in the survey? The numbers of students enrolled at the University, and the numbers responding (summed over the four categories A, B, C and D) are shown in Table 5.6, with the response rates.

Overall, only 26% of male and 17% of female students expressed an opinion by filling in a ballot paper. The response rate is somewhat higher than average for 4th-year males and 3rd-year females, and slightly lower than average for the other groups. We could model these response rates, treating the populations as samples, but the pattern of response is clear.

Given such low response rates, it seems difficult to draw any conclusion about the view of the student population. We have no way of knowing whether response is itself related to intensity of feeling, and therefore possibly to the policy chosen.

Table 5.5. Vietnam survey sample sizes

Year	A	B	C	D	Total
Males					
1	175	116	131	17	439
2	160	126	135	21	442
3	132	120	154	29	435
4	145	95	185	44	469
G	118	176	345	141	780
Females					
1	13	19	40	5	77
2	5	9	33	3	50
3	22	29	110	6	167
4	12	21	58	10	101
G	19	27	128	13	187

Table 5.6. Survey response rates

Year	Response	Enrolled	Response rate
Males			
1	439	1768	0.248
2	442	1792	0.247
3	435	1693	0.257
4	469	1522	0.308
G	780	3005	0.260
Total	2565	9780	0.262
Females			
1	77	487	0.158
2	50	326	0.153
3	167	772	0.216
4	101	608	0.166
G	187	1221	0.153
Total	582	3414	0.170

We can, however, investigate how the policy chosen varies by sex and year for those who responded. This is a different question; whether the conclusions are applicable to the population as a whole is unresolvable.

We show in Table 5.7 the observed proportions choosing each policy, with the year totals. For males, the proportions choosing policies A and B decrease, and those choosing C and D increase steadily with year. For females, policy C is consistently favoured but there is substantial variation with year.

Table 5.7. Observed proportions

Year	A	B	C	D	Total
Males					
1	0.399	0.264	0.298	0.039	439
2	0.362	0.285	0.305	0.048	442
3	0.303	0.276	0.354	0.067	435
4	0.309	0.203	0.394	0.094	469
G	0.150	0.226	0.442	0.181	780
Females					
1	0.169	0.247	0.519	0.065	77
2	0.100	0.180	0.660	0.060	50
3	0.132	0.174	0.659	0.036	167
4	0.119	0.208	0.574	0.099	101
G	0.102	0.144	0.684	0.070	187

We proceed to model the proportions as though the respondents were randomly sampled from the student population. The variable response rates noted above are irrelevant in the analysis since we are including sex and year in the model (see Section 2.9 for a discussion of unequal sampling fractions).

We give two analyses, one based on the multinomial logit model and one on the continuation-ratio model. In both analyses, we give a different analysis from that in the first edition, using a suggestion of M. Green (in Aitkin, 1996b), to deal with the awkward Graduate group—we want to model trends over year of enrolment, but Graduates are a mixed group, whose actual years of University study may range from 5 to 9 or more. Green suggested using year 7 as a compromise, and we derive the very simple model resulting from this approach. We give first the multinomial logit model.

5.8.1 *Multinomial logit model*

The data are read in as 40 counts classified by factors `policy`, `year` and `sex`. Note that this structure is different from that in the first edition (which was a set of four vectors of length 10).

```
$input 'vietnam' $
$yvar count $error p $link l
$eliminate year*sex $
$fit policy $
scaled deviance =  361.72 at cycle 4
    residual df =   27
```

This is the null model on the multinomial logit scale: the `policy` and (eliminated) `year*sex` terms are included to reproduce the marginal totals. The fit is very poor:

strong differences in response proportions are present. We now add the sex and
year interactions with response:

```
$fit policy.sex : policy.year : + policy.sex $
scaled deviance =  216.31 at cycle 4
    residual df =   24

scaled deviance =  153.94 at cycle 4
    residual df =   15

scaled deviance =  19.194 (change =    -134.7) at cycle 3
    residual df =  12     (change =     -3  )
```

Large changes in deviance show that both interactions are necessary. The only
term not yet fitted is the three-way interaction, the deviance for this term is 19.19
on 12 df. Although this value does not indicate a notably bad fit, it is possible that
a large contribution to this deviance comes from only a few degrees of freedom of
the interaction term. We have already seen a different response pattern for males
and females in the observed proportions and this should be investigated further.
We model the responses for each sex separately.

```
$calc male = 2 - sex $weight male $

-- model changed
```

Only males are now being analysed, so no sex terms need be fitted (if they are,
they will be aliased and ignored).

```
$fit policy + year : + policy.year $disp e $
scaled deviance =  203.05 at cycle 4
    residual df =   12     from 20 observations

scaled deviance =  2.043e-14 (change =    -203.0) at cycle 2
    residual df =      0     (change =    -12  ) from 20 observations
```

	estimate	s.e.	parameter
1	-0.4112	0.1197	POLICY(2)
2	-0.2896	0.1155	POLICY(3)
3	-2.332	0.2540	POLICY(4)
4	0.1723	0.1689	POLICY(2).YEAR(2)
5	0.3159	0.1739	POLICY(2).YEAR(3)
6	-0.0117	0.1782	POLICY(2).YEAR(4)
7	0.8110	0.1688	POLICY(2).YEAR(5)
8	0.1197	0.1643	POLICY(3).YEAR(2)
9	0.4437	0.1656	POLICY(3).YEAR(3)
10	0.5332	0.1602	POLICY(3).YEAR(4)

11	1.362	0.1572	POLICY(3).YEAR(5)
12	0.3009	0.3441	POLICY(4).YEAR(2)
13	0.8161	0.3265	POLICY(4).YEAR(3)
14	1.139	0.3069	POLICY(4).YEAR(4)
15	2.510	0.2830	POLICY(4).YEAR(5)

scale parameter 1.000

    eliminated term: YEAR.SEX

The large deviance for the interaction term (203.05) shows that proportions adopting each policy differ significantly over years. How can we describe these differences simply from the parameter estimates?

We have two orderings: by year and by policy. Within each policy, the logit increases with year, but not smoothly: there is a large jump in the logit in each case from year 4 to year 5, and for policy 2 (B) the logit for year 4 is small and negative, whereas it is large and positive for the other years. Thus a linear trend over year will not fit all years, though it might fit years 1–4. We include the graduates in the model by assigning them to year 7 in the linear modelling of years. The year pattern of logits becomes stronger with increasing policy, suggesting that a linear trend over policy might fit as well, that is, the regression coefficients on year are proportional over policy (see Fig. 5.12).

We try these two model simplifications.

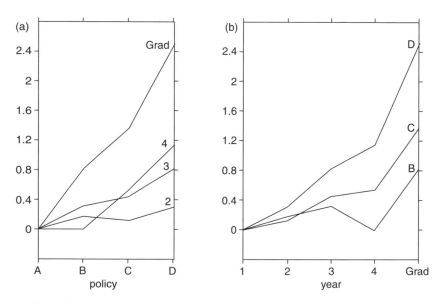

Fig. 5.12. Parameter estimates for interaction term POLICY.PYEAR for males in VIETNAM data; (a) ordered by POLICY (b) ordered by YEAR

```
$calc linp = policy $
$fit - policy.year + linp + linp.year $disp e $
scaled deviance =  13.914 (change =   +13.91) at cycle 3
    residual df =   8     (change =    +8  ) from 20 observations
```

	estimate	s.e.	parameter
1	-0.4130	0.0692	POLICY(2)
2	-0.3542	0.1035	POLICY(3)
3	-2.113	0.1625	POLICY(4)
4	0.000	aliased	LINP
5	0.0688	0.0715	YEAR(2).LINP
6	0.2274	0.0713	YEAR(3).LINP
7	0.3227	0.0700	YEAR(4).LINP
8	0.7378	0.0653	YEAR(5).LINP

```
scale parameter 1.000
    eliminated term: YEAR.SEX
```

The deviance shows a reasonable fit. We now assign year 7 to the graduates.

```
$calc liny = year : liny = %if(%eq(liny,5),7,liny) $
$fit - year.linp + liny.linp $disp e $
scaled deviance =  14.935 (change =   +1.021) at cycle 3
    residual df =  11     (change =    +3  ) from 20 observations
```

	estimate	s.e.	parameter
1	-0.5721	0.0614	POLICY(2)
2	-0.6727	0.0842	POLICY(3)
3	-2.589	0.1407	POLICY(4)
4	0.000	aliased	LINP
5	0.1267	0.0096	LINP.LINY

```
scale parameter 1.000
    eliminated term: YEAR.SEX
```

The linear model across years fits very well. We examine the residuals.

```
$print %rs $
```

```
 -0.542 -0.072  1.016 -0.675 -0.127  0.783 -0.164 -0.973
 -0.512  0.570  0.272 -0.600  1.578 -2.215  0.455  0.045
 -0.260  0.807 -0.977  0.974  0.     0.     0.     0.
  0.     0.     0.     0.     0.     0.     0.     0.
  0.     0.     0.     0.     0.     0.     0.     0.
```

There is a large residual of −2.22 at the 14th observation, where the observed number 95 of fourth-year students choosing policy B—the current policy—is not close to the model fitted value of 120.04. This feature is visible in the figure. We refit the above model excluding the 14th observation.

```
$calc w = 1 : w(14) = 0 : wt = male*w $
$weight wt $
$fit . $disp e $
-- model changed
scaled deviance =   6.4964 (change =   -8.438) at cycle 3
    residual df =  10       (change =   -1   ) from 19 observations


            estimate        s.e.      parameter
       1     -0.5164       0.0640      POLICY(2)
       2     -0.6869       0.0850      POLICY(3)
       3     -2.612        0.1419      POLICY(4)
       4      0.000        aliased     LINP
       5      0.1285       0.0097      LINP.LINY
scale parameter 1.000
    eliminated term: YEAR.SEX
```

The model fits very closely. Should we look for a further model which fits better, or is there a sensible interpretation of this model, and the outlying 14th observation?

A linear trend in policy means a linear increase in the log odds against policy A with each step along the policy scale: the policy options have been expressed in a smoothly ordered way. The outlying observation 14 does not fit this trend: fourth-year male students choose option B much less often than option A, instead of slightly less often as would be predicted from the linear increase in the odds against A.

We can give a plausible explanation for this. Fourth-year male students faced the draft and the possibility of serving in the war at the end of their academic year, unless they obtained deferment, for example by becoming graduate students. A policy of continuing the present conflict, rather than either bringing it to a presumably rapid end by invasion, or withdrawing to relatively safe positions, might have been viewed as the policy most likely to endanger their own lives. It is not surprising to find that this policy was much less favoured than A or C among fourth year students. Students in other undergraduate years did not face this immediate prospect.

We now turn to the females.

```
$calc female = 1 - male $
$weight female $
-- model changed

$fit policy + year $disp e $
scaled deviance =  13.263 at cycle 3
    residual df =  12      from 20 observations
```

	estimate	s.e.	parameter
1	0.3913	0.1536	POLICY(2)
2	1.648	0.1296	POLICY(3)
3	-0.6518	0.2026	POLICY(4)

scale parameter 1.000
eliminated term: YEAR.SEX

```
$print %rs $
```

```
  0.     0.     0.     0.     0.     0.     0.     0.     0.
  0.     0.     0.     0.     0.     0.     0.     0.     0.
  0.     0.     1.177  1.371 -1.262  0.047 -0.445 -0.007  0.231
 -0.100  0.361 -0.206  0.400 -1.417 -0.092  0.651 -0.754  1.412
 -0.798 -1.160  0.867  0.322
```

The null logit model gives a good fit; the largest residual is -1.42. There are no differences among the proportions choosing the four policies, other than those attributable to sampling variation. The fitted proportions from the model are obtained from

$$\hat{p}_j = \frac{e^{\hat{\theta}_j}}{\sum_{j=1}^{r} e^{\hat{\theta}_j}}, \quad \hat{\theta}_1 = 0,$$

and are 0.122, 0.180, 0.634 and 0.064. These are simply the marginal proportions giving each response, collapsed over year.

The device of fitting separate models for each sex is useful as the model structures are very different. The same result can be achieved by fitting the following single model to the complete data.

```
$calc v = male*liny*linp $
$weight w $
-- model changed
$fit policy + female + policy.female  + v $disp e $
scaled deviance =  19.759 at cycle 3
     residual df -  22    from 39 observations
```

	estimate	s.e.	parameter
1	-0.5164	0.0640	POLICY(2)
2	-0.6869	0.0850	POLICY(3)
3	-2.612	0.1419	POLICY(4)
4	0.000	aliased	FEMALE
5	0.1285	0.0097	V
6	0.9077	0.1664	POLICY(2).FEMALE
7	2.335	0.1550	POLICY(3).FEMALE
8	1.960	0.2474	POLICY(4).FEMALE

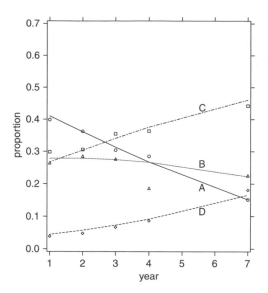

Fig. 5.13. Fitted model, males

```
scale parameter 1.000
     eliminated term: YEAR.SEX
```

The deviance of 19.76 on 22 df is the sum of the values 6.50 on 10 df for males and 13.26 on 12 df for females. The 14th observation is still excluded. The corresponding Pearson residual can be directly calculated: it is not listed by `$disp r $` since this observation is weighted out.

```
$calc (%yv(14) - %fv(14))/%sqrt(%fv(14)) $
```

```
-3.494
```

Note that this is a "jack-knife" residual, in the sense defined in Section 3.4, since this point has not been used in fitting the model.

We now construct the fitted proportions from the model. We change the first edition analysis by *not* including a term to reproduce the observed value for observation 14—it is left as a large residual from the model. The fitted models for each sex are shown in Figs 5.13 and 5.14.

```
$tab the %fv total for sex,year into t $
$calc fp = %fv/t(year + 5*female) $
   :    op = %yv/t(year + 5*female) $
   :    %re = male $
$graph (s = 1 h = 'year' v = 'proportion' y = 0,0.7)
         op,fp liny 5,2,7,8,10,11,12,13 policy,policy $
$calc %re = female $
```

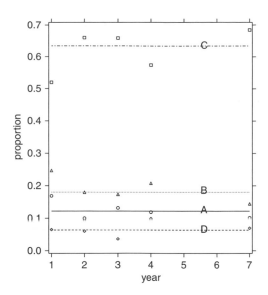

Fig. 5.14. Fitted model, females

```
$graph (s = 1 h = 'year' v = 'proportion' y = 0,0.7)
        op,fp liny 5,2,7,8,10,11,12,13 policy,policy $
```

The model fits the observed proportions closely except at observation 14.

In conclusion, 63% of female students responding choose policy C, while 18% choose B, 12% A and 6% D. These patterns are consistent over years. For males, support for policy A declines steadily with year, while that for policy B increases slightly and then decreases. Support for C increases rapidly, but that for D only slowly.

We now consider the continuation-ratio model.

5.8.2 *Continuation-ratio model*

We write the model in the form

$$\frac{n!}{n_A!(n_B + n_C + n_D)!}\, p_A^{n_A}(1 - p_A)^{n_B+n_C+n_D}.$$

$$\frac{(n_B + n_C + n_D)!}{n_B!(n_C + n_D)!}\left(\frac{p_B}{1 - p_A}\right)^{n_B}\left(\frac{p_C + p_D}{1 - p_A}\right)^{n_C+n_D}.$$

$$\frac{(n_C + n_D)!}{n_C!n_D!}\left(\frac{p_C}{1 - p_A - p_B}\right)^{n_C}\left(\frac{p_D}{1 - p_A - p_B}\right)^{n_D}.$$

We fit three binomial logit models by constructing two new variables of length 30 containing in blocks of 10 the counts for A versus (B + C + D), B versus (C + D) and C versus D. The five-category year factor is again replaced by its linear term, with Graduates coded to year 7.

```
$var 30 d1 d2 lsex lyear lres $
   : 10 a b c d i $
$calc i = %cu(1) $
   :    a = count(1 + 4*(i - 1)) $
   :    b = count(2 + 4*(i - 1)) $
   :    c = count(3 + 4*(i - 1)) $
   :    d = count(4 + 4*(i - 1)) $
   :   at = a + b + c + d $
   :   bt = b + c + d $
   :   ct = c + d $
   :    d1(i) = a $
: d1(10 + i) = b $
: d1(20 + i) = c $
   :    d2(i) = at $
: d2(10 + i) = bt $
: d2(20 + i) = ct $
   :  lsex = %gl(2,5) $
   : lyear = %gl(5,1) $
   : lres  = %gl(3,10) $
$factor lsex 2 lyear 5 lres 3 $
$yvar d1 $error b d2 $link g
$fit lres + lsex + lyear + lsex.lyear $
scaled deviance =   107.73 at cycle 3
    residual df =    18

$fit + lres.lsex  $disp e $
scaled deviance =   23.482 (change =   -84.25) at cycle 3
    residual df =  16      (change =    -2   )
```

	estimate	s.e.	parameter
1	-0.4720	0.0784	1
2	0.3436	0.0672	LRES(2)
3	2.438	0.0871	LRES(3)
4	-1.155	0.2316	LSEX(2)
5	-0.0922	0.1033	LYEAR(2)
6	-0.3259	0.1033	LYEAR(3)
7	-0.5124	0.1021	LYEAR(4)
8	-1.061	0.0936	LYEAR(5)
9	0.2855	0.1808	LRES(2).LSEX(2)
10	1.874	0.2327	LRES(3).LSEX(2)

```
    11       -0.3344          0.3402        LSEX(2).LYEAR(2)
    12        0.0731          0.2615        LSEX(2).LYEAR(3)
    13        0.1092          0.2873        LSEX(2).LYEAR(4)
    14        0.5005          0.2580        LSEX(2).LYEAR(5)
scale parameter 1.000
$fit + lres.lyear $disp e $
scaled deviance =  10.576 (change =    -12.91) at cycle 3
    residual df =    8      (change =    -8   )

             estimate          s.e.       parameter
     1       -0.4163          0.09533     1
     2        0.2115          0.1490      LRES(2)
     3        2.332           0.2476      LRES(3)
     4       -1.129           0.2305      LSEX(2)
     5       -0.1570          0.1361      LYEAR(2)
     6       -0.4104          0.1372      LYEAR(3)
     7       -0.4078          0.1357      LYEAR(4)
     8       -1.280           0.1340      LYEAR(5)
     9        0.2734          0.1830      LRES(2).LSEX(2)
    10        1.849           0.2354      LRES(3).LSEX(2)
    11        0.1517          0.2105      LRES(2).LYEAR(2)
    12        0.1537          0.2036      LRES(2).LYEAR(3)
    13       -0.1911          0.2071      LRES(2).LYEAR(4)
    14        0.4295          0.1919      LRES(2).LYEAR(5)
    15        0.1258          0.3437      LRES(3).LYEAR(2)
    16        0.2685          0.3207      LRES(3).LYEAR(3)
    17       -0.1592          0.3038      LRES(3).LYEAR(4)
    18        0.2824          0.2789      LRES(3).LYEAR(5)
    19       -0.3546          0.3414      LSEX(2).LYEAR(2)
    20        0.0488          0.2628      LSEX(2).LYEAR(3)
    21        0.1262          0.2852      LSEX(2).LYEAR(4)
    22        0.4683          0.2584      LSEX(2).LYEAR(5)
scale parameter 1.000
```

The three-way interaction deviance is 10.58 on 8 df. Inspection of the parameter estimates for the saturated model (not shown) shows that the three-way interactions all have z-statistics around 1 or less. We may eliminate them, and then examine the no-three-factor interaction model above. We first recode the Graduates to year 7 and define a variate with these values.

```
$calc linyear = lyear : linyear = %if(%eq(linyear,5),7,linyear) $
```

Examining the two-way interactions, we find that only appreciable terms are lres(3).lsex(2), lres(2).lyear(5) and lsex(2).lyear(5). We replace the five-year lyear factor by the linyear variate.

```
$fit - lres.lyear + lres.linyear $disp e $
scaled deviance =  18.862 (change =   +8.286) at cycle 3
    residual df =  14      (change =   +6    )
```

	estimate	s.e.	parameter
1	-0.3750	0.0996	1
2	0.1183	0.1247	LRES(2)
3	2.324	0.1866	LRES(3)
4	-1.127	0.2300	LSEX(2)
5	-0.0826	0.1077	LYEAR(2)
6	-0.3083	0.1214	LYEAR(3)
7	-0.4903	0.1389	LYEAR(4)
8	-1.041	0.1915	LYEAR(5)
9	0.2737	0.1812	LRES(2).LSEX(2)
10	1.869	0.2336	LRES(3).LSEX(2)
11	-0.3411	0.3397	LSEX(2).LYEAR(2)
12	0.0560	0.2611	LSEX(2).LYEAR(3)
13	0.0854	0.2872	LSEX(2).LYEAR(4)
14	0.4671	0.2585	LSEX(2).LYEAR(5)
15	-0.0322	0.0365	LRES(1).LINYEAR
16	0.0291	0.0364	LRES(2).LINYEAR
17	0.000	aliased	LRES(3).LINYEAR

```
scale parameter 1.000
```

The deviance increase is consistent with the df. We examine the residuals.

```
$print %rs $
```

```
 -0.041 -0.452 -0.489  1.906 -0.834 -0.195 -0.439 -0.150
  0.106  0.505 -0.129  0.564  0.474 -2.048  0.887  0.803
  0.241 -0.534  0.899 -0.823  0.362 -0.023  0.145 -0.132
 -0.131 -1.307  0.218  1.134 -1.558  0.587
```

The largest residual is −2.0—the variate model fits well. The interaction terms for lres.linyear look small. We remove them.

```
$fit - lres.linyear $disp e $
scaled deviance =  23.482 (change =   +4.620) at cycle 3
    residual df =  16      (change =   +2    )
```

	estimate	s.e.	parameter
1	-0.4720	0.0784	1
2	0.3436	0.0672	LRES(2)
3	2.438	0.0871	LRES(3)
4	-1.155	0.2316	LSEX(2)

5	-0.0922	0.1033	LYEAR(2)
6	-0.3259	0.1033	LYEAR(3)
7	-0.5124	0.1021	LYEAR(4)
8	-1.061	0.0936	LYEAR(5)
9	0.2855	0.1808	LRES(2).LSEX(2)
10	1.874	0.2327	LRES(3).LSEX(2)
11	-0.3344	0.3402	LSEX(2).LYEAR(2)
12	0.0731	0.2615	LSEX(2).LYEAR(3)
13	0.1092	0.2873	LSEX(2).LYEAR(4)
14	0.5005	0.2580	LSEX(2).LYEAR(5)

scale parameter 1.000

The deviance change is at the 18.8% point of χ_2^2. We continue the factor-variate replacement.

```
$fit - lsex.lyear + lsex.linyear $disp e $
scaled deviance =  25.298 (change =   +1.816) at cycle 3
    residual df =  19      (change =   +3    )
```

	estimate	s.e.	parameter
1	-0.3466	0.0898	1
2	0.3436	0.0672	LRES(2)
3	2.438	0.0870	LRES(3)
4	-1.420	0.1972	LSEX(2)
5	-0.0267	0.1028	LYEAR(2)
6	-0.1277	0.1179	LYEAR(3)
7	-0.2189	0.1380	LYEAR(4)
8	-0.4386	0.2148	LYEAR(5)
9	0.2829	0.1807	LRES(2).LSEX(2)
10	1.868	0.2324	LRES(3).LSEX(2)
11	-0.1064	0.0377	LSEX(1).LINYEAR
12	0.000	aliased	LSEX(2).LINYEAR

scale parameter 1.000

```
$fit - lyear + linyear $disp e $
scaled deviance =  25.591 (change =   +0.2928) at cycle 3
    residual df =  22      (change =   +3    )
```

	estimate	s.e.	parameter
1	-0.2486	0.0653	1
2	0.3437	0.0672	LRES(2)
3	2.437	0.0870	LRES(3)
4	-1.422	0.1960	LSEX(2)
5	-0.1829	0.0138	LINYEAR

```
      6        0.2830        0.1807        LRES(2).LSEX(2)
      7        1.869         0.2324        LRES(3).LSEX(2)
      8        0.1067        0.0376        LSEX(2).LINYEAR
scale parameter 1.000
```

The effect of linyear is small $(-0.183 + 0.107 = -0.076)$ for sex 2. We try to remove it.

```
$calc lsex1 = %eq(lsex,1) : lsex1ly = lsex1*linyear $
$fit - linyear - lsex.linyear + lsex1ly $disp e $
scaled deviance =   30.395 (change =   +4.804) at cycle 3
    residual df =   23      (change =   +1   )

           estimate        s.e.        parameter
      1     -0.2486         0.06529     1
      2      0.3437         0.06717     LRES(2)
      3      2.437          0.08700     LRES(3)
      4     -1.725          0.1425      LSEX(2)
      5     -0.1829         0.01376     LSEX1LY
      6      0.2776         0.1804      LRES(2).LSEX(2)
      7      1.837          0.2309      LRES(3).LSEX(2)
scale parameter 1.000
```

The linear trend for females is small, but very significant. We replace it for the final model.

```
$fit + linyear + lsex.linyear - lsex1ly $
scaled deviance =   25.591 (change =   -4.804) at cycle 3
    residual df =   22      (change =   -1   )
```

The final continuation-ratio model has a deviance of 25.59 on 22 df compared with the final multinomial logit model deviance of 19.76, also with 22 df. The deviance difference of 5.83 looks very substantial, with a corresponding likelihood ratio of 0.0542 against the continuation-ratio model. However, the deviance reported for the multinomial logit model excludes observation 14— if this is included the model deviance increases by 8.44 to 28.20, and the continuation-ratio model fits better, without any need to declare observation 14 an outlier!

It may be puzzling, however, that the continuation-ratio model finds a linear trend over years for females where the multinomial logit model does not. This difference occurs because we did not in fact fit the linear trend model for females, because the null model appeared to fit well. We re-examine this point below, but first present the final continuation-ratio model on the original scale.

We first tabulate the fitted probabilities (shown to 4 df) from the three logit models.

```
$calc fp = %fv/%bd $tab the fp mean for lres,lsex,lyear $
       LYEAR     1       2       3       4       5
LRES LSEX
    1    1   0.3938  0.3510  0.3106  0.2728  0.1781
         2   0.1484  0.1390  0.1302  0.1218  0.0994

    2    1   0.4781  0.4327  0.3885  0.3460  0.2341
         2   0.2459  0.2321  0.2188  0.2061  0.1712

    3    1   0.8814  0.8609  0.8375  0.8110  0.7125
         2   0.9281  0.9229  0.9173  0.9113  0.8910
```

The first two rows of the table also give the fitted probabilities for response A in the multinomial model. The remaining rows give the fitted conditional probabilities for responses B and C, conditional on $B + C + D$ and $C + D$ respectively. To obtain the fitted multinomial probabilities for responses B, C and D, we multiply the resp 2 entries by the corresponding probabilities $(1 - p_A)$ to give the multinomial probabilities for category B, and multiply the resp 3 entries by $(1 - p_A - p_B)$ to give the multinomial probabilities for category C; those for category D are obtained by difference from 1. These are easily calculated in GLIM, and are given below rounded to 3 dp from the GLIM output:

```
$var 10 fa fb fc fd $
$calc fa = fp(i) :
    fb = fp(i + 10)*(1 - fa) :
    fc = fp(i + 20)*(1 - fa - fb) :
    fd = 1 - fa - fb - fc $
$look 1 5 fa fb fc fd $
        FA       FB       FC       FD
 1    0.394    0.290    0.279    0.038
 2    0.351    0.281    0.317    0.051
 3    0.311    0.268    0.353    0.069
 4    0.273    0.252    0.386    0.090
 5    0.178    0.192    0.449    0.181

  : 6 10 fa fb fc fd $
        FA       FB       FC       FD
 6    0.148    0.209    0.596    0.046
 7    0.139    0.200    0.610    0.051
 8    0.130    0.190    0.623    0.056
 9    0.122    0.181    0.635    0.062
10    0.099    0.154    0.665    0.082
```

We note again the quite different models for females from the constant model given previously. In particular, the female proportions supporting the "hawk" policies A and B decline steadily with year, as they do for males. For the continuation-ratio model, the females behave quite similarly to the males, in terms of trend over year. Note also that the "discrepancy" we found in the under-use of Policy B by males in year 4 is not present in this model. Thus our interpretation of this discrepancy is speculative.

We finally return to the multinomial logit analysis of the females, and try the effect of linear trends across policies and years.

```
$newjob $
$input 'vietnam' $
$yvar count $error p $link l
$calc female = %eq(sex,2) $
$weight female $
$eliminate year $
$fit policy $
 scaled deviance =  13.263 at cycle 3
     residual df =  12      from 20 observations

$calc liny = year : liny = %if(%eq(liny,5),7,liny) $
$calc linp = policy $
$fit + policy.liny $disp e $
scaled deviance =  7.7048 (change =   -5.558) at cycle 3
     residual df =  9      (change =   -3    ) from 20 observations


          estimate        s.e.      parameter
     1      0.4099        0.3163     POLICY(2)
     2      1.269         0.2701     POLICY(3)
     3     -1.108         0.4389     POLICY(4)
     4     -0.1123        0.0940     POLICY(1).LINY
     5     -0.1172        0.0884     POLICY(2).LINY
     6     -0.0180        0.0789     POLICY(3).LINY
     7      0.000         aliased    POLICY(4).LINY
  scale parameter 1.000
      eliminated term: YEAR

$calc linp = policy $
$fit - policy.liny + linp.liny $disp e$
scaled deviance =   8.7522 (change =   +1.047) at cycle 3
     residual df =  11      (change =   +2    ) from 20 observations
```

```
          estimate          s.e.        parameter
    1      0.1883          0.1779        POLICY(2)
    2      1.229           0.2311        POLICY(3)
    3     -1.300           0.3667        POLICY(4)
    4      0.0530          0.0252        LINY.LINP
scale parameter 1.000
    eliminated term: YEAR


$fit - liny.linp $
  scaled deviance =  13.263 (change =   +4.511) at cycle 3
      residual df =  12     (change =   +1    ) from 20 observations

$fit + liny.linp $
scaled deviance  =  8.7522 (change =   -4.511) at cycle 3
      residual df =  11     (change =   +1    ) from 20 observations
```

The replacement of year by its linear component in the interaction fits quite well
($\chi^2_{3,5.56} = 0.765$) as does the replacement of policy by its linear component. The
single df for interaction is then important, giving the same pattern of response as
for males.

We show in Figs 5.15 and 5.16 the fitted proportions for males and females
from the continuation-ratio model. The GLIM directives are omitted.

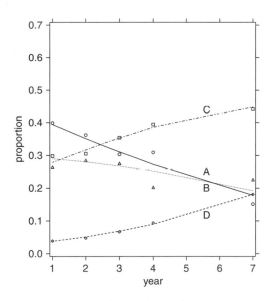

Fig. 5.15. Fitted CR model, males

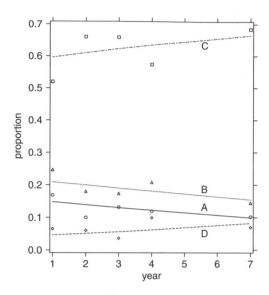

Fig. 5.16. Fitted CR model, females

5.9 Structured multinomial responses

In many studies with observational data, several categorical responses may be observed. The associations between these categorical response variables may be of interest, and particularly the effect of explanatory variables on the strength of these associations. The file toxaemia contains the number of women giving birth to their first child who showed toxaemic signs (hypertension and/or proteinurea, classified as Yes or No) during pregnancy. The data may be found in Brown *et al.* (1983) and were collected in Bradford between 1968 and 1977 on 13,384 United Kingdom women.

The four possible response patterns are cross-classified by social class (I–V), giving five categories of a factor class, and a three-level factor smoke, relating to the number of cigarettes smoked per day during pregnancy (0, 1–19, 20+). The counts in the four response categories are given by the variables hu, hn, nu and nn, representing those with hypertension and proteinurea, hypertension only, proteinurea only and neither symptom. We show these counts in Table 5.8, and the proportions in all four categories in Table 5.9.

How do the prevalence and association of these symptoms vary with class and smoke? We first consider each binary outcome separately, in greater detail than in the first edition, and then examine their association and the effect this has on the analysis.

5.9.1 *Independent outcomes*

We begin with hypertension, and tabulate the observed proportions with hypertension. The four outcome combinations are held in variates of length 15.

Table 5.8. Hypertension and proteinurea counts

Smoke class	1				2				3			
	HU	HN	NU	NN	HU	HN	NU	NN	HU	HN	NU	NN
1	28	82	21	286	5	24	5	71	1	3	0	13
2	50	266	34	785	13	92	17	284	0	15	3	34
3	278	1101	164	3160	120	492	142	2300	16	92	32	383
4	63	213	52	656	35	129	46	649	7	40	12	163
5	20	78	23	245	22	74	34	321	7	14	4	65

Table 5.9. Hypertension and proteinurea proportions

Smoke class	1				2				3			
	HU	HN	NU	NN	HU	HN	NU	NN	HU	HN	NU	NN
1	0.067	0.197	0.050	0.686	0.048	0.229	0.048	0.676	0.059	0.177	0.000	0.765
2	0.044	0.234	0.030	0.692	0.032	0.227	0.042	0.700	0.000	0.288	0.058	0.654
3	0.059	0.234	0.035	0.672	0.039	0.161	0.046	0.753	0.031	0.176	0.061	0.732
4	0.064	0.216	0.053	0.667	0.041	0.150	0.054	0.756	0.032	0.180	0.054	0.734
5	0.055	0.213	0.063	0.669	0.049	0.164	0.075	0.712	0.078	0.156	0.044	0.722

```
$input 'toxaemia' $
$calc h - hu + hn $
    :  u = hu + nu $
    :  t = hu + hn + nu + nn $
    : ph = h/t $
$tab the ph mean with t for class,smoke$
  SMOKE      1       2       3
  CLASS
      1   0.2638   0.2762   0.2353
      2   0.2784   0.2586   0.2885
      3   0.2932   0.2004   0.2065
      4   0.2805   0.1909   0.2117
      5   0.2678   0.2129   0.2333
```

The observed proportions are quite variable over the range 0.2–0.3 and show a generally lower hypertension rate for the smokers. Class differences are hard to identify. We fit models.

```
$yvar h $err b t $
$fit : + class : + smoke : + class.smoke $disp e $
 scaled deviance =   126.18 at cycle 3
     residual df =    14
```

```
scaled deviance =  117.70 (change =   -8.481) at cycle 3
    residual df =   10    (change =   -4    )
scaled deviance =  14.62 (change =   -103.1) at cycle 3
    residual df =    8    (change =   -2    )


scaled deviance =  1.364e-12 at cycle 2
    residual df =       0
```

	estimate	s.e.	parameter
1	-1.026	0.1111	1
2	0.0740	0.1294	CLASS(2)
3	0.1466	0.1156	CLASS(3)
4	0.0843	0.1318	CLASS(4)
5	0.0204	0.1621	CLASS(5)
6	0.0629	0.2449	SMOKE(2)
7	-0.1523	0.5822	SMOKE(3)
8	-0.1637	0.2779	CLASS(2).SMOKE(2)
9	0.2018	0.6611	CLASS(2).SMOKE(3)
10	-0.5670	0.2511	CLASS(3).SMOKE(2)
11	-0.3141	0.5930	CLASS(3).SMOKE(3)
12	-0.5649	0.2694	CLASS(4).SMOKE(2)
13	-0.2203	0.6091	CLASS(4).SMOKE(3)
14	-0.3647	0.2952	CLASS(5).SMOKE(2)
15	-0.0313	0.6442	CLASS(5).SMOKE(3)

```
scale parameter 1.000
```

The two-way interaction deviance is large and cannot be omitted. The interaction terms allow several different interpretations. We note that the standard errors for the smoke(3) category are much larger than those for the smoke(2) category because of the much smaller sample sizes. A careful inspection shows that the differences between the levels 2 and 3 of smoke in the interactions is nearly constant, between 0.253 and 0.366. This suggests that the interaction structure can be simplified by setting these differences equal in the model. To clarify the simplification, write s for smoke and c for class. The model

$$\beta_1 c2.s2 + \beta_2 c2.s3 + \beta_4 c3.s2 + \beta_5 c3.s3 + \cdots$$

under the restriction $\beta_2 - \beta_1 = \beta_4 - \beta_3 = \cdots = \delta$ reduces to

$$\beta_1 c2.(s2 + s3) + \beta_3 c3.(s2 + s3) + \cdots + \delta(c2 + c3 + \cdots).s3.$$

The variables $s2 + s3$ and $c2 + c3 + \cdots + c5$ are simply expressed though their logical complements:

```
$calc sn1 = %ne(smoke,1) : cn1 = %ne(class,1) $
      : s3 = %eq(smoke,3) $
$fit class + smoke + sn1 + cn1 + class.sn1 + cn1.s3 $disp e$
 scaled deviance =  0.26458 at cycle 3
     residual df =  3
```

	estimate	s.e.	parameter
1	-1.026	0.1111	1
2	0.0740	0.1294	CLASS(2)
3	0.1466	0.1156	CLASS(3)
4	0.0843	0.1318	CLASS(4)
5	0.0204	0.1621	CLASS(5)
6	0.0629	0.2449	SMOKE(2)
7	-0.1523	0.5825	SMOKE(3)
8	0.000	aliased	SN1
9	0.000	aliased	CN1
10	-0.1549	0.2753	CLASS(2).SN1
11	-0.5727	0.2509	CLASS(3).SN1
12	-0.5534	0.2670	CLASS(4).SN1
13	-0.3573	0.2917	CLASS(5).SN1
14	0.2909	0.6185	CN1.S3

```
scale parameter 1.000
```

The aliased terms are included as they are marginal to the interactions; without them GLIM reparametrizes the interactions. Further inspection shows that the interaction parameter estimates for classes 3, 4 and 5 are very similar. If this were true for class 2 as well we could replace the class variable in the interaction by cn1 defined above. The estimate for class 2 is considerably smaller, so we include a separate term in the model for the interaction of class 2 with sn1. The marginal main effect for class 2 is not needed.

```
$fit - class.sn1 + cn1.sn1 + c2.sn1 $disp e $
 scaled deviance =  1.9480 (change =   +1.683) at cycle 3
     residual df =  5       (change =   +2    )
```

	estimate	s.e.	parameter
1	-1.026	0.1111	1
2	0.0740	0.1294	CLASS(2)
3	0.1387	0.1152	CLASS(3)
4	0.0835	0.1245	CLASS(4)
5	0.1269	0.1382	CLASS(5)
6	0.0629	0.2449	SMOKE(2)

```
    7      -0.1523        0.5825        SMOKE(3)
    8       0.000         aliased       SN1
    9       0.000         aliased       CN1
   10      -0.5517        0.2495        CN1.SN1
   11       0.2921        0.6185        CN1.S3
   12       0.3967        0.1331        C2.SN1
scale parameter 1.000
```

The smoke terms are small and can be omitted. We continue with model simplification.

```
$fit - smoke $disp e $
  scaled deviance =  2.0751 (change =  +0.1271) at cycle 3
     residual df =  6       (change =  +1     )

          estimate        s.e.      parameter
    1       -1.026        0.1111     1
    2        0.0740       0.1294     CLASS(2)
    3        0.1387       0.1152     CLASS(3)
    4        0.0835       0.1245     CLASS(4)
    5        0.1269       0.1382     CLASS(5)
    6        0.0342       0.2321     SN1
    7        0.000        aliased    CN1
    8       -0.5230       0.2369     CN1.SN1
    9        0.0769       0.0894     CN1.S3
   10        0.3967       0.1331     C2.SN1
scale parameter 1.000


$fit - cn1.s3 $disp e $
  scaled deviance =  2.8074 (change =  +0.7323) at cycle 3
     residual df =  7       (change =  +1     )

          estimate        s.e.      parameter
    1       -1.026        0.1111     1
    2        0.0740       0.1294     CLASS(2)
    3        0.1383       0.1152     CLASS(3)
    4        0.0851       0.1245     CLASS(4)
    5        0.1272       0.1382     CLASS(5)
    6        0.0342       0.2321     SN1
    7        0.000        aliased    CN1
    8       -0.5106       0.2364     CN1.SN1
    9        0.3931       0.1330     C2.SN1
scale parameter 1.000
```

```
$fit - class $disp e $
 scaled deviance =   4.2377 (change =   +1.430) at cycle 3
     residual df =  10       (change =   +3    )

             estimate        s.e.        parameter
        1      -1.026       0.1111        1
        2       0.0342      0.2321        SN1
        3       0.1205      0.1141        CN1
        4      -0.5048      0.2361        CN1.SN1
        5       0.3409      0.1117        C2.SN1
 scale parameter 1.000


$fit - sn1 $disp e $
 scaled deviance =   4.2594 (change =   +0.02169) at cycle 3
     residual df =  11       (change =   +1       )

             estimate        s.e.        parameter
        1      -1.019       0.0976        1
        2       0.1127      0.1010        CN1
        3      -0.4706      0.0433        CN1.SN1
        4       0.3409      0.1117        C2.SN1
 scale parameter 1.000


$fit - cn1 $disp e $
 scaled deviance =   5.5233 (change =   +1.264) at cycle 3
     residual df =  12       (change =   +1    )

             estimate        s.e.        parameter
        1      -0.9136      0.02517       1
        2      -0.4629      0.04276       CN1.SN1
        3       0.3409      0.1117        C2.SN1
 scale parameter 1.000


$calc fp = %fv/%bd $
$tab the fp mean for class,smoke $


   SMOKE     1        2        3
   CLASS
       1   0.286    0.286    0.286
       2   0.286    0.262    0.262
       3   0.286    0.202    0.202
       4   0.286    0.202    0.202
       5   0.286    0.202    0.202
```

For non-smokers there are no social class differences, and for class 1 there are no smoking differences. For the other classes smokers have a slightly lower hypertension rate in class 2, and a much lower rate in classes 3–5. Note that for this hypertension response, the smoking categories can be collapsed to smokers/non-smokers, and classes 3–5 can be collapsed together.

We now examine the proteinurea response.

```
$calc pu = u/t $
$tab the pu mean with t for class,smoke $
  SMOKE      1      2       3
  CLASS
       1   0.1175  0.0952  0.0588
       2   0.0740  0.0739  0.0577
       3   0.0940  0.0858  0.0918
       4   0.1169  0.0943  0.0856
       5   0.1175  0.1242  0.1222
```

The proteinurea rate appears to decline with increasing smoking and shows some class differences. We fit models.

```
$yvar u $
$fit : + class : + smoke $disp e $
scaled deviance =   27.254 at cycle 3
     residual df =  14


scaled deviance =    6.2603 (change =    -20.99) at cycle 3
     residual df =  10        (change =    -4   )


scaled deviance =    3.0798 (change =     -3.180) at cycle 3
     residual df =   8        (change =    -2   )


              estimate        s.e.       parameter
         1      -2.054       0.1376      1
         2      -0.4512      0.1673      CLASS(2)
         3      -0.2043      0.1427      CLASS(3)
         4      -0.0427      0.1558      CLASS(4)
         5       0.1370      0.1721      CLASS(5)
         6      -0.1070      0.0642      SMOK(2)
         7      -0.1180      0.1231      SMOK(3)
scale parameter 1.000
```

The class main effect model is a good fit.

```
$fit - smoke $disp e $
scaled deviance =    6.2603 (change =     +3.180) at cycle 3
     residual df =  10        (change =    +2   )
```

```
          estimate        s.e.      parameter
     1      -2.077        0.1369     1
     2      -0.4576       0.1673     CLASS(2)
     3      -0.2263       0.1422     CLASS(3)
     4      -0.0750       0.1547     CLASS(4)
     5       0.0970       0.1706     CLASS(5)
scale parameter 1.000
```

Classes 1, 4 and 5 have very similar responses, while 2 and 3 are different. We combine the former classes, and display the fitted proportions.

```
$calc c3 = %eq(class,3) $
$fit - class + c2 + c3 $disp e $
scaled deviance =    8.1525 (change =   +1.893) at cycle 3
    residual df =   12       (change =   +2    )
```

```
          estimate        s.e.      parameter
     1      -2.094        0.05401    1
     2      -0.4407       0.1102     C2
     3      -0.2094       0.06618    C3
scale parameter 1.000
```

```
$calc  fp = %fv/%bd $
$tab the fp mean with t for class $
```

```
CLASS      1       2       3       4       5
 MEAN   0.110   0.073   0.091   0.110   0.110
```

Class 2 has the lowest rate, and class 3 is lower than the others.

We now consider the association between the two binary responses.

5.9.2 *Correlated outcomes*

In each of the 5 × 3 cell of the table classified by class and smoke, we have a 2 × 2 classification of H and U. We compute the log-odds ratio in each table, and present the table of "raw" log-odds ratios:

```
$calc lodds = %log(hu*nn/(hn*nu)) $
$tab the lodds mean for class,smoke $
```

```
--- invalid function/operator argument(s)
SMOKE   1    2    3
CLASS
    1   1.53 1.08   -
    2   1.47 0.86   -
```

```
3   1.58 1.37 0.73
4   1.32 1.34 0.87
5   1.00 1.03 2.09
```

The warning message is due to the two zero observed counts, which are both associated with very small sample sizes. The log-odds ratios appear fairly constant at about 1.2.

To perform a full analysis of these data with the four response categories we need to use the Poisson model representation. This requires the blocking of the variables into vectors of length 60, as described in Section 5.5. However, we now need to take account of the structure of our responses; hypertension and proteinurea are both defined as two-level factors: we do not have a single four-level response factor.

```
$var 60 count bclass bsmoke bh bu $
$assign count = nn,hn,nu,hu :
       bclass = class,class,class,class :
       bsmoke = smoke,smoke,smoke,smoke $
$calc bh = %gl(2,15) $
    : bu = %gl(2,30) $
$factor bclass 5 bsmoke 3 bh 2 bu 2 $
$slen 60 $
 -- standard length re-initialised

$yvar count $err p $link l
$eliminate bclass*bsmoke $
$fit bh*bu $disp e $
 scaled deviance =   179.03 at cycle 3
     residual df =    42

            estimate        s.e.      parameter
     1        -1.244       0.02178      BH(2)
     2        -2.772       0.04247      BU(2)
     3         1.365       0.06062      BH(2).BU(2)
 scale parameter 1.000
     eliminated term: BCLASS.BSMOKE
```

The bclass*bsmoke term reproduces the marginal totals of the women in each class/smoking combination. The bh*bu term reproduces the "saturated model" pattern of symptom dependence. Consequently this model is equivalent to the null model on the multinomial logit scale, the marginal totals being fixed for each explanatory variable combination and a common response pattern being fitted. The deviance is 179.03 on 42 df: the null model is a poor fit as we know and there are strong patterns of association to be modelled. The interaction between bh and bu is positive and substantial.

We proceed by removing the bh.bu interaction terms from the saturated model as far as possible, to identify the association structure between the symptoms, and then follow the analysis for the separate hypertension and proteinurea responses. To save space we suppress the parameter estimates in the early models.

```
$fit bh*bu*bclass*bsmoke - bh.bu.bclass.bsmoke $
scaled deviance =  12.681 (change =   +12.68) at cycle 4
    residual df =   8      (change =   +8   )

$fit - bh.bu.bclass $
scaled deviance =  15.648 (change =   +2.967) at cycle 4
    residual df =  12      (change =   +4   )

$fit - bh.bu.bsmoke $
scaled deviance =  22.286 (change =   +6.637) at cycle 4
    residual df =  14      (change =   +2   )
```

The deviance change is rather large. We refit this interaction and examine the parameter estimates for it.

```
$fit + bh.bu.bsmoke $disp e $
scaled deviance =  15.648 (change =   -6.637) at cycle 4
    residual df =  12      (change =   -2   )
...
    16       -0.2072      0.1315      BH(2).BU(2).BSMOKE(2)
    17       -0.5871      0.2576      BH(2).BU(2).BSMOKE(3)
...
```

The strength of the association appears to decrease with increased smoking. We try replacing bsmoke by its linear component.

```
$calc lsmoke = bsmoke $
$fit - bh.bu.bsmoke + bh.bu.lsmoke $disp e $
scaled deviance =  15.921 (change =   +0.2726) at cycle 4
    residual df =  13      (change =   +1    )

     3        1.752       0.1603      BH(2).BU(2)
...
    24       -0.2530      0.1000      BH(1).BU(1).LSMOKE
    25        0.000       aliased     BH(1).BU(2).LSMOKE
    26        0.000       aliased     BH(2).BU(1).LSMOKE
    27        0.000       aliased     BH(2).BU(2).LSMOKE
```

The linear trend term is clearly important, with a deviance change of 6.36 if it is omitted. The interaction is reparametrized to the first level of bh and bu because the lsmoke term is not included in the model with its single interactions.

The association between the two symptoms decreases with increasing level of smoking, from a log-odds ratio of 1.5 (1.75 − 0.25—an odds ratio of 4.5) for non-smokers to 1.00 (1.75 − 3 ∗ 0.25—an odds ratio of 2.7) for heavy smokers. We proceed with simplification of the single interaction terms with bh or bu.

```
$fit - bu.bclass.bsmoke $
scaled deviance =  19.388 (change =   +3.467) at cycle 3
    residual df =  21     (change =   +8    )

$fit - bu.bsmoke $
scaled deviance =  19.611 (change =   +0.2233) at cycle 3
    residual df =  22     (change =   +1    )
```

We clarify the model by including the marginal terms without changing the model fit.

```
$fit + lsmoke + bh.lsmoke + bu.lsmoke $disp e $
scaled deviance =  19.611 (change =   0.) at cycle 3
    residual df =  22     (change =   0 )
...
    13      0.0999      0.2456      BH(2).BSMOKE(2)
    14     -0.0799      0.5834      BH(2).BSMOKE(3)
    15      0.000       aliased     BH(2).LSMOKE
...
```

The bh.bsmoke interaction terms can be omitted, while the marginal bh.lsmoke interaction is retained.

```
$fit - bh.bsmoke $
scaled deviance =  19.611 (change =   0.) at cycle 3
    residual df =  22     (change =   0 )
...
    15      0.1511      0.2811      BH(2).BCLASS(1).BSMOKE(2)
    16      0.0224      0.6445      BH(2).BCLASS(1).BSMOKE(3)
    17     -0.0253      0.1905      BH(2).BCLASS(2).BSMOKE(2)
    18      0.1993      0.4174      BH(2).BCLASS(2).BSMOKE(3)
    19     -0.4222      0.1487      BH(2).BCLASS(3).BSMOKE(2)
    20     -0.3041      0.2981      BH(2).BCLASS(3).BSMOKE(3)
    21     -0.4150      0.1778      BH(2).BCLASS(4).BSMOKE(2)
    22     -0.2002      0.3288      BH(2).BCLASS(4).BSMOKE(3)
    23     -0.2091      0.1796      BH(2).BCLASS(5).BSMOKE(2)
    24      0.000       aliased     BH(2).BCLASS(5).BSMOKE(3)
...
```

The bclass.bsmoke interactions for bh show the same pattern as before so we define the same terms, and replace the bclass.bsmoke interaction as before.

```
$calc bcn1 = %ne(bclass,1) : bsn1 = %ne(bsmoke,1) $
   : bs3 = %eq(bsmoke,3) $
$fit - bh.bclass.bsmoke + bcn1 + bsn1 + bs3 + bclass.bsn1
+ bcn1.bs3 + bh.(bcn1 + bsn1 + bclass.bsn1 + bcn1.bs3) $disp e $
scaled deviance = 19.897 (change = +0.2855) at cycle 3
    residual df = 25    (change = +3    )
...
30      0.2866      0.6185      BH(2).BCN1.BS3

$fit - bh.bcn1.bs3 $disp e $
scaled deviance = 20.119 (change = +0.2222) at cycle 3
    residual df = 26    (change = +1    )
...
    26      -0.1298     0.2639      BH(2).BCLASS(2).BSN1
    27      -0.5404     0.2381      BH(2).BCLASS(3).BSN1
    28      -0.5146     0.2547      BH(2).BCLASS(4).BSN1
    29      -0.3124     0.2807      BH(2).BCLASS(5).BSN1
...
$calc bc2 = %eq(bclass,2) $
$fit - bh.bclass.bsn1 + bh.bcn1 + bh.bc2 + bcn1.bsn1
+ bc2.bsn1 + bh.bcn1.bsn1 + bh.bc2.bsn1 $disp e $
scaled deviance = 22.002 (change =  +1.883) at cycle 3
    residual df = 28    (change =  +2   )
...
    12      -0.4953     0.1715      BU(2).BCLASS(2)
    13      -0.2263     0.1465      BU(2).BCLASS(3)
    14      -0.03980    0.1599      BU(2).BCLASS(4)
    15       0.1381     0.1764      BU(2).BCLASS(5)
...

$calc bc3 = %eq(bclass,3) $
$fit - bu.bclass + bu.(bc2 + bc3) + bc3 $disp e $
scaled deviance = 23.926 (change =  +1.924) at cycle 3
    residual df = 30    (change =  +2    )
...
     9       0.1480     0.1303      BH(2).BCLASS(2)
    10       0.1764     0.1158      BH(2).BCLASS(3)
    11       0.0862     0.1245      BH(2).BCLASS(4)
    12       0.1299     0.1382      BH(2).BCLASS(5)
...
$fit - bh.bclass + bh.bcn1 $disp e $
scaled deviance = 26.374 (change =  +2.448) at cycle 3
    residual df = 32    (change =  +2    )
```

	estimate	s.e.	parameter
1	-1.325	0.1442	BH(2)
2	-2.785	0.1242	BU(2)
...			
8	1.737	0.1587	BH(2).BU(2)
9	0.0953	0.0894	BH(2).LSMOKE
10	0.1270	0.0663	BU(2).LSMOKE
11	0.1563	0.1150	BH(2).BCN1
12	-0.0340	0.2532	BH(2).BSN1
...			
18	0.0109	0.0732	BH(1).BC2
...			
20	-0.4945	0.1128	BU(2).BC2
...			
22	-0.2203	0.0666	BU(2).BC3
23	-0.2454	0.0991	BH(2).BU(2).LSMOKE
24	-0.5226	0.2364	BH(2).BCN1.BSN1
25	0.3932	0.1331	BH(2).BC2.BSN1

scale parameter 1.000

The aliased terms are suppressed for clarity.

We conclude the reduction at this point, though some unnecessary terms remain, and could be removed by further definitions of explicit variables for these terms and other terms to which they are marginal.

As for the separate symptom analyses, for hypertension there are no significant social class differences for non-smokers, and for class 1 there are no significant smoking differences. For the other classes smokers have a slightly lower hypertension rate in class 2, and a much lower rate in classes 3–5.

For proteinurea, class 2 has a lower incidence than the others.

We give finally separate fitted models for proteinurea and hypertension from this model. We "unblock" the fitted values into vectors for the two symptoms:

```
$var 15 i fh fnh fu fnu fph fpu $
$calc i = %cu(1) $
    : fh = %fv(15+i) + %fv(45+i) $
    : fnh = %fv(i) + %fv(30+i) $
    : fph = fh/(fh + fnh) $
    : fu = %fv(30+i) + %fv(45+i) $
    : fnu = %fv(i) + %fv(15+i) $
    : fpu = fu/(fu + fnu) $
$tab the fph mean for class,smok $
```

```
SMOKE      1       2       3
CLASS
    1   0.264   0.269   0.281
    2   0.279   0.260   0.275
    3   0.288   0.198   0.209
    4   0.295   0.203   0.213
    5   0.295   0.203   0.213
```

: the fpu mean for class $

```
CLASS     1       2       3       4       5
 MEAN  0.113   0.073   0.089   0.109   0.109
```

6
Survival data

6.1 Introduction

Over the last 30 years there has been a rapid development of probability models and statistical analysis for technological and medical survival data. Many studies have been made of the length of life or of periods of remission of animal or human subjects being treated for serious diseases. These studies have used probability models for duration of life which originated in engineering reliability studies of the operating lifetimes of electrical and mechanical systems. We will use the terms "death" and "failure" interchangeably to represent the events and "survival time" to represent the times to these events. In this chapter we give an account of the main probability models and their use in data analysis. The recognition of the special form of the likelihood in an important class of these distributions allows their simple fitting as generalized linear models. Many books are now available on survival analysis; a concise account can be found in Cox and Oakes (1984), detailed discussions in Kalbfleisch and Prentice (1980) and Lawless (1982), and more applied treatments in Collett (2003) and Hosmer and Lemeshow (1999).

This chapter is restricted to a discussion of single events: we do not deal with models for *recurrent events*, like repeated spells of remission or of unemployment.

6.2 The exponential distribution

The exponential distribution plays a central role in survival analysis. Although few systems have exponentially distributed lifetimes, most of the useful survival distributions are directly related to the exponential distribution. We first consider its properties as a survival distribution. In this chapter we use the notation T for a non-negative random variable.

The density function of the exponential random variable T for survival time t is

$$f(t) = \Pr(t < T < t + dt)/dt$$
$$= (1/\mu)e^{-t/\mu} \quad \mu, t > 0,$$

where μ is the mean of the distribution; the variance is μ^2. The cumulative distribution function (cdf) is

$$\Pr(T \leq t) = F(t) = 1 - e^{-t/\mu}, \quad \mu, t > 0.$$

A fundamental concept in survival analysis is that of the *hazard function* $h(t)$ which is the conditional density function at time t given survival up to time t:

$$h(t)\,dt = \Pr(t < T \le t + dt \mid T > t)$$
$$= f(t)\,dt/[1 - F(t)].$$

It is convenient to introduce a notation for the upper tail probability: we define the *survivor function*

$$S(t) = \Pr(T > t) = 1 - F(t).$$

Then

$$h(t) = f(t)/S(t).$$

The hazard function can be interpreted as the instantaneous failure rate at time t. Any continuous probability distribution can be specified equivalently by its density, survivor or hazard function. In particular the density and survivor functions can be obtained from the hazard function by

$$S(t) = \exp(-H(t))$$
$$f(t) = h(t)\exp(-H(t))$$

where the function

$$H(t) = \int_0^t h(u)\,du$$

is the *integrated hazard function*. The usefulness of the hazard function is clear in the servicing and repair of mechanical or electrical systems. When the system has been in operation for some time, the appropriate servicing or repair schedule depends on the probability of failure, conditional on the previous operating history. Historically, the effect of explanatory variables on survival time has been expressed by modelling the hazard function; as we shall see, such models are models for functions of the parameters of the probability distribution.

For the exponential distribution

$$h(t) = 1/\mu$$
$$H(t) = t/\mu.$$

Thus, the exponential distribution has *constant hazard*: the probability of death at time t is not dependent on the length of previous lifetime. The exponential distribution represents the lifetime distribution of an item which does not age or wear: the instantaneous probability of failure is the same no matter how long the item has already survived. In most applications this strong property does not hold, and so the exponential distribution has limited application as a survival distribution.

In a more general context, if "death" is replaced by an event which can occur repeatedly in time, then these events occur in a time-homogeneous Poisson process with rate $\lambda = 1/\mu$ if the times between events are independent exponential variables with the same mean μ. Thus, the probability of r events in time t has the Poisson distribution

$$\Pr(r \mid \lambda, t) = e^{-\lambda t} (\lambda t)^r / r! \quad r = 0, 1, 2, \ldots.$$

This relation between the exponential and Poisson distributions is used to advantage in Section 6.7 for ML estimation with censored data.

6.3   Fitting the exponential distribution

The exponential distribution is available as a standard distribution in GLIM as it is a special case of the *gamma* distribution, considered in Section 6.9. The error distribution is specified as g, and the scale parameter has to be fixed at 1.

We illustrate with a simple example from Feigl and Zelen (1965). The file feigl contains the survival times (the variable time) in weeks of 33 patients suffering from acute myelogeneous leukaemia, and the values of two explanatory variables, white blood cell count wbc in thousands and a positive or negative factor ag, positive values (ag = 1) being defined by the presence of Auer rods and/or significant granulation of the leukaemic cells in the bone marrow at diagnosis, and negative values (ag = 2) if both Auer rods and granulation are absent.

We first graph time against wbc in Fig. 6.1.

```
$input 'feigl' $
$graph (s = 1 h = 'wbc' v = 'time')
        time wbc 5 $
```

There is a heavy concentration of points along both the horizontal and vertical axes. We try the log scales, and use a different plotting symbol for the two levels of ag (circles for ag+, crosses for ag-) in Fig. 6.2.

```
$calc lt = %log(time)   :   lwbc = %log(wbc) $
$graph (s = 1 h = 'log wbc' v = 'log time')
        lt lwbc 5,6 ag $
```

There is a general decline in log survival time with increasing lwbc, and the negative ag group appears to have shorter survival times. We try fitting the exponential distribution using a log link function for the mean with lwbc as the explanatory variable interacting with the factor ag.

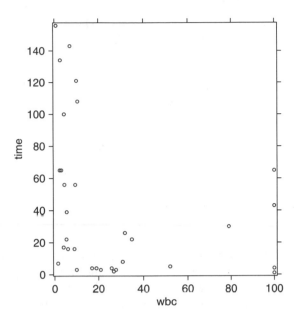

Fig. 6.1. Feigl data: survival time vs wbc

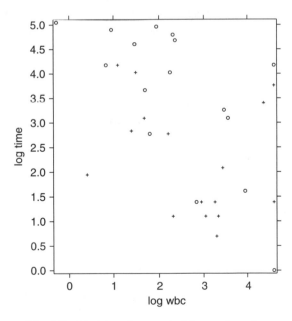

Fig. 6.2. Feigl data: log survival time vs log wbc

Thus the model is $\log \mu = -\log h(t) = \boldsymbol{\beta}'\mathbf{x}$, so a log-linear model for the mean is also a log-linear model for the hazard function.

```
$yvar time $error g $scale 1 $
$link 1 $fit ag*lwbc $disp e $
  scaled deviance =  38.555 at cycle 6
     residual df =  29
```

	estimate	s.e.	parameter
1	5.150	0.501	1
2	-1.874	0.785	AG(2)
3	-0.482	0.174	LWBC
4	0.329	0.267	AG(2).LWBC

```
scale parameter 1.000
```

It might be thought that the residual deviance from the model could be used for testing the goodness of fit, since the standard application of large sample theory would suggest that the deviance is distributed as χ^2_{n-p}. However, as in Section 4.4, as $n \to \infty$ the number of parameters in the saturated model also goes to infinity, and this violates one of the regularity conditions for the validity of this asymptotic distribution.

The interaction appears non-significant and might be omitted from the model. We note also that the slope of the regression in the second ag group is $(-0.482 + 0.329)$, that is, -0.153, so that an alternative simplification of the model might be achieved by setting this slope to zero, as in the solv example in Section 3.1.

```
$fit - ag.lwbc $disp e $
scaled deviance =  40.319 (change =   +1.764) at cycle 6
    residual df =  30      (change =   +1    )
```

	estimate	s.e.	parameter
1	4.730	0.412	1
2	-1.018	0.349	AG(2)
3	-0.304	0.132	LWBC

```
scale parameter 1.000
$calc ag1 = %eq(ag,1)  : z = ag1*lwbc $
$fit ag + z $disp e $
scaled deviance =  39.373 at cycle 6
    residual df =  30
```

	estimate	s.e.	parameter
1	5.150	0.501	1
2	-2.263	0.560	AG(2)
3	-0.482	0.174	Z

```
scale parameter 1.000
```

The deviances of the two reduced models differ by only 0.95. We cannot choose between these two models on the basis of the data, because the slope of the regression in the second ag group is poorly defined.

The log link, while a natural choice for positive survival times, is not the only possible choice. The mean survival time can also be related to the linear predictor using the *reciprocal link*, giving a *linear hazard model*

$$\mu^{-1} = \lambda = \boldsymbol{\beta}'\mathbf{x}.$$

Here λ is the mean rate of dying, or failure rate (per unit time). This choice of link function is sometimes thought natural because sufficient statistics exist for the model parameters on this scale. This property does not, however, assist the model-fitting in any way, and it does not prevent the occurrence of negative fitted values from the linear model.

```
$link r $fit ag*lwbc $disp e $
  scaled deviance =   39.779 at cycle 5
     residual df =   29
```

	estimate	s.e.	parameter
1	0.00598	0.00359	1
2	0.02196	0.02638	AG(2)
3	0.00585	0.00238	LWBC
4	0.00581	0.01116	AG(2).LWBC

```
scale parameter 1.000
```

The interaction is small and can be omitted.

```
$fit - ag.lwbc $disp e $
scaled deviance =   40.044 (change =   +0.2650) at cycle 5
    residual df =   30      (change =   +1      )
```

	estimate	s.e.	parameter
1	0.00580	0.00340	1
2	0.03441	0.01461	AG(2)
3	0.00611	0.00232	LWBC

```
scale parameter 1.000
```

With the reciprocal link an equally good fit is obtained if wbc instead of lwbc is used as the explanatory variable.

Finally, we consider the identity link. The original analysis by Feigl and Zelen used this link, with

$$\mu = 1/\lambda = \boldsymbol{\beta}'\mathbf{x}.$$

```
$link i $fit ag*lwbc $disp e $
  scaled deviance =   40.505 at cycle 7
     residual df =   29
```

	estimate	s.e.	parameter
1	110.5	36.00	1
2	-80.77	37.31	AG(2)
3	-20.03	8.653	LWBC
4	15.69	9.305	AG(2).LWBC

scale parameter 1.000

The deviance is 40.51, and the interaction, though large, does not appear significant. Omitting the interaction gives an error message:

```
$fit - ag.lwbc $disp e $
```

```
** fitted mean out of range for error distribution, at [ ag.lwbc ] on level 2
from channel 1. At cycle 3, the fitted mean for unit 32 is -1.7115, which is
less than or equal to zero when the error is G. This indicates an
inappropriate model.
```

The negative fitted value corresponds to observation 32 with the largest value of wbc in the ag negative group and a small value of survival time. This difficulty is avoided with the log link because with this link the fitted values can never be negative.

The three interaction models have very similar deviances, so it is clear that we cannot choose among them from statistical considerations. All three models can be expressed in a Box–Cox form for their link functions:

$$\beta'\mathbf{x} = \eta = (\mu^\theta - 1)/\theta, \quad \mu = (1 + \theta\eta)^{1/\theta}$$

with $\theta = -1$, 0 and 1 for the reciprocal, log and identity links. The corresponding deviances are 39.78, 38.56 and 40.51: the log link has the smallest deviance, but all three fit the data equally well.

The choice of a final model is made difficult, as in the solv example of Section 3.1, by the small sample size. We choose the log link and the parallel slopes model ag + lwbc as a reasonable description.

We now consider model criticism for the exponential distribution.

6.4 Model criticism

How do we know that the exponential distribution is a reasonable choice? Probability plotting of approximately standardized exponential variables provides a guide, as for the normal distribution. For many of the probability distributions in this chapter, the survivor function has a simple form which facilitates plotting. We therefore use, in preference to the empirical cdf, the *empirical survivor function*

$\hat{S}(t)$, defined by

$$\hat{S}(t) = \frac{\text{no. of } t_i > t}{n}.$$

If T_i is exponential with mean $\mu_i = 1/\lambda_i$, then $T_i/\mu_i = \lambda_i T_i$ is standard exponential with mean 1. If $\mu_i = g(\eta_i)$ with η_i the linear predictor, define

$$u_i = t_i/\hat{\mu}_i$$

with

$$\hat{\mu}_i = g(\hat{\boldsymbol{\beta}}' \mathbf{x}_i).$$

The u_i have approximately standard exponential distributions, though they are not independent, as with normal residuals.

The survivor function for the standard exponential is

$$S(u) = e^{-u}$$

so that

$$\log S(u) = -u$$

and

$$-\log(-\log S(u)) = -\log u.$$

We first graph the empirical survivor function $\hat{S}(u)$ of the u_i against u, using the interaction model.

```
$link l $fit aq*lwbc $
$calc u = %yv/%fv $
```

The u_i are closely related to the Pearson residuals. These are

$$e_i = \frac{t_i - \hat{\mu}_i}{\hat{\mu}_i} = u_i - 1.$$

We calculate the empirical survivor function, first sorting the scaled survival times in increasing order.

```
$sort ou u $
$calc s = 1 - %cu(1)/(%nu + 1) $
```

The survivor function estimate uses $(n + 1)$ in the denominator to avoid a zero value for the last observation (see Section 2.3 for a discussion of plotting positions). We graph $\hat{S}(u)$ against ou, together with the fitted survivor function exp(-ou) in Fig. 6.3.

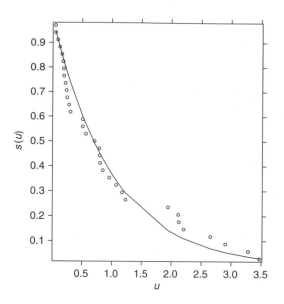

Fig. 6.3. Feigl data: residual survivor functions

```
$calc sexp = %exp(-ou) $
$graph (s = 1 h = 'u' v = 'S(u)')
          s,sexp ou 5,10 $
```

The bends in the graph at the edges of sets of data points are due to GLIM's linear interpolation. As with other cdf and probability plots, and especially with small samples, it is difficult to decide whether a real failure of the probability model is occurring. Since $\hat{S}(u)$ is essentially an estimated binomial probability for each u, 95% simultaneous confidence bounds can be placed on the graph in Fig. 6.4 using the macro binocisim as in Section 2.3. We ignore here the dependence between the residuals caused by the model-fitting; the bounds are used only qualitatively to indicate model and data agreement. A formal consideration of the dependence between observations would suggest that the bounds should be widened somewhat but we do not consider this further. We suppress in this graph, and the two following, the empirical survivor function to focus attention on the adequacy of the model fit.

```
$input 'binocisim' $
$calc r = %nu + 1 - %cu(1) : n = %nu + 1 $
$use binocisim r n $
$graph (s = 1 h = 'u' v = 'S(u)')
          sexp,pl__,pu__ ou 10,9,9 $
```

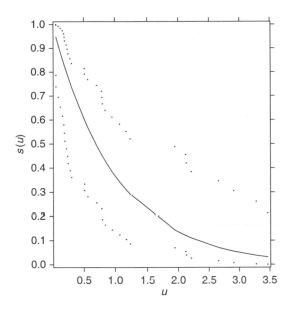

Fig. 6.4. Feigl data: fitted survivor function and bounds

The bounds on the true survivor function in Fig. 6.4 are very wide because of the small sample size, and the exponential model clearly fits well. We now log transform the scales of $S(u)$ and u, since the graphs on these scales are linear for the exponential distribution, and may indicate the form of the departure from the exponential distribution if this is incorrect (see Section 6.14).

```
$calc lsexp = %log(sexp) $
      : lpl = %log(pl__) : lpu = %log(pu__) $
$graph (s = 1 h = 'u' v = 'log S(u)')
        lsexp,lpl,lpu ou 10,9,9 $
```

The bounds in Fig. 6.5 are very condensed for small values of u. We now repeat the log transformations on both scales in Fig. 6.6. (The transformation of $\log S(u)$ is changed slightly from that in the first edition to the *negative* of $\log S(u)$, to ensure that all the survival graphs show a *decreasing* relation with increasing time.)

```
$calc llsexp = -%log(-lsexp) $
      : lou = %log(ou) $
      : llpl = -%log(-lpl) : llpu = -%log(-lpu) $
$graph (s = 1 h = 'log u' v = '-log[-log S(u)]')
        llsexp,llpl,llpu lou 10,9,9 $
```

Figure 6.6 is more informative, with a more nearly constant boundwidth, and is more useful in detecting departures from the exponential distribution, particularly

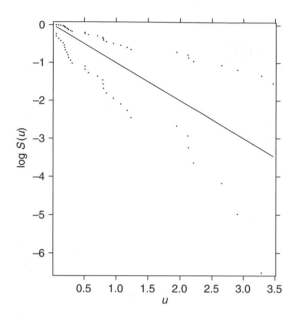

Fig. 6.5. Feigl data: fitted log survivor function and bounds

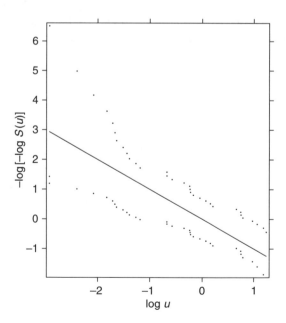

Fig. 6.6. Feigl data: fitted log–log survivor function and bounds

for the Weibull distribution (Section 6.10). The figure shows a good agreement of
the fitted straight line with the bounds. The exponential distribution assumption
seems satisfactory.

Now we assess the possibility that particularly influential observations have
affected the fitted model. The assessment of influence is based on the hat matrix
from the iteratively reweighted least squares algorithm, as in Section 4.3. The
diagonal elements of this matrix are stored in the system vector %lv.

```
$extract %lv $
$calc i = %ind(%lv) $
$graph (s = 1 h = 'observation number' v = 'leverage')
       %lv i 5 $
```

The leverage values are shown in Fig. 6.7. Using the value $2(p + 1)/n = 0.242$
as a rough guide, we find that observations $2(h_2 = 0.297)$ and $21(h_{21} = 0.282)$
are potentially influential. Both are on the edge of the variable space, with the two
smallest values of wbc. Deleting these observations in turn makes no substantial
difference to the fitted model. Returning to the main-effect model, we find only one
potentially influential observation, (2), which has no substantial effect on the fitted
model. We conclude that the data are represented adequately by the exponential
distribution with parallel regressions on the log scale (although other regression
models also fit the data adequately).

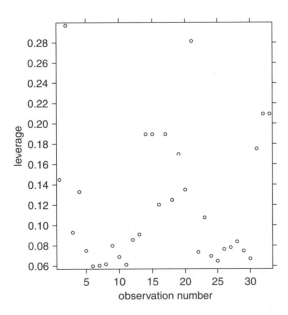

Fig. 6.7. Feigl data: leverage values, interaction model

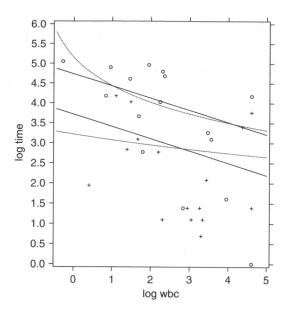

Fig. 6.8. Feigl data: fitted log (solid) and reciprocal (dotted) link function models

Figure 6.8 shows the complete data on the log t–log wbc scale and the parallel slopes models for survival time using the log (straight lines) and reciprocal (curves) links. The two sets of fitted values are computed from the fitted models over a grid using $predict.

```
$link l $fit ag + lwbc $
$var 110 x $calc x = -0.5 + 0.05*%cu(1) $
$predict lwbc = x ag = 1 $calc pl1 = %plp $
$predict lwbc = x ag = 2 $calc pl2 = %plp $
$link r $fit . $
$predict lwbc = x ag = 1 $calc pr1 = %log(%pfv) $
$predict lwbc = x ag = 2 $calc pr2 = %log(%pfv) $
$graph (s = 1 h = 'log wbc' v = 'log time' y = 0,6 x = -0.5,5,5)
       lt,pl1,pl2,pr1,pr2 lwbc,x,x,x,x 5,6,10,10,11,11 ag,,,, $
```

(Note the use of the factor argument to define different plotting characters for ag+ and ag- ; for the fitted values different characters are not needed and so the factor arguments are empty for these values.)

The fitted models agree closely relative to the random variation in the data. One aspect of this figure may seem strange: the fitted models for each group seem to lie above most of the points for that group. This is a consequence of the exponential assumption: on the log scale of time the distribution is extreme value

(Section 6.12) with scale parameter 1. This distribution has negative skew: large negative and small positive values are characteristic of it (see Fig. 6.19).

6.5 Comparison with the normal family

The Box–Cox normal transformation family discussed in Chapter 3 provides an alternative family of possible survival distributions. Does this family provide a better representation than the exponential distribution? We first explore the Box–Cox transformation without the log transformation of wbc:

```
$accuracy 5 $
$input %plc boxcox $
$error n $link i $
$macro yvar time $endmac
$macro model ag*wbc $endmac
$use boxcox $
```

Specifying a grid for λ of $-2(0.5)2$, we find the minimum of the disparity $-2p\ell(\lambda)$ near $\lambda = 0$. Replotting over a finer grid we find the minimum disparity of 289.42 occurs at $\lambda = 0.11$, with an approximate 95% confidence interval of $(-0.15, 0.38)$. The value at $\lambda = 0$ is 290.12. The log scale for time is strongly indicated. (The disparity values are different from the values of $-2p\ell(\lambda)$ in the first edition as all the constants are included in the disparity macro computation.)

Should wbc be log transformed as well?

```
$macro model ag*lwbc $endmac
$use boxcox $
```

The minimum disparity of 286.26 now occurs at $\lambda = 0.19$, and the value at $\lambda = 0$ is 288.37. The log transformation of wbc at $\lambda = 0$ decreases the disparity by 1.75: there is not much to choose between the models, but the lwbc model is slightly preferred. We now fit this model to log time.

```
$use boxfit $
Value of lambda?
$DIN? 0
```

The normal quantile plot of the raw residuals shows a reasonable straight-line fit. How do we choose between the exponential and the lognormal distributions for time? This problem does not fit into the standard likelihood ratio test theory because the two distributions are not in the same family and have different numbers of parameters. We discuss the problem further in Section 6.20, where the two distributions are embedded in a larger family: here we simply compare the disparities for the two models.

Further problems arise in this comparison, because for the exponential model there are constants omitted from the calculation of the deviance which have to be included if comparisons of disparity are to be made between different probability models.

For the exponential distribution, the GLIM deviance is

$$D = -2\log[L(\hat{\boldsymbol{\beta}})/L(\hat{\mu}_i)],$$

where

$$L(\hat{\mu}_i) = \prod_i \frac{1}{\hat{\mu}_i} e^{-t_i/\hat{\mu}_i}$$

and $\hat{\mu}_i = t_i$ giving

$$L(\hat{\mu}_i) = e^{-n} \Big/ \prod_i t_i.$$

Thus

$$-2\ell(\hat{\boldsymbol{\beta}}) = -2\log L(\hat{\boldsymbol{\beta}})$$
$$= D + 2\left(n + \sum \log t_i\right).$$

For the exponential model with log link, $D = 38.56$, and we calculate the disparity using

```
$calc %c = %cu(%log(time)) : 38.56 + 2*(%nu + %c) $
```

which gives a disparity of 291.32, compared with 288.37 for the lognormal.

Thus the exponential distribution provides a slightly worse fit, possibly because it has no free scale-parameter. For the lognormal distribution the fitted model is

$$5.425 - 2.525\,\mathrm{AG}(2) - 0.818\,\mathrm{LWBC} + 0.583\,\mathrm{AG}(2).\mathrm{LWBC}$$
$$(0.603)\quad(0.945)\qquad\quad(0.209)\qquad\qquad(0.321)$$

Again the interaction term can be omitted (a disparity increase of 3.56), giving a final model

$$4.802 - 0.988\,\mathrm{AG}(2) - 0.571\,\mathrm{LWBC}$$
$$(0.514)\quad(0.436)\qquad\quad(0.165)$$

The slope estimate is greater in magnitude for the lognormal than for the exponential model (-0.304), but the ag difference is very similar. Survival appears to decline more rapidly with wbc under the lognormal distribution than under the exponential, but since these distributions fit nearly equally well, we cannot be precise about this rate of decline.

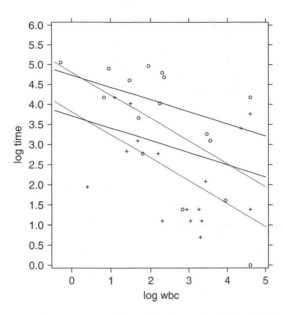

Fig. 6.9. Feigl data: fitted lognormal (dotted) and exponential log link (solid) models

The fitted models for the lognormal (dotted lines) and exponential (solid lines) distributions are shown in Fig. 6.9.

```
$yvar lt $err n $link i $
$fit ag + lwbc $disp e $
$predict lwbc = x ag = 1 $calc pln1 = %plp $
$predict lwbc = x ag = 2 $calc pln2 = %plp $
$graph (s = 1 h = 'log wbc' v = 'log time' y = 0,6 x = -0.5,5.5)
        lt,pl1,pl2,pln1,pln2 lwbc,x,x,x,x 5,6,10,10,11,11 ag,,,, $
```

We consider this dataset further in Sections 6.9, 6.11 and 6.17, where we compare these models with those produced assuming gamma, Weibull and piecewise exponential distributions.

6.6 Censoring

A characteristic feature of survival data is the occurrence of *censored* observations, that is, observations on which the complete lifetime or survival time is not observed. Censoring occurs for a number of different reasons. In industrial experiments on destructive testing of components, testing is usually terminated after a fixed time, or after the failure of some fixed proportion of components on test. The remaining components have not failed and their complete lifetimes are not observed. In medical studies censoring arises from the withdrawal of patients from the study because they have moved away or have stopped returning for follow-up, and from the need to analyse the data at various stages of the study before complete lifetimes are observed on all patients. Censoring is generally on the right; that is, the observed time is less than the actual survival time. Much less commonly it may be on the left; that is, the observed time is *greater* than the actual survival time. We will assume that censoring is on the right for most of this chapter; left-censoring is considered in Section 6.18.

Throughout this chapter we shall assume that censoring is *uninformative*, that is, that censoring of a lifetime does not occur because of factors associated with the survival time which would have been observed. This assumption is similar to that of data missing at random in Section 3.17. Censoring would be informative if, for example, patients were withdrawn from a drug treatment study because they were suffering unexpected or severe effects of the treatment which might have reduced their survival times.

As with missing data, it is prudent to compare by logistic regression the group of censored observations with the group of uncensored observations to see if censoring is systematically related to the explanatory variables. Systematic effects would be a warning of the possible failure of the assumption of uninformative censoring, though they do not imply it logically.

6.7   Likelihood function for censored observations

When some observations have censored lifetimes, the likelihood function is changed, and, in general, maximum likelihood estimation can no longer be carried out using the standard GLIM fitting algorithm (Section 2.9.3). Nevertheless, for survival distributions related to the exponential (and also for the logistic and log–logistic distributions in Section 6.18) it is still possible to use GLIM model-fitting procedures. We illustrate with the exponential distribution.

For each individual in the sample we observe the vector of explanatory variables $\mathbf{x}_i$, and a pair of variables (t_i, w_i). The *censoring indicator* w_i takes the value 1 if the survival time t_i for the i-th observation is *uncensored* and zero if it is censored. Thus when $w_i = 1$, the true survival time $T_i = t_i$, while when $w_i = 0$, it is known only that $T_i > t_i$. The contribution of a censored survival time to the likelihood is thus the probability $S(t_i)$.

The joint distribution of T_i and the Bernoulli censoring indicator W_i can be expressed as

$$\Pr(T_i, W_i) = \Pr(T_i \mid W_i) \Pr(W_i)$$
$$= \Pr(W_i \mid T_i) \Pr(T_i).$$

The critical assumption of uninformative censoring is that $\Pr(W_i \mid T_i) = \Pr(W_i)$ is independent of T_i, and that the Bernoulli distribution of W_i is free of the model parameters in the distribution of T_i. In this case the censoring process can be omitted from the likelihood, which can then be expressed as

$$L(\boldsymbol{\beta}) = \prod_{i=1}^{n} [f(t_i)]^{w_i} [S(t_i)]^{1-w_i}$$

$$= \prod_{i=1}^{n} [h(t_i)]^{w_i} S(t_i).$$

For the exponential distribution with mean μ_i, we have

$$h(t_i) = 1/\mu_i = \lambda_i$$

$$S(t_i) = e^{-\lambda_i t_i}$$

so that

$$L(\boldsymbol{\beta}) = \prod_{i} \lambda_i^{w_i} e^{-\lambda_i t_i}$$

$$= \prod_{i} (\lambda_i t_i)^{w_i} e^{-\lambda_i t_i} \bigg/ \prod_{i} t_i^{w_i}.$$

The term in the denominator is not a function of the parameter vector $\boldsymbol{\beta}$ and can be omitted for ML calculations, though it must be retained for the computation of the disparity. The remaining term is identical to the likelihood function for a set of n observations w_i having independent Poisson distributions with mean $\lambda_i t_i$ (the term $1/w_i!$ is missing from the Poisson likelihood, but this is a constant and can also be omitted). Since w_i is only either 0 or 1, this result looks strange. Where does the Poisson distribution come from?

The Poisson result is easily understood if we visualize a set of n independent time-homogeneous Poisson processes as in Section 6.2, the i-th having rate λ_i. The number of events for the i-th process in a time-interval of length t_i then has a Poisson distribution with mean $\lambda_i t_i$. We observe the i-th process until either the first event occurs, or a fixed time has elapsed without the event occurring. In the first case, one event occurs at time t_i; in the second, no event has occurred by time t_i.

The Poisson likelihood for the censored and uncensored exponential survival times allows us to fit models to the hazard rate λ using the Poisson model (Aitkin and Clayton, 1980). Write θ_i for the Poisson mean:

$$\theta_i = \lambda_i t_i.$$

If the *linear hazard model* is used, then

$$\lambda_i = \boldsymbol{\beta}' \mathbf{x}_i$$

and

$$\theta_i = \boldsymbol{\beta}'(t_i \mathbf{x}_i).$$

To fit the model, each explanatory variable $\mathbf{x}_i$ (including the unit vector $\mathbf{1}$) has to be multiplied by t_i, w_i is declared as the response variable with Poisson error and identity link function, and the intercept term is omitted.

If the *log-linear hazard model* is used, then

$$-\log \mu_i = \log \lambda_i = \boldsymbol{\beta}'\mathbf{x}_i$$

and

$$\log \theta = \log \lambda_i + \log t_i$$
$$= \boldsymbol{\beta}'\mathbf{x}_i + \log t_i.$$

This model is particularly simple to fit: the same error and response variable specifications are used, but now with log link function and an offset of $\log t_i$.

This approach can also be used for uncensored observations, by defining the censoring indicator as a vector of 1s. We illustrate with the example of Section 6.3:

```
$input 'feigl' $
$calc lt = %log(time) : lwbc = %log(wbc) : w = 1 $
$yvar w $error p $link l $offset lt $
$fit ag*lwbc  $disp e $
 scaled deviance =  38.56 at cycle 5
     residual df =  29
```

	estimate	s.e.	parameter
1	-5.150	0.514	1
2	1.871	0.732	AG(2)
3	0.482	0.180	LWBC
4	-0.328	0.246	AG(2).LWBC

We obtain the same parameter estimates and deviance as in Section 6.3 (to within the accuracy set by GLIM's convergence criterion), though with opposite signs for the parameters, but the standard errors are different (e.g. 0.732 for the s.e. of AG(2), rather than 0.789). This is because the expected information matrix for the Poisson model treats t_i as a constant explanatory variable, while in the exponential model t_i is treated as a random variable and is replaced in the expected information matrix by its expected value μ_i. Thus the standard errors reported from the Poisson model are based on the observed, rather than the expected, information matrix for the censored exponential model. See Section 2.9.3 for details.

It may seem strange that we can fit a model to a constant response variable, but the offset introduces the variation to be modelled by the explanatory variables.

If the *reciprocal hazard model* is used, then

$$\lambda_i = 1/\boldsymbol{\beta}'\mathbf{x}_i$$

giving

$$\mu_i = \boldsymbol{\beta}'\mathbf{x}_i$$

and the model is a linear model for the *mean*. Then

$$\theta_i = \lambda_i t_i = t_i/\boldsymbol{\beta}'\mathbf{x}_i,$$

so that

$$\theta_i^{-1} = \boldsymbol{\beta}'(\mathbf{x}_i/t_i).$$

The model can be fitted by dividing each explanatory variable (including the unit vector **1**) by t_i, and using the reciprocal link function without an intercept.

The Poisson representation of the likelihood applies more generally to *proportional hazard* models, as we shall see in Section 6.15.

6.8 Probability plotting with censored data: the Kaplan–Meier estimator

The presence of censoring complicates our assessment of the probability distribution assumption for the response variable. If T_i is exponential with mean μ_i, then T_i/μ_i is standard exponential, but if the true lifetime is censored at t_i, then t_i/μ_i is a censored value from the standard exponential distribution. Thus, when a model has been fitted to mixed censored and uncensored observations, the $\mu_i = t_i/\hat{\mu}_i$ are themselves mixed censored and uncensored values from the (approximately) standard exponential distribution. The u_i are the Poisson fitted values for any link: $u_i = \theta_i = \hat{\lambda}_i t_i = t_i/\hat{\mu}_i$. However, it is not clear how to construct the empirical survivor function of the u_i to assess the correctness of the probability distribution. The *product-limit* or *Kaplan–Meier* (1958) estimator of the survivor function is used for this purpose.

We begin by considering the set of all discrete times $0 < a_1 < a_2 < \cdots < a_N$ at which deaths or censorings may occur. There is no real loss of generality in assuming that these times are discrete, because the finite precision of measurement means that the values of survival time actually recorded can take only a finite (though possibly large) number of values (see Chapter 2 for a discussion of measurement precision). Let A_j denote the time-interval $(a_{j-1}, a_j]$ with $a_0 = 0$. For the continuous survival distribution with density $f(t)$, survivor function $S(t)$ and hazard $h(t)$, we define the corresponding functions for the resulting grouped discrete distribution:

$$f_j = \Pr(T \in A_j)$$
$$= S(a_{j-1}) - S(a_j) = s_j - s_{j+1},$$

where

$$s_j = \Pr(T > a_{j-1}) = f_j + f_{j+1} + \cdots + f_N,$$

and

$$h_j = \Pr(T \in A_j \mid T > a_{j-1}) = f_j/s_j.$$

Then

$$h_j = (s_j - s_{j+1})/s_j$$

so that

$$s_{j+1}/s_j = 1 - h_j$$

whence

$$s_{r+1} = \prod_{j=1}^{r} s_{j+1}/s_j = \prod_{j=1}^{r}(1 - h_j)$$

since $s_1 = 1$. The survivor function at time t can therefore be expressed as

$$S(t) = \prod_{j:a_j<t} (1 - h_j).$$

We estimate the survivor function by considering the passage of each individual through time, shown in the table below.

The circles represent censoring and the crosses death. The i-th individual survives to the beginning of the interval A_{j_i} and is either finally censored or dies at the end of the interval. Let w_{ij} be the censoring indicator for the i-th individual at a_j; the hazard in A_j is h_j, and the survivor function at a_{j-1} is s_j, and at a_j is s_{j+1}.

The contribution L_i to the likelihood for the i-th individual is f_{j_i} if the individual dies in A_{j_i}, and s_{j_i+1} if he survives beyond the end a_j of the interval. Thus

$$L_i = f_{j_i}{}^{w_{ij_i}} s_{j_i+1}{}^{1-w_{ij_i}}$$

$$= (h_{j_i} s_{j_i})^{w_{ij_i}} [s_{j_i}(1-h_{j_i})]^{1-w_{ij_i}}$$

$$= s_{j_i} h_{j_i}{}^{w_{ij_i}} (1-h_{j_i})^{1-w_{ij_i}}$$

$$= h_{j_i}{}^{w_{ij_i}} (1-h_{j_i})^{1-w_{ij_i}} \prod_{j=1}^{j_i-1}(1-h_j)$$

$$= \prod_{j=1}^{j_i} h_j{}^{w_{ij}} (1-h_j)^{1-w_{ij}}$$

since all the w_{ij} for $j < j_i$ are zero. The likelihood function over all individuals is then

$$L = \prod_{i=1}^{n} L_i = \prod_{j=1}^{N}\prod_{i\in R_j} h_j{}^{w_{ij}} (1-h_j)^{1-w_{ij}},$$

where R_j is the *risk set* of individuals in the j-th interval, that is the set of individuals not already dead or withdrawn (censored).

The likelihood function is the product of a set of N binomial likelihoods, one for each interval. In the j-th interval, the number of individuals at risk is

$$\sum_{i\in R_i} 1 = r_j,$$

and the number dying is

$$\sum_{i\in R_j} w_{ij} = d_j.$$

The constant probability of death for each individual is h_j, so we have immediately that the MLE of h_j is

$$\hat{h}_j = d_j/r_j$$

and therefore that of $S(t)$ is

$$\hat{S}(t) = \prod_{j:a_j<t} (1-d_j/r_j).$$

This is the *product-limit* or *Kaplan–Meier* estimator of the survivor function. When $d_j = 0$, $\hat{h}_j$ is zero: a change in the estimated survivor function occurs only at an observed death time. Thus in computing the Kaplan–Meier estimator we need to consider only the times at which deaths occur: all other times can be omitted from the set of a_j. When there is no censoring, the Kaplan–Meier estimator reduces to the usual empirical survivor function. Thus the usual empirical cumulative distribution function is in fact the non-parametric maximum likelihood estimator of the cumulative distribution function $F(t)$ when no assumption is made about the form of F.

The Kaplan–Meier estimator is a *step-function*, with steps only at the observed death times. It is usually graphed as a step-function, but we will follow the convention of graphing it as a point function only at the observed death times.

We illustrate with a simple example. The file gehan contains 42 observations of remission times in weeks of child patients with acute leukaemia. A randomized treatment group was treated with 6-mercaptopurine, the other group was a control. The data are presented and analysed by Freireich *et al.* (1963) and Gehan (1965). We first examine the treatment group, in which censoring is heavy. The observations t for this group are given below, in increasing order, with the censoring indicator w.

```
------------------------------------------------------------------
t   6   6   6   6   7   9  10  10  11  13  16  17  19  20  22  23  25  32  32  34  35
w   1   1   1   0   1   0   1   0   0   1   1   0   0   0   1   1   0   0   0   0   0
------------------------------------------------------------------
```

The first censored observation at 6 weeks we take to just *exceed* 6 weeks, and similarly for that at 10 weeks. The survivor function takes the value 1 at $t_0 = 0$ and changes value only at the observed remission times $t_j = 6, 7, 10, 13, 16, 22$ and 23 weeks. We construct a table of t_j, r_j, d_j, h_j and $\hat{S}(t)$.

j	t_j	r_j	d_j	h_j	$1 - h_j$	$\prod(1 - h_j)$
0	0	21	0	0.0	1.0	1.0
1	6	21	3	0.1429	0.8571	0.8571
2	7	17	1	0.0588	0.9412	0.8067
3	10	15	1	0.0667	0.9333	0.7529
4	13	12	1	0.0833	0.9167	0.6902
5	16	11	1	0.0909	0.9091	0.6275
6	22	7	1	0.1429	0.8571	0.5378
7	23	6	1	0.1667	0.8333	0.4482

The Kaplan–Meier estimator can be computed using the macro kaplan, stored in the file kaplan. The macro requires as arguments the name of the variable containing the censored and uncensored survival times, and the censoring indicator which must be coded 1 for uncensored and 0 for censored observations. It produces a graph of the Kaplan–Meier estimator against the survival variable named as the macro argument. The survivor function estimate is held in the variate surv_ and the *uncensored* values of the survival variable in the variate time_. If the largest observation is uncensored, the survivor function estimate at this point will be zero. When the log of the survivor function is calculated, this will give a value of zero which will appear out of place in the graph. The system vector %re can be used to suppress the plotting of this point, using for example $calc %re = %ne(surv_,0) $. We suppress for the same reason the point $t = 0, S(t) = 1$.

We illustrate with the full set of data from gehan in which the response variable is t, the censoring indicator is w and the treatment group factor is g. We first graph the survivor functions in the two treatment groups, by copying each group data into new variables and graphing them together.

```
$units 42 $
! Remission times in acute leukemia
$data t $read
1 1 2 2 3 4  4  5  5  8  8  8  8 11 11 12 12 15 17 22 23
6 6 6 6 7 9 10 10 11 13 16 17 19 20 22 23 25 32 32 34 35
$data w $read
1 1 1 1 1 1 1 1 1 1 1 1 1 1 1 1 1 1 1 1 1
1 1 1 0 1 0 1 0 0 1 1 0 0 0 1 1 0 0 0 0 0
$calc g = %gl(2,21) $factor g 2 $
!          variables:
!            t  remission time in weeks
!            w  censor variate (1 = uncensored, 0 = censored)
!            g  group treatment (1 = placebo, 2 = 6-mercaptopurine)
$input 'kaplan' $
$var 21 t1 t2 w1 w2 index $
$calc index = %cu(1) $
        : t1 = t(index) : t2 = t(21 + index) $
        : w1 = w(index) : w2 = w(21 + index) $
$use kaplan t1 w1 $
$calc surv1 = surv_  : time1 = time_ $
$use kaplan t2 w2 $
$calc surv2 = surv_  : time2 = time_ $
$graph (s = 1 h = 't' v = 'S(t)')
        surv1,surv2 time1,time2 5,7 $
```

The survivor functions shown in Fig. 6.10 are very different in the two groups. The treatment has a major effect on survival.

For later purposes we transform to the (negative) log–log scale for $S(t)$ and the log scale for t, and redraw the graph (Fig. 6.11). We restrict plotting of the

SURVIVAL DATA

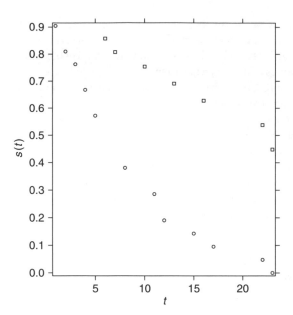

Fig. 6.10. Gehan data: survivor functions

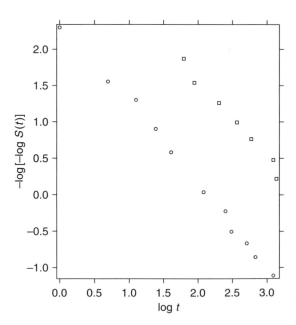

Fig. 6.11. Gehan data: log–log survivor functions

transformed zero for $S(t)$ using weights:

```
$calc llsl = -%log(-%log(survl)) : lls2 = -%log(-%log(surv2)) $
    :    lt1 = %log(timel) : lt2 = %log(time2) $
    :    ww1 = %ne(survl,0) : ww2 = %ne(surv2,0) $
$graph (s = 1 h = 'log t' v = '-log[-log S(t)]')
            llsl/ww1,lls2/ww2 lt1,lt2 5,7 $
```

It is striking that the two survivor functions are nearly linear and parallel on these scales with common slope nearly 1.0, consistent with an exponential distribution for survival in each group (Fig. 6.11). This parallelism is a distinguishing feature of *proportional hazard models*, discussed in general in Section 6.15.

We now fit an exponential distribution with the two-group model g. The fitted values are then the standardized exponential values u_l.

```
$yvar w $error p $link l $
$calc ofs = %log(t) $offset ofs $
$fit g $disp e $
scaled deviance =   38.017 at cycle 4
     residual df =   40
```

	estimate	s.e.	parameter
1	-2.159	0.218	1
2	-1.527	0.398	G(2)

scale parameter 1.000

The hazard in the second group is significantly lower than in the first, with an estimated group difference of 1.527 with standard error 0.398. We now examine the survivor function for the pooled residuals. As with the feigl example we use the binocisim macro to place 95% confidence bounds on the estimated survivor function. The simultaneous bounds for the Kaplan–Meier estimator were given by Hollander *et al.* (1997). The number of "trials" is accessed from the kaplan macro as rj_ and the number of "successes" is computed as rj_*surv_. The (negative) exponential survivor function is computed over a fine grid as the variate exp to give a smoother fitted curve.

```
$use kaplan %fv w $
$input 'binocisim' $
$calc r = rj_*surv_ $
$var 260 sexp ind $calc ind = 0.01*%cu(1) : sexp = %exp(-ind) $
$calc ww = %ne(surv_,0) $
$use binocisim r rj_ $
$graph (s = 1 h = 'u' v = 'S(u)' y = 0,1)
            surv_/ww,pl_ _/ww,pu_ _/ww,sexp time_,time_,time_,ind
            5,6,6,10 $
```

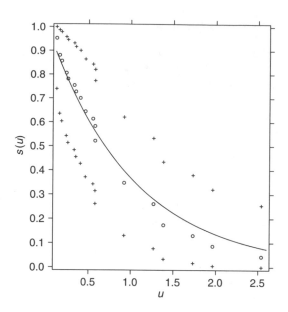

Fig. 6.12. Gehan data: survivor function and bounds

The survivor function in Fig. 6.12 appears to decrease exponentially; the bounds on the true survivor function increase in width with time, but there is little evidence of non-exponential curvature.

Transformations of the proportion scale may be used as in the `feigl` example to examine this issue further. As the fitted models are now straight lines the fine grid is not necessary.

```
$calc  ls = %log(surv_) $
   : lsexp = %log(sexp) $
   :   lpl = %log(pl__) : lpu = %log(pu__) $
$graph (s = 1 h = 'u' v = 'log S(u)')
         ls/ww,lsexp,lpl/ww,lpu/ww time_,ind,time_,time_ 5,10,9,9 $
$calc lls = -%log(-ls) $
   :llsexp = -%log(-lsexp) $
   :  lind = %log(ind) $
   :    lu = %log(time_) $
   :  llpl = -%log(-lpl) : llpu = -%log(-lpu) $
$graph (s = 1 h = 'log u' v = '-log[-log S(u)]')
         lls/ww,lllsexp,llpl/ww,llpu/ww lu,lind,lu,lu 5,10,9,9 $
```

The points in Fig. 6.13 appear to curve away from the line for large values of scaled time, and their slope in Fig. 6.14 is slightly greater than 1 in magnitude—the fitted line is below the data points on the left, but above them on the right. However, the

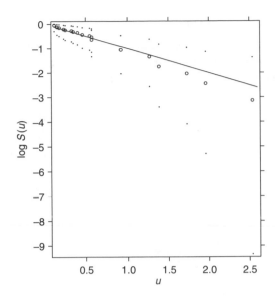

Fig. 6.13. Gehan data: log survivor function and bounds

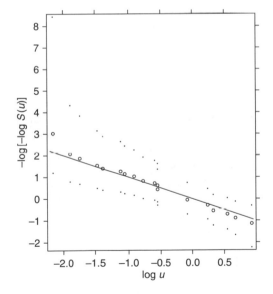

Fig. 6.14. Gehan data: log–log survivor function and bounds

line is well within the bounds so it is not clear whether the exponential distribution is satisfactory. In Section 6.9 we consider the gamma, and in Section 6.11 the Weibull distribution as alternative models for the survival time.

We now consider survival distributions related to the exponential.

6.9 The gamma distribution

Suppose that an observed lifetime is a *sum* of times for r separate stages of life, each of which is exponentially distributed with the same mean θ. Then the total lifetime has a gamma (r, θ) distribution with parameters r and θ:

$$f(t) = \frac{1}{\Gamma(r)\theta^r} e^{-t/\theta} t^{r-1}, \quad t, r, \theta > 0.$$

The value of r, the *shape* parameter, can more generally be any positive real number.

The mean of the distribution is $r\theta$, and we adopt the GLIM parametrization using $\mu = r\theta$, writing

$$f(t \mid r, \mu) = \frac{r^r}{\Gamma(r)\mu^r} e^{-rt/\mu} t^{r-1}, \quad t, r, \mu > 0.$$

The variance is μ^2/r, and the exponential family *scale* parameter is $\phi = 1/r$. The cdf is

$$F(t \mid r, \mu) = \int_0^t f(u \mid r, \mu)\, du$$

$$= \frac{1}{\Gamma(r)} \int_0^{rt/\mu} e^{-z} z^{r-1}\, dz$$

$$= F_r(rt/\mu),$$

where $F_r(t)$ is the cdf of the "standardized" gamma distribution with mean 1 and shape parameter r. This is available in GLIM as the system function %gp(t,r). Graphs of the density are shown in Fig. 6.15 for $\mu = 5$ and $r = 0.5, 1, 2, 5$. Skewness decreases as r increases.

The gamma distribution does not have an analytic hazard function, and does not have a proportional hazard function except for $r = 1$, and so modelling is generally of the (log) mean rather than the (log) hazard. Graphs of the hazard function are shown in Fig. 6.16 for the same values of μ and r.

For $r < 1$ the hazard is monotone decreasing. For $r = 1$ the hazard is constant, since the distribution is exponential. For $r > 1$ the hazard is monotone increasing to an asymptote.

The gamma distribution can be fitted straightforwardly for a constant shape parameter r. We assume that the explanatory variables do not affect the shape parameter; Smyth (1989) has developed double modelling of both μ and r, as for the normal distribution in Chapter 3. We give details in Section 6.9.3.

We first consider uncensored observations, with a log-linear model $\log \mu_i = \boldsymbol{\beta}' \mathbf{x}_i$.

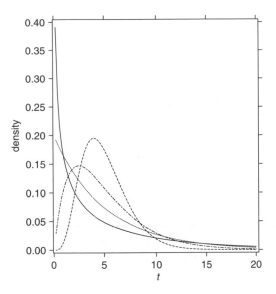

Fig. 6.15. Gamma density functions

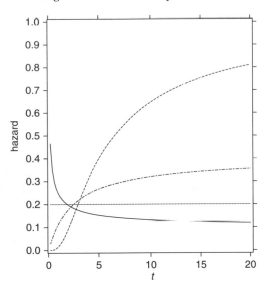

Fig. 6.16. Gamma hazard functions

6.9.1 *Maximum likelihood with uncensored data*

The log-likelihood is

$$\log L = nr \log r - n \log \Gamma(r) - r \sum \log \mu_i - r \sum (t_i/\mu_i) + (r-1) \sum \log t_i,$$

and

$$\frac{\partial \log L}{\partial \boldsymbol{\beta}} = r \sum \{(t_i/\mu_i) - 1\} \mathbf{x}_i$$

$$\frac{\partial \log L}{\partial r} = n(1 + \log r) - n\psi(r) - \sum \log \mu_i - \sum (t_i/\mu_i) + \sum \log t_i$$

$$\frac{\partial^2 \log L}{\partial \boldsymbol{\beta} \partial \boldsymbol{\beta}'} = -r \sum (t_i \mathbf{x}_i \mathbf{x}_i'/\mu_i)$$

$$\frac{\partial^2 \log L}{\partial \boldsymbol{\beta} \partial r} = \sum \{(t_i/\mu_i) - 1\} \mathbf{x}_i$$

$$\frac{\partial^2 \log L}{\partial r^2} = n/r - n\psi'(r),$$

where $\psi(r)$ and $\psi'(r)$ are the digamma and trigamma functions respectively. Since $E[T_i] = \mu_i$, the expectation of the cross-derivative is zero, and so is its sample value at the MLE $\hat{\boldsymbol{\beta}}$ (since the cross-derivative is just the score for $\boldsymbol{\beta}$ divided by r). The fact that r is estimated, rather than known, has no effect asymptotically on the precision of $\hat{\boldsymbol{\beta}}$. The MLE of $\boldsymbol{\beta}$ does not depend on r, since r cancels in the score equation for $\boldsymbol{\beta}$, but the observed and expected informations contain r as a multiplying factor. The expected information is $I(\boldsymbol{\beta}) = r \sum \mathbf{x}_i \mathbf{x}_i'$, independent of $\hat{\boldsymbol{\beta}}$.

Thus the MLE of $\boldsymbol{\beta}$ for the gamma distribution will be (in this parametrization) the same as that for the exponential distribution, and the standard errors will be $1/\sqrt{r}$ times those for the exponential. The MLE $\hat{r}$ of r is the solution of

$$n(1 + \log r) - n\psi(r) + \sum \log(t_i/\hat{\mu}_i) - \sum (t_i/\hat{\mu}_i) = 0.$$

The macro gamma in the file gammaml finds the MLE $\hat{r}$ (held in the scalar r_) by a simple Newton–Raphson step starting from the GLIM estimate below, using $\partial \log L/\partial r$ and $\partial^2 \log L/\partial r^2$, and gives the standard error from the expected information.

A simple approximation is accurate for large r. Using Stirling's formula for the gamma function,

$$\log \Gamma(r) \doteq \tfrac{1}{2} \log(2\pi) - r + (r - \tfrac{1}{2}) \log r$$

and correspondingly

$$\psi(r) \doteq \log r - 1/(2r)$$
$$\psi'(r) \doteq 1/r + 1/(2r^2).$$

So for large r the MLE $\hat{r}$ is approximately the solution of

$$n(1 + \log r) - n\left(\log r - \frac{1}{2r}\right) - \sum(t_i/\hat{\mu}_i) + \sum \log(t_i/\hat{\mu}_i) = 0,$$

which gives

$$\hat{r} \doteq \frac{n}{2}\left(\sum(t_i/\hat{\mu}_i) - \sum \log(t_i/\hat{\mu}_i) - n\right)^{-1} = n/\text{dev},$$

where dev is the GLIM deviance, with approximate standard error $r\sqrt{2/n}$. This gives the simple GLIM approximation for $\hat{r}$ as the reciprocal of the mean deviance. However, for small r the approximation is quite inaccurate.

The LR "goodness of fit" test for the comparison of nested models needs care. If r were *known*, the "saturated" model would have $\hat{\mu}_i = t_i$, with the corresponding log-likelihood

$$\log L_{\text{sat}} = nr \log r - n \log \Gamma(r) - r \sum \log t_i - nr + (r - 1) \sum \log t_i,$$

and the LRTS for a given model relative to the saturated model would be

$$LRTS = -2(\log L_{\text{model}} - \log L_{\text{sat}})$$
$$= 2r\left\{\sum(t_i/\hat{\mu}_i) - n - \sum \log(t_i/\hat{\mu}_i)\right\}.$$

But this test statistic can be used only when the scale factor r is known and equal under both models. When no scale parameter has been explicitly set by the user, GLIM calculates its deviance using the above expression with $r = 1$ (the value corresponding to the exponential distribution). GLIM uses the reciprocal of the mean deviance as an estimate of r only for calculating standard errors for the parameters. (In fact GLIM uses the residual degrees of freedom $n - p - 1$ rather than the sample size n in calculating the mean deviance.)

When the scale-parameter is *known*, it can be set using the $scale directive (as in normal variance modelling in Chapter 3). In this case GLIM reports *scaled* deviances which are correct, so that scaled deviance differences between nested models are asymptotically χ^2 under the null hypothesis. If the scale-parameter is *not* set, the GLIM (unscaled) deviances between nested models are those for the exponential distribution, and not the gamma. When r is estimated by ML under each model, disparity differences are again distributed as χ^2 under the null hypothesis. The gamma macro prints out the disparity for the fitted model and holds it in the scalar d_.

We illustrate with the feigl data of Section 6.3. We first give the standard GLIM analysis.

```
$newjob $
$input 'feigl' $
```

```
$calc lwbc = %log(wbc) $
$yvar time $err g $link l $
$fit ag*lwbc $disp e $
   deviance =   38.555 at cycle 6
residual df =   29
```

	estimate	s.e.	parameter
1	5.150	0.577	1
2	-1.874	0.905	AG(2)
3	-0.482	0.200	LWBC
4	0.329	0.308	AG(2).LWBC

```
scale parameter 1.329
```

As no argument is specified for $scale, GLIM estimates the scale-parameter $\phi = 1/r$ from the mean deviance as $38.555/29 = 1.329$, giving $\tilde{r} = 0.752$. For the full ML analysis, we specify the model in a model macro, and use the gamma macro with argument the time variable:

```
$input 'gammaml' $
$macro model ag*lwbc $endm
$use gamma time $
Gamma model fitted with ML estimate of shape parameter
  -2 log Lmax is   291.314,
   deviance relative to saturated exponential model is   38.552
```

	estimate	s.e.	parameter
1	5.150	0.504	1
2	-1.874	0.789	AG(2)
3	-0.482	0.175	LWBC
4	0.329	0.268	AG(2).LWBC

```
scale parameter 1.011

Gamma shape parameter r =   0.989 (s.e. =   0.214)
```

The MLE of r is very close to 1.0, the exponential distribution. The disparity change relative to the exponential model is 0.003: within the class of gamma distributions, the exponential is almost the best fit (see Section 6.11 for a similar result for the Weibull distribution, and Section 6.17 for the piecewise exponential distribution for these data.)

We do not proceed further with this example at this point, since it can be treated as exponential. However we re-consider this example with *double modelling* of both parameters in Section 6.9.3. Note that the GLIM mean deviance estimate $\tilde{r}$ of 0.752 differs considerably from the MLE.

We now consider the case of censored observations, again restricting our analysis to the log link.

6.9.2 *Maximum likelihood with censored data*

The likelihood as constructed in Section 6.7 requires the survivor function which is expressed in terms of the incomplete gamma function.

We solve the likelihood equations indirectly, using the EM algorithm, and considering the censored observations as incomplete. The complete data log-likelihood is linear in t_i and $\log t_i$, and so in the E-step of the algorithm the censored values of t and $\log t$ are replaced by their conditional expectations given the observed data and the current parameter estimates.

For an observation right-censored at a, the conditional distribution of T given $T > a$ is

$$f(t \mid T > a) = f(t)/S(a), \quad t > a.$$

The conditional expectation of T for a right-censored observation is therefore

$$\mathrm{E}[T \mid T > a] = \int_a^\infty t f(t)\, \mathrm{d}t \Big/ S(a)$$

$$= \mu \frac{[1 - F_{r+1}(ra/\mu)]}{[1 - F_r(ra/\mu)]}.$$

To find the conditional expectation of $\log T$, we first find its conditional moment generating function. Write

$$M(\theta) = \mathrm{E}[e^{\theta \log T} \mid T > a]$$

$$= \mathrm{E}[T^\theta \mid T > a]$$

$$= \int_a^\infty t^\theta f(t)\, \mathrm{d}t / S(a)$$

$$= \frac{\mu^\theta}{r^\theta} \frac{\Gamma(r+\theta)}{\Gamma(r)} \frac{[1 - F_{r+\theta}(ra/\mu)]}{[1 - F_r(ra/\mu)]}.$$

The cumulant function of $\log T$ is (omitting terms not involving θ)

$$K(\theta) = \log M(\theta) = \theta(\log \mu - \log r) + \log \Gamma(r+\theta) + \log[1 - F_{r+\theta}(ra/\mu)].$$

Differentiating w.r.t. θ and setting $\theta = 0$ gives

$$K'(0) = \mathrm{E}[\log T \mid T > a]$$

$$= \log \mu - \log r + \psi(r) + \frac{g(0 \mid ra/\mu)}{[1 - F_r(ra/\mu)]},$$

where

$$g(\theta \mid ra/\mu) = \frac{\partial}{\partial \theta}[-F_{r+\theta}(ra/\mu)].$$

We evaluate $g(0 \mid ra/\mu)$ by numerical differentiation as

$$g(0 \mid ra/\mu) = [F_{r-\delta}(ra/\mu) - F_{r+\delta}(ra/\mu)]/2\delta$$

for some small δ.

Thus in the E-step of the EM algorithm, we replace the censored values of t_i by their conditional expectations

$$t_i^* = \mu_i \frac{[1 - F_{r+1}(rt_i/\mu_i)]}{[1 - F_r(rt_i/\mu_i)]},$$

and the censored values of $\log t_i$ by their conditional expectations

$$(\log t_i)^* = \log \mu_i - \log r + \psi(r) + \frac{g(0 \mid rt_i/\mu_i)}{[1 - F_r(rt_i/\mu_i)]}.$$

In the M-step the score equations for β and r are solved using these "expected data".

The macro cgamma in the file gammaml fits the gamma distribution to right-censored data using this approach. The macro requires two arguments, the response variable and the censoring indicator. A macro model must be specified containing the model to be fitted; the link function is reset to log. The standard errors for the $\hat{\beta}_j$ are underestimated (being based on the complete data information matrix); correct standard errors can be calculated as described in Chapter 2, by equating the Wald test statistic for the parameter to the disparity change on omitting the explanatory variable. The disparity $-2 \log L_{max}$ is calculated and printed in the macro from the censored and uncensored observations, using the ML estimates.

We illustrate with the gehan example.

```
$newjob $
$input 'gehan' : 'gammaml' $
$macro model g $endmac
$yvar t $error g $link l $
$use cgamma t w $

-2 log Lmax is  212.884,
          estimate        s.e.      parameter
     1       2.159       0.1695       1
     2       1.281       0.2397       G(2)
  scale parameter 0.6031


gamma shape parameter r =  1.658 (s.e. =  0.332)


WARNING - all standard errors are underestimated
```

```
$macro model 1 $endm
$use cgamma $

 -2 log Lmax is   232.354,
           estimate            s.e.        parameter
      1         2.837        0.1359         1
scale parameter 0.7756


gamma shape parameter r =   1.289 (s.e. =   0.253)


WARNING - all standard errors are underestimated
```

The treatment effect (on the log mean scale) is 1.28; the fitted mean survival under the treatment group is $e^{1.28} = 3.6$ times that in the control group, and the 95% confidence interval for the ratio of means is (2.0–6.4). The underestimated standard error of 0.240 is corrected by using the disparity change of 19.47 on omitting the treatment effect:

$$SE \doteq 1.281/\sqrt{19.47} = 0.290;$$

the "complete data" SE underestimates by 17%. These results are qualitatively similar to those for the exponential distribution, though the treatment effect has the opposite sign, since the mean rather than the hazard is being modelled.

The disparity for the exponential model is 217.05 (see Section 6.10), so the disparity change for the gamma shape parameter is 4.17 on 1 df, moderately strong evidence for the gamma over the exponential distribution.

We now consider the modelling of both gamma parameters; we follow the approach in Section 3.19.

6.9.3   Double modelling

We restrict the discussion to uncensored data. We generalize the regression model to the form

$$\log \mu_i = \boldsymbol{\beta}'\mathbf{x}_i, \quad \log r_i = \boldsymbol{\gamma}'\mathbf{z}_i.$$

As for the normal distribution in Chapter 3, this ensures positivity of the scale-parameter r. The variables $\mathbf{z}_i$ may be a subset or superset of $\mathbf{x}_i$, or may be unrelated. The parameters $\boldsymbol{\beta}$ and $\boldsymbol{\gamma}$ are assumed to be unrelated. The log-likelihood is

$$\ell = \sum_i \{r_i \log r_i - \log \Gamma(r_i) - r_i \log \mu_i - r_i t_i/\mu_i + (r-1)\log t_i\}$$

and

$$\frac{\partial \ell}{\partial \boldsymbol{\beta}} = \sum_i r_i \mathbf{x}_i [t_i/\mu_i - 1]$$

$$\frac{\partial^2 \ell}{\partial \boldsymbol{\beta} \partial \boldsymbol{\beta}'} = -\sum_i r_i \mathbf{x}_i \mathbf{x}_i' t_i/\mu_i$$

$$E\left[\frac{\partial^2 \ell}{\partial \boldsymbol{\beta} \partial \boldsymbol{\beta}'}\right] = -\sum_i r_i \mathbf{x}_i \mathbf{x}_i'$$

$$\frac{\partial \ell}{\partial \boldsymbol{\gamma}} = \sum_i r_i \mathbf{z}_i [1 + \log r_i - \psi(r_i) + \log(t_i/\mu_i) - t_i/\mu_i]$$

$$\frac{\partial^2 \ell}{\partial \boldsymbol{\gamma} \partial \boldsymbol{\gamma}'} = \sum_i r_i \mathbf{z}_i \mathbf{z}_i' [1 - r_i \psi'(r_i) + 1 + \log r_i - \psi(r_i) + \log(t_i/\mu_i) - t_i/\mu_i]$$

$$\frac{\partial^2 \ell}{\partial \boldsymbol{\beta} \partial \boldsymbol{\gamma}'} = \sum_i r_i \mathbf{x}_i \mathbf{z}_i' [t_i/\mu_i - 1].$$

For the expected information, we have immediately

$$E\left[\frac{\partial^2 \ell}{\partial \boldsymbol{\beta} \partial \boldsymbol{\gamma}'}\right] = 0.$$

For the expectation of $\log(T_i)$, we have from the censored case above, letting $a \to \infty$,

$$E[\log T_i] = \psi(r_i) + \log(\mu_i) - \log r_i,$$

and hence

$$E\left[\frac{\partial^2 \ell}{\partial \boldsymbol{\gamma} \partial \boldsymbol{\gamma}'}\right] = \sum_i r_i \mathbf{z}_i \mathbf{z}_i' [1 - r_i \psi'(r_i)].$$

Since the cross-derivative has zero expectation as in the normal model, the scoring algorithm reduces to separate algorithms for $\boldsymbol{\beta}$ and $\boldsymbol{\gamma}$, conveniently applied using successive relaxation.

For given $\boldsymbol{\gamma}$, we take r_i as a prior weight in the gamma model for $\boldsymbol{\beta}$ with known scale 1, and for given $\boldsymbol{\beta}$, we formulate the scoring algorithm as an iterative weighted normal regression with adjusted dependent variate

$$z_i = \mu_i - [1 + \log r_i - \psi(r_i) + \log(t_i/\mu_i) - t_i/\mu_i]/[1 - r_i \psi'(r_i)]$$

and iterative weight variate

$$w_i = r_i[1 - r_i\psi'(r_i)].$$

The algorithm begins conveniently with the fit of the constant-r gamma model to estimate β, followed by alternate γ and β steps until the disparity converges. At this point the standard errors for both models (based on the expected information) are correct.

This double modelling procedure is implemented in the macro doubgamm. It requires two model macros, modm and mods, for the mean and scale respectively. Log-linear models are assumed for both of these. The name of the response variable is required as an argument to doubgamm. We illustrate with the feigl data. We fit interaction models to both mean and scale, and reduce first the scale model, and then the mean model, by backward elimination as in Chapter 3.

```
$input 'feigl' : 'doubgamm' $
$macro modm ag*lwbc $endmac
$macro mods ag*lwbc $endmac
$use doubgamm time$
  Mean log-linear model
          estimate        s.e.      parameter
    1       4.972         0.212      1
    2      -1.579         0.601      AG(2)
    3      -0.383         0.123      LWBC
    4       0.185         0.237      AG(2).LWBC
  scale parameter 1.000


  -- model changed
  Scale log-linear model
          estimate        s.e.      parameter
    1       2.311         0.667      1
    2      -1.926         1.012      AG(2)
    3      -0.734         0.218      LWBC
    4       0.587         0.333      AG(2).LWBC
  scale parameter 1.000


  -2 log Lmax =    283.045
```

The interaction can be omitted from the scale model. Further simplification leads to the final model:

```
$macro modm ag + lwbc $endm
$macro mods lwbc $endm
$use doubgamm $
  Mean log-linear model
```

	estimate	s.e.	parameter
1	4.897	0.292	1
2	-1.181	0.307	AG(2)
3	-0.343	0.120	LWBC

scale parameter 1.000

-- model changed
 Scale log-linear model

	estimate	s.e.	parameter
1	1.214	0.500	1
2	-0.416	0.165	LWBC

scale parameter 1.000

 -2 log Lmax =   286.658

The disparity change from the full double interaction model to the final model is 3.61 on 3 df, and from the final model to the constant-r model is 4.66 on 1 df. There is some evidence, though it is not very strong, of a decrease in scale r, and therefore an increase in variance, with lwbc. The "t"-statistic for lwbc in the scale model does not correspond well to the disparity change: the squared t-statistic is $(0.4163/0.1646)^2 = 6.40$. This is to be expected in a small sample with a non-identity link function.

Over the observed range of wbc from 0.75 to 100, the fitted value of log r decreases from 1.334 to -0.703, so that r decreases from 3.80 to 0.50; the effect on the variance of T is a reduction by a factor of $1/3.8 = 0.26$ at the smallest value of wbc and an increase by a factor of 2 at the largest value. An increase in variance is consistent with the appearance of the data in Fig. 6.1, where the survival times are remarkably variable in both ag groups at wbc = 100.

The variance heterogeneity does not, however, much affect the mean model parameters in this example: the estimates for ag and lwbc in the double model are about 15% greater than those in the constant r model, though the standard errors in the former model are reduced because of the modelling of the heterogeneity. There is no doubt of the importance of both variables.

6.10 The Weibull distribution

The Weibull distribution was first used for the strength of materials corresponding to a form of "weakest link" model: specifically, if $Z_1, \ldots, Z_r$ are independently distributed on $(0, \infty)$ and $V = \min(Z_j)$ then the distribution of V, suitably standardized, approaches the Weibull distribution as r increases.

The Weibull distribution can be expressed as a power transform of the exponential. Suppose U has an exponential distribution with mean μ and constant hazard $\lambda = 1/\mu$, and $T = U^{1/\alpha}$ where $\alpha > 0$. Then T has a Weibull distribution

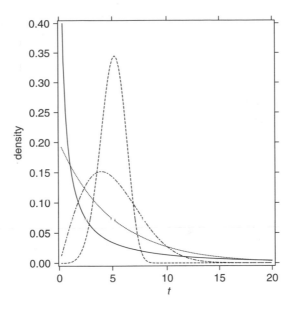

Fig. 6.17. Weibull density functions

with density, survivor and hazard functions (expressed in terms of the exponential hazard λ)

$$f(t) = \alpha \lambda t^{\alpha-1} e^{-\lambda t^{\alpha}}, \quad t > 0$$

$$S(t) = e^{-\lambda t^{\alpha}}$$

$$h(t) = \alpha \lambda t^{\alpha-1}.$$

The mean of the distribution is $\Gamma(1 + 1/\alpha)\mu^{1/\alpha}$, the median is $(\mu \log 2)^{1/\alpha}$, and the variance is $[\Gamma(1+2/\alpha) - \Gamma^2(1+1/\alpha)]\mu^{2/\alpha}$. If $\alpha = 1$, then the hazard function is constant and T has an exponential distribution; for $\alpha > 1$ the hazard function is monotone increasing with t, while for $\alpha < 1$ it is monotone decreasing. The shape of the density varies considerably with the *shape parameter* α. Graphs of the density are shown in Fig. 6.17 for a mean of 5 and $\alpha = 0.5, 1, 2, 5$.

The density is generally right-skewed, but for large α relative to the mean it can be negatively skewed (as it is for mean 5 and $\alpha = 5$). Graphs of the hazard are shown in Fig. 6.18 for the same values.

If the explanatory variables **x** do not affect the shape parameter, then the Weibull distribution has the important property of a *proportional hazard function*. The effect of the explanatory variables in any regression model for λ is simply to multiply the hazard by some constant: the functional form in t remains the same. This is true regardless of the way the explanatory variables affect the parameter λ.

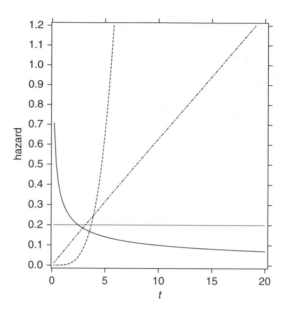

Fig. 6.18. Weibull hazard functions

The Weibull distribution has the further property of being an *accelerated failure* (or life) *time distribution*. If T_1 and T_2 have Weibull distributions with the same shape parameter α but different parameters λ_1 and λ_2, then

$$S_1(t) = e^{-\lambda_1 t^\alpha}$$

$$S_2(t) = e^{-\lambda_2 t^\alpha}$$

$$= S_1(\phi t),$$

where $\phi = (\lambda_2/\lambda_1)^{1/\alpha}$. Thus the survivor function for T_2 is the same as that for T_1 if time for T_1 is scaled by the factor ϕ: death or failure is *accelerated* for T_2 relative to T_1 if $\phi > 1$.

6.11   Maximum likelihood fitting of the Weibull distribution

The Poisson likelihood representation for the exponential distribution can be easily generalized to the Weibull distribution (Aitkin and Clayton, 1980). Given observations $(t_i, w_i, \mathbf{x}_i)$ with w_i the censoring indicator, the likelihood function is, from Section 6.7,

$$L(\boldsymbol{\beta}, \alpha) = \prod_{i=1}^{n} [h(t_i)]^{w_i} S(t_i).$$

Assuming a constant shape parameter, we have for the Weibull distribution

$$h(t_i) = \alpha \lambda_i t_i^{\alpha-1},$$

$$S(t_i) = e^{-\lambda_i t_i^{\alpha}},$$

and we write

$$\theta_i = \lambda_i t_i^{\alpha} = H(t_i),$$

where H is the integrated hazard function. Then

$$L(\boldsymbol{\beta}, \alpha) = \prod_l [\alpha \theta_i / t_i]^{w_i} e^{-\theta_i}$$

$$= \alpha^{\sum w_i} \prod_i \theta_i^{w_i} e^{-\theta_i} \Big/ \prod_i t_i^{w_i}.$$

Again the term $\prod t_i^{w_i}$ is a constant, and the term $\prod \theta_i^{w_i} e^{-\theta_i}$ is a Poisson likelihood, but there is now an additional term $\alpha^{\sum w_i}$ in the full likelihood. The log of the Weibull hazard function is

$$\log h(t_i) = \log \alpha + (\alpha - 1) \log t_i + \log \lambda_i,$$

and if λ_i itself has a log-linear model:

$$\log \lambda_i = \boldsymbol{\beta}' \mathbf{x}_i$$

then the hazard is log-linear in the explanatory variables:

$$\log h(t_i) = \log \alpha + (\alpha - 1) \log t_i + \boldsymbol{\beta}' \mathbf{x}_i.$$

As with the exponential distribution, it is possible to fit linear or reciprocal hazard models to λ by scaling the variables by t^{α} or $t^{-\alpha}$, respectively, but the fitted values are not guaranteed to be positive (as in Section 6.5) and we do not discuss this possibility further.

Since

$$\log \theta_i = \alpha \log t_i + \boldsymbol{\beta}' \mathbf{x}_i,$$

if α were known the log-linear hazard model could again be fitted using the log-linear Poisson model for w with a known offset $\alpha \log t$. On the other hand if $\boldsymbol{\beta}$ were known, α could be estimated from the likelihood equation

$$\frac{\partial \ell}{\partial \alpha} = 0.$$

Writing $n_1 = \sum w_i$, the number of uncensored observations, we have

$$\frac{\partial \ell}{\partial \alpha} = n_1/\alpha + \sum (w_i - \theta_i) \log t_i$$

$$\frac{\partial^2 \ell}{\partial \alpha \partial \boldsymbol{\beta}} = -\sum \theta_i \log t_i \mathbf{x}_i$$

and hence

$$\hat{\alpha}^{-1} = \sum (\hat{\theta}_i - w_i) \log t_i / n_1.$$

The last equation defines $\hat{\alpha}$ only recursively, and the information matrix is not block diagonal in α and $\boldsymbol{\beta}$ in any parametrization, so estimation of α affects the precision of the estimate of $\boldsymbol{\beta}$. Joint ML estimation of $\boldsymbol{\beta}$ and α can be carried out by the method of successive relaxation. First α is fixed at $\alpha_0 = 1$, giving an exponential distribution, and $\boldsymbol{\beta}$ is estimated from the Poisson model taking $\log t_i$ as an offset. Then α_0' is estimated from the above likelihood equation using the fitted values $\hat{\theta}_i$. A new estimate $\alpha_1 = (\alpha_0 + \alpha_0')/2$ is then used to define the offset $\alpha_1 \log t_i$ for a new fit of the Poisson model, and this process is continued until convergence. The damped estimate α_1 is used rather than α_0' because the successive estimates α_j and α_j' oscillate about the MLE, and damping substantially accelerates the rate of convergence.

The library macro `weibull` (Aitkin and Francis, 1980) fits the Weibull distribution using this approach. The macro requires three arguments: the response variable, the censoring indicator, and a scalar which takes the value 1 if a Weibull fit is required (an exponential fit being the first iteration) or 0 if only the exponential fit is required. A `model` macro must also be specified containing the model to be fitted. Standard errors for the parameter estimates $\boldsymbol{\beta}$ are slightly underestimated because the fitting algorithm treats α as known in the estimation of $\boldsymbol{\beta}$. This can be corrected by dropping each variable in turn and equating the disparity change to the squared Wald statistic. (Roger and Peacock (1982), and Roger (1985) gave direct standard error calculations using the device of extra data.)

We illustrate with the gehan example.

```
$input %plc weibull $
$input 'gehan' $
$calc %s = 1 $
$macro model g $endmac
$use weibull t w %s $
```

```
                    -- Model is g
```

```
Exponential fit

  Deviance    shape       df
              parameter
   217.05     1.0000      40
Weibull fit
   213.16     1.3650      39
-- Standard errors of estimates given below are underestimated
            estimate        s.e.      parameter
      1        -3.069       0.2172       1
      2        -1.730       0.3963       G(2)
scale parameter 1.000


$macro model 1 $endmac
$use weibull $


  -- model changed


                      -- Model is 1
  Weibull fit


  Deviance    shape       df
              parameter
   232.81     1.1420      40
-- Standard errors of estimates given below are underestimated
            estimate        s.e.      parameter
      1        -3.295       0.1818       1
scale parameter 1.000
```

The deviance reported by the macro is equal to the disparity, which for the exponential model is 217.05, and that for the Weibull is 213.16, with an estimate $\hat{\alpha}$ of 1.365. Hazard is increasing with time, and the hazard in the treatment group is substantially *reduced* relative to that in the control group, by the factor $e^{-1.73} = 0.177$. The change in disparity (or deviance) of 3.89 relative to the exponential is just on the 5% point of χ_1^2: there is marginal evidence to support a Weibull distribution rather than an exponential (but see Section 6.17 for further discussion of this example). Omitting the group variable gives a disparity change of 19.65, so the likelihood ratio test-based standard error is

$$\text{SE} = |\beta|/\sqrt{\text{disp.change}} = 1.730/4.433 = 0.390$$

compared with the fixed-α estimate of 0.396. There is essentially no underestimation.

Note also that the gamma distribution (Section 6.9) fits equally well, with a disparity of 212.88. Comparing the two treatment parameters is simplified if we transform the Weibull parameter to the log mean scale, which requires reflecting the sign and scaling by α, giving $1.730/1.365 = 1.267$ with SE $0.396/1.365 = 0.290$, compared with the gamma value of 1.281 with SE 0.290. We cannot choose between the distributions (which have the same number of parameters) for these data.

In Section 6.5 we compared the lognormal and exponential distributions on the feigl data, and found that the lognormal gave a slightly better fit. Does the extra shape parameter of the Weibull improve its fit compared to the lognormal?

```
$newjob $
$input 'feigl' : %plc weibull $
$calc %s = 1 : w = 1 : lwbc = %log(wbc) $
$macro model ag*lwbc $endmac
$use weibull time w %s $
                        -- Model is ag*lwbc
 Exponential fit

  Deviance    shape      df
            parameter
   291.32    1.0000      29
Weibull fit
   291.29   0.97887      28
-- Standard errors of estimates given below are underestimated
           estimate         s.e.      parameter
     1       -5.045        0.514      1
     2        1.842        0.733      AG(2)
     3        0.476        0.180      LWBC
     4       -0.325        0.246      AG(2).LWBC
 scale parameter 1.000
```

The disparity for the exponential fit is 291.32, exactly that calculated in Section 6.5. The disparity for the Weibull is almost the same, and the shape parameter estimate is almost 1. Thus within the Weibull family the exponential provides almost the best fit. In Section 6.5 we found that the lognormal distribution for the same model gave a disparity of 288.38; this is 2.91 less than for the Weibull and an unimportant difference. (See Chapter 2 for interpretation of disparity differences for models with the same number of parameters.)

6.12   The extreme value distribution

If T has a Weibull distribution with parameters α and μ as defined in Section 6.10, then $Y = \log T$ has the extreme value distribution with scale-parameter $\sigma = 1/\alpha$

and location-parameter $\theta = \sigma \log \mu$. The density, survivor and hazard functions are

$$f(y) = \frac{1}{\sigma} \exp\left[\frac{y-\theta}{\sigma} - \exp\left(\frac{y-\theta}{\sigma}\right)\right], \quad -\infty < y < \infty, -\infty < \theta < \infty, \sigma > 0$$

$$S(y) = \exp\left[-\exp\left(\frac{y-\theta}{\sigma}\right)\right]$$

$$h(y) = \frac{1}{\sigma} \exp\left(\frac{y-\theta}{\sigma}\right).$$

The mean and variance are $\theta + \sigma\psi(1+\sigma)$ and $\sigma^2\psi'(1+\sigma)$, where ψ and ψ' are the digamma and trigamma functions respectively. The hazard function increases exponentially, which limits the usefulness of the distribution as a model for survival data. The standard form of the density ($\theta = 0, \sigma = 1$) is plotted in Fig. 6.19. Since the extreme value is a location and scale family, all distributions have the same shape: the density has a long left tail and short right tail—it has negative skew.

The extreme value distribution can be fitted using the Poisson representation of the likelihood as for the Weibull distribution. The likelihood is, for general location

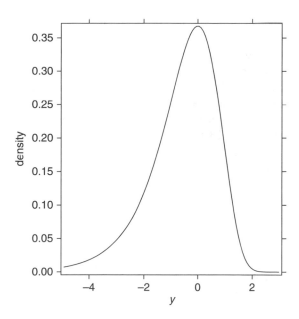

Fig. 6.19. Extreme value density function

parameters θ_i,

$$L(\theta_1, \ldots, \theta_n, \sigma) = \prod_{i=1}^{n} [h(y_i)]^{w_i} S(y_i)$$

$$= \prod_{i=1}^{n} [H(y_i)]^{w_i} e^{-H(y_i)} \prod_{i=1}^{n} \left[\frac{h(y_i)}{H(y_i)} \right]^{w_i},$$

where $H(y_i)$ is the integrated hazard function. The first part of the above expression is a Poisson likelihood with mean

$$H(y_i) = \exp\left(\frac{y_i - \theta_i}{\sigma} \right)$$

so that

$$\log H(y_i) = y_i/\sigma - \theta_i/\sigma = \alpha y_i - \alpha \theta_i.$$

The second term in the likelihood expression is

$$\prod_{i=1}^{n} \frac{1}{\sigma^{w_i}} = \prod_{i=1}^{n} \alpha^{w_i} = \alpha^{n_1},$$

where n_1 is the number of uncensored observations.

The likelihood can be seen to be similar to that given in Section 6.11 for the Weibull distribution, with t_i replaced by $y_i = \log(t_i)$. The parameter estimates from the Poisson fit correspond to the extreme value model $-\alpha\theta_i = \boldsymbol{\beta}'\mathbf{x}_i$, or $\theta_i = -\boldsymbol{\beta}'\mathbf{x}_i/\alpha$, a *linear* (and proportional hazard) model for the location-parameter θ. To fit the extreme value distribution in GLIM, we therefore need to replace $\log t_i$ in the offset by t_i. No other change is required, though for convenience a separate macro extval (not in the GLIM macro library) is used to fit the extreme value distribution.

We illustrate with the gehan example.

```
$newjob $
$input 'extval' : 'gehan' $
$macro model g $endmac
$use extval t w $
Extreme value fit
 -2 log L      shape      df
              parameter
  240.015      0.116    39.
  ***WARNING
```

```
    Standard errors of estimates given below are underestimated
            estimate          s.e.      parameter
      1        -1.314        0.218      1
      2        -2.177        0.398      G(2)
   scale parameter 1.000


$macro model 1 $endm
$use extval$
Extreme value fit

   -2 log L      shape      df
                parameter
    266.251      0.084      40.
   ***WARNING
      Standard errors of estimates given below are underestimated
            estimate          s.e.      parameter
      1        -1.769        0.183      1
   scale parameter 1.000
```

The corrected standard error for the group effect is $2.177/\sqrt{26.244} = 0.425$. The disparity is 240.01, much larger than the Weibull disparity of 213.16, and the estimate of alpha is 0.116 (corresponding to an estimate of σ of 8.64). This distribution is clearly inappropriate: the extreme value hazard increases exponentially with t ($e^{0.116t}$) compared to the Weibull power hazard ($t^{0.365}$).

For the feigl data, it is similarly found that the disparity for the model ag*lwbc is 323.70 for the extreme value distribution, much larger than the value of 291.29 for the Weibull distribution. The extreme value distribution is again inappropriate.

6.13   The reversed extreme value distribution

The extreme value distribution is asymmetric, and is defined on the infinite range $(-\infty < y < \infty)$. The distribution of $-Y = \log(1/T)$ is the *reversed* extreme value distribution, which has positive skew. The density, survivor and hazard functions are

$$f(y) = \frac{1}{\sigma} \exp\left[-\frac{y-\theta}{\sigma} - \exp\left(-\frac{y-\theta}{\sigma}\right)\right]$$

$$S(y) = 1 - \exp\left[-\exp\left(-\frac{y-\theta}{\sigma}\right)\right]$$

$$h(y) = f(y)/S(y).$$

The hazard function is monotone increasing and approaches an asymptote of $1/\sigma$. This form of the hazard function arises because for large y, $f(y)$ is essentially exponential, with constant hazard. The likelihood function for this distribution does not have the Poisson form, because the survivor function is not exponential. It is not easily fitted in GLIM and is not considered further in this book.

6.14   Survivor function plotting for the Weibull and extreme value distributions

In Section 6.8 we described the use of the Kaplan–Meier estimator of the survivor function to validate the exponential distribution. We can extend this procedure to assess the validity of the Weibull and extreme value distributions.

For the exponential distribution, we used the Poisson fitted values to construct the Kaplan–Meier estimate which was then plotted against u using a suitable scale. For the Weibull distribution,

$$S(t) = e^{-\lambda t^{\alpha}},$$

and the Poisson fitted values are

$$\hat{\theta}_i = \hat{\lambda}_i t_i^{\hat{\alpha}}$$

which are again approximately (censored or uncensored) standard exponential variables. Thus the same plotting procedures are used for these values based on the Weibull distribution.

In the same way for the extreme value distribution

$$S(y) = \exp\{-\exp[(y - \theta)/\sigma]\}$$

and the Poisson fitted values are $\exp[(y-\hat{\theta})/\hat{\sigma}]$, so again these have approximately standard exponential distributions, censored or uncensored.

We illustrate in Figs 6.20 and 6.21 with the Weibull distribution fitted to the Gehan data, repeating the directives from the previous Kaplan–Meier fit. We omit the survivor function graph and give only the two transformed graphs. The approximation for the lower limit pl_ of the simultaneous confidence band fails (becomes negative) for small pl_ and these values (which would otherwise be set to zero by the log function) are deleted by the weight variate w1. The band is unhelpfully wide here.

```
$newjob $
$input %plc weibull $
$input 'gehan' : 'kaplan' $
$macro model g $endmac
```

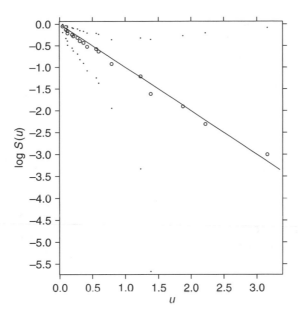

Fig. 6.20. Gehan data: log survivor function and bounds, Weibull

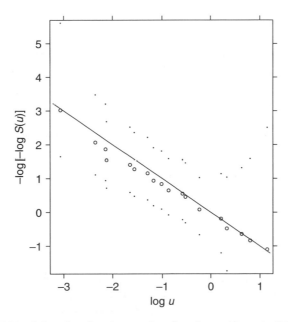

Fig. 6.21. Gehan data: log–log survivor function and bounds, Weibull

```
$calc %s = 1 $use weibull t w %s $
$use kaplan %fv w $
$input 'binocisim' $
$calc r = riset_*surv_ $
$var 332 sexp ind $
$calc ind = 0.03 + 0.01*%cu(1) : sexp = %exp(-ind) $
      : ww = %ne(surv_,0) $
$use binocisim r riset_ $
$calc ls = %log(surv_) $
  : lsexp = %log(sexp) $
  :   lpl = %log(pl__) : lpu = %log(pu__) : wl = %ne(lpl,0) $
$graph (s = 1 h = 'u' v = 'log S(u)')
        ls/ww,lsexp,lpl/wl,lpu/ww time_,ind,time_,time_ 5,10,9,9 $
$calc lls = -%log(-ls) $
  :llsexp = -%log(-lsexp) $
  :  lind = %log(ind) $
  :    lu = %log(time_) $
  :  llpl = -%log(-lpl) : llpu = -%log(-lpu) $
$graph (s = 1 h = 'log u' v = '-log[-log S(u)]')
        lls/ww,lllsexp,llpl/wl,llpu/ww lu,lind,lu,lu 5,10,9,9 $
```

The graph in Fig. 6.20 is very close to a straight line: the curvature from the exponential model visible in Figure 6.13 has been removed. However, in Fig. 6.21 the straight line appears slightly out of place, with the magnitude of the (negative) slope overestimated because of the extreme point on the left. There is still some question of the appropriateness of the Weibull model. We consider this further in Section 6.17.

6.15   The Cox proportional hazards model and the piecewise exponential distribution

A further application of the Poisson likelihood representation enables us to fit an important model developed by Cox (1972). In this *proportional hazards model* the hazard function is

$$h(t) = \lambda_0(t)g(\mathbf{x}, \boldsymbol{\beta}),$$

where $\lambda_0(t)$ is an arbitrary, unspecified function of t—the *base-line hazard* when $\mathbf{x} = \mathbf{0}$. The effect of the covariates $\mathbf{x}$ will be assumed to be through a log-linear model, $g(\mathbf{x}, \boldsymbol{\beta}) = e^{\boldsymbol{\beta}'\mathbf{x}}$, so that the individual covariates x_j themselves affect the hazard proportionately. The likelihood function is

$$L(\boldsymbol{\beta}, \lambda_0) = \prod_i [h(t_i)]^{w_i} S(t_i)$$

$$= \prod_i [h(t_i)]^{w_i} e^{-H(t_i)}$$

$$= \prod_i [H(t_i)]^{w_i} e^{-H(t_i)} \prod_i [h(t_i)/H(t_i)]^{w_i},$$

where $H(t)$ is the integrated hazard function. Because of the proportional hazards assumption, the second product in the likelihood does not depend on $\boldsymbol{\beta}$, and the first is again a Poisson likelihood, with

$$\log H(t_i) = \log \Lambda_0(t_i) + \boldsymbol{\beta}' \mathbf{x}_i,$$

where

$$\Lambda_0(t) = \int_0^t \lambda_0(u) \, du$$

is the *integrated baseline hazard*. Since $\lambda_0(t)$ is arbitrary, it is not clear how to fit the model without parametrizing $\lambda_0(t)$ in some way.

In his ground-breaking paper, Cox (1972) used a conditioning argument to eliminate the baseline hazard in a *partial likelihood*. This partial likelihood could then be maximized without reference to the unknown baseline hazard.

We use a different semi-parametric approach based on the *piecewise exponential distribution* due to Breslow (1974); see also Whitehead (1980), and Aitkin *et al.* (1983). This approach retains the hazard but in a (semi-) parametric form which is modelled explicitly and so can be examined using the same methods as for the parametric models already discussed.

We define a set of disjoint time intervals $(a_{j-1}, a_j]$ defined by "cut-points" $0 = a_0 < a_1 < a_2 < \cdots < a_N < a_{N+1} = \infty$. In the j-th interval $(a_{j-1}, a_j]$, we model the hazard function as *constant*, with parameter λ_j. That is, the survival time distribution $f(t)$ in $(a_{j-1}, a_j]$ is exponential with mean $1/\lambda_j$, but the mean may be different in each piece of the time axis—hence the name piecewise exponential.

The semi-parametric model has the same flexibility as the model in which $\lambda_0(t)$ is unspecified. We replace the assumption of an arbitrary hazard by the assumption of a step-function hazard with steps at $a_1, a_2, \ldots, a_N$.

We can thus write the baseline hazard as

$$\lambda_0(t) = \lambda_j = \exp(\phi_j), \quad a_{j-1} < t \le a_j, \ j = 1, \ldots, N+1.$$

where the ϕ_j are a set of $N+1$ time-interval constants.

The proportional hazard assumption can then be written

$$h_j = \lambda_j \exp \boldsymbol{\beta}' \mathbf{x} = \exp(\phi_j + \boldsymbol{\beta}' \mathbf{x}).$$

To construct the likelihood function, we consider as in Section 6.8 the survival experience of each individual through time, shown in the table below.

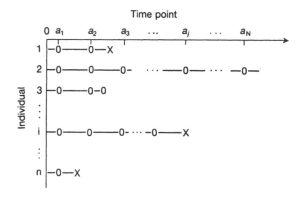

The circles represent censoring and the crosses death. The i-th individual experiences a sequence of censorings at $a_1, a_2, \ldots$ and either final censoring or death at t_i, defined to fall in the N_i-th interval, so that $a_{N_{i-1}} < t_i < a_{N_i}$. Let h_{ij} be the hazard function for the i-th individual in the j-th interval $(a_{j-1}, a_j]$, with

$$h_{ij} = \exp(\phi_j + \boldsymbol{\beta}' \mathbf{x}_i),$$

and let w_{ij} be the censoring indicator for the i-th individual in the j-th interval. The survivor function for the i-th individual in the j-th interval is

$$S_{ij}(t) = \exp[-H_{ij}(t)] = \exp(-h_{ij}t).$$

Define e_{ij} to be the exposure time of the i-th individual in the j-th interval, so that

$$e_{ij} = \begin{cases} a_j - a_{j-1}, & j = 1, \ldots, N_i - 1 \\ t_i - a_{N_{i-1}}, & j = N_i. \end{cases}$$

Then the contribution of the i-th individual to the likelihood is

$$L_i = \prod_{j=1}^{N_i} h_{ij}^{w_{ij}} S_{ij}(e_{ij}).$$

The full likelihood is thus

$$L(\boldsymbol{\beta}, \phi_1, \ldots, \phi_{N+1}) = \prod_{i=1}^{n} \prod_{j=1}^{N_i} h_{ij}^{w_{ij}} \exp(-h_{ij}e_{ij})$$

$$= \prod_i \prod_j \theta_{ij}^{w_{ij}} \exp(-\theta_{ij}) \Big/ \prod_i \prod_j e_{ij}^{w_{ij}}$$

with

$$\theta_{ij} = h_{ij}e_{ij}$$
$$= e_{ij} \exp(\phi_j + \boldsymbol{\beta}'\mathbf{x}_i).$$

The denominator in the likelihood is a constant, and the numerator is again a Poisson likelihood, with a log-linear model for θ_{ij}:

$$\log \theta_{ij} = \log e_{ij} + \phi_j + \boldsymbol{\beta}'\mathbf{x}_i.$$

6.16   Maximum likelihood fitting of the piecewise exponential distribution

We now consider the maximization of the likelihood in all the parameters $\boldsymbol{\beta}$, $\phi_1, \ldots, \phi_{N+1}$. We have

$$\ell = \log L = \sum_{i=1}^{n} \sum_{j=1}^{N_i} (w_{ij} \log \theta_{ij} - \theta_{ij}) - \sum_{i=1}^{n} \sum_{j=1}^{N_i} w_{ij} \log e_{ij}$$

$$\frac{\partial \ell}{\partial \phi_j} = \sum_{i \in R_j} (w_{ij} - \theta_{ij}),$$

where R_j is the risk set of individuals in the j-th interval. For a maximum of the likelihood,

$$d_j = \sum_{i \in R_j} w_{ij} = \sum_{i \in R_j} \hat{\theta}_{ij} = \hat{\lambda}_j \sum_{i \in R_j} e_{ij} \exp(\hat{\boldsymbol{\beta}}'\mathbf{x}_i),$$

where d_j is the number of deaths in the j-th interval. If there are no deaths in the j-th interval, then $d_j = 0$. Since $\hat{\theta}_{ij}$ is non-negative, this equality is only possible if $\hat{\lambda}_j = 0$, whatever the model $\boldsymbol{\beta}'\mathbf{x}$. That is, the term in the maximized likelihood for this interval is 1. Thus, in comparing two models, intervals with no deaths make no contribution to the maximized log-likelihood under both models. The second term in the log-likelihood depends only on t_i and the a_j and is the same under both models.

The choice of interval cut-points determines the extent of *smoothing* of the estimated piecewise hazard function. *Minimal* smoothing is achieved by choosing the distinct death times as cut-points (Breslow, 1974), so that each interval contains just one death (or more if there are tied death times). The resulting estimate for $\hat{\beta}$ corresponds closely to that obtained from the conditional likelihood for the Cox model (Breslow, 1974). Other choices of cut-points will in general give different estimates of β, since the estimated hazard parameters $\hat{\phi}_j$ are not independent of $\hat{\beta}$.

Fitting of the model in GLIM4 is greatly simplified (relative to the relaxation method described for GLIM3 in the first edition) by the use of $eliminate. The cut-point nuisance parameters ϕ_j are eliminated from the model-fitting (in the variable t_ints), and only the β parameter vector is reported in the output. The censoring variable is declared as the y-variate, the Poisson model is declared with an offset of $\log e_{ij}$, and the fitted model is **x** with the ϕ parameters eliminated. However the cut-point parameters can still be recovered, and may be graphed against the time variable to examine the form of the hazard. This is done more informatively using the integrated hazard function. The piecewise exponential model is fitted using GLIM4 library macros in the subfile cox.

6.17   Examples

We consider first the gehan data.

```
$newjob $
$input 'gehan' : %plc cox $
```

The cox subfile contains a number of macros. The macro coxmodel sets up the vector of distinct death times which is used for fitting the piecewise exponential distribution, and expands the censoring indicator and the exposure time variable to length $\sum_i N_i$. It requires as arguments the survival time and the censoring indicator, and the name of the variable to contain the distinct death times.

```
$use coxmodel t w ddt $
There are 17 distinct event times.
```

Any specified set of cut-points can be set up instead of the distinct death times, using the macro inib instead of coxmodel in the same way. These macros set up the model-fitting structure, and to fit the model we use $fit. The macros allow *time-dependent* covariates, which are discussed further in Section 6.22. Time-constant covariates have to be indexed by the variate ind_, to "expand" them to the same length as the censoring and time variables. We fit the group model.

```
$fit g(ind_) $disp e $
scaled deviance =  202.37 at cycle 7
```

```
    residual df =   407

          estimate           s.e.        parameter
    1        -1.521         0.410        G(IND_)(2)
scale parameter 1.000
    eliminated term: T_INTS
```

The scaled deviance reported by the macro is equal to the disparity. The group effect estimate of -1.521 is very similar to that from the exponential model, and noticeably smaller in magnitude than the Weibull estimate (-1.731). The disparity is 202.37; the df reported from the output is based on the expanded data set used in fitting the model, of survival experience of each individual through the time intervals; the correct df is 23, reflecting the estimation of 1 parameter in the model and 18 hazard parameters for the 17 distinct death times. The disparity change relative to the exponential is 13.68 on 17 df, very weak evidence against the exponential.

To examine the form of the estimated hazard parameters $\exp(\hat{\phi}_j)$, we convert them to the estimated *baseline survivor function*, using

$$\Lambda_j = \sum_i^j \lambda_i, \quad S_{0j} = \exp(-\Lambda_j).$$

To check the similarity of the baseline hazard to Weibull, we use as before the graph of $-\log(-\log S_{0j})$ against $\log t_j$.

We graph it in Fig. 6.22 against the log of the death times using the macro bsf. This is a slight variation of the library macro phaz which graphs the estimated hazard function against log death time. The bsf macro graphs both λ_j and $-\log(-\log S_{0j})$ against $\log t_j$. We show only the second graph; the log-hazard parameters are so variable as to provide little information about the hazard.

The estimated baseline survivor function is held in the vector bsf_; the death times are in ddt and the logged death times are in the vector lt_. The bsf macro requires as argument the name of the cut-point variate.

```
$input 'bsf' $
$use bsf ddt $
```

The log–log survivor function is nearly linear, and therefore close to Weibull or exponential, but has some curvature. The group effect estimate is much closer to the exponential estimate than to the Weibull estimate.

Thus the piecewise exponential distribution provides important information about the hazard function, and may suggest a suitable parametric distribution.

We repeat the analysis with the feigl data.

```
$newjob $
$input 'feigl': 'bsf' : %plc cox $
```

SURVIVAL DATA

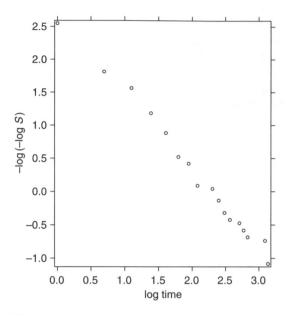

Fig. 6.22. Gehan data: baseline log–log survivor function estimate

```
$calc w = 1 $
$use coxmodel time w ddt $
There are 22 distinct event times.
$calc lwbc = %log(wbc) $
$fit ag(ind_)*lwbc(ind_) $disp e $
scaled deviance =   268.05 at cycle 6
     residual df =   312


           estimate          s.e.        parameter
     1        2.435          0.909        AG(IND_)(2)
     2        0.642          0.211        LWBC(IND_)
     3       -0.495          0.276        AG(IND_)(2).LWBC(IND_)
 scale parameter 1.000
     eliminated term: T_INTS
```

The disparity is 268.05 on 8 df.

```
$use bsf ddt $
```

The hazard graph (not shown) has a decline in the middle and a steep rise at the end—the hazard does not appear consistent with the exponential distribution. The survivor function graph in Fig. 6.23 shows a corresponding double bend, and is definitely not Weibull, though the overall slope is close to -1.

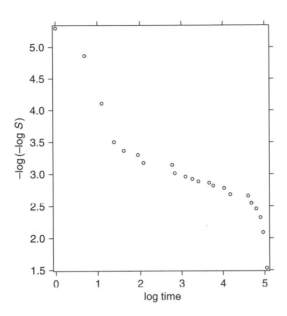

Fig. 6.23. Feigl data: baseline log–log survivor function estimate

The parameter estimates and standard errors from the interaction model are somewhat larger in magnitude than those for the exponential distribution, but less than those for the lognormal distribution (Section 6.5), and again the interaction can be omitted.

```
$fit - ag(ind_).lwbc(ind_) $disp e $
scaled deviance =  271.35 (change =   +3.302) at cycle 5
     residual df =  313    (change =  +1    )


            estimate        s.e.      parameter
      1       1.017        0.418      AG(IND_)(2)
      2       0.360        0.134      LWBC(IND_)
scale parameter 1.000
     eliminated term: T_INTS
```

The disparity increases by 3.30. Note that there is no constant term—this is eliminated with the hazard constants.

The parameter estimates and standard errors for the main-effect model are close to those from the exponential model, and are also closely equivalent to those reported in Cox and Oakes (1984, p. 100) obtained using the conditional likelihood, though Cox and Oakes use a different parametrization of the model, with ag reversed in sign and lwbc centred to an origin.

6.18   The logistic and log–logistic distributions

The logistic distribution is very similar to the normal but has slightly heavier tails. Its advantage in survival analysis is that both right- *and* left-censoring can be easily handled, since the likelihood can be expressed as a special form of binomial likelihood (Bennett and Whitehead, 1981).

The density, survivor and hazard functions for the general location- and scale-parameter logistic distribution are

$$f(y) = \frac{1}{\sigma} \exp\left(\frac{y - \mu}{\sigma}\right) \bigg/ \left[1 + \exp\left(\frac{y - \mu}{\sigma}\right)\right]^2, \quad -\infty < y < \infty$$

$$S(y) = 1 \bigg/ \left[1 + \exp\left(\frac{y - \mu}{\sigma}\right)\right]$$

$$h(y) = \frac{1}{\sigma} \exp\left(\frac{y - \mu}{\sigma}\right) \bigg/ \left[1 + \exp\left(\frac{y - \mu}{\sigma}\right)\right];$$

the distribution mean and variance are μ and $\sigma^2 \pi^2/3$, respectively.

The hazard function is logistic, and monotone increasing; the distribution does not have a proportional hazard function, but does have an accelerated failure time, under the model $\mu_i = \boldsymbol{\beta}' \mathbf{x}_i$, σ constant.

Let c_i and b_i be dummy indicators for right- and left-censoring. Left-censoring is uncommon, but may occur if individuals fail before the first measurement period occurs. For example, if survival is recorded in weeks and an individual dies within the first week, the value is left-censored at one week. (In general, recording in this way will lead to a *grouped* survival distribution, but this has little effect on inference unless the grouping is very coarse. In this case the discrete time approach of Section 6.23 should be used.)

For each individual, we have the data $(y_i, \mathbf{x}_i, b_i, c_i)$. The likelihood function is

$$L(\boldsymbol{\beta}, \sigma) = \prod_i [f(y_i)]^{1 - b_i - c_i} [S(y_i)]^{c_i} [F(y_i)]^{b_i}.$$

Write

$$F(y_i) = p_i;$$

then

$$S(y_i) = 1 - p_i$$
$$f(y_i) = p_i(1 - p_i)/\sigma$$

and

$$L(\boldsymbol{\beta}, \sigma) = \frac{1}{\sigma^m} \prod_i p_i^{1 - c_i} (1 - p_i)^{1 - b_i},$$

where m is the number of uncensored observations. The product has the form of the likelihood from n binomial observations with $r_i = (1 - c_i)$ successes in $n_i = (2 - b_i - c_i)$ trials, and the success probability is

$$p_i = \exp\left(\frac{y_i - \mu_i}{\sigma}\right) \Big/ \left[1 + \exp\left(\frac{y_i - \mu_i}{\sigma}\right)\right]$$

so that

$$\operatorname{logit} p_i = (y_i - \mu_i)/\sigma = y_i/\sigma - \beta' x_i/\sigma.$$

(This parametrization is different from that in Bennett and Whitehead who absorb σ into β and use $1/\sigma$ as the scale-parameter.)

Maximization of the likelihood can be achieved by a relaxation method using the binomial error model to estimate β, and an additional step for the estimation of σ. For fixed σ, the logit model is fitted to the number of successes $1 - c_i$ with binomial denominator $2 - b_i - c_i$, using an offset of y_i/σ. The quantities $(y_i - \hat{\mu}_i)$ are obtained by multiplying the linear predictor by σ. A new estimate of σ is then obtained by solving the likelihood equation $\partial \log L/\partial \sigma = 0$, which reduces to

$$m\hat{\sigma} = \sum_i (n_i \hat{p}_i - r_i)(y_i - \hat{\mu}_i).$$

Alternate estimation of β and σ continues until convergence, with damping of successive estimates of σ. As with the Weibull macro, the standard errors of the estimated parameters in $\hat{\beta}$ are slightly underestimated because σ is taken as known in each iteration. This can be corrected using the approach of Roger and Peacock (1982), or by dropping each variable in turn as in Chapter 2.

The macro `logist` in the file `clogistic` fits the logistic distribution. The parameter estimates and standard errors differ by a scale factor σ from those given by Bennett and Whitehead's macro, but are comparable with those for the normal distribution in Section 6.19.

Before considering an example, we describe the log–logistic distribution. If Y is logistic, then $T = e^Y$ is log–logistic. The density, survivor and hazard functions are

$$f(t) = \frac{\alpha}{\theta} \frac{(t/\theta)^{\alpha-1}}{[1 + (t/\theta)^\alpha]^2}, \quad t, \theta, \alpha > 0$$

$$S(t) = \frac{1}{1 + (t/\theta)^\alpha}$$

$$h(t) = \frac{\alpha}{\theta} \frac{(t/\theta)^{\alpha-1}}{1 + (t/\theta)^\alpha},$$

where $\theta = e^\mu$, $\alpha = 1/\sigma$.

For $\alpha \leq 1$ the hazard is monotone decreasing, while for $\alpha > 1$ it has a single maximum. The log-logistic distribution is a convenient approximation to the lognormal since it possesses similar positive skew. It is fitted using the macro llogist in the file clogistic, which gives estimates of β and σ, where $\log(\theta_i) = \beta'x_i$. The parameter estimates again differ from those in Bennett and Whitehead by the factor σ, and the disparity is different from Bennett and Whitehead's deviance as they omit the term $2\sum(1 - b_i - c_i)\log t_i$ required to make the logistic and log-logistic comparable.

We illustrate with the gehan data. Both macros require as arguments the response variable and a single censoring indicator w_i, defined as $w_i = 0$ for right-censoring, $w_i = 2$ for left-censoring and $w_i = 1$ for no censoring. The fitted model is specified in a macro called model, as for the Weibull distribution.

```
$newjob $
$input 'gehan' : 'clogistic' $
$macro model g $endmac
$use logistic t w $
                    -- Model is g


  LOGISTIC fit


   Deviance    shape      df
              parameter
    236.92     5.5978      39
  -- Standard errors of estimates given below are underestimated
             estimate        s.e.     parameter
      1        8.289        1.936     1
      2        13.98        3.428     G(2)
  scale parameter 31.34
```

The disparity is 236.92 with an estimated σ of 5.60.

```
$use llogistic t w $
                    -- Model is g


  LOG-LOGISTIC fit


   Deviance    shape      df
              parameter
    215.32    0.54646      39
  -- Standard errors of estimates given below are underestimated
             estimate        s.e.     parameter
      1        1.893        0.207     1
      2        1.265        0.319     G(2)
  scale parameter 0.2986
$macro model 1 $endm
```

```
$use llogistic $
 -- model changed


                    -- Model is 1


  LOG-LOGISTIC fit

    Deviance    shape       df
              parameter
     230.70   0.63759       40
 -- Standard errors of estimates given below are underestimated
            estimate         s.e.      parameter
      1         2.470        0.175        1
  scale parameter 0.4065
```

The disparity is 215.32, and the estimated group difference is 1.265, with (underestimated) standard error 0.319; the estimated σ is 0.546. Fitting the null model gives a disparity of 230.70, and a corrected standard error for the group difference parameter of $1.265/\sqrt{15.38} = 0.323$. The disparity is greater than that for the Weibull (213.2) in Section 6.11, and much lower than that for the logistic, as we should expect from the skewness of the survival times.

6.19 The normal and lognormal distributions

The density, survivor and hazard functions for the normal (μ, σ) distribution for Y are

$$f(y) = \frac{1}{\sqrt{2\pi}\sigma} \exp\left\{ -\frac{1}{2} \frac{(y - \mu)^2}{\sigma^2} \right\}$$

$$S(y) = 1 - \Phi\left(\frac{y - \mu}{\sigma} \right)$$

$$h(y) = f(y)/S(y).$$

The hazard function is monotone increasing, and $h(y) \to (y - \mu)/\sigma$ for large y.

The density, survivor and hazard functions for the lognormal distribution for T (a *positive* random variable) with parameters (μ, σ) are

$$f(t) = \frac{1}{\sqrt{2\pi}\sigma t} \exp\left\{ -\frac{1}{2} \frac{(\log t - \mu)^2}{\sigma^2} \right\}$$

$$S(t) = 1 - \Phi\left(\frac{\log t - \mu}{\sigma} \right)$$

$$h(t) = f(t)/S(t).$$

The hazard function first increases and then decreases to zero.

In Section 6.5 we compared the exponential and lognormal distributions for the `feigl` data, where there were no censored observations. As with the gamma distribution, censoring complicates the fitting of the normal model because the survivor function does not have a simple analytic form. However, it is possible to fit the normal model with both left- and right-censoring; for simplicity we consider only right-censoring (see Wolynetz, 1979 for the general case). For right-censoring with censoring indicator w_i we have

$$L(\boldsymbol{\beta}, \sigma) = \prod_i [f(y_i)]^{w_i} [S(y_i)]^{1-w_i},$$

where

$$f(y_i) = \frac{1}{\sqrt{2\pi}\sigma} \exp\left[-\frac{1}{2\sigma^2}(y_i - \mu_i)^2\right]$$

$$\mu_i = \boldsymbol{\beta}' \mathbf{x}_i.$$

Maximization of the log-likelihood can be achieved using an EM algorithm (Dempster *et al.*, 1977; Wolynetz, 1979; Aitkin, 1981).

In the normal model the log-likelihood is linear in y_i and y_i^2. In the E-step, the values of y_i and y_i^2 for the censored observations are replaced by their conditional expectations given the current parameter estimates. In the M-step, the likelihood equations are solved to give new parameter estimates using the conditional expectations obtained in the E-step as real data.

For an observation right-censored at a, the conditional density is

$$f(y \mid Y > a) = f(y) \big/ S\left(\frac{a-\mu}{\sigma}\right), \quad y > a.$$

The conditional expectation of Y is

$$E[Y \mid Y > a] = \int_a^\infty y f(y) \, dy \big/ S\left(\frac{a-\mu}{\sigma}\right)$$

$$= \mu + \sigma h\left(\frac{a-\mu}{\sigma}\right)$$

and that of Y^2 is

$$E[Y^2 \mid Y > a] = \int_a^\infty y^2 f(y) \, dy \big/ S\left(\frac{a-\mu}{\sigma}\right)$$

$$= \mu^2 + \sigma^2 + \sigma(\mu + a) h\left(\frac{a-\mu}{\sigma}\right),$$

where $h(y)$ is the hazard function.

Thus in the E-step, we construct from the current parameter estimates the expected observations

$$\tilde{y}_i = w_i y_i + (1 - w_i) \left[\mu_i + \sigma h \left(\frac{y_i - \mu_i}{\sigma} \right) \right]$$

$$\tilde{y}_i^2 = w_i y_i^2 + (1 - w_i) \left[\mu_i^2 + \sigma^2 + \sigma (\mu_i + y_i) h \left(\frac{y_i - \mu_i}{\sigma} \right) \right].$$

In the M-step, we estimate the parameters β and σ. Thus for the identity link, we have the estimates

$$\hat{\beta} = \left(\sum_i \mathbf{x}_i \mathbf{x}_i' \right)^{-1} \left(\sum \mathbf{x}_i \tilde{y}_i \right)$$

$$\hat{\sigma}^2 = \sum_i (\tilde{y}_i^2 - 2\mu_i \tilde{y}_i + \mu_i^2)/n,$$

where $\mu_i = \beta' \mathbf{x}_i$.

The normal and lognormal distributions can be fitted using the macros norm and lnorm in the GLIM macro library. Both macros require as arguments the response variable and the censoring indicator. The fitted model is again specified in a macro called model as for the Weibull distribution.

The standard errors based on $\hat{\sigma}^2 (\sum \mathbf{x}_i \mathbf{x}_i')^{-1}$ may be seriously in error. Correct standard errors can be obtained by dropping each variable in turn from the model.

We illustrate with the gehan data.

```
$newjob $
$input 'gehan' : %plc cnormal $
$macro model g $endm
$use norm t w $
                     -- Model is g
  Normal Fit
  -2log Lmax  sigma    DF
    235.94   9.5452    40.
-- Standard errors of estimates given below are underestimated
            estimate        s.e.      parameter
     1        8.667        2.071        1
     2        13.88        2.929        G(2)
  scale parameter 90.09
$use lnorm t w $
                     -- Model is g
  Lognormal Fit
  -2log Lmax  sigma    DF
    213.41   0.92361   40.
```

```
-- Standard errors of estimates given below are underestimated
            estimate            s.e.        parameter
     1        1.825           0.202        1
     2        1.349           0.286        G(2)
scale parameter 0.8556
```

The disparity for the lognormal is slightly greater than that for the Weibull (213.16) in Section 6.11, slightly less than for the log–logistic, and much lower than that for the normal, as we might expect.

```
$macro model 1 $endm
$use lnorm t w $
                        -- Model is 1
   Lognormal Fit
        -2log Lmax  sigma    DF
          230.79    1.1130   41.
-- Standard errors of estimates given below are underestimated
            estimate            s.e.        parameter
     1        2.463           0.170        1
scale parameter 1.217
```

The disparity increases by 17.38, giving a standard error of $1.349/\sqrt{17.38} = 0.324$. The standard error of g is underestimated by 12%.

Note that the variance estimates for the lognormal and log–logistic distributions of T are very similar. The estimated variance of the log–logistic is $\hat{\sigma}^2\pi^2/3$, that is, 0.981 with $\hat{\sigma} = 0.546$, while that of the lognormal is 0.856. It is very difficult to distinguish between these distributions in small samples.

The empirical survivor function of the residuals from the normal or lognormal model can be constructed as in Section 6.8 using the vector %rs as the first argument of kaplan; a Q–Q plot may then be constructed using the %nd transformation of surv_.

6.20 Evaluating the proportional hazard assumption

We illustrate the importance of the proportional hazard assumption with a complex example of medical survival data. The data are adapted from Prentice (1973); Prentice's data are reproduced in Kalbfleisch and Prentice (1980, pp. 223–4), and are held in the file prentice. They consist of survival times t in days of 137 lung cancer patients from a Veteran's Administration Lung Cancer trial, together with explanatory variables: performance status status, a measure of general medical status on a continuous scale 1–9.9, with 1–3 completely hospitalized, 4–6 partial confinement to hospital, 7–9.9 able to care for self; age in years age; time in months from diagnosis mfd to starting on the study; a factor prior therapy prior

(1 no, 2 yes); a factor treatment treat (1 standard, 2 test) and a factor tumour type type (1 squamous, 2 small, 3 adeno, 4 large). There are three censored observations; the censoring indicator is w.

We first examine the survivor functions for the four different cell types. This is achieved by selecting the appropriate subsets of data with $pick, using the Kaplan–Meier macro and saving the output survivor function and time variables in four sets of variates. These are then log–log transformed as described in Section 6.8 and plotted on the same graph, with suppression of points with a zero survivor function. If the other explanatory variables have little effect and the hazard functions are proportional, the transformed survivor functions should be approximately parallel curves; if they are linear as well, a Weibull distribution with a common shape parameter is well-supported.

```
$newjob $
$input 'prentice': 'kaplan' $
$calc i1 = %eq(type,1) $
$pick t1,w1 t,w i1 $
$use kaplan t1 w1 $
$calc s1 = surv_ : time1 = time_ : ww1 = %ne(s1,0) $
  : lls1 = -%log(-%log(s1)) : lt1 = %log(time1) $
...
$calc i4 = %eq(type,4) $
$pick t4,w4 t,w i4 $
$use kaplan t4 w4 $
$calc s4 = surv_ : time4 = time_ : ww4 = %ne(s4,0) $
  : lls4 = -%log(-%log(s4)) : lt4 = %log(time4) $
$graph (s = 1 h = 'log t' v = '-log[-log S(t)]')
       lls1/ww1,lls2/ww2,lls3/ww3,lls4/ww4
       lt1,lt2,lt3,lt4 2,4,5,7 $
```

The survivor functions in Fig. 6.24 show an interesting pattern, which is more clearly visible in colour than in the monotone graph printed here. Three of the cell types show nearly linear, and closely parallel, log–log survivor functions. The fourth (triangles—the squamous cell type) is also linear but appears to have a different slope. Thus all four appear to follow Weibull survival, but the squamous cell type has a shape parameter different from the other three.

This complicates the analysis, since our Weibull analysis assumes the shape parameter is unaffected by the explanatory variables. We deal with this by separating the squamous cell type data from the others, and analysing them separately. We begin with the squamous cell type, fitting first the piecewise exponential distribution with the full two-way interaction model and examining the hazard. A graph of survival time against mfd (not shown) shows considerable skew in this variable, which is removed by a log transformation lmfd. We use this transformed variable

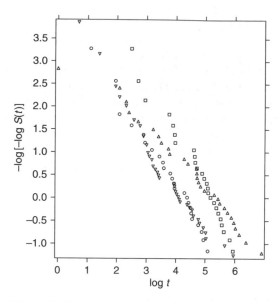

Fig. 6.24. Prentice data: cell type survivor functions

in subsequent modelling. We first extract the explanatory variable values for this cell type.

```
$input %plc cox weibull $
$calc lmfd = %log(mfd) $
$pick stat1,lmfd1,age1,prior1,treat1 status,lmfd,age,prior,treat i1 $
$factor prior1 2 treat1 2 $
$use coxmodel t1 w1 ddt1 $
$fit stat1(ind_) + lmfd1(ind_) + age1(ind_)
   + prior1(ind_) + treat1(ind_)  $disp e $


 scaled deviance =  346.95 at cycle 6
     residual df =  484


            estimate          s.e.       parameter
        1    -0.333          0.112        STAT1(IND_)
        2     0.517          0.248        LMFD1(IND_)
        3     0.035          0.025        AGE1(IND_)
        4    -0.372          0.427        PRIOR1(IND_)(2)
        5    -0.133          0.409        TREAT1(IND_)(2)
 scale parameter 1.000
     eliminated term: T_INTS


$use bsf ddt1 $
```

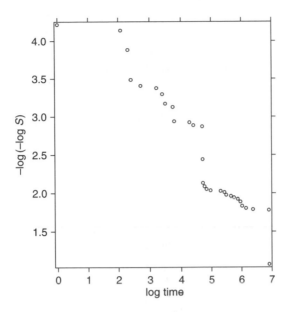

Fig. 6.25. Prentice data: type1 survivor function

The log–log survivor function in Fig. 6.25 is roughly linear in $\log t$ but with the two extreme points well out of line. Hazard increases with (log) months from diagnosis and decreases with status, but treatment shows no effect for this cell type. We extend the model to include all two-way interactions. Backward elimination gives a final model with the interaction of prior and status as well as lmfd:

```
scaled deviance =  343.88 at cycle 6
    residual df =  485


           estimate        s.e.       parameter
   1       -0.206         0.131       STAT1(IND_)
   2        0.495         0.246       LMFD1(IND_)
   3        2.872         1.422       PRIOR1(IND_)(2)
   4       -0.540         0.239       PRIOR1(IND_)(2).STAT1(IND_)
scale parameter 1.000
    eliminated term: T_INTS
```

Hazard declines with increasing status much more rapidly for those with prior therapy, but is higher in the prior therapy group for patients with low status, and lower for patients with high status, than in the group without prior therapy. The crossover occurs at status = 2.872/0.540 = 5.3. A Weibull distribution gives a similar model:

```
$calc %s = 1 $
$macro model lmfd + stat1*prior1 $endm
$use weibull t1 w1 %s$


                        -- Model is lmfd1 + stat1*prior1
  Exponential fit

   Deviance    shape      df
             parameter
    376.06    1.0000      30
  Weibull fit
    376.06    1.0011      29
  -- Standard errors of estimates given below are underestimated
             estimate          s.e.      parameter
        1      -4.586          0.983      1
        2       0.455          0.223      LMFD1
        3      -0.224          0.127      STAT1
        4       2.473          1.370      PRIOR1(2)
        5      -0.444          0.222      PRIOR1(2).STAT1
  scale parameter 1.000
```

The disparity is increased by 32.18, with 29 df. The Weibull hazard is exponential, with a shape parameter estimate of 1.00. The actual hazard however is not exponential, though this does not affect the final model, nor much affect the parameter estimates.

Now we analyse the other three cell types. Cell type (type0 below) is defined as a four level factor, with level 2 as the reference category. Since there are no data for Type 1, the parameter for this level is intrinsically aliased. This allows Types 3 and 4 to be consistently identified by their original labelling. The expansion indicator ind_ has first to be deleted, as do the structures in the bsf macro, for the different length of the distinct death times. We begin again with the main-effect model, and then extend it with all two-way interactions.

```
$calc i0 = %ne(type,1) $
$pick t0,w0 t,w i0 $
  : stat0,lmfd0,age0,prior0,type0,treat0
    status,lmfd,age,prior,type,treat i0 $
$factor prior0 2 treat0 2 type0 4 (2) $
$del ind_ $
$use coxmodel t0 w0 ddt0 $
$fit stat0(ind_) + lmfd0(ind_) + age0(ind_) +
    prior0(ind_) + typ0(ind_) + treat0(ind_) $disp e $
```

```
scaled deviance =    986.19 at cycle 7
   residual df =  3379
```

	estimate	s.e.	parameter
1	-0.340	0.064	STAT0(IND_)
2	-0.265	0.157	LMFD0(IND_)
3	-0.019	0.010	AGE0(IND_)
4	0.422	0.313	PRIOR0(IND_)(2)
5	0.000	aliased	TYPE0(IND_)(1)
6	0.217	0.279	TYPE0(IND_)(3)
7	-0.657	0.276	TYPE0(IND_)(4)
8	0.590	0.246	TREAT0(IND_)(2)

```
scale parameter 1.000
   eliminated term: T_INTS


$del cuhaz_ bsf_ lls_ $
$use bsf ddt0 $
```

The log–log survivor function in Fig. 6.26 curves smoothly—it is not linear. Hazard decreases with `status`, but unexpectedly *increases* with the new treatment. We extend the model to include all two-way interactions. Backward elimination

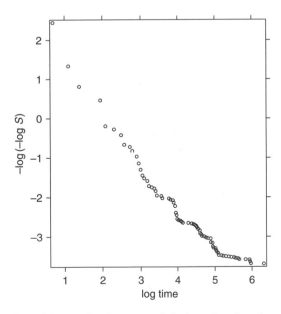

Fig. 6.26. Prentice data: types 2–3–4 survivor function

gives the final model:

```
scaled deviance =   988.22 at cycle 7
     residual df =  3381

          estimate         s.e.      parameter
     1     -0.232         0.065      STAT0(IND_)
     2      1.712         0.770      PRIOR0(IND_)(2)
     3      0.609         0.223      TREAT0(IND_)(2)
     4     -0.635         0.254      TYPE4(IND_)
     5     -0.285         0.134      PRIOR0(IND_)(2).STAT0(IND_)
scale parameter 1.000
     eliminated term: T_INTS
```

The status.prior interaction is similar to that for Type 1, but there is no lmfd effect and the adverse treatment effect remains. The hazard for Type 4 cells is lower than for Types 2 and 3 (type4 is a dummy for this cell type).

We fit the same model using the Weibull distribution.

```
$macro model stat0*prior0 + treat0 + type4 $endmac
$use weibull t0 w0 %s $

                      -- Model is stat0*prior0+treat0+type4
Exponential fit

  Deviance    shape      df
            parameter
   1043.3    1.0000      96
Weibull fit
   1037.7    1.2036      95
   1037.7    1.2068      95
-- Standard errors of estimates given below are underestimated
          estimate         s.e.      parameter
     1     -6.616         1.017      1
     2      0.070         0.167      STAT0
     3      1.806         0.751      PRIOR0
     4      0.593         0.205      TREAT0
     5     -0.645         0.245      TYPE4
     6     -0.297         0.130      STAT0.PRIOR0
scale parameter 1.000
```

The deviance increases by 49.5, for 69 fewer degrees of freedom. It would appear from this comparison alone that the Weibull should be a good fit, but the piecewise hazard is not well represented by a Weibull hazard, and the deviance change between the models is also not well represented by χ^2_{69}, because the piece-wise hazard parameters increase with n and so asymptotic theory does not apply.

The Weibull parameter estimates are similar to those above, except that for the status variable, which is very small and positive for those with no prior therapy, instead of significantly large and negative.

The evaluation of the proportional hazards assumption can clearly have an important bearing on the interpretation of the data: the conclusions from this re-analysis are importantly different from those from the first-edition analysis which treated all types as having the same Weibull shape parameter, and which consequently mis-stated the treatment effect, the form of the hazard and the importance of prior therapy and months from diagnosis.

6.21 Competing risks

The survival time modelling of this chapter can be extended to an important class of processes in which failure may be from one of several causes. In the example considered below of heart transplantation patients, death of the patient may occur by rejection of the heart or from other causes not related to rejection (e.g. from infections due to lowered resistance caused by immuno-suppressant drugs). If treatment or patient background variables affect the hazard differently for different causes of death, then an analysis which does not distinguish the different causes may misrepresent both the importance of the explanatory variables and the nature of the hazard function.

Consider in general the case of n individuals on whom we observe $(t_i, w_i, j_i, \mathbf{x}_i)$ for $i = 1, \ldots, n$. Here t_i, w_i and $\mathbf{x}_i$ are the survival time, censoring indicator and explanatory variables as in previous sections, and j_i is the cause of failure, taking one of the values $1, 2, \ldots, k$ for the k possible causes of failure. The associated random variable will be denoted by J.

We define the *cause-specific hazard function* or *sub-hazard function* (Crowder, 2001) $h_j(t)$ by

$$h_j(t) \, dt = \Pr(t < T \leq t + dt, J = j \mid T > t).$$

That is, $h_j(t)$ is the instantaneous failure rate for failure from the j-th cause at time t, given survival to time t. The overall hazard function is then

$$h(t) = \sum_{j=1}^{k} h_j(t)$$

since failure must be from one of the k given causes. The survivor function is then

$$S(t) = e^{-H(t)}$$

with

$$H(t) = \int_0^t h(u)\,du = \sum_{j=1}^k H_j(t),$$

where $H_j(t)$ is the cause-specific integrated hazard function.

The *cause-specific density function* $f_j(t)$ for survival time for the j-th cause is then given by

$$f_j(t)\,dt = \Pr(t < T \le t + dt, J = j)$$
$$= \Pr(T > t)\Pr(t < T < t + dt, J = j \mid T > t)$$
$$= S(t)h_j(t)\,dt$$

so that

$$f_j(t) = h_j(t)S(t), \quad j = 1,\ldots,k.$$

To construct the likelihood function, define a set of k dummy indicators d_{ij} for the i-th individual by

$$d_{ij} = \begin{cases} 1 & \text{if failure for the } i\text{-th individual is from cause } j \\ 0 & \text{otherwise } (j = 1,\ldots,k). \end{cases}$$

Then $\sum_j d_{ij} = w_i$.

The likelihood function can be written

$$L = \prod_{i=1}^n [f_{j_i}(t_i)]^{w_i}[S(t_i)]^{1-w_i}$$
$$= \prod_i [h_{j_i}(t_i)]^{w_i} S(t_i)$$
$$= \prod_{i=1}^n \prod_{j=1}^k [h_j(t_i)]^{d_{ij}} e^{-H_j(t_i)}.$$

Interchanging the products, we see that the likelihood is a product of k factors, the j-th being the likelihood obtained by treating death from cause j as the outcome, and deaths from any other cause as censoring. Since no assumptions have been made about $h_j(t)$, we can fit completely unrelated models to each cause of death very simply.

We illustrate with a much-analysed set of data from the Stanford Heart Transplantation programme (Crowley and Hu, 1977). For a discussion of the data and a detailed analysis see Aitkin *et al.* (1983); slightly different data from the same study were presented and analysed in Kalbfleisch and Prentice (1980) and Cox and Oakes (1984). A more extensive data set was given by Miller and Halpern (1982) but without cause of death.

The file `stan` contains the data on 65 transplanted patients, consisting of the patient's age at transplantation `age`, prior open-heart surgery `surg` ($1 =$ yes, $0 =$ no), a censoring indicator `died` ($1 =$ yes, $0 =$ no), the survival time in days after transplant `surv`, a score `mm` representing the mismatch between the patient's and the donor's tissue type (values range from 0.00 to 3.05), and an indicator `rej` for death by rejection ($1 =$ yes, $0 =$ no). One zero survival time is recoded to 0.5. There are 41 deaths and 24 censored survivals, with 39 distinct death times. We begin by fitting the piecewise exponential distribution, without distinguishing the causes of death.

```
$newjob $
$input 'stan' : %plc cox $
$use coxmodel surv died ddt $
$fit (age + surg + mm)(ind_) $disp e $
 There are 39 distinct event times.
 scaled deviance =    504.84 at cycle 7
     residual df =   1527

          estimate        s.e.        parameter
    1       0.056        0.022        AGE(IND_)
    2      -0.863        0.483        SURG(IND_)
    3       0.427        0.291        MM(IND_)
scale parameter 1.000
    eliminated term: T_INTS
```

Age is clearly important: risk increases with age as would be expected. The importance of surgery and mismatch score is not clearly established (we obtain the same conclusions by looking at disparity changes due to omitting each variable in turn from the model). We now graph the log hazard and log–log survivor functions against log survival time.

```
$use bsf ddt $
```

Figure 6.27 (of the log–log survivor function) shows a peculiar feature. There is a sudden dramatic fall in the survivor function around 55 days (4 on the log time scale) followed by another dramatic change in slope.

These changes can be identified in the ordered death times, where there are 13 deaths between 44 and 68 days, and then a gap to 127 days (around 5 on the log scale). Of these 13 deaths, 11 are by rejection, compared with 29 out of 41 overall. Could this peculiar behaviour be due to different hazards for death by rejection and other causes?

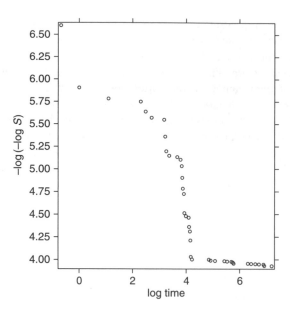

Fig. 6.27. Transplant survival, all causes

We first consider death by rejection. We need to delete the variable ind_ which has to change its length.

```
$calc drej = rej*died $
$del ind_ cuhaz_ lls_ lt_ bsf_ $
$use coxmodel surv drej rejddt $
$fit (age + surg + mm)(ind_) $disp e $
 There are 27 distinct event times.
 -- number of units changed
 -- offset unset
 scaled deviance =  359.43 at cycle 7
     residual df =  947


          estimate       s.e.      parameter
     1      0.099       0.031      AGE(IND_)
     2     -1.053       0.618      SURG(IND_)
     3      0.919       0.357      MM(IND_)
 scale parameter 1.000
     eliminated term: T_INTS
```

The standard errors have increased because the effective sample size—the number of deaths—has decreased. However, all three variables have increased

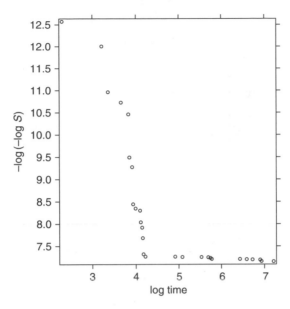

Fig. 6.28. Transplant survival, rejection

in importance, and mismatch is now clearly significant, though surgery appears not, with a Wald test statistic of -1.70. However if we drop surgery from the model, the deviance increases by 3.71, so the importance of surgery is still equivocal.

```
$use bsf rejddt $
```

Figure 6.28 shows a precipitous fall in the log–log survivor function from 10 to 65 days, followed by a very slow decline from 130 days onward. This behaviour does not correspond to any standard survival distribution. (The lognormal distribution does not provide an adequate fit, with a disparity of 407.67 compared with 359.42 for the piecewise exponential, a change of 48.24 for 26 df.)

We now analyse the deaths from other causes.

```
$calc nrej = (1 - rej)*died $
$del ind_ cuhaz_ lls_ lt_ bsf_ $
$use coxmodel surv nrej nrejddt $
$fit (age + surg + mm)(ind_) $disp e $
 There are 12 distinct event times.
 -- number of units changed
 -- offset unset
 scaled deviance =   165.56 at cycle 8
     residual df =   641
```

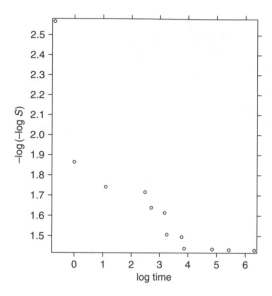

Fig. 6.29. Transplant survival, other causes

```
            estimate          s.e.        parameter
      1       -0.008          0.032        AGE(IND_)
      2       -0.397          0.787        SURG(IND_)
      3       -0.449          0.538        MM(IND_)
  scale parameter 1.000
      eliminated term: T_INTS
```

The standard errors have increased further because of the small number of deaths, and *none* of the variables appears important. We examine the hazard and survivor function.

```
$use bsf nrejddt $
```

The log–log survivor function in Fig. 6.29 decreases nearly linearly with log time apart from the first point, suggesting a Weibull distribution for survival time for deaths from other causes.

We fit the null model.

```
$fit $disp e $
  scaled deviance =   166.60 at cycle 8
      residual df =   644

  -- No parameters to display
  scale parameter 1.000
      eliminated term: T_INTS
```

The disparity increases by 1.04 on 3 df. We fit the Weibull model.

```
$input %plc weibull $
$calc %s = 1 $
$macro model 1 $endm
$use weibull surv nrej %s $
                        -- Model is 1
 -- number of units changed
 -- offset unset
 -- model changed
Exponential fit

  Deviance    shape     df
              parameter
    207.66    1.0000     64
Weibull fit
    188.63    0.63914    63
 ...
    182.03    0.35847    63
 -- Standard errors of estimates given below are underestimated
           estimate      s.e.    parameter
       1      -3.619     0.289    1
scale parameter 1.000
```

The Weibull disparity is an increase of 15.44 on 11 df compared to the piecewise exponential. The MLE of the shape parameter is 0.358: risk declines rapidly with time. Fitting the previous model with age, surgery and mismatch confirms their irrelevance for the Weibull distribution, with the same deviance change of 1.04 on 3 df.

Thus in this example the competing risk framework reveals the importance of mismatch and the very rapid increase in hazard in the first 60 days for deaths by rejection, and the rapidly declining hazard for deaths from other causes which is unrelated to the explanatory variables, but reflects the normal recovery process following major surgery.

The cause-specific hazard analysis combines deaths from causes other than the one being examined with "true" censoring from survival into a single censored group. An alternative analysis is possible in which the censored observations are regarded as coming from a *mixture* distribution (Chapter 7), since the cause of death for the survivors is unknown.

Larson and Dinse (1985) applied this analysis to the Stanford Heart Transplant data; their mixture model for the survivors used a logistic regression model for the probability of death by rejection, with the same covariates used in the logistic model as in the cause-specific piecewise exponential hazard models. The covariates were mismatch and age (both scaled to have mean zero and variance 1), and waiting time for transplant, but not prior surgery.

Waiting time was irrelevant in all the models, and the logistic model could be reduced to the null model, giving a constant probability of death by rejection.

The practical conclusions are the same as in the analysis above: age and mismatch are important for death by rejection, but not for death by other causes.

6.22   Time-dependent explanatory variables

All of the analysis in this chapter has been based on the assumption that the explanatory variables **x** are *constant over time*, that is that they do not change during the lifetime of the individual. Many medical variables are of this type, but others may change their values during the individual's lifetime. For example, measures of physical or physiological status may be available during the course of treatment; these may be important predictors of survival time, and more relevant than the same measures taken before treatment begins. Variables of this type are called *time-dependent*.

Such variables can be incorporated into a proportional hazards model using the piecewise exponential distribution. We need first two notational changes. The time-varying explanatory variables may be changing their values continuously in time, but in practice they are measured only at follow-up or interview times, which will generally not correspond to the times a_j at which the hazard function changes. We assume that these variables can be taken as constant over the time intervals between measurements, and we extend the set of cut-points a_j to include the measurement times for each individual.

We now write $\mathbf{x}_{ij}$ for the values of the explanatory variables, measured at a_{j-1}, for the i-th individual in the j-th time-interval $a_{j-1} < t \leq a_j$. In practice many, perhaps most, components of $\mathbf{x}_{ij}$ will be constant over time intervals.

The piecewise exponential model of Section 6.15 now applies with

$$h_{ij} = \exp(\phi_j + \boldsymbol{\beta}'\mathbf{x}_{ij}).$$

We do not illustrate the use of time-varying variables here as these are usually associated with *repeated event* times, for example, with several recurrences of a disease. The analysis of repeated events requires *variance component* or *frailty* models, which are not discussed in this book.

6.23   Discrete time models

The discussion so far in this chapter has assumed that time is measured on a continuous scale, or at least that the discreteness of the recording of time is small compared to the range of possible time values.

Time may, however, be grouped into quite broad categories; for example, when deaths of laboratory animals are recorded at relatively long intervals, or when human studies use follow-up periods which are long in comparison to the progress of the disease.

In such cases the likelihood construction of Section 6.15 is not appropriate, and a model for discrete time is needed. We follow the development in Prentice and Gloeckler (1978; see also Kalbfleisch and Prentice, 1980, p. 98 and Thompson, 1981).

Suppose the continuous distribution with density $f(t)$, survivor function $S(t)$ and hazard $h(t)$ is grouped into s intervals $A_j = (t_{j-1}, t_j]\ j = 1, \ldots, s$, with $t_0 = 0, t_s = \infty$. Write f_j, s_j and h_j for the probability mass, survivor and hazard functions for the resulting discrete distribution.

Then

$$
\begin{aligned}
f_j &= \Pr(T \in A_j) \\
&= S(t_{j-1}) - S(t_j) \\
&= s_j - s_{j+1}, \\
h_j &= \Pr(T \in A_j \mid T > t_{j-1}) \\
&= f_j / (f_j + f_{j+1} + \cdots + f_s) \\
&= f_j / s_j
\end{aligned}
$$

so that

$$
s_j = \prod_{k=1}^{j-1} (1 - h_k)
$$

as in Section 6.8. Consider the survival experience of each individual through time as in Section 6.15. The i-th individual experiences a sequence of censorings at $t_1, t_2, \ldots$ and either dies or is finally censored in the j_i-th interval. Let w_{ij} be the censoring indicator and h_{ij} the hazard for the i-th individual in the j-th interval; the i-th contribution to the likelihood is then

$$
\begin{aligned}
L_i &= \prod_{j=1}^{j_i} h_{ij}^{w_{ij}} s_{ij} \\
&= \prod_{j=1}^{j_i} h_{ij}^{w_{ij}} (1 - h_{ij}) \\
&= \prod_{j=1}^{j_i} h_{ij}^{w_{ij}} (1 - h_{ij})^{1 - w_{ij}}.
\end{aligned}
$$

For the proportional hazards model,

$$h(t, \mathbf{x}) = \lambda_0(t) e^{\boldsymbol{\beta}' \mathbf{x}}$$

$$S(t, \mathbf{x}) = \exp(-\Lambda_0(t) e^{\boldsymbol{\beta}' \mathbf{x}}).$$

Thus

$$s_j = \exp(-\Lambda_0(t_{j-1}) e^{\boldsymbol{\beta}' \mathbf{x}})$$

$$h_j = 1 - s_{j+1}/s_j$$

$$= 1 - \exp[-e^{\boldsymbol{\beta}' \mathbf{x}} \{\Lambda_0(t_j) - \Lambda_0(t_{j-1})\}]$$

so that

$$h_{ij} = 1 - \exp(-e^{\boldsymbol{\beta}' \mathbf{x}_i + \psi_j}),$$

where

$$\psi_j = \log\{\Lambda_0(t_j) - \Lambda_0(t_{j-1})\}.$$

Since nothing is assumed about $\lambda_0(t)$, the ψ_j are unrelated, unknown constants as in the proportional hazards model. The likelihood function over all individuals is then

$$L = \prod_{i=1}^{n} \prod_{j=1}^{j_i} h_{ij}^{w_{ij}} (1 - h_{ij})^{1-w_{ij}} = \prod_{j=1}^{s} \prod_{i \in R_j} h_{ij}^{w_{ij}} (1 - h_{ij})^{1-w_{ij}},$$

where R_j is the set of individuals at risk in the k-th time-interval. This is a product of $\sum j_i$ Bernoulli likelihoods, one for each individual in each interval. In each interval the individuals currently at risk either die or survive to the next interval. The death probability in the j-th interval for the i-th individual is h_{ij}, and

$$\log\{-\log(1 - h_{ij})\} = \boldsymbol{\beta}' \mathbf{x}_i + \psi_j.$$

Thus the discrete time proportional hazard model can be fitted by treating the observations in each time-interval as independent across intervals, with the censoring indicator as the response variable with a Bernoulli distribution, and a complementary log–log link function (Section 4.2); the regression model contains the explanatory variables and the interval parameters ψ_j. If the explanatory variables are themselves categorical, the data have the form of a contingency table, and the response variable is the number of deaths in the interval, which has a binomial distribution with the number at risk in the interval as the binomial denominator.

As for the Poisson likelihood for the continuous time proportional hazards model, the interval parameters ψ_j can be eliminated from the binomial likelihood using `eliminate`.

We illustrate with the gehan example. Suppose that remission times are recorded in months (units of 4 weeks) instead of weeks. The grouped data are presented below as a contingency table, with the number at risk and the number dying in each month for each treatment group. Censoring within an interval is treated as censoring at the end of the previous interval.

		Time					
		1-4	5-8	9-12	13-16	17-20	21-24
G1	*d*	7	6	4	1	1	2
	r	21	14	8	4	3	2
G2	*d*	0	4	1	2	0	2
	r	21	20	13	12	7	7

The analysis is straightforward.

```
$newjob $
$unit 12 $
$data d r $
$read
 7 21  6 14  4  8  1  4  1 3  2 2
 0 21  4 20  1 13  2 12  0 7  2 7
$factor g  2 month 6 $
$calc g = %gl(2,6)  : month = %gl(6,1) $
$yvar d $error b r $link c $
$eliminate month $
$fit : + g $disp e $
 scaled deviance =  24.884 at cycle 4
     residual df =  6

 scaled deviance =  6.8019 (change =   -18.08) at cycle 4
     residual df =  5       (change =   -1  )

          estimate         s.e.      parameter
     1      -1.698        0.4255       G(2)
 scale parameter 1.000
     eliminated term: MONTH
```

The estimate of g is reasonably close to that (−1.521) from the piecewise exponential model (Section 6.16), and is between those for the exponential and Weibull distributions.

An alternative to fitting the grouped time-variable as a factor in the model is to construct a conditional likelihood which depends only on the common group difference. We condition on the total number at risk in each time-interval, in the same way as for the 2×2 table in Section 4.6. Various tests based on the resulting conditional hypergeometric distribution are possible; the best known is the *log rank* test (Mantel, 1966).

It is not necessary to have categorical explanatory variables to fit the discrete time model. In general the model can be fitted using the macros in cox. As for the proportional hazards model in continuous time with time-dependent covariates, the data vector w_{ij} is expanded to length $\sum N_i$, as are the explanatory variables $\mathbf{x}_i$. These expansions are achieved by indexing the variables by the index vector ind_. The initializing macros coxmodel or inib are used to set up the individuals at risk in each interval and expand the censoring indicator.

Before fitting the model it is necessary to declare the model as Bernoulli with complementary log–log link. The censoring indicator is already declared as the y-variable. Finally, the offset must be turned off as the macros set this in readiness for the standard Poisson fit of the piecewise exponential distribution.

We illustrate with the feigl data, using the distinct death times as discrete cut-points. Note that both the explanatory variables and the binomial denominator (a vector of 1s) have to be indexed by the expanding vector ind_.

```
$newjob $
$input 'feigl' : %plc cox $
$calc w = 1 $
$use coxmodel time w ddt $
$calc lwbc = %log(wbc) : n = 1 $
$error b n(ind_) $link c $offset $
$fit (ag*lwbc)(ind_) $disp e $


 There are 22 distinct event times.
 -- number of units changed
 scaled deviance =   184.85 at cycle 6
     residual df =   312
```

	estimate	s.e.	parameter
1	2.871	0.937	AG(IND_)(2)
2	0.746	0.222	LWBC(IND_)
3	-0.597	0.281	AG(IND_)(2).LWBC(IND_)

```
 scale parameter 1.000
     eliminated term: T_INTS
$macro title Discrete hazard $endm
$use phaz ddt title $
```

The hazard estimates in the plot are not $\log h_{ij}$ but $\log[-\log(1 - h_{ij})]$, which is approximately $\log h_{ij}$ if h_{ij} is small (see below). The estimates (not shown here) are initially fairly constant but show a steady increase with cut-point number from cut-point 14 on. The behaviour of these estimates does not however in general provide information about the form of the hazard function, since it depends on both the spacing of the cut-points and on $\lambda(t)$. The interaction appears significant compared to its standard error. We try omitting it from the model.

```
$fit - ag(ind_).lwbc(ind_) $disp e $
 scaled deviance =  188.28 (change =   +3.431) at cycle 4
     residual df =  313    (change =   +1    )

           estimate         s.e.      parameter
     1        1.118        0.417      AG(IND_)(2)
     2        0.391        0.135      LWBC(IND_)
 scale parameter 1.000
     eliminated term: T_INTS
```

The disparity change is not significant, and the main-effect parameter estimates are similar to those from the piecewise exponential model; the observed survival times in weeks are not heavily grouped relative to the variation in individual survival.

It is easily seen that if $t_j - t_{j-1}$ is small,

$$\psi_{rj} = \log[\Lambda_0(t_j) - \Lambda_0(t_{j-1})]$$
$$\approx \log[(t_j - t_{j-1})\lambda_0(t_j)]$$
$$= \log(e_j) + \phi_j$$

from the piecewise exponential. As h_{ij} will also be very small,

$$\log\{-\log(1 - h_{ij})\} \approx \log h_{ij},$$

so we obtain the previous piecewise exponential model as the Poisson limit of the binomial model for precise measurement in continuous time.

7
Finite mixture models

7.1 Introduction

The exponential family of distributions and its extensions discussed in earlier chapters are extremely useful in statistical analysis, but they cannot represent all types of data of scientific interest.

In this chapter we consider a general extension of the exponential family to *mixtures* of distributions from this family. The need for this extension will become clear in subsequent chapters in which we discuss *random effect models* and their fitting by ML.

Mixtures are extensively discussed in the books by Everitt and Hand (1981), Maritz and Lwin (1989), Titterington *et al.* (1985), McLachlan and Basford (1988), Lindsay (1995), Böhning (1999) and McLachlan and Peel (2000). Our treatment is much more limited.

We restrict our discussion to the *mixed* exponential family distribution

$$m(y \mid \theta, \phi) = \int f(y \mid z, \phi) g(z \mid \theta) \, dz$$

where $f(y \mid z, \phi)$ is a "kernel" exponential family density or mass function depending on a parameter ϕ and a random variable Z, and $g(z \mid \theta)$ is the "mixing" distribution of Z, depending on the parameter θ. The value of z is unobservable, and we can observe only y.

Two different classes of problems arise:

1. We know or specify the form of the mixing distribution $g(z \mid \theta)$ up to the unknown value of θ, and want to draw likelihood inferences about θ and ϕ.
2. The distribution $g(z \mid \theta)$ is unknown, and we want to draw inferences about ϕ, without making any assumption about $g(z)$; we may also want to draw inferences about $g(z)$.

We deal with the first class of problems in subsequent chapters, as no new problems arise from them. In this chapter we consider in detail the second class, in the framework of *finite mixtures*.

We first illustrate the need for mixture distributions with the girl birthweight example in Chapter 2.

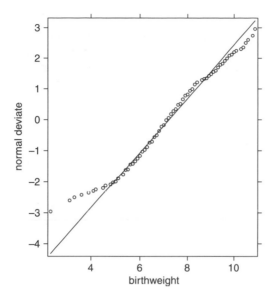

Fig. 7.1. Birthweight: normal deviate of cumulative proportion, girls

7.2 Example—girl birthweights

We saw in Chapter 2 that the cdf plot for the girl birthweights was not normal, with two notable bends in the Q–Q plot (reproduced in Fig. 7.1).

Such changes in slope of the Q–Q plot are a good indication of the presence of a mixture; the observations from the separate components, if we could identify them, would each have a straight-line plot if the component distributions were normal, but the Q–Q plot of the mixture of distributions has to bend to accommodate the different means (and possibly variances) of the component distributions.

7.3 Finite mixtures of distributions

If the distribution of Z is *discrete*, on the set of values (support points) $z_1, \ldots, z_K$ with probabilities $\pi_1, \ldots, \pi_K$, then the distribution of Y is a *finite mixture*. In this case we write the *mixed* probability distribution for the observed Y in the form

$$m(y \mid z_1, \ldots, z_K, \pi_1, \ldots, \pi_{K-1}, \phi) = \sum_{k=1}^{K} \pi_k f_k,$$

where $f_k = f(y \mid z_k, \phi)$, $k = 1, \ldots, K$ is an exponential family density or mass function depending on a parameter z_k unique to f_k, and a parameter ϕ common to all the K component densities. The proportions π_k of each component in the mixture are non-negative and sum to 1, so there are only $K - 1$ distinct proportion parameters. (To remove unnecessary generality, we will in fact require the π_k to be *positive*, so that the mixture has K non-empty components.)

An appealing interpretation of mixtures is that they arise because of the omission or suppression of a K-category group identifier G which, if observed, would allow us to fit a one-way classification model. We make use of this interpretation implicitly in the ML fitting of the model.

7.4   Maximum likelihood in finite mixtures

Given a random sample $y_1, \ldots, y_n$ from the mixture distribution, the likelihood is

$$L = L(z_1, \ldots, z_K, \pi_1, \ldots, \pi_{K-1}, \phi) = \prod_{i=1}^{n} m_i = \prod_i \left(\sum_k \pi_k f_{ik} \right),$$

where

$$m_i = m(y_i \mid z_1, \ldots, z_K, \pi_1, \ldots, \pi_{K-1}, \phi), \quad f_{ik} = f(y_i \mid z_k, \phi).$$

The log-likelihood is

$$\ell = \ell(z_1, \ldots, z_K, \pi_1, \ldots, \pi_{K-1}, \phi) = \sum_i \log m_i = \sum_i \log \left(\sum_k \pi_k f_{ik} \right)$$

and the score for z_k is

$$\frac{\partial \ell}{\partial z_k} = \sum_i \frac{\pi_k}{m_i} \frac{\partial f_{ik}}{\partial z_k}$$

$$= \sum_i \frac{\pi_k f_{ik}}{m_i} \frac{\partial \log f_{ik}}{\partial z_k}.$$

Write

$$w_{ik} = \frac{\pi_k f_{ik}}{m_i};$$

then

$$\frac{\partial \ell}{\partial z_k} = \sum_i w_{ik} s_{ik}(z_k)$$

where

$$s_{ik}(z_k) = \frac{\partial \log f_{ik}}{\partial z_k}$$

is the score for z_k in the k-th component distribution. Similarly

$$\frac{\partial \ell}{\partial \phi} = \sum_i \sum_k w_{ik} s_{ik}(\phi).$$

Thus the likelihood (score) equations

$$\frac{\partial \ell}{\partial z_k} = 0, \quad \frac{\partial \ell}{\partial \phi} = 0$$

are *weighted* forms of the single-distribution score equations for these para-meters, with weights w_{ik}. We now apply Bayes' theorem to the hypotheses that the observation y_i comes from component k. The prior probabilities are π_k from the original mixture formulation, and

$$\Pr(y_i \mid G = k) = f_{ik},$$

so the posterior probability that y_i comes from component k is

$$\Pr(G = k \mid y_i) = \frac{\pi_k f_{ik}}{\sum_\ell \pi_\ell f_{i\ell}} = w_{ik},$$

and the weights w_{ik} are just these posterior probabilities.

Now consider the mixture probabilities π_k. Differentiating the log-likelihood subject to the constraint $\sum \pi_k = 1$ using a Lagrange multiplier λ, we have

$$\ell^* = \ell - \lambda \left(\sum \pi_k - 1 \right)$$

$$\frac{\partial \ell^*}{\partial \pi_k} = \sum_i \frac{f_{ik}}{m_i} - \lambda$$

$$= \sum_i \frac{w_{ik}}{\pi_k} - \lambda.$$

Solving the score equations gives immediately

$$\hat{\pi}_k = \sum_i w_{ik} / \lambda,$$

and multiplying by λ and summing over k gives

$$\hat{\lambda} = \sum_k \sum_i w_{ik} = \sum_i \sum_k w_{ik} = n,$$

so

$$\hat{\pi}_k = \sum_i w_{ik} / n.$$

ML estimation of the model parameters can now proceed iteratively by an EM algorithm (Day, 1969; Dempster *et al.*, 1977; Aitkin, 1980; McLachlan and Krishnan, 1997). In the M step, the z_k and ϕ are estimated by solving the weighted score equations with given weights w_{ik}, while in the E step, the weights w_{ik} are updated using the new parameter estimates in the previous M step.

The weights w_{ik} can be interpreted as *conditional expectations* of the missing or unobserved K-category group identifier G described in Section 7.3. Define $G_{i1}, \ldots, G_{iK}$ to be the binary indicators for membership of the i-th observation in groups $1, 2, \ldots, K$. If these were observed, the "complete data" likelihood would be

$$L^* = \prod_i \prod_k (\pi_k f_{ik})^{G_{ik}},$$

and the conditional expectation of the log-likelihood used in the E step would replace G_{ik} by $E[G_{ik} \mid y_i] = \Pr(G_{ik} = 1 \mid y_i) = w_{ik}$.

Multiple maxima are a common feature of mixture likelihoods and different starting values for the EM algorithm need to be used to locate these maxima. Finch *et al.* (1989) discussed strategies for this. In the EM algorithm which we use for fitting these models, initial estimates of the mass-point locations are obtained from the *Gaussian quadrature* mass-points used for the normal distribution (see Section 8.3 for details). Searching for multiple maxima is assisted through a scaling parameter `tol_` which scales in or out these initial estimates of the mass-point locations. This is not a completely general method for locating multiple maxima. We give an example in Section 7.8.

An important point is that the above results for ML estimation are not restricted to exponential family models: they apply quite generally. Further the densities $f_k(y)$ do not even have to be of the same form—we could for example have a mixture of a gamma and a lognormal distribution, though we will not in fact consider such mixtures.

7.5   Standard errors

A general feature of the EM algorithm is that it gives ML estimates, but not their standard errors. These require additional computation beyond EM itself. The most direct method to obtain standard errors is via computation of the observed information matrix. We write the density or mass function of the observed data as $m(y \mid \boldsymbol{\psi})$, depending on the parameter $\boldsymbol{\psi}$ in the joint distribution of Y and Z. Then

$$m(y \mid \boldsymbol{\psi}) = \int f(y \mid z, \boldsymbol{\psi}) g(z \mid \boldsymbol{\psi}) \, \mathrm{d}z.$$

Writing $m(y_i \mid \boldsymbol{\psi}) = m_i, f(y_i \mid z_i, \boldsymbol{\psi}) = f_i$ and $g(z_i \mid \boldsymbol{\psi}) = g_i$, the contribution of a single observation y_i to the log-likelihood and its derivatives for a sample are:

$$\ell_i = \ell_i(\boldsymbol{\psi}) = \log m(y_i \mid \boldsymbol{\psi})$$

$$= \log \int f_i g_i \, dz_i$$

$$s_i(\boldsymbol{\psi}) = \frac{\partial \ell_i}{\partial \boldsymbol{\psi}} = \frac{(\partial/\partial \boldsymbol{\psi}) \int f_i g_i \, dz_i}{m_i}$$

$$= \frac{\int \left[(\partial f_i/\partial \boldsymbol{\psi}) g_i + f_i (\partial g_i/\partial \boldsymbol{\psi}) \right] dz_i}{m_i}$$

$$= \int \frac{f_i g_i}{m_i} \left[\frac{\partial \log f_i}{\partial \boldsymbol{\psi}} + \frac{\partial \log g_i}{\partial \boldsymbol{\psi}} \right] dz_i$$

$$= \int h_i \left[\frac{\partial \log f_i}{\partial \boldsymbol{\psi}} + \frac{\partial \log g_i}{\partial \boldsymbol{\psi}} \right] dz_i.$$

Here

$$h_i = \frac{f_i g_i}{m_i}$$

is the conditional distribution of Z_i given y_i. The term $(\partial \log f_i)/\partial \boldsymbol{\psi} + (\partial \log g_i)/\partial \boldsymbol{\psi}$ is called the "complete data score", since it corresponds to observing the "complete data" Y and Z. Then

$$s_i(\boldsymbol{\psi}) = \mathrm{E}_{Z_i \mid y_i} \left[\frac{\partial \log f_i}{\partial \boldsymbol{\psi}} + \frac{\partial \log g_i}{\partial \boldsymbol{\psi}} \right]$$

$$= \mathrm{E}_{Z_i \mid y_i}[s_{ci}(\boldsymbol{\psi})],$$

where the subscript c indicates "complete data". Here $s_{ci}(\boldsymbol{\psi})$ is the i-th component of the complete-data score, and we assume sufficient regularity in the density f to allow the interchange of differentiation and integration. Then, summing over all n observations, *the observed-data score is equal to the conditional expectation of the complete-data score.*

Before taking the second derivative, we note that

$$\log h_i = \log f_i + \log g_i - \log m_i$$

and hence

$$\frac{\partial \log h_i}{\partial \boldsymbol{\psi}} = \frac{\partial \log f_i}{\partial \boldsymbol{\psi}} + \frac{\partial \log g_i}{\partial \boldsymbol{\psi}} - \frac{\partial \log m_i}{\partial \boldsymbol{\psi}}$$

$$= s_{ci}(\boldsymbol{\psi}) - s_i(\boldsymbol{\psi}),$$

the difference between the complete-data and the observed-data score. Then the observed data Hessian is

$$H(\psi) = \sum_i H_i(\psi) = \sum_i \frac{\partial^2 \ell_i}{\partial \psi \partial \psi'}$$

$$= \sum_i \frac{\partial}{\partial \psi'} \int h_i s_{ci}(\psi) \, dz_i$$

$$= \sum_i \int \left[h_i H_{ci}(\psi) + s_{ci}(\psi) \frac{\partial h_i}{\partial \psi'} \right] dz_i$$

$$= \sum_i \int [H_{ci}(\psi) + s_{ci}(\psi)[s_{ci}(\psi) - s_i(\psi)]'] h_i \, dz_i.$$

Interchanging the order of summation and integration, the second term in the integral, the matrix $\sum_i s_{ci}(\psi)[s_{ci}(\psi) - s_i(\psi)]'$, is the *sample covariance matrix* of the complete-data score. The first term in the integral is the complete-data Hessian, and so (Louis, 1982; Oakes, 1999) *the observed-data information matrix is equal to the conditional expectation of the complete-data observed information matrix minus the conditional covariance matrix of the complete-data score.* In many incomplete-data problems in which the EM algorithm is used with a package which analyses complete-data, standard errors reported by the package are those for the complete-data information matrix. These are always too small, since they correspond to treating the conditional expectations of the unobserved data as though they were known, but can be corrected by computing the conditional covariance matrix of the complete-data score. Since GLIM lacks the matrix facilities necessary for the computation of this matrix, we give an alternative approach, decribed in Chapter 2 which is often useful (Aitkin, 1994; Dietz and Böhning, 1995).

In the applications considered in the next chapters, the common parameter ϕ is frequently a regression coefficient vector β and is the parameter of principal interest. Standard errors (SE) for an individual coefficent β_j can be obtained by omitting the corresponding explanatory variable x_j from the regression model, recording the disparity change Δdisp_j, and equating the Wald test statistic for the hypothesis $\beta_j = 0$ to the LRTS, giving

$$\left[\frac{\hat{\beta}_j}{\text{SE}(\hat{\beta}_j)} \right]^2 = \Delta\text{disp}_j$$

whence

$$\text{SE}(\hat{\beta}_j) \doteq \frac{|\hat{\beta}_j|}{\sqrt{\Delta\text{disp}_j}}.$$

In small samples from non-linear models the likelihood may be far from normal and there may be poor agreement between the likelihood ratio and Wald tests. In this case any standard error is less useful, as likelihood-based confidence intervals for the parameter will not be symmetric around the MLE. Nevertheless the approximate standard error is still useful in this case as it gives the correct impression of the importance of the variable in the model; the usual standard error is misleading if interpreted in a Wald significance test sense.

7.6 Testing for the number of components

We have assumed above that the number of components K in the mixture is specified. However even with well-separated components it may be unclear how many components are needed to provide an adequate fit to the data. It might be expected that this question could be resolved straightforwardly by increasing the number of components until the decrease in disparity becomes non-significant. Two theoretical properties of ML in finite mixtures however prevent us from using the standard asymptotic properties of the LR test:

1. The mixture model is non-regular, in the sense that the $(K-1)$-component mixture is not a simple restriction of the K-component mixture: for example restricting the component-specific parameters so that $z_{K-1} = z_K$ leaves these two components indistinguishable, with π_{K-1} and π_K not separately identifiable. Equivalently, if π_K is set to zero, z_K is not identifiable.
2. In single-parameter models like the binomial and Poisson, the disparity does not continue to decrease indefinitely as K increases. Surprisingly, at some value K_0 the disparity stabilizes, and increasing K further gives the same disparity, the fitted $(K_0 + 1)$-component mixture degenerating to a K_0-component mixture, and similarly for larger numbers of components.

This unusual result can be expressed quite generally, following Kiefer and Wolfowitz (1956), Laird (1978) and Lindsay (1983). Under certain conditions (discussed at length in, e.g. Maritz and Lwin, 1989), the mixing distribution can be consistently estimated, that is the estimate $\hat{g}(z)$ converges to the true $g(z)$ as $n \to \infty$. This was established by Kiefer and Wolfowitz (1956). The form of the MLE $\hat{g}(z)$ of $g(z)$ was established by Laird (1978) and Lindsay (1983), and is a *discrete* distribution on $\hat{K}$ points of support, with probability masses $\hat{\pi}_1, \ldots, \hat{\pi}_K$ at mass-points $\hat{z}_1, \ldots, \hat{z}_K$. The number, location and masses of these mass-points have to be determined computationally.

The one-parameter exponential family distributions satisfy the conditions for consistent estimation of $g(z)$, except for the Bernoulli distribution. It is easy to see why the latter distribution fails: if $Y \mid Z$ is Bernoulli $b(1, Z)$ and Z has some distribution $g(z)$ on $(0,1)$ with mean μ, then the distribution of Y is

$$p(y) = \int z^y (1-z)^{1-y} g(z)\, dz.$$

So

$$p(1) = \int z g(z)\, dz = \mu,$$

and $p(0) = 1 - \mu$. Thus Y again has a Bernoulli distribution with success probability μ. The distribution of Y is the same for all distributions $g(z)$ with the same mean, and so $g(z)$ cannot be consistently estimated.

For binomial distributions $b(n, Z)$ with $n > 1$, the marginal probabilities from the compound binomial distribution are not binomial: for $n = 2$ for example they are $\mu^2 + \sigma^2$ for $r = 2$, $2\mu(1-\mu) - 2\sigma^2$ for $r = 1$ and $(1-\mu)^2 + \sigma^2$ for $r = 0$, where μ and σ^2 are the mean and variance of Z. These probabilities are binomial only if $\mathrm{Var}[Z] = 0$. Thus the distribution of Z is identified, but only up to its first two moments. For general n the first n moments of Z are identified; this does not define uniquely the distribution of Z.

The form of the general MLE of $g(z)$ means that the estimated distribution of Y is a *finite mixture of exponential families*:

$$\hat{m}(y) = \sum_{k=1}^{\hat{K}} f(y \mid \hat{z}_k) \hat{\pi}_k.$$

The number $\hat{K}$, locations $\hat{z}_k$ and masses $\hat{\pi}_k$ have to be determined computationally to maximize the likelihood. This brings us back to the finite mixture problem discussed at the beginning of the chapter: we have a straightforward EM algorithm for maximizing finite mixture likelihoods, and in a more general form than for the simple example in Section 7.2. The MLE $\hat{K}$ of the number of components K in the one-parameter family is found by increasing K until the maximized likelihood stabilizes. We make extensive use of this approach in subsequent chapters. In the two-parameter family there is no MLE $\hat{K}$. The maximizing $\hat{g}(z)$, where it exists, is called the *non-parametric maximum likelihood (NPML) estimate* of $g(z)$—non-parametric because no parametric model is assumed for $g(z)$. (In models with additional regression parameters, this estimate of $g(z)$ is often called *semi-parametric*, because of the additional model parameters. However these additional parameters do not affect the nature of the estimate, so we retain the term *non-parametric* even when there are additional model parameters.)

Since no general distributional results are available for the distribution of the LRTS in the general mixture problem (though we give some results for special cases in subsequent chapters), *bootstrap* methods are the only general methods available, though they are not very satisfactory either (McLachlan and Peel, 2000, p. 198).

For the special case of testing a two-component normal mixture against a single normal distribution, Thode *et al.* (1988) gave tables of simulated percentage points of the LRTS.

A common alternative approach is through the Akaike or Bayesian Information Criteria (AIC or BIC), in which the disparity for the K-component mixture is penalized by a measure of model complexity depending on the number of model parameters. The penalty functions for these criteria are derived by assuming the large-sample normality of the likelihood function around the parameter MLEs. Since this assumption fails for mixture likelihoods with poorly defined components, the applicability of these criteria to mixture likelihoods is questionable. A more detailed discussion of criteria for assessing the number of components is given in McLachlan and Peel (2000), Chapter 6.

In the bootstrap approach, to test the null hypothesis of a K-component mixture against the alternative of a $(K + 1)$-component mixture, we simulate a sample of n observations from the fitted K-component mixture, and fit the K- and $(K + 1)$-component mixture models to the simulated sample. The value of the LRTS for the null hypothesis is then computed for this sample. This sampling procedure and model fitting are replicated to give R independent samples, and the value of the LRTS for the observed sample is compared with the R simulation values.

If the null hypothesis is true, the probability that the observed value is larger than any of the R simulation values is approximately $1/(R + 1)$. So for a 5% level test, we need 19 simulation values, and the hypothesis is rejected if the observed sample value is larger than all 19 simulation values. For a 1% level test, we need 99 simulated values.

The test size is only approximate because we are treating the MLEs of the parameters from the given sample data as the true values in the simulations. McLachlan and Peel (1997) showed that the P-values from the test tend to be non-conservative, giving overstatements of significance.

The bootstrap test procedure was proposed by Hope (1968); an illustration of its use in mixture models was given in Aitkin *et al.* (1981). McLachlan (1987) provided FORTRAN code for the simulations assuming a multivariate normal mixture. We use the GLIM macro `bootlrts` for this purpose.

We illustrate with the girl birthweight example.

7.6.1   *Example*

We first fit a mixture of normals with different means μ_k but a common variance σ^2. Applying the earlier approach gives the score equations

$$\frac{\partial \ell}{\partial \mu_k} = \sum_i w_{ik}(y_i - \mu_k) = 0$$

$$\frac{\partial \ell}{\partial \sigma} = \sum_i \sum_k w_{ik} \left[-\frac{1}{\sigma} + \frac{1}{\sigma^3}(y_i - \mu_k)^2 \right] = 0$$

giving

$$\hat{\mu}_k = \sum_i w_{ik} y_i \bigg/ \sum_i w_{ik}$$

$$\hat{\sigma}^2 = \sum_i \sum_k w_{ik} (y_i - \mu_k)^2 / n$$

and

$$\hat{\pi}_k = \sum_i w_{ik}/n.$$

The EM algorithm is extremely simple, alternating weighted mean and sums of squares calculations with recalculation of the weights using

$$w_{ik} = \frac{\pi_k f_{ik}}{\sum_\ell \pi_\ell f_{i\ell}}$$

with

$$f_{ik} = \frac{1}{\sqrt{2\pi}\sigma} \exp\left\{ -\frac{1}{2\sigma^2} (y_i - \mu_k)^2 \right\}.$$

The algorithm is implemented in the set of macros in the file `alldist`. The macro `remnp` fits the general mixture model. For simple mixtures we specify the "fixed" and "random" macros as null, with a simple intercept term. The number of components is specified in the `mass`k macros, with $k = 1, 2, \ldots, 10, 20$ mass-points (components) implemented. The normal distribution is the default, and the y-variate is specified as the argument to `remnp`.

```
$input 'gbirthwt' : 'alldist' $
$macro fixed 1 $endm
$macro random 1 $endm
$use mass1 $
$use remnp bwt $
```

```
            estimate        s.e.      parameter
       1       7.223       0.04501      MASS_(1)
    scale parameter 1.313

    MLE of sigma =    1.145
    mixture proportions :   1.000
    -2 log L = 2014.39 , deviance =   2014.39, after 3 iterations
```

With only one component, there is no mixture, and we obtain as parameter estimates the mean birthweight of 7.22 pounds (the mean MASS_(1) of the "first" component) and the MLE $\hat{\sigma} = 1.145$ pounds. (GLIM provides the unbiased estimate $s = \sqrt{1.313} = 1.146$ from the scale parameter. This should be ignored, since for $K > 1$ it has no connection to the MLE $\hat{\sigma}$.) We now increase the number of components.

```
$use mass2 $use remnp $

           estimate       s.e.      parameter
     1        7.370      0.04492     MASS_(1)
     2        7.080      0.04439     MASS_(2)
  scale parameter 0.6460


MLE of sigma =    1.136
mixture proportions :  0.4940  0.5060
-2 log L = 2014.43 , deviance =  2014.43, after 69 iterations

$use mass3 $use remnp $

           estimate       s.e.      parameter
     1        9.135      0.06305     MASS_(1)
     2        7.078      0.02086     MASS_(2)
     3        3.893      0.1500      MASS_(3)
  scale parameter 0.2497


MLE of sigma =   0.8649
mixture proportions :  0.0970  0.8859  0.0171
-2 log L = 1972.06 , deviance =  1972.06, after 62 iterations

$use mass4 $use remnp $

           estimate       s.e.      parameter
     1        9.168      0.05479     MASS_(1)
     2        7.256      0.02554     MASS_(2)
     3        6.919      0.02469     MASS_(3)
     4        3.857      0.1299      MASS_(4)
  scale parameter 0.1817


MLE of sigma =   0.8520
mixture proportions :  0.0934  0.4299  0.4600  0.0166
-2 log L = 1971.97 , deviance =  1971.97, after 37 iterations
```

The disparity for the two-component mixture is slightly greater than for the one-component model, showing that the EM algorithm has not quite converged to the boundary one-component value. The two components differ in mean by 0.29, but this is only 0.26 (estimated) standard deviations, and a two-component mixture with such a small separation between the components is indistinguishable from a single normal distribution even in this large sample.

For the three-component mixture the disparity decreases by 42.33, a value so large that a formal test is hardly needed (though we give one below). The four-component model reduces the disparity by only 0.09, and it is clear that the second and third components are simply a partition of the second component in the three-component model, with means differing only by 0.34, which is 0.4 standard deviations.

Increasing the number of components to six leaves the disparity essentially unchanged, and the cdf of the fitted three-component mixture distribution fits uniformly inside the cdf confidence band, as was shown in Chapter 2, so we conclude that the three-component mixture provides a satisfactory model for the girl birthweights.

The three-component model is made up of a central sub-population of about 89% with mean 7.1 pounds, a "high birthweight" sub-population of about 10% with mean 9.1 pounds, and a "low birthweight" sub-population of about 2% with mean 3.9 pounds; the common standard deviation is 0.86 pounds.

We finally consider the LRT for the existence of a three-component mixture. We generate $M = 19$ independent samples of size 648 from the single normal distribution with mean 7.22 and standard deviation rounded to 1.15, and fit the single- and three-component mixture distributions, saving the disparity changes (the macro holds the disparity in the scalar f__). Then we compare the disparity change in the real data with that from the simulations.

```
$macro boot
   $calc yy = 7.22 + 1.15*%nd(%sr(0)) $
   $use mass1 $use remnp yy $calc %a = f__ $
   $use mass3 $use remnp $calc lrt(%i) = %a - f__ $
   $calc %i = %i + 1 : %n = %n - 1 $
$endm

$var 19 lrt $
$var 648 yy $
$calc %n = 19 : %i = 1 $
$while %n boot $
$sort slrt lrt $
$print slrt $
-0.0331   0.0165   0.0867   0.1318   0.1992   0.5094   0.5625
 0.5684   0.8823   0.8901   0.9403    1.566    2.299    2.326
 4.442    4.456    6.245    6.880
```

The observed data test statistic of 42.33 greatly exceeds the largest, 6.88, of the 19 simulated values. (Note that one of the LRT values is negative, though small; the three-component model gives a slightly larger deviance than the one-component model because the convergence criterion stops the three-component solution converging to the boundary.) The hypothesis of a single normal is firmly rejected in favour of the three-component mixture.

In Chapter 2 we raised the possibility that the variation was caused by other variables, like mother's weight and age. We examine this possibility, by fitting these variables explicitly in a regression model and refitting the mixture model.

```
$macro fixed agem + mwt $endm
$use mass1 $use remnp $

            estimate        s.e.        parameter
    1        0.01112      0.007301       AGE(I_)
    2        0.01295      0.001963       MWT(I_)
    3        5.219        0.2843         MASS_(1)
scale parameter 1.212


MLE of sigma =     1.099
mixture proportions :    1.000
-2 log L = 1960.77 , deviance =    1960.77, after 3 iterations
```

Mother's weight is very important, with a t-statistic of 6.60, but mother's age is not, with a t-statistic of 1.52. We drop mother's age and refit the model with one and three components.

```
$macro fixed mwt $endm
$use remnp $

            estimate        s.e.        parameter
    1        0.01378      0.001888       MWT(I_)
    2        5.425        0.2503         MASS_(1)
scale parameter 1.215


MLE of sigma =     1.101
mixture proportions :    1.000
-2 log L= 1963.09 , deviance =    1963.09, after 3 iterations


$use mass3 $use remnp $

            estimate        s.e.        parameter
    1        0.01334      0.0008339      MWT(I_)
    2        7.225        0.1270         MASS_(1)
    3        5.356        0.1104         MASS_(2)
    4        2.147        0.1836         MASS_(3)
scale parameter 0.2363
```

```
MLE of sigma =    0.8410
mixture proportions :   0.0972   0.8853   0.0175
-2 log L = 1919.33 , deviance =   1919.33, after 65 iterations
```

The three-component model appears clearly again, with a disparity change of 43.76. The mixture cannot be attributed to mother's weight. Though highly significant, the effect of mother's weight is not very large: a 10-pound increase in mother's weight is associated with an increase in the mean baby's weight of only 0.133 pounds: even a 50-pound increase in mother's weight corresponds to an increase of only 0.66 pounds, less than the standard deviation of 0.84 pounds. The "unexplained" variability about the regression remains large.

Note that the GLIM standard error for mother's weight is too small, being based on the complete-data conditional expected information. The indirect estimate of the standard error by equating the disparity change and the squared Wald statistic is

$$ SE = \frac{|\hat{\beta}|}{\sqrt{\Delta \text{disp}}} = \frac{0.01334}{\sqrt{52.73}} = 0.0018, $$

almost unchanged from the (one-component) normal regression model. The "complete data" value is a very substantial underestimate.

7.7   Likelihood "spikes"

A computational difficulty occurs in normal mixtures with different variances. If one observation is remote from the others, it may be identified as a mixture component with one observation, with a variance which goes to zero. This will invalidate the computation of the likelihood, since the "spike" density for the degenerate component will go to infinity, which is obviously invalid.

This difficulty arises from the failure to represent the measurement precision (discussed in Chapter 2) properly in this case. Consider a *single* observation y drawn from a population with a normal distribution model $N(\mu, \sigma^2)$ for Y. Let the measurement precision be δ. The likelihood for the single observation is then

$$ L(\mu, \sigma) = \Pr(y - \delta/2 < Y < y + \delta/2 \mid \mu, \sigma) $$
$$ = \Phi([y + \delta/2 - \mu]/\sigma) - \Phi([y - \delta/2 - \mu]/\sigma) $$

where $\Phi(z)$ is the standard normal cdf. If the measurement precision δ is small compared with σ, then as in Chapter 2 we may approximate the likelihood by the normal density at y multiplied by δ. But if σ is small, as we are considering if it may tend to zero, this approximation cannot be used, and the exact expression above must be retained. The MLE of μ may be shown by differentiation to be

$\hat{\mu} = y$. Substituting in the likelihood gives the profile likelihood in σ:

$$P(\sigma) = \Phi(\delta/[2\sigma]) - \Phi(-\delta/[2\sigma]) = 2\Phi(\delta/[2\sigma]) - 1.$$

This does not depend on the data value y at all, only the measurement precision δ. Thus the single observation is uninformative about σ, unless we have independent information about μ. This is not surprising—to obtain information about variability from data, we need to have observations which *show* variability!

It can be verified that the profile likelihood in σ is maximized at $\sigma = 0$, when $P(\sigma) = 1$. (If the invalid normal density approximation is used for the likelihood, again the MLE of σ is zero, but at this value the likelihood appears to be infinite, a sure sign that the calculation is invalid: since the likelihood is a probability, how can it be infinite?) This result is often presented as a criticism of direct likelihood inference—it is interpreted to mean that the single observation somehow provides misleading information that σ is very small—indeed, if $\sigma = 0$, that the observation y is *certain* to occur! In response, we note that if we are only ever allowed a single observation from a population, then any model for it will be irrelevant. But if, given the model, only one observation is to be taken, and in the absence of any other information about the population, what is more natural than to suppose that the observation we draw is certain to occur? We find this unreasonable because we have very strong prior information that degenerate populations with zero variances do not occur. It is this information, not that provided by the data, that leads us to the conclusion.

However, we have to deal with this difficulty in the computation of the likelihood for the mixture model. The usual solution is to bound the component standard deviations below by some small σ_0. This is easy to implement computationally (as we have done in the macro normvar), but the maximized likelihood resulting will depend on the value of σ_0. We can adopt a different solution.

So long as we do *not* have single observations belonging to mixture components, the problem does not arise, and the likelihood can be computed in its usual mixed-density form. Suppose now that one and only one observation belongs to component k uniquely, and no other observation has non-negligible probability of belonging to it. Then the likelihood for this observation should be maximized at 1, not at ∞. This can be achieved simply by removing this observation from the likelihood computation by assigning it weight zero. The maximized likelihood will then be correct for this mixture model, allowing an extra degenerate component for the excluded observation. The macro normvar allows the specification of a weight variable which can be used to weight out observations with their own mixture components.

However, if we consider further the general model with different variances, for n sample observations it is possible to fit a mixture model with n components with different means and variances, with all the variances equal to zero. The same model with *equal* variances would also fit the data with n components with zero variance. These "mixture" models in fact give the *empirical mass function*—the non-parametric estimate of the density when no assumption is made about it.

The empirical mass function is not a mixture in any real sense—any two-parameter distribution would give the same result as the normal. However the possibility of "pushing the mixture to the limit" means that the non-parametric ML estimate of the mixing distribution is well-defined only for one-parameter kernel distributions whose variances cannot be zero. For normal and other two-parameter distributions we need to rely on the bootstrap LRT to determine reasonable numbers of components.

7.8   Galaxy data

We discuss a second example: the well-known "galaxy" data (Roeder, 1990; Richardson and Green, 1997; Aitkin, 2001). The data are the recession velocities, in units of 10^3 km/s, of 82 galaxies receding from our own, sampled from six well-separated conic sections of space. The astronomers Postman *et al.* (1986) gave the full data by region.

One question of scientific interest is whether these galaxies form distinct super-clusters surrounded by voids in space. If so, some clustering of velocities would be expected, with the overall distribution of velocity being multimodal. Following all authors, including the astronomers, we do not analyse separately the data from the six regions: the individual regions have very small data sets, from which not much can be learned about clustering among or within regions.

The data are shown below in increasing order, and are held in the file `galaxy` in the variable v.

```
    Recession velocities of 82 galaxies

 9.172   9.350   9.483   9.558   9.775  10.227  10.406  16.084  16.170
18.419  18.552  18.600  18.927  19.052  19.070  19.330  19.343  19.343
19.440  19.473  19.529  19.541  19.547  19.663  19.846  19.856  19.863
19.914  19.918  19.973  19.989  20.166  20.175  20.179  20.196  20.215
20.221  20.415  20.629  20.795  20.821  20.846  20.875  20.986  21.137
21.492  21.701  21.814  21.921  21.960  22.185  22.209  22.242  22.249
22.314  22.374  22.495  22.746  22.747  22.888  22.914  23.206  23.241
23.263  23.484  23.538  23.542  23.666  23.706  23.711  24.129  24.285
24.289  24.366  24.717  24.990  25.633  26.960  26.995  32.065  32.789
34.279
```

There is a gap, or jump, between the seven smallest observations around 10 and the large central body of observations between 16 and 26, and another gap between 27 and 32, for the three largest observations. So we might expect to find at least three components in the data.

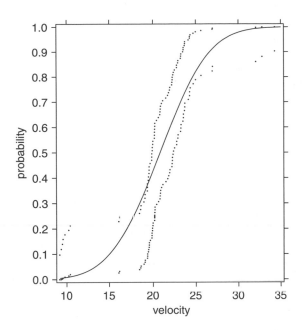

Fig. 7.2. Galaxy data cdf bounds and fitted normal

Figure 7.2 shows the 95% simultaneous confidence band for the population cdf and the superimposed fitted normal cdf. The GLIM code is given below:

```
$input 'galaxy' $
$calc i = %cu(1) : n = 83 $
$use binocisim i n $
$var 104 vv $calc vv = 9 + 26*%cu(1)/104 $
$calc  np = %np((vv - 20.83)/4.54)$
$graph (s = 1 h = 'velocity' v = 'probability')
       pl__,pu__,np v,v,vv 9,9,10 $
```

It is clear that a single normal distribution is not appropriate.

An immediate question with all mixture analyses is—what kernel density should be used? We follow previous discussions of these data and fit mixtures of normals with both equal and unequal variances; the latter are fitted by the same approach, but the common standard deviation σ is replaced by component-specific standard deviations σ_k. A different macro normvar is used for the model fitting, but the model specification is otherwise the same:

```
$input 'alldist' $
$macro fixed 1 $endm
$macro random 1 $endm
$use mass1 $
```

```
$use remnp v $
$use mass2 $use remnp $
...
$use mass10 $use remnp $
$input 'normvar' $
$use mass1 $
$use remnp v $
$use mass2 $use remnp $
...
$use mass8 $use remnp $
```

Table 7.1 gives the component means, standard deviations and proportions, and the disparity, for up to eight components.

Identifying the global maximum for each K requires careful searching. To do this we need different starting values of the mass-points for different K; these are set by varying the scalar `tol_` between 0.2 and 1.

For the unequal variance case, two of the component standard deviations approach zero at $K = 6$ (they correspond to two nearly identical observations in each component), and the same "spikes" appear for larger K. As we increase the number of components beyond six, these two components, and the extreme components with means 9.71 and 33.04, remain stable up to $K = 8$, while the main body of observations is split into further closely-spaced components with very little change in maximized likelihood. Beyond $K = 8$ the extreme groups are split further into components containing single observations. For the unequal variance model, at most five components (with 14 parameters) appear meaningful.

For the equal variance case, for $K \geq 6$ the same mean values for the components appear, though with slightly different masses. The six-component equal-variance model has 12 parameters and fits only slightly worse than the five-component unequal-variance model. For all models with $K > 2$, the smallest 7 and largest 3 observations form stable components (note that $\frac{7}{82} = 0.085$, $\frac{3}{82} = 0.037$). For $K > 3$ the central group of observations is split into successively smaller groups, with the largest subgroup stable at a mean of around 19.8. The evidence for at least three components looks quite strong.

The equal- and unequal-variance models can be compared by the LRT to assess the variance heterogeneity. Although the mixture models are non-regular in their mean and proportion parameters, the hypothesis of equal variances is interior to the parameter space and does not possess any singularity (though by analogy with standard tests of variance heterogeneity, the test may be affected badly by departures from the specified null mixture distribution). Comparing the deviances for the two models for each K, the unequal-variance model appears to be needed: the homogeneity of variance test gives a deviance difference of 18.40 on 2 df for $K = 3$, 11.45 on 3 df for $K = 4$ and 18.58 on 4 df for $K = 5$.

Table 7.1. Mixture estimates

		Equal variances				Unequal variances			
K	k	Mean	Prop.	sd	Disparity	Mean	Prop	sd	Disparity
1	1	20.83	1	4.54	480.83	20.83	1	4.54	480.83
2	1	21.49	0.533	4.49	480.83	21.35	0.740	1.88	440.72
	2	20.08	0.467			19.36	0.260	8.15	
3	1	32.94	0.037	2.08	425.36	33.04	0.037	0.92	406.96
	2	21.40	0.877			21.40	0.878	2.20	
	3	9.75	0.086			9.71	0.085	0.42	
4	1	33.04	0.037	1.32	416.50	33.05	0.037	0.92	395.43
	2	23.50	0.352			21.94	0.665	2.27	
	3	20.00	0.526			19.75	0.213	0.45	
	4	9.71	0.085			9.71	0.085	0.42	
5	1	33.04	0.037	1.07	410.85	33.05	0.036	0.92	392.27
	2	26.38	0.037			22.92	0.289	1.02	
	3	23.04	0.366			21.85	0.245	3.05	
	4	19.76	0.475			19.82	0.344	0.63	
	5	9.71	0.085			9.71	0.085	0.42	
6	1	33.04	0.037	0.81	394.58	33.04	0.037	0.92	365.15
	2	26.24	0.044			26.98	0.024	0.018	
	3	23.05	0.357			22.93	0.424	1.20	
	4	19.93	0.453			19.79	0.406	0.68	
	5	16.14	0.025			16.13	0.024	0.043	
	6	9.71	0.085			9.71	0.085	0.42	
7	1	33.04	0.037	0.66	388.87	33.04	0.037	0.92	363.04
	2	26.60	0.033			26.98	0.024	0.018	
	3	23.88	0.172			23.42	0.300	0.99	
	4	22.31	0.221			22.13	0.085	0.25	
	5	19.83	0.427			19.89	0.444	0.73	
	6	16.13	0.024			16.13	0.024	0.043	
	7	9.71	0.085			9.71	0.085	0.42	
8	1	33.04	0.037	0.59	388.18	33.04	0.037	0.92	360.99
	2	26.57	0.035			26.98	0.024	0.018	
	3	23.90	0.169			23.61	0.039	0.086	
	4	22.35	0.214			23.34	0.267	1.11	
	5	20.21	0.236			22.15	0.085	0.25	
	6	19.44	0.201			19.87	0.439	0.72	
	7	16.13	0.024			16.13	0.024	0.043	
	8	9.71	0.085			9.71	0.085	0.42	

However, bootstrapping this test gives a quite different picture. Generating 99 samples from the fitted three-component equal-variance model and then fitting the three-component equal- and unequal-variance models gives the following ordered deviance differences:

```
-14.56 -14.25 -13.05 -12.14 -10.32  -9.43  -8.80  -6.34
- 0.54 - 0.54   0.16   0.28   0.62   0.71   1.64   1.65
```

```
 1.74    2.20    2.33    2.35    2.38    2.59    2.71    2.92
 3.07    3.16    3.71    4.05    4.87    5.95    5.99    6.16
 6.30    7.92    8.68    8.86    9.48    9.69    9.73    9.97
10.37   10.86   11.29   11.89   11.96   12.14   12.32   14.11
14.27   14.46   17.02   17.41   17.43   17.83   18.50   18.69
18.87   19.79   19.79   20.01   20.10   22.86   23.81   24.39
24.70   25.57   27.24   27.50   28.23   28.33   28.76   29.07
29.33   30.60   30.92   31.28   32.20   32.81   32.98   33.13
33.42   33.43   37.41   37.83   38.49   39.54   40.06   40.74
41.23   43.49   43.80   49.57   53.38   57.32   57.55   79.88
80.59   82.85   90.50
```

The large negative values reflect the inability of the unequal-variance algorithm to find the point in the subspace of equal variances with much higher likelihood from the Gaussian quadrature-based starting values. This could be corrected by starting the EM algorithm for the unequal-variance model from the estimates of the equal-variance model.

Nevertheless it is clear that large values of the LRTS for variance homogeneity are quite likely: the observed value of 18.40 is the 55-th largest in the above table and so has a P-value of 0.46. This does not provide any convincing evidence of variance heterogeneity.

To assess the number of components needed, we again use the bootstrap test. We first compare the single normal model with the three-component equal variance model. The 19 simulated LRT statistics are computed for $K = 1$ to 3, using the estimated parameters are given below:

```
0.18    0.74    0.98    1.25    1.49    1.83    2.23    2.27
2.52    2.65    2.70    2.93    2.96    4.90    5.08    5.50
8.12    8.65   10.97
```

The observed LRTS is 55.47, far beyond the largest value. We reject the hypothesis $K = 1$ in favour of $K = 3$. We repeat the simulations using the model parameters for $K = 3$ and fit the three- and four-component equal-variance models, using 99 bootstrap samples instead of 19. The bootstrap code is a little more complicated in this case; the GLIM directives are shown below. The value of tol_ is set at 1, half of the within-component standard deviation.

```
$macro boot3
  $calc z = %sr(0) $
      : z1 = %le(z,0.086) $
      : z2 = %gt(z,0.086)*%le(z,0.963)   $
      : z3 = %gt(z,0.963) $
      :  u = 2.08*%nd(%sr(0)) $
      : yy = u + 9.75*z1 + 21.4*z2 + 32.94*z3 $
```

```
$use mass3 $use remnp yy $calc %a = f__ $
$use mass4 $use remnp $calc lrt(%i) = %a - f__ $
$calc %i = %i + 1 : %n = %n - 1 $
$endm

$var 99 lrt $
$var 82 yy $
$calc %n = 99 : %i = 1 $num tol_ = 1 $
$while %n boot3 $
$sort slrt lrt $
$print slrt $
 -0.005 -0.005 -0.004 -0.003 -0.003 -0.002 -0.002 -0.001
 -0.001 -0.001 -0.001 -0.001 -0.001 -0.000 -0.000 -0.000
 -0.000  0.001  0.033  0.066  0.080  0.155  0.071  0.719
  0.872  0.931  1.213  1.240  1.502  1.557  1.959  2.122
  2.228  2.401  2.554  2.585  2.697  2.745  3.065  3.223
  3.252  3.272  3.377  3.564  3.592  3.687  3.706  3.947
  3.981  4.084  4.711  4.946  4.979  5.148  5.440  5.453
  5.502  5.992  6.244  6.362  6.367  6.400  6.593  6.776
  6.916  6.990  7.205  7.361  7.436  7.500  7.826  7.965
  8.104  8.108  8.170  8.556  8.998  9.004 10.39  10.82
 11.54  11.75  12.78  14.23  15.05  17.59  18.81  19.72
 19.93  22.25  22.97  26.28  29.85  30.47  31.28  38.37
 39.41  40.31  51.39
```

The observed LRTS of 8.86 is the 77-th largest in the full set of 100, so it has a P-value of 0.24. Extending the bootstrapping further, we find the P-values (based on 99 bootstrap samples) for the 4–5, 5–6 and 6–7 comparisons are 0.10, 0.05 and 0.16. For a direct 3–6 comparison, the P-value of the observed LRTS of 30.88 is 0.11. Thus for the equal-variance model, the bootstrap LRT supports three components, with very weak evidence for six, given the over-statement of significance for larger numbers of components (McLachlan and Peel, 1997).

For unequal variances, which allow the same component mean separation but smaller standard deviations in one component and hence a higher likelihood, the bootstrap LRT was applied to the galaxy data by McLachlan and Peel in the discussion of Richardson and Green (1997); they reported P-values for 1–2, 2–3, 3–4, 4–5, 5–6 and 6–7 components of 0.01, 0.01, 0.01, 0.04, 0.02 and 0.22. So the bootstrap test with unequal variances supports four components, with weak evidence for six.

The fitted three-component equal-variance (solid curve) and unequal-variance (dotted curve) model cdfs are shown in Fig. 7.3 with the 95% simultaneous confidence band, while Fig. 7.4 gives the corresponding densities.

FINITE MIXTURE MODELS

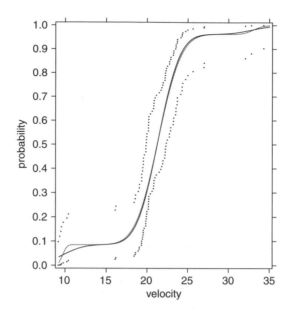

Fig. 7.3. Galaxy data cdf bounds and fitted three-component normal cdfs

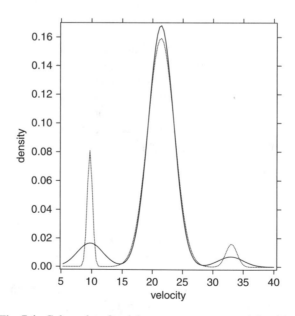

Fig. 7.4. Galaxy data fitted three-component normal densities

The GLIM code is given below:

```
$calc np3ev = 0.086*%np((vv -  9.75)/2.08) +
              0.877*%np((vv - 21.40)/2.08) +
              0.037*%np((vv - 32.94)/2.08) $
    : np3uv = 0.085*%np((vv -  9.71)/0.42) +
              0.878*%np((vv - 21.40)/2.20) +
              0.037*%np((vv - 33.04)/0.92) $
    : nd3ev = (0.086*%exp(-0.5*((vv - 9.75)/2.08)**2) +
              0.877*%exp(-0.5*((vv - 21.4)/2.08)**2) +
              0.037*%exp(-0.5*((vv - 32.94)/2.08)**2))
              /(2.08*%sqrt(2*%pi)) $
    : nd3uv = (0.085*%exp(-0.5*((vv -  9.71)/0.42)**2)/0.42 +
              0.878*%exp(-0.5*((vv - 21.4)/2.20)**2)/2.20 +
              0.037*%exp(-0.5*((vv - 33.04)/0.92)**2)/0.92)
              /%sqrt(2*%pi) $
$graph (s = 1  h = 'velocity' v = 'probability')
       pl__,pu__,np3ev,np3uv v,v,vv,vv 9,9,10,11 $
$qraph (s = 1  h = 'velocity' v = 'density')
       nd3ev,nd3uv vv 10,11 $
```

The fitted models are very similar, with the unequal-variance model fitting the small velocities slightly better. Increasing the number of components beyond four does not materially affect the fitted model: the additional complexity of the model is used only to interpolate, more and more precisely, the sample fluctuations in the empirical cdf (Aitkin, 2001).

Further discussion of mixture fitting for the galaxy data can be found in McLachlan and Peel (2000, pp. 194–196) and Aitkin (2001) who also compared several Bayesian analyses of the galaxy data.

7.9   Kernel density estimates

The normal mixture we have presented in this chapter is an alternative to the widely used *kernel density estimate* of a density, using a normal (Gaussian) kernel (see, e.g. Wand and Jones, 1994). Suppose we have observations $y_1, \ldots, y_n$ from an unknown continuous density (model) $f(y)$. We want to give a smooth estimate of the density, without a strong model assumption about it. The kernel density estimate $\tilde{f}(y)$ of f is defined by

$$\tilde{f}(y) = \frac{1}{nh} \sum_{i=1}^{n} K\left(\frac{y - y_i}{h}\right),$$

where $K(x)$ is the *kernel* density function. This is frequently taken as the normal $N(0, 1)$ density, though other kernels are also used; the choice of kernel is much less critical than the choice of the scale parameter h. We will consider only the normal kernel. The scale parameter (standard deviation) h is generally called the "bandwidth" parameter or "tuning constant". From our discussion of normal mixtures, the kernel density estimate can be immediately recognized as an equally-weighted n-component normal mixture, with component means equal to the observed values y_i, and common standard deviation h. It can be viewed alternatively as a *smoothing* of the NPML estimate of $f(y)$, which is just the *empirical mass function* with mass $1/n$ at y_i. The empirical mass function is smoothed out by assigning to y_i a normal density with positive standard deviation scaled by mass $1/n$ instead of a degenerate spike mass.

The difficulty in using this estimate is in the choice of the bandwidth h. From our discussion of the normal mixture with different variances, we can see that the same problem arises with the kernel density estimate. If we try to estimate h by ML, the maximized likelihood increases monotonically as $h \to 0$, and in the limit we recover the empirical mass function again. Considerable research has therefore been devoted to other ways of choosing the bandwidth, and this remains an active area of research.

In Fig. 7.5 we show the kernel density estimate for the galaxy data, for a range of values of the smoothing parameter h: 0.5, 1, 2, 4, together with the data. These were produced by the macro `kernel` which requires a specification of the bandwidth parameter—it does not compute an estimate of it. The macro stores

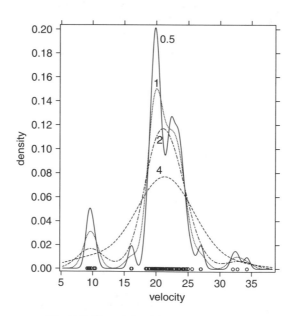

Fig. 7.5. Kernel densities for galaxy data

the fitted kernel density in f_ and the grid values of the variable in x_. The data are plotted at the zero value of the density.

```
$input 'kernel' $
$num  h = 0.5 $use kernel v h $calc f05 = f_ $
$calc h = 1    $use kernel      $calc f1  = f_ $
$calc h = 2    $use kernel      $calc f2  = f_ $
$calc h = 4    $use kernel      $calc f4  = f_ $
$var %sl zero $calc zero = 0 $
$graph (s = 1 h = 'velocity' v = 'density')
       f05,f1,f2,f4,zero x_,x_,x_,x_,v 10,11,12,13,5 $
$gtext 0.19 20 '0.5' : 0.15 20 '1' : 0.11 20 '2' : 0.075 20 '4' $
```

The density estimates become progressively more "peaked" as the bandwidth h decreases. At $h = 4$, near the MLE of σ for $K = 1$, the density is unimodal. At $h = 2$, near the MLE of σ for $K = 3$, a three-mode structure is clearly visible. At $h = 1$, near the MLE of σ for $K = 5$, the density is still trimodal, but with "shoulders". At $h = 0.5$, near the MLE of σ for $K = 7$, the density has seven modes. The choice of bandwidth is clearly critical if the density estimate is to be used to indicate multi-modality consistent with a mixture.

In the next chapter we consider the application of mixtures in *random effect models*.

8
Random effect models

8.1 Overdispersion

In Chapter 5 we found overdispersion in the fabric fault data; the Poisson GLM did not fit or represent the data adequately. The failure of a GLM to fit may be due to several causes. The distribution of Y may not be the specified exponential family member, or the regression model fitted may be mis-specified. A fruitful way of expressing the problem, which unifies these two possibilities, is through *omitted variables* from the regression model: the full model we have fitted is missing one or more important variables. If these have been recorded then they can be added to the model and fitted in the usual way (e.g. interactions). The common reality however, especially in survey or observational studies, is that we do not know what variables *should* have been recorded or measured to construct an adequate model.

We now make the (model) assumption that there is another set (vector) of unobserved variables $\mathbf{z}$, in addition to the observed set $\mathbf{x}$ in our model, and that the "correct" linear predictor is $\boldsymbol{\beta}'\mathbf{x} + \boldsymbol{\gamma}'\mathbf{z}$, where $\boldsymbol{\gamma}$ is the vector of regression coefficients for the unobservable variables. Since $\mathbf{z}$ was not measured, we know nothing about it: it is varying over the dataset in an unknown way, and is therefore a *random* vector as far as we are concerned. Since $\boldsymbol{\gamma}$ is also an unknown vector, the term $\boldsymbol{\gamma}'\mathbf{z}$ is in fact a *single* (scalar) unobserved random variable or *random effect* Z, and so we may write the linear predictor without any loss of generality as $\boldsymbol{\beta}'\mathbf{x} + Z$, where the scalar variable Z is random with an unknown distribution. This model including Z is called a *random effect model*. It should be noted that we *have* imposed a model restriction, namely that Z appears *additively* in the model on the scale of the linear predictor, that is, that Z does not interact with $\mathbf{x}$. We will relax this restriction later. We call the model for Y an *overdispersed* GLM.

We have also assumed that the distribution of Z does not depend on $\mathbf{x}$, for example, through a regression of Z on $\mathbf{x}$. Under the assumption of linear dependence, however, we can still proceed. Suppose that the distribution of Z does in fact depend on $\mathbf{x}$ through a location parameter, and so can be written as $g(z - \boldsymbol{\delta}'\mathbf{x})$. Then we may write

$$m(y) = \int f(y \mid z)g(z - \boldsymbol{\delta}'\mathbf{x})\,\mathrm{d}z = \int f(y \mid z^*)g(z^*)\,\mathrm{d}z^*,$$

where the distribution of $Z^* = Z - \delta'\mathbf{x}$ does not depend on $\mathbf{x}$, and the dependence of Y on Z^* is through the linear predictor

$$\eta = \beta'\mathbf{x} + Z = \beta'\mathbf{x} + Z^* + \delta'\mathbf{x} = \beta^{*'}\mathbf{x} + Z^*,$$

where $\beta^* = \beta + \delta$. So if the distribution of Z does have a regression on $\mathbf{x}$ of this form, we can still fit the model assuming that Z^* is independent of $\mathbf{x}$, but the regression coefficient β will be affected by the dependence. A substantial change in the estimate of the regression coefficient β, when we allow for overdispersion, is a warning that the omitted variable(s) represented by Z may be correlated with the explanatory variables $\mathbf{x}$, and so the regression of Y on $\mathbf{x}$ integrated over Z may be quite different from the conditional regression of Y on $\mathbf{x}$ given Z.

We now restrict consideration to the case of "homogeneous" Z, with distribution independent of $\mathbf{x}$. Since Z is unobserved, we can observe only the distribution of Y, which is now a *compound* or *mixed* exponential family distribution. We have lost the original distribution by compounding. Formally,

$$m(y) = \int f(y \mid z)g(z)\,\mathrm{d}z,$$

where $g(z)$ is the density or mass function of Z. Consequently

$$E[Y] = E[E[Y \mid Z]]$$
$$\mathrm{Var}[Y] = E[\mathrm{Var}[Y \mid Z]] + \mathrm{Var}[E[Y \mid Z]].$$

Write $\mu(Z)$ and $V(Z)$ for the mean and variance of Y given Z; then

$$E[Y] = E[\mu(Z)]$$
$$\mathrm{Var}[Y] = E[V(Z)] + \mathrm{Var}[\mu(Z)].$$

The variance of Y is always *inflated* by the compounding with Z, and the mean/variance relationship of the original ("kernel") distribution of Y is lost by the compounding. For example, if $Y \mid Z$ has a Poisson distribution with $V(Z) = \mu(Z)$ and log-linear model $\log \mu(Z) = \beta'\mathbf{x} + Z = \eta + Z$, then

$$E[Y] = E[e^{\eta+Z}] = e^{\eta}M(1)$$
$$\mathrm{Var}[Y] = E[e^{\eta+Z}] + \mathrm{Var}[e^{\eta+Z}]$$
$$= e^{\eta}M(1) + e^{2\eta}[M(2) - M^2(1)],$$

where $M(t)$ is the moment-generating function of Z. Thus

$$\mathrm{Var}[Y] = E[Y] + \phi E^2[Y],$$

where $\phi = \left[M(2)/M^2(1)\right] - 1$, for any distribution of Z. These results can be expressed equivalently, and very simply, in terms of the distribution of

$W = e^Z$: $M(1)$ is the mean, and $M(2) - M^2(1)$ is the variance of W, so ϕ is the *squared coefficient of variation* of W. (We use $g(z)$ for the pdf of the random effect distribution and $g(\mu)$ for the link function. Since these rarely appear in the same expression we hope the reader will not be confused.)

In general, for non-linear link functions g, the mean of the marginal distribution of Y no longer has the same link to the explanatory variables. Writing h for the inverse link function, using a second-order Taylor expansion we have for the random effect model

$$\mu(Z) = h(\eta + Z) \approx h(\eta) + Zh'(\eta) + \tfrac{1}{2}Z^2h''(\eta).$$

So for the compound model

$$E[Y] = E[\mu(Z)] \approx h(\eta) + \tfrac{1}{2}\sigma^2h''(\eta),$$

where σ^2 is the variance of Z and the mean of Z is taken as zero without any loss of generality (since a non-zero mean could be included with the intercept term in the regression model). Then $g(E[Y]) = h^{-1}(E[Y]) \neq \eta$ in general. However, if σ is small relative to the variation in η the link will still hold approximately, and for two special links the original link holds exactly: for the identity link we have immediately

$$E[Y] = \eta + E[Z] = \eta,$$

and for the log link

$$E[Y] = E[e^{\eta+Z}] = e^\eta M(1) = e^{\eta+\delta},$$

where $\delta = \log M(1)$. In this case, the intercept β_0 of the regression model is affected but the regression coefficients for the explanatory variables are not: the mean of Y still has a log-linear model in these variables.

Without a specific distributional assumption for Z, we cannot proceed directly to fit the model by ML, and are limited to *quasi-likelihood* (QL) analyses (Wedderburn, 1974; McCullagh and Nelder, 1989) which use only the mean and variance of Y. We do not discuss this approach further in this book, since we emphasize the use of specific models for Z. We note, however, that the QL approach is not sensitive to variations in the distribution of Z: for example, in the Poisson model above, the normal and the log-gamma distributions for Z, scaled to have the same mean and variance, will have the same value of ϕ, though the marginal distributions of Y and therefore the likelihoods will be different.

8.1.1 *Testing for overdispersion*

In the general framework of random effect modelling in this chapter, we use the LRT discussed below for model simplification, including testing for the presence of overdispersion. Earlier treatments of this model generally used the score test, since as we noted in Chapter 2 this does not require the fitting of the overdispersed model, only that of the null hypothesis model without overdispersion. We illustrate the score test with a brief example of heterogeneity in a simple Poisson model.

Poisson heterogeneity
Consider the heterogeneous Poisson model $Y_i \sim P(\mu_i)$, where there is no specific model for the heterogeneity in terms of explanatory variables. We want a test for heterogeneity, against the null homogeneous model $Y_i \sim P(\mu)$. We write $\log \mu_i = \theta + \delta_i$; the null hypothesis is $H_0 : \delta_i = 0$ for all i, and the alternative hypothesis is H_1 : not all $\delta_i = 0$. Here θ is the "nuisance parameter" of Chapter 2 and the δ_i are the parameters of interest. This model is over-parametrized with $n + 1$ parameters for n observations. We set $\delta_1 = 0$ for identifiability.

The log-likelihood and its derivatives are

$$\ell = \sum_i [-\mu_i + y_i \log \mu_i]$$

$$\frac{\partial \ell}{\partial \theta} = \sum_i [-\mu_i + y_i]$$

$$\frac{\partial \ell}{\partial \delta_i} = -\mu_i + y_i$$

$$\frac{\partial^2 \ell}{\partial \theta^2} = -\sum_i \mu_i$$

$$\frac{\partial^2 \ell}{\partial \theta \partial \delta_i} = -\mu_i$$

$$\frac{\partial^2 \ell}{\partial \delta_i^2} = -\mu_i.$$

The score components evaluated at the MLE $\hat{\mu}_i = e^{\hat{\theta}_0} = \bar{y}$ under H_0 give

$$\mathbf{s}_0' = (0, y_2 - \bar{y}, \ldots, y_n - \bar{y}),$$

and the expected information is

$$\mathcal{I} = \begin{bmatrix} \mu_+ & \mu_2 & \mu_3 & \cdots & \mu_n \\ \mu_2 & \mu_2 & 0 & \cdots & 0 \\ \mu_3 & 0 & \mu_3 & \cdots & 0 \\ \cdots & \cdots & \cdots & \cdots & \cdots \\ \mu_n & 0 & 0 & \cdots & \mu_n \end{bmatrix},$$

where $\mu_+ = \sum_1^n \mu_i$. Evaluating this under H_0 gives

$$
\mathcal{I}_0 = \begin{bmatrix}
n\bar{y} & \bar{y} & \bar{y} & \cdots & \bar{y} \\
\bar{y} & \bar{y} & 0 & \cdots & 0 \\
\bar{y} & 0 & \bar{y} & \cdots & 0 \\
\bar{y} & 0 & 0 & \cdots & \bar{y}
\end{bmatrix}
$$

with inverse matrix

$$
\mathcal{I}_0^{-1} = \frac{1}{\bar{y}} \begin{bmatrix}
1 & -n\mathbf{1}' \\
-n\mathbf{1} & \mathbf{I} - \mathbf{11}'
\end{bmatrix},
$$

where $\mathbf{1}$ is the unit vector of length $n - 1$, and the score test statistic is

$$
\mathbf{s}_0' \mathcal{I}_0^{-1} \mathbf{s}_0 = \frac{\sum_{i=1}^n (y_i - \bar{y})^2}{\bar{y}}.
$$

Under H_0 the score statistic has an asymptotic χ_{n-1}^2 distribution.

Since we concentrate on the formulation and fitting of the alternative model, we use the LRT where it is available. While the score test is sometimes *locally* most powerful (i.e. in the near neighbourhood of the null hypothesis), when departures from the null hypothesis are large, the score test may have substantially lower power than the LRT, and in any case we need to use the alternative model.

The normal model, and any other *specific* parametric model for Z, allows the use of the LRT for testing the null hypothesis $\sigma = 0$ with a *known* asymptotic distribution, given originally by Chernoff (1954) and frequently re-derived in various forms. For the general class of overdispersed models with a single overdispersion parameter, the LRTS is distributed asymptotically as an equal mixture of a degenerate distribution with mass 1 at 0, and a χ_1^2 distribution. The degenerate distribution arises because if the null exponential family model is correct but the overdispersed model is fitted, in large samples the estimate of σ will frequently be zero, on the boundary of the parameter space, and so the disparities for the null and overdispersed models will be the same.

Thus, a size α test can be obtained (asymptotically) as follows: if the LRTS is zero, accept H_0. If it is non-zero, compare it with the $100(1 - 2\alpha)\%$ point of the χ_1^2 distribution.

A comprehensive discussion of the LR and score tests for variance component models can be found in Verbeke and Molenberghs (2003).

We now consider several parametric distributions for the random effect Z.

8.2 Conjugate random effects

Each of the exponential family distributions has its own *conjugate* distribution for the canonical parameter θ of the distribution. The conjugate distribution for Z has the same analytic form in Z as the likelihood in Z from the distribution of

$Y \mid Z$, giving a simple form for the marginal distribution of Y. In some cases the conjugate distribution for Z can be used to model the extra variation.

8.2.1 Normal kernel: the t-distribution

For the random effect model for the normal mean, the conjugate distribution is also normal, but the resulting model is *unidentifiable*: if $Y \mid Z \sim N(\theta + Z, \sigma^2)$ conditionally and $Z \sim N(\mu, \phi^2)$ marginally, then $Y \sim N(\theta + \mu, \sigma^2 + \phi^2)$ marginally. If all the parameters are unknown, then μ is aliased with θ and ϕ^2 with σ^2: we simply have another unknown normal distribution marginally. So the normal distribution cannot be extended by conjugate compounding on the mean.

However, it *can* be extended by conjugate compounding on the *variance*. Let $Y \mid Z \sim N(\mu, \sigma^2/Z)$ with Z having a conjugate gamma $(r/2, 2/r)$ distribution with mean 1 as in Section 6.9. The marginal distribution of Y is then

$$
\begin{aligned}
m(y) &= \frac{(r/2)^{r/2}}{\sqrt{2\pi}\,\sigma\,\Gamma(r/2)} \int z^{(r-1)/2} \exp\left\{-z\left[\frac{r}{2} + \frac{(y-\mu)^2}{2\sigma^2}\right]\right\} dz \\
&= \frac{(r/2)^{r/2}\Gamma((r+1)/2)}{\sqrt{2\pi}\,\sigma\,\Gamma(r/2)} \left[\frac{r}{2} + \frac{(y-\mu)^2}{2\sigma^2}\right]^{-(r+1)/2} \\
&= \frac{\Gamma((r+1)/2)}{\sqrt{\pi r}\,\sigma\,\Gamma(r/2)} \left[1 + \frac{(y-\mu)^2}{r\sigma^2}\right]^{-(r+1)/2}
\end{aligned}
$$

which is a t-distribution of $(y - \mu)/\sigma$ with r degrees of freedom. Here Y has mean μ (provided $r > 1$) and variance $r\sigma^2/(r-2)$ (provided $r > 2$). The *Cauchy* distribution is the special case with $r = 1$: it is a *heavy-tailed* distribution with infinite mean and variance.

Thus, the t-distribution arises naturally as a conjugate extension of the normal— it is an *inverse gamma mixture of normals*, mixed on the variance. It is a candidate for modelling in the presence of *outliers*. We will illustrate this with an example, but first we show that the t-distribution can be fitted as a GLM with an additional iterative component to the algorithm. ML for a linear model $\mu_i = \boldsymbol{\beta}'\mathbf{x}_i$ in the t-distribution is readily achieved (Lange *et al.*, 1989). To simplify the log-likelihood derivatives, we first re-parametrize the distribution to $(\boldsymbol{\beta}, \phi, r)$ where $\phi = r\sigma^2$, and write

$$
e_i = y_i - \mu_i, \quad w_i = \left[1 + \frac{(y_i - \mu_i)^2}{\phi}\right]^{-1}.
$$

The log-likelihood for a regression model with n observations is:

$$
\ell = n\left[\log\Gamma\left(\frac{r+1}{2}\right) - \frac{1}{2}\log\pi - \log\Gamma\left(\frac{r}{2}\right) - \frac{1}{2}\log\phi\right] + \frac{r+1}{2}\sum_{i=1}^{n}\log w_i.
$$

The derivatives involve the expressions

$$\frac{\partial w_i}{\partial \boldsymbol{\beta}} = \frac{2}{\phi} w_i^2 e_i \mathbf{x}_i$$

$$\frac{\partial w_i}{\partial \phi} = \frac{1}{\phi^2} w_i^2 e_i^2,$$

while w_i is not a function of r. Then

$$\frac{\partial \ell}{\partial \boldsymbol{\beta}} = \frac{r+1}{\phi} \sum_i w_i e_i \mathbf{x}_i$$

$$\frac{\partial \ell}{\partial \phi} = -\frac{n}{2\phi} + \frac{r+1}{2\phi^2} \sum_i w_i e_i^2$$

$$\frac{\partial \ell}{\partial r} = \frac{n}{2} \left[\psi\left(\frac{r+1}{2}\right) - \psi\left(\frac{r}{2}\right) \right] + \frac{1}{2} \sum_i \log w_i,$$

where ψ is the digamma function.

The first two score equations can be expressed as

$$\hat{\boldsymbol{\beta}} = \left(\sum_i w_i \mathbf{x}_i \mathbf{x}_i' \right)^{-1} \sum_i w_i \mathbf{x}_i y_i$$

$$= (X'WX)^{-1} X'W\mathbf{y}$$

and

$$\hat{\phi} = (\hat{r} + 1) \sum_i \frac{w_i e_i^2}{n},$$

giving

$$\hat{\sigma}^2 = \left(\frac{\hat{r}+1}{\hat{r}} \right) \sum_i \frac{w_i (y_i - \hat{\boldsymbol{\beta}}' \mathbf{x}_i)^2}{n}$$

$$= \left(\frac{\hat{r}+1}{\hat{r}} \right) \frac{\text{RSS}}{n},$$

where RSS is the *weighted* residual sum of squares from the fitted model. The MLE of $\boldsymbol{\beta}$ is a weighted least squares estimate with weights w_i, though since the weights are themselves functions of $\boldsymbol{\beta}$ the weighting has to be done iteratively. The MLE of σ^2 depends on $\hat{r}$, which is the solution of

$$\psi\left(\frac{r+1}{2}\right) - \psi\left(\frac{r}{2}\right) = -\sum_i \frac{\log w_i}{n}.$$

The effect of the weights w_i is to downweight the observations with large residuals e_i from the fitted model, so reducing the effect of outliers: the t-distribution provides a parametric model for *robust regression*.

By direct calculation of the second derivatives and their expected values, Lange *et al.* (1989) showed that the expected information matrix is block diagonal, with a zero β, (ϕ, r) off-diagonal block. Thus, in large samples the estimation of ϕ and r does not affect the precision of $\hat{\beta}$, and the iterative procedure gives the correct standard errors for $\hat{\beta}$.

Fitting the t-distribution with known r is particularly simple; we iterate until convergence between fitting the model with weights w_i and recomputing the weights from the current estimates of β and ϕ. This is implemented in the macro tmac. For unknown r, we can compute the profile likelihood over a grid of values and hence obtain the joint MLEs of β, ϕ and r. Alternatively, we can introduce an additional level to our procedure, iterating between the estimation of β and ϕ for known r, and solving the score equation in r directly.

The iteratively weighted least squares algorithm described above is closely related to the EM algorithm: if we regard the random effect Z as unobserved data, we can proceed by EM, maximizing the conditional expected complete-data log-likelihood. It is straightforward to show that the EM estimates of β and σ^2 are also weighted least squares estimates, with weights $(1 + 1/r)w_i$, which are a constant multiple of the w_i. This multiple does not affect the estimates.

Example—the Brownlee stack-loss data

Here we reproduce the analysis of Brownlee's (1965) stack-loss data from Lange *et al.* (1989). The stack-loss data have been analysed many times for outliers: the observations numbered 1, 3, 4 and 21 have frequently been identified as outliers. A detailed study of the data, with a full list of previous analyses, can be found in Dodge (1999). The data consist of 21 observations on stack-loss y (the loss of acid through the stack) in a chemical plant for the conversion of ammonia to nitric acid, with three explanatory variables: air flow x_1, cooling water inlet temperature x_2 and acid concentration x_3. The (coded) data are listed in Table 8.1, and are held in the file stackloss.

```
$input 'stackloss' $
$yvar y $fit x1 + x2 + x3 $disp e $
    deviance =   178.83
 residual df =    17
```

```
         estimate            s.e.     parameter
  1       -39.92            11.90         1
  2         0.716            0.135        X1
  3         1.295            0.368        X2
  4        -0.152            0.156        X3
scale parameter 2.918
```

The disparity is

$$n\{1 + \log(2\pi) + \log(\text{RSS}/n)\} = 104.58.$$

The variables x_1 and x_2 appear important but x_3 does not. Increasing levels of the important variables increases the stack-loss. We check the leverage of the individual observations.

```
$extract %lv $print %lv $
0.302   0.318   0.175   0.129   0.052   0.078   0.219   0.219   0.140
0.200   0.155   0.217   0.158   0.206   0.191   0.131   0.412   0.161
0.175   0.080   0.285
```

The average leverage, $(p+1)/n$, is 0.19, and twice this is 0.38. Only the 17th observation exceeds this value.

We now fit the t model with fixed r, over a grid of r using the macro tmac.

```
$input 'tmac' $
$macro model x1 + x2 + x3 $endm
$num r_ = 30 $use tmac y $
$num r_ = 20 $use tmac $
...
$num r_ = 0.3 $use tmac $
$num r_ = 0.2 $use tmac $
```

Table 8.1. Stack-loss data

Obs.	y	x_1	x_2	x_3	Obs.	y	x_1	x_2	x_3
1	42	80	27	89	12	13	58	17	88
2	37	80	27	88	13	11	58	18	82
3	37	75	25	90	14	12	58	19	93
4	28	62	24	87	15	8	50	18	89
5	18	62	22	87	16	7	50	18	86
6	18	62	23	87	17	8	50	19	72
7	19	62	24	93	18	8	50	19	79
8	20	62	24	93	19	9	50	20	80
9	15	58	23	87	20	15	56	20	82
10	14	58	18	80	21	15	70	20	91
11	14	58	18	89					

Table 8.2. Stack-loss estimates

r	1	x_1	x_2	x_3	σ	Disparity
∞	−39.9	0.716	1.30	−0.152	2.92	104.58
30	−40.2	0.740	1.21	−0.145	2.81	104.50
20	−40.3	0.754	1.17	−0.142	2.75	104.43
10	−40.7	0.793	1.03	−0.133	2.56	104.12
8	−40.7	0.811	0.968	−0.129	2.46	103.92
6	−40.7	0.835	0.876	−0.124	2.30	103.56
5	−40.5	0.848	0.817	−0.120	2.19	103.27
4	−40.1	0.857	0.746	−0.115	2.03	102.85
3	−39.1	0.854	0.657	−0.104	1.76	102.14
2	−38.1	0.848	0.557	−0.089	1.34	100.63
1.1	−38.4	0.852	0.491	−0.071	0.93	99.14
1	−38.6	0.852	0.489	−0.068	0.88	99.16
0.5	−40.8	0.840	0.536	−0.044	0.38	101.09
0.4	−40.1	0.834	0.562	−0.055	0.18	101.912
0.3	39.9	0.833	0.567	−0.058	0.043	100.528
0.2	−39.7	0.834	0.566	−0.060	0.000	67.625

The estimates and disparities are shown in Table 8.2. The likelihood is fairly flat for large r but more peaked near $\hat{r} = 1.1$. The change in disparity between the normal model at $r = \infty$ and that at $\hat{r}$ is 5.44. The disparity is not monotone for $r < 1$ – it increases as r decreases from $r = 1.1$ to $r = 0.4$, and then decreases again with further decrease in r. With decreasing r, x_1 becomes more important and x_2 and x_3 less important. We now examine the weights and leverage values for $r = 1.1$.

```
$calc r_ = 1.1 $use tmac y $
 -2 log Lmax =   99.138 with sigma =    0.929
       and scale parameter    1.100
    deviance =   9.4870 (change =  -0.01071)
 residual df =   17      (change =   0     )

            estimate        s.e.      parameter
     1       -38.44        3.255        1
     2        0.8520       0.04589      X1
     3        0.4910       0.1180       X2
     4       -0.07128      0.04387      X3
  scale parameter 0.5581


$print w_ $
0.032  0.914  0.028  0.014  0.491  0.301  0.762  0.817  0.451
0.987  0.771  0.997  0.096  0.250  0.342  0.982  0.882  0.979
0.643  0.234  0.010
```

```
$extract %lv $print %lv $
0.025   0.756   0.014   0.004   0.058   0.049   0.292   0.312   0.118
0.320   0.230   0.422   0.026   0.092   0.114   0.219   0.502   0.230
0.183   0.028   0.007
```

The weights on observations 1, 3, 4 and 21 are very low, below 0.033, and that on observation 13 is 0.096. As small weights correspond to large residuals, these points are identified as outliers, downweighted and effectively excluded from this robust regression model fit. An issue of concern is that the total weight may be substantially less than n: here it is 10.98. Nearly half the data have been "lost" in the downweighting, and the parameter estimates become more strongly dependent on the remaining observations with high weights and leverages, particularly 2, 12 and 17.

For smaller values of r these effects become extreme. The weights and leverage values for $r = 0.5$ are:

```
0.003   0.954   0.002   0.001   0.047   0.023   0.066   1.000   0.036
1.000   0.310   0.846   0.008   0.018   0.053   0.995   0.703   0.791
0.145   0.024   0.001


0.002   0.868   0.001   0.001   0.009   0.006   0.043   0.658   0.017
0.404   0.142   0.535   0.003   0.011   0.027   0.339   0.577   0.284
0.068   0.004   0.001
```

The regression is now being determined only by observations 2, 8, 10, 12, 16, 17 and 18—the total weight is only 7.03. At $r = 0.2$ only four weights are 1.0, on observations 2, 8, 12 and 18, and all the other weights are zero. The four-parameter model fits almost exactly the total observation weight of four, with a variance estimate of almost zero and an almost undefined disparity.

We consider the stack-loss data further below (Section 8.7.4) with a different random effect model, and come to a quite different conclusion.

8.2.2   Poisson kernel: the negative binomial distribution

For the Poisson distribution of $Y \mid Z$, with $\log \mu(Z) = \eta + Z$, the conjugate distribution of Z is the log-gamma. It is simpler to work with $W = e^Z$, so that $\log \mu(W) = \eta + \log W$. W has a gamma distribution, which we take in the standard form (Section 6.9)

$$g(w) = \frac{r^r}{\Gamma(r)} e^{-rw} w^{r-1}, \quad w, r > 0,$$

so that W has mean 1 and variance $1/r$. The mean of the corresponding log-gamma distribution of Z is $\psi(r) - \log r$ which is approximately $-1/2r$, and the variance is $\psi'(r)$ which is approximately $1/r + 1/2r^2$.

The marginal distribution of Y is then negative binomial, a *gamma mixture of Poissons*:

$$
\begin{aligned}
m(y) &= \frac{r^r}{\Gamma(r)y!} \int_0^\infty e^{-w e^\eta} (w\, e^\eta)^y e^{-rw} w^{r-1}\, dw \\
&= \frac{r^r e^{\eta y}}{\Gamma(r)y!} \int_0^\infty e^{-w(e^\eta + r)} w^{y+r-1}\, dw \\
&= \frac{r^r e^{\eta y}}{\Gamma(r)y!} \frac{\Gamma(y+r)}{(e^\eta + r)^{y+r}} \\
&= \frac{\Gamma(y+r)}{\Gamma(r)y!} \left(\frac{e^\eta}{e^\eta + r}\right)^y \left(\frac{r}{e^\eta + r}\right)^r,
\end{aligned}
$$

with mean $\mu = e^\eta$ and variance $\mu + \mu^2/r$. To fit the negative binomial as a GLM we write

$$
m(y) = \frac{\Gamma(y+r)}{\Gamma(r)y!} p^y (1-p)^r
$$

with

$$
p = \frac{e^\eta}{e^\eta + r}
$$

so that

$$
\operatorname{logit} p = \eta - \log r.
$$

The intercept is altered in this formulation, but the other regression coefficients are unaffected.

The likelihood for a sample $(y_1, \mathbf{x}_1), \ldots, (y_n, \mathbf{x}_n)$ is then

$$
L(\boldsymbol{\beta}, r) = \prod_{i=1}^n \frac{\Gamma(y_i + r)}{\Gamma(r)y_i!} p_i^{y_i} (1-p_i)^r
$$

with

$$
\operatorname{logit} p_i = \boldsymbol{\beta}' \mathbf{x}_i - \log r.
$$

To fit the model, if r were known $\boldsymbol{\beta}$ could be estimated directly from a binomial logit model for p_i, treating y_i as the number of successes in $y_i + r$ Bernoulli trials. The intercept would then be an estimate of $\beta_0 - \log r$. The correct intercept is achieved by using an offset of $-\log(r)$ in the model. For the estimation of r, we first consider the information matrix for β and r. From the log-likelihood

$$\ell(\beta, r) = \sum \{\log \Gamma(y_i + r) - \log \Gamma(r) - \log y_i! + y_i \eta_i + r \log r$$
$$- (y_i + r) \log(e^{\eta_i} + r)\}$$

we have

$$\frac{\partial \ell}{\partial r} = \sum \left\{ \psi(y_i + r) - \psi(r) + 1 + \log r - \left(\frac{y_i + r}{e^{\eta_i} + r} + \log(e^{\eta_i} + r) \right) \right\}$$

$$\frac{\partial^2 \ell}{\partial r^2} = \sum \left\{ \psi'(y_i + r) - \psi'(r) + \frac{1}{r} - \left(\frac{e^{\eta_i} - y_i}{(e^{\eta_i} + r)^2} + \frac{1}{e^{\eta_i} + r} \right) \right\}$$

$$\frac{\partial^2 \ell}{\partial \beta \partial r} = -\sum \left\{ -\frac{y_i + r}{(e^{\eta_i} + r)^2} e^{\eta_i} + \frac{e^{\eta_i}}{e^{\eta_i} + r} \right\} \mathbf{x}_i.$$

To evaluate the expected information, we note that

$$E[Y_i] = E[E[Y_i \mid W_i]] = E[\mu_i(Z_i)] = E[W_i e^{\eta_i}] = e^{\eta_i}$$

and hence

$$E\left[\frac{\partial^2 \ell}{\partial \beta \partial r} \right] = 0$$

$$E\left[\frac{\partial^2 \ell}{\partial r^2} \right] = \sum \left\{ E[\psi'(Y_i + r)] - \psi'(r) + \frac{1}{r} - \frac{1}{e^{\eta_i} + r} \right\}.$$

Since the expected information is diagonal, full ML estimation is easily attained by successive relaxation, alternating between estimating β for fixed r and r for fixed β, the latter by a simple Newton algorithm as for the gamma distribution in Chapter 6. The expectation of the trigamma function in the second equation has no standard form and it is therefore simpler to work with the observed information in this parameter. This is implemented in the macro negbinom. Note also that the estimation of r does not affect the precision of $\hat{\beta}$ asymptotically, because of the diagonal form of the information matrix.

Example: the fabric fault data
In Chapter 5 we considered the data on fabric faults and modelled them by the Poisson distribution. We saw that the residual deviance from the model was quite large (64.54 on 30 df), suggesting that other relevant factors were varying over the data. We now fit the Poisson, and the negative binomial distribution using the macro negbinom, and save the fitted values and the fitted

two-standard deviation bands around the fitted values, for both distributions. The fitted values from the negative binomial macro are stored in the vector nbfv_. The macro prints out a deviance relative to a saturated Poisson model and can thus be used to compare the negative binomial to a Poisson model. The standard deviation printed out is the approximate standard deviation of Z, that is, $\sqrt{1/r + 1/2r^2}$.

```
$input 'faults' : 'negbinom' $
$calc ll = %log(l) $
$macro model ll $endm
$yvar n $err p $link l $
$fit #model $disp e$

scaled deviance =  64.537  at cycle 4
   residual df =  30

          estimate        s.e.      parameter
      1     -4.173        1.135       1
      2      0.9969       0.1759      LL
scale parameter 1.000


$calc pfv = %fv $
$sort syv,spfv,sl %yv,pfv,l l $
$calc psd = %sqrt(spfv) $
    : pul = spfv + 2*psd $
    : pll = spfv - 2*psd $
$use negbinom $

scaled deviance =  30.402 at cycle 3
   residual df =  30

          estimate        s.e.      parameter
      1     -3.793        1.424       1
      2      0.9374       0.2225      LL
scale parameter 1.000


 - 2 log Lmax =     48.30
Deviance relative to saturated Poisson =     52.09
ML estimate of r =    8.521
Standard deviation of random effect on the log scale is   0.3529

$sort snbfv nbfv_ l $
$calc nbsd = %sqrt(snbfv + snbfv**2/r_) $
    : nbul = snbfv + 2*nbsd $
    : nbll = snbfv - 2*nbsd $
```

```
$macro model 1 $endm
$use negbinom $

  scaled deviance =   32.660 at cycle 3
      residual df =   31

            estimate          s.e.       parameter
      1          2.183        0.1056        1
scale parameter 1.000

  - 2 log Lmax =       63.97
Deviance relative to saturated Poisson =       67.76
ML estimate of r =     4.091
Standard deviation of random effect on the log scale is    0.5260
```

The fitted negative binomial model has deviance 52.09 (the GLIM scaled deviance should be ignored), a reduction of 12.45 compared to the Poisson model. The parameter estimates are slightly affected by the overdispersion modelling, with a lower slope and a higher intercept. The standard error of $\hat{\beta}_1$ is increased by 30%. (Note that we have two standard error estimates here: dropping the roll-length variable and fitting the null negative binomial model gives a deviance change of $67.76 - 52.09 = 15.67$, and the corresponding standard error of $\hat{\beta}_1$ is $0.937/\sqrt{15.67} = 0.237$, slightly larger than the information-based standard error of 0.223. The profile likelihood in β_1 is slightly skewed.)

In Fig. 8.1 we show the fitted Poisson and negative binomial models, together with 2SD limits around the fitted values. We restrict the vertical scale to non-negative values; the negative values within the variability bounds reflect the asymmetry of the Poisson and negative binomial distributions, and the unsuitability of 2SD limits for these distributions for small means.

The fitted values agree closely, despite the differences in the parameter estimates, but the variability bounds for the negative binomial are substantially wider than those for the Poisson.

```
$gstyle   5 line 0 sym 5 col 1 $
$gstyle   1 1     1 s    0 c    1 $
$gstyle   2 1     2 s    0 c    1 $
$gstyle   3 1     3 s    0 c    1 $
$graph (s = 1 h = 'length' v = 'faults' y = 0,30)
        syv,spfv,snbfv,spul,spll,nbul,nbll sl 5,1,1,2,2,3,3 $
```

For this example, the deviance for the Poisson model is 64.54 on 30 df, and that for the negative binomial model is 52.09. The difference of 12.45 is the LRTS, which is very large compared with $\chi^2_{1,0.90} = 2.71$. There is no doubt about the presence of overdispersion.

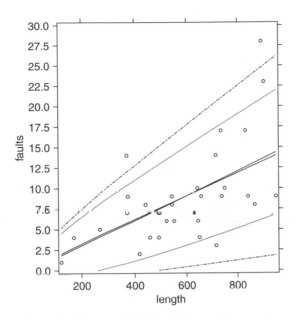

Fig. 8.1. Poisson and NB fits and variability bounds

8.2.3 *Binomial kernel: beta-binomial distribution*

Consider first the natural parametrization of the binomial distribution in terms of the logit parameter $\theta = \log[\,p/(1-p)]$. The probability of y successes in n trials is

$$P(y \mid n, p) = \binom{n}{y} p^{y}(1-p)^{n-y}$$

$$= \binom{n}{y} \frac{e^{\theta y}}{(1 + e^{\theta})^{n}}.$$

For a simple null random effect model $\theta = Z$ the conjugate distribution for Z is the logistic, equivalent to the beta distribution on the p scale, giving the marginal *beta-binomial* distribution for Y as

$$m(y) = \binom{n}{y} \frac{1}{B(a,b)} \int_{-\infty}^{\infty} \frac{e^{z(y+a)}}{(1 + e^{z})^{n+a+b}} \, dz$$

$$= \binom{n}{y} \frac{B(y + a, n - y + b)}{B(a,b)},$$

a hypergeometric distribution. However, for any non-null model $\theta = \eta + Z$, with η the linear predictor and Z having the above logistic distribution, we have

$$m(y) = \binom{n}{y} \frac{1}{B(a,b)} \int_{-\infty}^{\infty} \frac{e^{(\eta+z)y}}{(1+e^{\eta+z})^n} \cdot \frac{e^{az}}{(1+e^z)^{a+b}} \, dz$$

and the integral cannot be evaluated explicitly. Thus conjugate random effect modelling is limited to the trivial null model in this distribution.

This difficulty can be circumvented by using a different model for the extra variation (Crowder, 1978, 1979). Suppose that $Y_i \mid P_i \sim b(n_i, P_i)$ conditionally, while $P_i \sim \text{beta}(a_i, b_i)$ marginally, with mean $\mu_i = a_i/(a_i + b_i)$ and variance $\phi_i \mu_i(1 - \mu_i)$, where $\phi_i = (a_i + b_i + 1)^{-1}$. Then marginally Y_i has the beta-binomial (hypergeometric) distribution as before, with mean $E[Y_i] = n_i \mu_i$ and variance

$$V_i = n_i^2 \text{Var}[P_i] + E[n_i P_i(1 - P_i)]$$

$$= n_i^2 \phi_i \mu_i(1 - \mu_i) + n_i(1 - \phi_i)\mu_i(1 - \mu_i)$$

$$= n_i \mu_i(1 - \mu_i)(1 + (n_i - 1)\phi_i).$$

Without some restriction on the a_i and b_i, this model is unidentifiable. We therefore set the ϕ_i equal to ϕ. The model with logistic link for μ_i can now be fitted by QL through the mean–variance relationship provided that ϕ can be estimated in some way. ML fitting of the model is however very difficult since the beta distribution parameters appearing in the likelihood have a complex dependence on β and ϕ. We do not consider this approach further.

8.2.4   Gamma kernel

For a gamma distribution we have

$$f(y \mid \theta, r) = \frac{\theta^r}{\Gamma(r)} e^{-\theta y} y^{r-1},$$

where the mean of the distribution is $\mu = r/\theta$. Assuming the random variation is in the mean of the gamma distribution, with the parameter r constant, the conjugate distribution for the canonical parameter θ is again the gamma distribution. For a simple null random effect model $\theta = Z$, we have for the conjugate distribution

$$g(z) = \frac{\phi^s}{\Gamma(s)} e^{-\phi z} z^{s-1}$$

and the marginal distribution of Y is

$$
\begin{aligned}
m(y \mid r, s, \phi) &= \frac{y^{r-1}}{\Gamma(r)} \frac{\phi^s}{\Gamma(s)} \int_0^\infty e^{-z(\phi+y)} z^{r+s-1} \, \mathrm{d}z \\
&= \frac{1}{y} \frac{\Gamma(r+s)}{\Gamma(r)\Gamma(s)} \left(\frac{y}{\phi+y}\right)^r \left(\frac{\phi}{\phi+y}\right)^s
\end{aligned}
$$

which is essentially a beta distribution of $y/(\phi+y)$. We do not consider this further.

8.2.5 Difficulties with the conjugate approach

The conjugate random effect approach is commonly used with the Poisson distribution: the negative binomial is a popular model in many application areas, for example, consumer purchasing (Chatfield and Goodhart, 1970). The t-distribution is also used increasingly as a long-tailed distributional model. However, the conjugate approach does not work in binomial models and other approaches are necessary. Because of this lack of generality of the conjugate approach, we turn in the next section to using the normal distribution for the random effect. The resulting normal random effect models are widely used.

8.3 Normal random effects

The non-conjugate normal distribution for Z means that the likelihood of $Y \mid Z$ cannot be integrated analytically (we exclude the normal likelihood, as discussed in Section 8.2.1). The need for numerical integration of the likelihood by Gaussian quadrature has in the past discouraged the use of normal random effect models, but Bock and Aitkin (1981) and Hinde (1982) showed that the numerical integration can be expressed as a simple finite mixture of the type considered in Section 7.1, and so these models can be fitted straightforwardly using the EM algorithm.

As before, we have an unobserved random effect Z_i for the i-th observation, but now the Z_i are independently normally distributed $N(0, 1)$, and conditional on Z_i, Y_i has a GLM with linear predictor $\eta_i = \boldsymbol{\beta}'x_i + \sigma Z_i$. The likelihood is then

$$
L(\boldsymbol{\beta}, \sigma) = \prod_{i=1}^n \int f(y_i \mid \boldsymbol{\beta}, \sigma, z_i)\phi(z_i) \, \mathrm{d}z_i,
$$

where $\phi(z)$ is the standard normal density function.

Since the integral is not analytic except for $Y \mid z$ normal, we approximate it by Gaussian quadrature: we replace the integral over the normal Z by a finite sum over K Gaussian quadrature mass-points z_k with masses π_k; the z_k and π_k are given in standard references (Abramowitz and Stegun, 1970; Stroud and Sechrest, 1966).

458 RANDOM EFFECT MODELS

The likelihood is then

$$L(\boldsymbol{\beta}, \sigma) \approx \prod_{i=1}^{n} \sum_{k=1}^{K} \pi_k f(y_i \mid \boldsymbol{\beta}, \sigma, z_k).$$

The likelihood is thus (approximately) the likelihood of a finite mixture of exponential family densities with *known* mixture proportions π_k at *known* mass-points z_k, with the linear predictor for the i-th observation in the k-th mixture component being

$$\eta_{ik} = \boldsymbol{\beta}'x_i + \sigma z_k.$$

We may also regard this as the exact likelihood for this discrete mixing distribution for Z. This is inherently of interest because the NPML estimate of the mixing distribution, when it exists (Chapter 7) is a discrete distribution on a finite number of mass-points. In Section 8.6 we consider the joint estimation of $\boldsymbol{\beta}$, the π_k and the mass-points z_k, but for the moment consider the latter quantities as fixed.

ML in this model follows immediately from the results of Section 7.4, with some simplifications. We identify the ϕ parameter there with $\boldsymbol{\beta}$ and σ, and there are no unknown component-specific parameters θ_k. The mixture proportions π_k are known and do not have to be estimated.

The score component for $\boldsymbol{\beta}$ is

$$\frac{\partial \ell}{\partial \boldsymbol{\beta}} = \sum_i \frac{\sum_k \pi_k f_{ik}(\partial \log f_{ik}/\partial \boldsymbol{\beta})}{\sum_\ell \pi_\ell f_{i\ell}} = \Sigma_i \Sigma_k w_{ik} s_{ik}(\boldsymbol{\beta}),$$

where w_{ik} is the posterior probability that observation y_i comes from component k:

$$w_{ik} = \frac{\pi_k f_{ik}}{\sum_\ell \pi_\ell f_{i\ell}}$$

and $s_{ik}(\boldsymbol{\beta})$ is the $\boldsymbol{\beta}$-component of the score for observation i in component k:

$$s_{ik}(\boldsymbol{\beta}) = \frac{y_i - \mu_{ik}}{V_{ik} g'_{ik}} x_i$$

(Chapter 2). Similarly for σ,

$$s_{ik}(\sigma) = \frac{y_i - \mu_{ik}}{V_{ik} g'_{ik}} z_k.$$

Equating the score to zero gives likelihood equations which are simple weighted sums of those for an ordinary GLM with weights w_{ik}; alternately solving these equations for given weights w_{ik}, and updating these weights from the current parameter estimates, is an EM algorithm. The double summation over i and k is conveniently (if inefficiently) handled in GLIM by expanding the data vectors to length $K \times n$ by replicating y and $\mathbf{x}$ K times, with each block of values corresponding to a specific quadrature point, and in this expanded version each quadrature value is repeated n times (Hinde, 1982; Anderson and Aitkin, 1985). Model fitting is then identical to that of a single sample of size Kn with prior weight vector w. Initial estimates for the first E-step for β are conveniently obtained from the ordinary GLM fit, and for σ by arbitrary specification other than zero (e.g. $\sigma = 1$).

This approach is implemented in the macro `alldist`, mentioned in Chapter 7 for normal mixtures, and discussed at length in Aitkin and Francis (1995). We give examples of its use below. We note one peculiarity of Gaussian quadrature: the disparity does not reduce monotonically as we increase the number of mixture components K. This is because all the models have the same number of unknown parameters, since the mass-point locations and masses are known. Variations in the disparity are due to the better or worse representation of the extra variation by the discrete "normal" distributions with different numbers of components. As the number of mass-points increases these variations decrease, but it is not easy to give a value of the disparity for the "normal" random effect distribution, without using a very large number of mass-points. Crouch and Spiegelman (1990) found that for the logistic model with a normal random effect, even 20-point Gaussian quadrature may not give the integral accurately if the variance of Z is large. Lesaffre and Spiessens (2001) found variation in the integral with even larger numbers of mass-points, and recommended adaptive quadrature and routine investigation of this variation with the number of mass-points.

We address this problem below.

8.3.1 *Predicting from the normal random effect model*
While the fitted random effect model correctly assesses the effects of the explanatory variables, it cannot be used *predictively*, to predict the mean response for an individual given the values of the explanatory variables, since the random effect for the individual is unobservable. For this purpose we need the fitted *marginal* model, integrating the *conditional* (on the random effect) model over the random effect.

For the Poisson distribution the mean and variance can be expressed analytically, since

$$E[Y] = E[E[Y \mid Z]] = E[e^{\beta'\mathbf{x}+\sigma Z}] = e^{\beta'\mathbf{x}+\frac{\sigma^2}{2}},$$

$$\mathrm{Var}[Y] = E[Y] + (e^{\sigma^2} - 1)E^2[Y],$$

but for binomial models the logit and other transformations do not give an analytic mean or variance function.

The fitted mean values can be computed numerically from the conditional model (assuming a normal random effect) and are the π_k-weighted sums of the means

of the component distributions. These can be obtained directly from the output of
`alldist`, using

```
$tab the %fv mean with p_ for i_ into mfv $
```

The vector p_ holds the posterior probabilities π_k (expanded to length $n \times K$), and
the vector i_ holds the observation number (also expanded). The resultant vector
mfv holds the fitted mean values (of length n) for the marginal distribution of Y.
This vector can be graphed against the explanatory variables, but the linearity of
the original linear predictor is lost.

The variability of the prediction is much harder to assess, except for simple
models like the Poissson, and we do not give results here.

We now give several examples of Gaussian quadrature for normal random effect
models.

8.4   Gaussian quadrature examples

8.4.1   *Overdispersion model fitting*

To fit an overdispersion model, we give the usual $error specification, but only
certain links are available, namely those giving an unbounded range for the linear
predictor. We specify a `fixed` macro which contains the regression variables to
be fitted, and a `random` macro which for simple overdispersion models contains
just the intercept 1 (we consider more general models in Section 8.8). The number
of components K is set by the appropriate mass-point macro, as described in
Chapter 7. Gaussian quadrature is carried out using the macro `remz`, which requires
one or two arguments—the response variable Y, and the binomial denominator if
a binomial model is being fitted.

8.4.2   *Poisson—the fabric fault data*

We have already seen the negative binomial analysis of the fabric fault data.
We extend this analysis with the Poisson/normal model (Hinde, 1982; Aitkin,
1996a). The results are summarized in Table 8.3. Standard errors for $\hat{\beta}_1$ are given
beside the estimates. They are calculated as described in Chapter 7, by omitting
the roll-length variable, refitting the model and equating the deviance change to
the squared Wald statistic.

```
$input 'faults' : 'alldist' $
$macro fixed ll $endm
$macro random 1 $endm
$use mass10 $
$yvar n $err p $link l
$use remz n $
$use mass20 $use remz $
```

Table 8.3. Estimates and deviances for fabric fault data

K	β_0	β_1	SE(β_1)	σ	Deviance
1	−4.173	0.997	0.176	0.000	64.54
2	−4.417	1.034	0.206	0.339	52.28
3	−3.308	0.849	0.233	0.358	50.96
4	−3.750	0.921	0.237	0.322	51.91
5	−3.823	0.933	0.239	0.348	51.82
6	−3.737	0.919	0.235	0.344	51.61
10	−3.751	0.922	0.241	0.339	51.74
20	−3.749	0.921	0.240	0.340	51.72
NB	−3.793	0.937	0.223	0.353	52.09
QL	−3.86	0.94	0.23	0.36	

Table 8.3 also gives the negative binomial estimates, and Nelder's (1985) QL estimates and standard errors for these parameters, based on the variance/mean relation $V(\mu) = \mu + \phi\mu^2$, where $\phi = \sigma^2$ is estimated from the mean residual deviance. (Note that for the normal random effect model, ϕ, the coefficient of variation of $W = e^Z$, is $e^{\sigma^2} - 1 = 0.123$ for 20-point quadrature compared to $\phi = 0.36^2 = 0.13$ for the QL method.)

The normal mass-point models show reasonable consistency in the regression slope estimates, apart from that for $K = 3$, which is somewhat smaller than for other values of K. They appear to be converging at $K = 20$. The deviance is not monotone in K (as we noted above, it need not be). The deviances for the normal and negative binomial models are very close in this small sample: here we cannot distinguish the normal distribution from the log-gamma for the unobserved variation.

In Fig. 8.2 we graph the fitted values and 2SD limits for the negative binomial and Poisson/normal ($K = 20$) compound models. They are almost identical.

```
$tab the %fv mean with p_ for i_ into pnfv $
$sort syv,spnfv,sl n,pnfv,l 1 $
$calc pnsd = %sqrt(spnfv + (%exp(0.34**2) - 1)*spnfv**2) $
     : pnul = spnfv + 2*pnsd $
     : pnll = spnfv - 2*pnsd $
$gstyle 5 line 0 sym 5 col 2 $
$gstyle 1 1    1 s    0 c    1 $
$gstyle 2 1    2 s    0 c    1 $
$gstyle 3 1    3 s    0 c    1 $
$graph (s = 1 h = 'length' v = 'faults' y = 0,30)
        syv,spnfv,snbfv,pnul,pnll,nbul,nbll sl 5,1,1,2,2,3,3 $
```

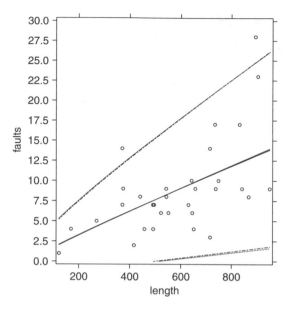

Fig. 8.2. Poisson/normal and NB fits and variability bounds

Note that the residual deviance for the compound model is substantially above the "degrees of freedom" of 29. This result has often caused confusion, and has been interpreted to mean that the compound model also does not "fit" the data. This misunderstanding arises because of the difference between one- and two-parameter families. In the former the "saturated model" has a parameter for every observation, and provides a "baseline" against which we assess the current model. In two-parameter families however (with an extra variability parameter) the model with a parameter for each observation cannot be used to reproduce the data, as the estimate of the scale parameter becomes zero and the model degenerates. So in two-parameter families obtained by extending one-parameter families by overdispersion, there is no interpretation of the "residual deviance" relative to the saturated model as a "goodness-of-fit" test of the overdispersion model—only the change in deviance is of inferential importance, representing the importance of the extra variation.

8.4.3 *Binomial—the toxoplasmosis data*

We consider the overdispersed logistic regression model for incidence of the parasitic condition toxoplasmosis in 34 cities of El Salvador, discussed by Efron (1986) and in the GLIM4 manual, and re-analysed in Aitkin (1996a). Table 8.4 shows the number (n) of men tested and the number (r) with a positive test for toxoplasmosis in each city, together with the annual rainfall (x) in metres in the city. These data are in the file `toxoplas`.

Table 8.4. Toxoplasmosis data

r	n	x	r	n	x	r	n	x
2	4	1.735	3	10	1.936	1	5	2.000
2	2	1.750	3	5	1.800	2	8	1.750
7	19	2.077	3	6	1.920	8	10	1.800
7	24	2.050	0	1	1.830	15	30	1.650
4	22	2.200	0	1	2.000	6	11	1.770
0	1	1.920	33	54	1.770	4	9	2.240
5	18	1.620	0	1	1.650	8	11	2.250
41	77	1.796	24	51	1.890	7	16	1.871
46	82	2.063	9	13	2.100	23	43	1.918
53	75	1.834	8	13	1.780	3	10	1.900
1	6	1.976	23	37	2.292			

We fit polynomial models in rainfall to the logit of the proportion with toxoplasmosis. We suppress the output.

```
$input 'toxoplas' : 'alldist' $
$yvar r $err b n $
$fit : + x<1> : + x<2> : + x<3> $
```

The regression shows a strong cubic trend, with deviances of 74.21, 74.09, 74.09 and 62.64 for the null, linear, quadratic and cubic models respectively. We graph the fitted cubic and the data on the proportion scale in Fig. 8.3.

```
$calc op = r/n $
$assign xx = 1.62, 1.63, ..., 2.3 $
$predict x = xx $calc fp = %pfv $
$sort sop,sx op,x x $
$graph (s = 1 h = 'rainfall' v = 'proportion')
        sop,fp sx,xx 5,10 $
```

It is difficult to understand the scientific significance of the cubic model. We do not show the variability bounds around the fitted values as these vary considerably with sample size and give a very complex picture.

Overdispersion appears substantial; the residual deviance of 62.64 with 30 df casts doubt on the existence of the cubic trend. (This interpretation implicitly assumes the residual deviance has a χ^2_{30} distribution, which is not correct, as noted in Section 2.10.4.) A simple scale correction for the overdispersion (McCullagh and Nelder, 1989) would set a scale parameter ϕ equal to the mean deviance of 2.09, scaling the change in the deviance for the cubic term to 5.48.

We give in Table 8.5 the deviances for 2 to 10 and 20 mass-points for the cubic model using Gaussian quadrature. The deviances show substantial changes between odd and even numbers of mass-points when K is small. In all cases

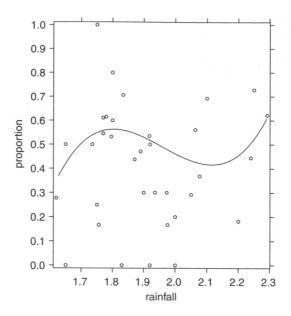

Fig. 8.3. Observed proportions and fitted cubic

Table 8.5. Deviances for toxoplasmosis data

K	Deviance	σ
1	62.64	0.000
2	55.05	0.360
3	51.46	0.480
4	55.15	0.338
5	52.51	0.510
6	54.12	0.387
10	53.80	0.426
20	53.84	0.421

the reduction in deviance relative to the binomial model is substantial, between 9 and 12. This greatly exceeds the 10% point of χ_1^2: there is real extra variation.

Refitting successively the null, linear and quadratic models with overdispersion using 20-point quadrature gives deviances of 59.76, 59.76 and 59.58. The deviance change for the cubic term relative to the quadratic is still rather large (5.74 on 1 df), but the deviance change for the cubic relative to the null model is almost the same (5.92 on 3 df). A plausible interpretation is that the cubic model is unnecessary and that there is no strong evidence of *any* trend with rainfall, but strong evidence of variation with other unmeasured factors.

Efron (1986) used the *double exponential family* model to analyse these data; this introduced an extra "variability" parameter into the generalized binomial distribution, to unlink the mean and variance of the distribution. However, the normalizing constant in the resulting distribution is not available analytically, and this affects the parameter estimation. A discussion of this problem in the Poisson distribution was given in Aitkin (1995).

8.5 Other specified random effect distributions

We may use the Gaussian quadrature approach to fit other specified random effect distributions with very small changes in the procedure. Suppose we wish to use a specific distribution $g(z)$ for Z, for example, a t_r-distribution for specified r. One simple approach is still to use the Gaussian mass-point locations z_k (rather than the mass-point locations for the t_r distribution) and to scale the masses π_k by the density ratio $t_r(z_k)/\phi(z_k)$. This is closely related to *importance sampling* (Tanner, 1996). The resulting masses are not however *optimal* for the t_r-distribution because the mass-points z_k are optimized for Gaussian quadrature assuming the normal distribution.

8.6 Arbitrary random effects

We now consider the most general form of random effect model, in which we make *no* assumption about the distribution of the random effects. It may seem impossible to proceed if we know nothing about this distribution, but as we signalled in Section 7.5, the identification of general mixtures provides a very general framework for random effect models (Clayton and Kaldor, 1987).

Consider again the general overdispersion model of Section 8.1, in which a random effect Z_i with a distribution $g(z)$ is added to the linear predictor $\boldsymbol{\beta}'\mathbf{x}_i$ for the i-th observation. We observe n observations $y_1, \ldots, y_n$ from the marginal or compound distribution

$$m(y) = \int f(y \mid z) g(z) \, \mathrm{d}z.$$

Now we make no assumption about the form of $g(z)$, but instead regard it as another unknown, which we wish to estimate together with $\boldsymbol{\beta}$. The likelihood is

$$L(\boldsymbol{\beta}, g(z)) = \prod_{i=1}^{n} \int f(y_i \mid z_i, \boldsymbol{\beta}) g(z_i) \, \mathrm{d}z_i,$$

where the linear predictor is

$$\eta_i = \boldsymbol{\beta}'\mathbf{x}_i + z_i.$$

As discussed in Chapter 7, the NPML estimate of $g(z)$, when it exists, is known (Laird, 1978; Lindsay, 1983) to be a discrete distribution on a finite number $\hat{K}$ of mass-points $\hat{z}_k$, with masses $\hat{\pi}_k$. Thus, the *profile* likelihood in β, maximized over $g(z)$, is the $\hat{K}$-component finite mixture likelihood

$$P(\beta) = \prod_{i=1}^{n} \left\{ \sum_{k=1}^{\hat{K}} f(y_i \mid \hat{z}_k, \beta) \hat{\pi}_k \right\},$$

where $\hat{K}, \hat{z}_k$ and $\hat{\pi}_k$ are functions of β.

In order to maximize this profile likelihood, we can reformulate the problem as the maximization of the *joint* likelihood

$$L(\beta, K, \pi_1, \ldots, \pi_{K-1}, z_1, \ldots, z_K) = \prod_{i=1}^{n} \left\{ \sum_{k=1}^{K} f(y_i \mid z_k, \beta) \pi_k \right\}$$

over β and all of the parameters of the mixture. The number of components K is unknown, so the likelihood has to be maximized over this as well. This maximization problem is exactly that for the general finite mixture, discussed in Chapter 7. So we have a very general solution to the problem of modelling overdispersion, without any parametric model assumption being required for the random effects.

The estimated components in the mixture may have a direct interpretation. A simple example is the existence of *outliers*: an observation remote from the main body may be split off, and identified as a single component with mass $\hat{\pi}_k = 1/n$. Less commonly, the components may correspond to actual *latent classes* of observations, corresponding to an omitted factor. The posterior probabilities of component membership are frequently informative: outliers can be immediately identified as described above, and latent classes can be identified by a high proportion of 1s and 0s. In general however no substantive meaning can be attached to the components—they merely represent in a discrete form the variation which we usually think of as continuous.

Testing for overdispersion in the more general mixture model is greatly complicated by the estimation of the mixing distribution. As we noted in Chapter 7, there is no satisfactory general theory for the distribution of the LRTS in this model, and for semi-formal tests we need to resort to bootstrapping. It is clear that the distribution of the LRTS must be stochastically larger than it is in the normal random effect model case, and so if the deviance change is not significant by this test, it will certainly not be so by a formal test. This result is however not very helpful in assessing the reality of larger deviance changes.

We now re-examine the previous examples for which we made normal random effect model assumptions, using the NPML approach. For this purpose we use the macro file `alldist`, previously used for normal random effect modelling. The same macro file provides NPML estimation. We need only use the macro `remnp` instead of the macro `remz`; otherwise the model fitting is identical, though the NPML macro requires many more EM iterations for convergence, since the distribution $g(z)$ has to be estimated instead of being fixed by the parametric model assumption. The posterior probabilities are held in the weight vector w_; they can be tabulated against observation index i_ and component index k_ using $tprint w_ i_,k_ $.

8.7 Examples

8.7.1 *The fabric fault data*

We fit the model with one, two and three mass-points:

```
$input 'faults' : 'alldist' $
$macro fixed 11 $endm
$macro random 1 $endm
$use mass1 $use remnp n $
 -- data list abolished
           estimate        s.e.       parameter
      1      -4.173        1.135       MASS_(1)
      2       0.9969       0.1759      LL(I_)
 scale parameter 1.000


 mixture proportions :    1.000
 -- change to data values affects model
 -2 log L =  187.84 , deviance =     64.54, after 3 iterations

$use mass2 $use rcmnp $
 -- data list abolished
           estimate        s.e.       parameter
      1      -2.383        1.101       MASS_(1)
      2      -3.147        1.074       MASS_(2)
      3       0.8019       0.1677      LL(I_)
 scale parameter 1.000


 mixture proportions :   0.2048   0.7952
 -- change to data values affects model
 -2 log L =  172.66 , deviance =     49.36, after 25 iterations

$use mass3 $use remnp $
 -- data list abolished
           estimate        s.e.       parameter
      1      -2.358        1.101       MASS_(1)
      2      -3.120        1.075       MASS_(2)
```

```
   3        -3.146        1.089        MASS_(3)
   4         0.7986       0.1678       LL(I_)
scale parameter 1.000


mixture proportions :   0.2019  0.6663   0.1318
-- change to data values affects model
-2 log L =  172.66 , deviance =    49.37, after 30 iterations
```

Further increases in K leave the deviance essentially unaltered, with replication of the mass-points and split masses. The NPML estimate is a two-point distribution, a somewhat surprising result. The deviance is 49.37 compared to 51.72 for 20-point Gaussian quadrature. For comparability with the results from Gaussian quadrature we rescale the mass-point locations to give the mixing distribution mean zero; the resulting mass-points are 0.608 with mass 0.205 and -0.156 with mass 0.795. The standard deviation of the estimated mixing distribution is 0.308, slightly less than that for the normal models. With the rescaling of the mixture distribution the estimate of the intercept parameter is $\hat{\beta}_0 = -2.383 - 0.608 = -2.991$. The slope estimate, $\hat{\beta}_1 = 0.802$, is substantially below those for the simple Poisson and Poisson/normal models, and the negative binomial estimate. As before we can obtain a standard error for $\hat{\beta}_1$ by omitting the term and equating the deviance change $(65.18 - 49.36)$ to the squared Wald statistic, giving a value of 0.202. We note that the slope is still quite consistent with a true slope of 1. Fitted values and 2SD bounds for the Poisson/non-parametric mixture may be calculated as for the Poisson/normal, with the slight change that the value of ϕ has to be computed directly from the moment-generating function of the two-point mixing distribution (from Section 8.1 $\phi = [M(2)/M^2(1)] - 1)$:

```
$use mass2 $use remnp $
$calc 0.795*%exp(-0.156) + 0.205*%exp(0.608) $
   1.057
$calc 0.795*%exp(2*-0.156) + 0.205*%exp(2*0.608) $
   1.274
$calc 1.274/(1.057**2) - 1 $
   0.1405
$tab the %fv mean with p_ for i_ into npfv $
$sort snpfv npfv l $
$calc pnpsd = %sqrt(snpfv + 0.1405*snpfv**2) $
    : pnpul = snpfv + 2*pnpsd $
    : pnpll = snpfv - 2*pnpsd $
$graph (s = 1 h = 'length' v = 'faults' y = 0,30)
       syv,snpfv,snbfv,pnpul,pnpll,nbul,nbll sl
       5,1,1,2,2,3,3 $
```

The fitted values and 2SD bounds are shown in Fig. 8.4 for the two-point mixture model and the negative binomial model. Despite the substantial differences in

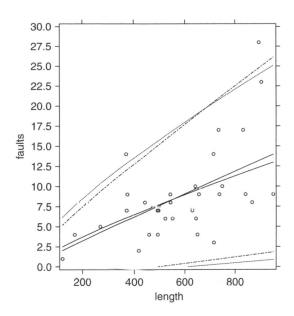

Fig. 8.4. Poisson/non-parametric mixture and NB fits and variability bounds

the estimates, the fitted values are not very different, but the two-component mixture bounds (dot-dashed curves) are slightly narrower than those for the negative binomial (dotted curves), reflecting the smaller standard deviation of the two-point distribution.

In Fig. 8.5 the posterior probabilities of membership in component 1 are plotted as a function of the standardized residuals from the Poisson fit. The posterior probability increases almost monotonically with the residual, with only five observations having a posterior probability of more than 0.5 of being in component 1; these observations have standardized residuals 1.255, 2.530, 3.541, 1.773 and 3.946. Thus component 1 contains the observations with large positive residuals; these are "downweighted" in their effect on the fitted model by the intercept term $\hat{z}_1 = 0.608$, compared with $\hat{z}_2 = -0.156$ for the second component with small positive, or negative, residuals. Since these large observations are mainly at the upper end of the range of fabric length, the regression coefficient $\hat{\beta}_1$ for this variable is reduced. We referred to this possibility in Section 8.1: if Z is correlated with $\mathbf{x}$, then fitting the overdispersion model will change the regression coefficient on $\mathbf{x}$.

The deviance change on fitting the general mixture compared with the simple Poisson model is 15.18. The improvement in deviance over the normal random effect model (2.36 for two extra parameters) is quite small, suggesting that the normal assumption for the random effect is reasonable. Comparing the NPML fit with the Gaussian quadrature fit appears to raise a difficult model comparison

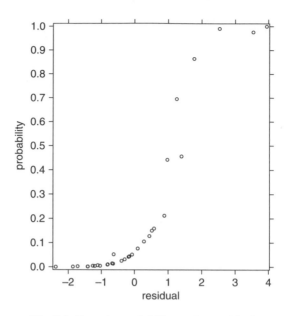

Fig. 8.5. Posterior probability against residual

problem, as we are comparing a normal model for Z with an unspecified one. However, we are actually comparing two discrete distributions for the random effect; in one the masses are fixed, and the mass-points are fixed up to a scale parameter, while in the other the mass-points and masses are both estimated. Since the parameter estimates are not on a boundary, we can invoke standard asymptotic theory to compare the deviances for the two models by the usual LRT. The difference of 2.36 is far from significant compared to χ_2^2, and we conclude that the normal model (*or* the negative binomial) is quite satisfactory.

8.7.2   *The toxoplasmosis data*

Refitting the cubic model with `remnp`, the NPML estimate is a three-point distribution which is close to the mass-points and masses for three-point Gaussian quadrature. Mass-points (centered at 0) for the three-point estimate are 0.780, −0.020 and −1.120 with masses of 0.171, 0.724 and 0.105 respectively. Deviances for two and three mass-point models are 52.42 and 51.14 (compared to 51.46 for three-point Gaussian quadrature). There is again a substantial reduction in the significance of the cubic term, with the deviances for the null, linear, quadratic and cubic models 56.17, 55.92, 55.46 and 51.14, respectively. A plot (not shown) of the posterior probabilities of component membership against the residuals from the simple logit model shows that the large positive residuals are in the first component, the large negative residuals in the third, and the intermediate residuals (between −1.5 and +1) are predominantly in the second.

As for the normal random effect analysis, a plausible interpretation is that the cubic model is unnecessary and that there is no strong evidence of *any* trend with rainfall, but strong evidence of variation with other unmeasured factors.

The deviance improvement from three-point Gaussian quadrature is very small: the estimated mixing distribution is almost exactly the same as that for the quadrature, apart from the scale factor.

8.7.3 *Leukaemia remission data*

We briefly consider the exponential survival example of Feigl and Zelen (1965) discussed in Chapter 6. Recall that the exponential model ag + lwbc with a log link gave a deviance of 40.32 on 30 df. Weibull and gamma models gave deviances and parameter estimates very close to the exponential values.

For any number of mass-points, Gaussian quadrature gives the same deviance of 40.32, with an estimated σ close to zero. The three-point NPML estimate gives a deviance of 39.72, only 0.60 less. This deviance change would not be significant if the mixing distributions were *fixed*, and so is certainly not significant when it is estimated as well. We again conclude that there is no evidence of extra variation beyond the exponential model.

8.7.4 *The Brownlee stack-loss data*

We conclude with a re-analysis of the stack-loss data. As noted by many critics of outlier analyses of these data, the data do not come from a randomized experiment and so conventional modelling approaches are suspect, but we present here for interest the results of the finite mixture estimation. Aitkin and Tunnicliffe Wilson (1980) presented a detailed investigation of normal mixture models for outliers, with suspected outliers assigned to specific mixture components. These components had their own means, unrelated to the regression structure modelled for the "good" observations. Our approach here (following Aitkin, 1996a) is different since the "outliers" still contribute to the estimation of the regression model parameters.

For all the mixture models, parameter estimates and the disparity are sensitive to starting values. For the two-component model the best solution gives a disparity of 100.39 (using tol_ = 3); parameter estimates are shown for this model, and for three- (tol_ = 3) and four-component (tol_ = 1.9) models, in Table 8.6. We abbreviate MASS to M. The mixture models are fitted without intercepts.

Posterior probabilities of component membership are given for the same three models in Table 8.7. The two-component model splits the observations into two groups, with each observation having posterior probability of at least 0.907 of falling in one component. The three-component model isolates observations 1, 3 and 4 in the first component, and observation 21 in the third. (Observations 1, 3, 4 and 21 are those repeatedly identified as outliers in previous analyses.)

Table 8.6. Mixture fits for stack-loss data

K	x_1	x_2	x_3	σ	M_(1)	M_(2)	M_(3)	M_(4)	Disparity
1	0.716	1.30	−0.152	2.92	−39.9				104.58
2	0.605	1.83	−0.280	1.38	−30.5	−36.2			100.39
3	0.766	0.62	−0.061	1.03	−30.0	−37.2	−45.5		85.72
4	0.874	0.91	0.033	0.51	−50.9	−55.5	−59.5	−67.4	72.96

Table 8.7. Posterior probabilities of component membership

	K								
	2		3			4			
Obs.	1	2	1	2	3	1	2	3	4
1	0.999	0.001	1	0	0	0	1	0	0
2	0.000	1.000	0	1	0	0	0	1	0
3	1.000	0.000	1	0	0	0	1	0	0
4	1.000	0.000	1	0	0	1	0	0	0
5	0.002	0.998	0	1	0	0	0	1	0
6	0.000	1.000	0	1	0	0	0	1	0
7	0.000	1.000	0	1	0	0	0	1	0
8	0.002	0.998	0	1	0	0	0	1	0
9	0.000	1.000	0	1	0	0	0	1	0
10	0.993	0.007	0	1	0	0	1	0	0
11	1.000	0.000	0	1	0	0	1	0	0
12	1.000	0.000	0	1	0	0	1	0	0
13	0.093	0.907	0	1	0	0	0	1	0
14	0.987	0.013	0	1	0	0	0	1	0
15	1.000	0.000	0	1	0	0	1	0	0
16	0.971	0.029	0	1	0	0	1	0	0
17	0.000	1.000	0	1	0	0	1	0	0
18	0.009	0.991	0	1	0	0	1	0	0
19	0.002	0.998	0	1	0	0	1	0	0
20	0.910	0.090	0	1	0	0	1	0	0
21	0.000	1.000	0	0	1	0	0	0	1

For the four-component model the MLE of the residual SD is 0.5, half a measurement unit. Aitkin and Tunnicliffe Wilson (1980) noted a similar phenomenon in their mixture modelling: with four components the residual standard deviation was so small that the normal density representation of the likelihood breaks down unless measurements can be recorded to another decimal place. This model splits the observations into four groups, each observation having posterior probability at least 0.999 of falling in one component. Observation 4 falls in the first component, and 21 in the fourth. The other observations are split in several long sequences between the second and third components. This result strongly suggests a real latent class model, with a "missing factor" from the regression model. The appearance of the successive observations 5–9 in the third component and observations 15–20 in the second component suggests a five-day week effect.

We define a week factor with five levels: week 1 contains observations 1–4, week 2 observations 5–9, week 3 observations 10–14, week 4 observations 15–19 and week 5 observations 20 and 21. Adding week to the three-variable normal model with no mixture structure gives a deviance of 76.28 which is close to that for the four-component mixture model.

```
$calc i = %cu(1) $
 : week = 1 + %gt(i,4) + %gt(i,9) + %gt(i,14) + %gt(i,19) $
$factor week 5 $
$use mass1 $
$macro fixed x1 + x2 + x3 + week $endm
$use remup $
```

	estimate	s.e.	parameter
	0.371	0.128	X1
	0.884	0.592	X2
	-0.026	0.102	X3
	-10.88	2.01	WEEK(2)
	-10.38	4.10	WEEK(3)
	-13.05	3.81	WEEK(4)
	-11.80	3.34	WEEK(5)
	-12.01	15.34	MASS_(1)
sigma	1.488		
deviance	76.28		

The deviance change between the normal regression model and that including week is 28.30 on 4 df. Equivalently, the F-statistic for the contribution of week is 9.25 for $F_{4,13}$; these are both significant beyond the 0.1% level. There is no question of the importance of this factor.

The importance of X_2 has decreased in this model and it appears that airflow X_1 is the only important variable when the weekly time sequence of the data is taken into account. The weeks after week 1 (which contains the "outlier" observations 1, 3 and 4) show a dramatically lower stack-loss, with week 4 lower than weeks 2, 3 and 5.

"Week" however is not an explanatory variable in any real sense: the need for this factor in the model is a strong indication that there are other very important omitted process variables which apparently changed their values in successive operating weeks, as did the variables x_1 and x_2. (This analysis was repeated with six- and seven-day weeks, and for five-day weeks starting with different observations in the sequence, with totally negative results: there is no evidence at all for a week effect longer than five days, or of the sequence starting at a different day.)

We conclude that the persistent identification of outliers in these data may be a result of model mis-specification: the data are consistent with consecutive daily observations of a five-day working week process, in which changes in the process were made at the end of some working weeks. These changes were apparently in addition to the substantial changes in the level of airflow and water temperature between weeks. When the week is included as a factor in the model, the evidence for a mixture disappears.

The detailed discussion of the data in Dodge (1999) appears to bear out the near-daily time sequencing of the observations, though no full list of the dates of the observations appears to exist.

We now consider more general versions of random effect models, in which the model regression coefficients as well as the intercept are random.

8.8   Random coefficient regression models

The overdispersion model can be interpreted as a *random intercept* model. The random effect Z_i can be combined with the model intercept term β_0 to give a random intercept B_{0i}, with mean β_0 and variance σ^2, the variance of Z.

This formulation of the random effect model can be extended to more general *random coefficient models*. Consider a simple example with a single variable x_{1i} whose coefficient β_1 varies across the data, in addition to the intercept. We represent it by $B_{1i} = \beta_1 + U_i$, where U_i corresponds to variation in the regression coefficient about a "mean" β_1. The remaining regression coefficients $\boldsymbol{\beta}_2$ are fixed. Then conditional on U_i *and* Z_i, the regression model is

$$\eta_i = \beta_0 + \beta_1 x_{1i} + \boldsymbol{\beta}_2' \mathbf{x}_{2i} + Z_i + U_i x_{1i},$$

while Z_i and U_i have some joint distribution $g(z, u)$. The likelihood is then

$$L(\boldsymbol{\beta}) = \prod_i \int \int f(y_i \mid z, u) g(z, u) \, dz \, du.$$

If the distribution $g(z, u)$ is assumed Gaussian with unknown covariance matrix, we need to numerically integrate over both parameters. This can be achieved by two independent single-parameter numerical integrations by using parameter orthogonalization. We rewrite the model as

$$\eta_i = \beta_0 + \beta_1 x_{1i} + \boldsymbol{\beta}_2' \mathbf{x}_{2i} + \sigma_Z Z_i + \sigma_U U_i x_{1i},$$

where Z_i and U_i are bivariate normal with means zero, variances 1 and $\text{corr}(Z_i, U_i) = \rho$. The distribution of (U, Z) can be re-expressed as the independent

$N(0, 1)$ distributions of U and Z^*, the residual from the regression of Z on U:

$$Z^* = \frac{Z - \rho U}{\sqrt{1 - \rho^2}}.$$

The model above can then be re-expressed as

$$\eta_i = \beta_0 + \beta_1 x_{1i} + \boldsymbol{\beta}_2' \mathbf{x}_{2i} + \sigma_Z \left(\rho U_i + \sqrt{1 - \rho^2} Z_i^* \right) + \sigma_U U_i x_{1i}$$

$$= \beta_0 + \beta_1 x_{1i} + \boldsymbol{\beta}_2' \mathbf{x}_{2i} + \lambda_0 Z_i^* + (\lambda_1 + \lambda_2 x_{1i}) U_i.$$

Approximating the double integral by two finite sums over the independent Z_i^* and U_i, we have

$$L(\boldsymbol{\beta}) \approx \prod_i \sum_k \sum_\ell f(y_i \mid z_k, u_\ell) \pi_k \pi_\ell$$

with linear predictor

$$\eta_{ik\ell} = \beta_0 + \beta_1 x_{1i} + \boldsymbol{\beta}_2' \mathbf{x}_{2i} + \lambda_0 z_k + (\lambda_1 + \lambda_2 x_{1i}) u_\ell.$$

The unknown variances and covariance of Z and U are replaced by the three regression coefficients of z_k, u_ℓ and the interaction $x_{1i} u_\ell$.

The double integration still requires K^2 terms in the summation, and with R random parameters we would need K^R terms, making this approach impractical for a large number of random parameters. We have not implemented Gaussian quadrature for random coefficient models, only for random intercept models.

However, when we estimate the joint distribution of both Z_i and U_i non-parametrically, we again obtain the NPML estimate of $g(z, u)$ as a discrete distribution on a finite number $\hat{K}$ of points in the (z, u) plane, with an estimated mass $\hat{\pi}_k$ and estimated mass-points $(\hat{z}_k, \hat{u}_k)$ for the k-th component. The likelihood maximized over $g(z, u)$ can again be expressed as a finite mixture, of regressions with different intercept and slope for x_1 in each component of the mixture, and common slopes for $\mathbf{x}_2$.

In the GLIM implementation of this model, the components are indexed by the "component factor". The random coefficients can be handled by including in the regression model, both the main effect of the factor z_k, and the interaction of this factor with the explanatory variable x_1. Again the number of mass-points has to be determined by sequential increase from 1, and, in general, more mass-points may be required than for the case of fixed β_1. This process is quite general, and is invoked simply by specifying in the random macro the terms which are to have

random regression coefficients (the intercept term is always included). We stress
that this approach is valid only for NPML estimation: if remz is used instead of
remnp with a random coefficient model, the resulting analysis is *not* a full Gaussian
quadrature over all the random parameters.

We now illustrate this extension with an example.

8.8.1   *Example—the fabric fault data*

We extend the model to a random linear regression on log length:

```
$inp 'fabric' : 'alldist' $
$yvar n $err p $link l
$macro fixed ll $endm
$macro random ll $endm
$use mass2 $use remnp $
-- data list abolished
              estimate        s.e.       parameter
       1       -4.553         1.634        MASS_(1)
       2       -2.115         1.443        MASS_(2)
       3        1.123         0.2498       LL(I_)
       4       -0.4873        0.3369       MASS_(2).LL(I_)
scale parameter 1.000


mixture proportions :   0.2702   0.7298
-- change to data values affects model
-2 log L =   171.58 , deviance =     48.29, after 17 iterations


$calc tol_= 1 $
$use mass3 $use remnp $
-- data list abolished
              estimate        s.e.       parameter
       1      -10.39         14.25         MASS_(1)
       2      -13.59          2.742        MASS_(2)
       3       -0.8364        1.295        MASS_(3)
       4        2.198         2.416        LL(I_)
       5        0.2666        2.452        MASS_(2).LL(I_)
       6       -1.755         2.425        MASS_(3).LL(I_)
scale parameter 1.000


mixture proportions :   0.0260   0.3184   0.6556
-- change to data values affects model
-2 log L =   169.37 , deviance =     46.07, after 27 iterations


$calc tol_ = 0.3 $
$use mass4 $use remnp $
-- data list abolished
```

```
          estimate        s.e.       parameter
   1       -0.5198        3.147       MASS_(1)
   2      -13.80          2.917       MASS_(2)
   3       -1.205         1.344       MASS_(3)
   4       -1.249         4.964       MASS_(4)
   5        0.5157        0.4991      LL(I_)
   6        1.980         0.6656      MASS_(2).LL(I_)
   7       -0.01693       0.5420      MASS_(3).LL(I_)
   8       -0.01078       0.9270      MASS_(4).LL(I_)
scale parameter 1.000


mixture proportions :   0.0525   0.2798   0.6220   0.0457
-- change to data values affects model
-2 log L =  168.73 , deviance =    45.44, after 23 iterations
```

Identification of the NPML estimate for different K requires some searching over values of tol_. For $K = 2, 3$ or 4 the deviance is 48.29 with the default setting of tol_ = 0.5, while for $K = 3$ it is 46.07 with tol_ = 1 and for $K = 4$ it is 45.44 with tol_ = 0.3, though the third and fourth components are essentially identical. For $K > 4$ the deviance stabilizes at 45.44. With this small sample and complex model, several local maxima appear.

The change in deviance of 3.93 relative to the simple overdispersion model is quite small; if this change were for a *fixed* distribution of (Z, U) it would barely be significant compared to $\chi^2_{1, 0.90}$. There is no real evidence of non-constant regression slope.

8.9 Algorithms for mixture fitting

We use the EM algorithm for fitting mixture models. It has the great virtues of relatively easy programming and guaranteed convergence to at least a local maximum, but locating the global maximum may require extensive searching over starting points for the algorithm. Other algorithms are available which are able to locate where additional mass-points should be placed to increase the likelihood, given the current mass-point configuration. These are *vertex* algorithms based on optimal design theory, and are discussed in Lindsay (1995) and Böhning (1999), among others.

These approaches do not necessarily locate the same maximum, as illustrated in the following example.

The trypanosome data
Follman and Lambert (1989) gave an example of a logistic regression with a varying intercept term. The data consist of numbers y of trypanosomes killed out of n treated at a treatment dose x. The data are shown in Table 8.8, and are graphed

Table 8.8. Trypanosome data

x	n	y
4.7	55	0
4.8	49	8
4.9	60	18
5.0	55	18
5.1	53	22
5.2	53	37
5.3	51	47
5.4	50	50

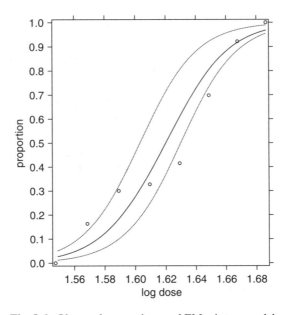

Fig. 8.6. Observed proportions and EM mixture model

in Fig. 8.6. The logistic regression on log x is a very poor fit, with deviance 24.66 with 6 df. (A CLL link gives a much better fit, with a deviance of 15.83, but we do not pursue this here.)

For comparison with Follman and Lambert we fit the random intercept model using log dose, although a slightly better fit is obtained with dose x.

```
$inp 'trypanos' : 'alldist' $
$calc lx = %log(x) $
$yvar y $err b n
$macro fixed lx $endm
$macro random 1 $endm
```

```
$use mass1 $use remnp y n $
$use mass2 $use remnp $
...
```

The NPML estimate of the mixing distribution found by this approach has two mass-points with location $(0.898, -0.546)$ and masses $(0.379, 0.621)$ and deviance 16.41. Parameter estimates are $\hat{\beta}_0 = -87.37$, $\hat{\beta}_1 = 53.92$. The fitted proportions from the marginal mixed logistic model are obtained from the fitted values, and are shown in Fig. 8.6. We use `predict` to get a smoother fitted curve.

```
$use mass2 $use remnp $
$tab the w_ mean for i_,k_ $
$ass px = 4.65,4.66,...,5.45 $
$calc px = %log(px) $
$predict lx = px k__ = 1 $calc p1 = %pfv $
$predict lx = px k__ = 2 $calc p2 = %pfv $
$calc mfp = pp_(1)*p1 + pp_(2)*p2 $
$calc  op = r/n $
$graph (s = 1  h = 'log dose' v = 'proportion')
        op,mfp,p1,p2 lx,px,px,px 5,10,11,11 $
```

For the EM estimates, the posterior probabilities of membership in the upper component are

$$(0.074, 0.997, 1.000, 0.001, 0.000, 0.000, 0.175, 0.782).$$

Thus, the EM estimates separate the second and third observations off from the others, with the last observation having high weight for this component as well. The dashed curves in the figure are the fitted component models.

Using an optimal design algorithm, Follman and Lambert found a two-point distribution for the intercept with location $(6.27, -3.23)$, masses $(0.34, 0.66)$ and a deviance of 3.38. Parameter estimates are $\hat{\beta}_0 = -202.47, \hat{\beta}_1 = 124.8$. Tabulation over a grid of β_0 and β_1 shows that the deviance evaluated at these parameter values is indeed the minimum of 3.38, but the EM algorithm as implemented here cannot find this point, despite searching over starting values using `tol_`. Fig. 8.7 shows the two logistic components and the fitted model.

```
$var 100 fl1 fl2 $
$calc fl1 = -196.2 + 124.8*lpx $
    : fl1 = %exp(fl1) $
    : pl1 = fl1/(1 + fl1) $
    : fl2 = -205.7 + 124.8*lpx $
    : fl2 = %exp(fl2) $
    : pl2 = fl2/(1 + fl2) $
    : pl  = 0.34*pl1 + 0.66*pl2 $
```

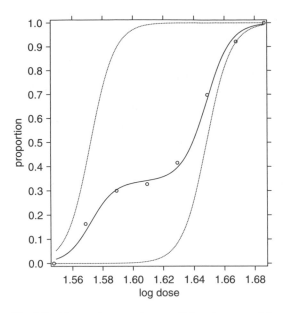

Fig. 8.7. Observed proportions and FL mixture model

```
$graph (s = 1  h = 'log dose' v = 'proportion')
        op,pl,pl1,pl2 lx,lpx,lpx,lpx 5,10,11,11 $
```

The fitted probabilities nearly interpolate the observed proportions, with a very low deviance. The posterior probabilities are 1 for all observations except the fifth—the first four fall in the first component, and the last three in the second.

For the second mixture, the standard deviation of the estimated mixing distribution is 4.50, compared with 0.70 for the first. The two-components are completely separated in the second mixture, with a "low-dose" sub-population of 34% which is killed almost completely by a log-dose of 1.62; the remaining "high-dose" sub-population of 66% requires higher doses for mortality.

A peculiarity of the EM algorithm here is that if it is started from the Follman/Lambert estimates, it jumps in one iteration to the NPML estimates above. Further study of this example, and others, would be rewarding. Aitkin (2000) discussed a second example.

8.10 Modelling the mixing probabilities

We have assumed that the mixing distribution of the random effect Z does not depend on the explanatory variable **x** except through a location shift, which does not affect the model-fitting, though it may change its interpretation.

An alternative approach, widely used in computer science, is to model Z as discrete, with component proportions π_k modelled as multinomial logit functions of the explanatory variables, in addition to the model for the kernel parameter. These *double* models are known as *mixtures of experts* models in computer science literature (see, e.g. Jacobs *et al.* 1991), an "expert" being a regression model within a single component of the mixture. The usual form of the model allows a different regression model for the kernel density parameter within each mixture component. The double model can be fitted straightforwardly by successive relaxation, by alternating the mixture fitting of the kernel models and the fitting of the multinomial logit model. This is similar to the double modelling in Chapter 3 of the normal mean and variance parameters. The discrete mixture model likelihood of Section 8.6:

$$L(\boldsymbol{\beta}, K, \pi_1, \ldots, \pi_{K-1}, z_1, \ldots, z_K) = \prod_{i=1}^{n} \left\{ \sum_{k=1}^{K} f(y_i \mid z_k, \boldsymbol{\beta}) \pi_k \right\}$$

is generalized using the multinomial model

$$\theta_{ik} = \log(\pi_{ik}/\pi_{i1}) = \boldsymbol{\delta}_k' \mathbf{x}_i$$

and the component-specific regression parameter vector $\boldsymbol{\beta}_k$, to give the mixture of experts likelihood

$$L(\boldsymbol{\beta}_1, \ldots, \boldsymbol{\beta}_K, \boldsymbol{\delta}_1, \boldsymbol{\delta}_2, \ldots, \boldsymbol{\delta}_K, K) = \prod_{i=1}^{n} \left\{ \sum_{k=1}^{K} f(y_i \mid \boldsymbol{\beta}_K) \pi_{ik} \right\},$$

where $\boldsymbol{\delta}_1 = 0$, with log-likelihood

$$\ell = \log L = \sum_i \log \left\{ \sum_k \pi_{ik} f_{ik} \right\} = \sum_i \log f_i.$$

The derivatives of the log-likelihood with respect to $\boldsymbol{\beta}_k$ follow from Section 8.3:

$$\frac{\partial \ell}{\partial \boldsymbol{\beta}_k} = \sum_i w_{ik} \mathbf{s}_{ik}(\boldsymbol{\beta}_k),$$

and those with respect to $\boldsymbol{\delta}_k (k > 1)$ are

$$\frac{\partial \ell}{\partial \boldsymbol{\delta}_k} = \sum_i \frac{1}{f_i} \left\{ \frac{\partial \pi_{ik}}{\partial \boldsymbol{\delta}_k} f_{ik} - \frac{\partial \pi_{iK}}{\partial \boldsymbol{\delta}_K} f_{iK} \right\}$$

$$= \sum_i \left\{ \frac{\pi_{ik} f_{ik}}{f_i} \mathbf{x}_i - \frac{\pi_{iK} f_{iK}}{f_i} \mathbf{x}_i \right\}$$

$$= \sum_i (w_{ik} - w_{iK}) \mathbf{x}_i.$$

We do not give details of algorithm implementation.

The model is heavily parametrized, and as for the double normal model there may be identifiability difficulties in assessing the effects of variables in the two sets of models. The expected information matrix may not be a reliable indicator of effect importance.

An important special case of the mixture of experts model is one in which the component-specific regression models are the same: only the logistic regression models are different for each component. This special case was proposed by Aitkin and Foxall (2003) as an alternative to the *feed-forward neural network* or *multi-layer perceptron*.

8.11 Mixtures of mixtures

An interesting question is whether mixtures of exponential family distributions can be generalized further by mixing. Can we develop and fit mixtures of mixtures?

Two issues arise immediately. The first is that for exponential family distributions, mixing with a conjugate random effect distribution gives in some cases closed-form distributions with unlinked mean and variance parameters, giving greater generality than the original exponential family distribution. However, the resulting compound distribution may not have a conjugate mixing distribution, and numerical integration will then be necessary, as it is for the normal mixing distribution with Poisson and binomial distributions.

The second is that the possible improvement in likelihood, by mixing the mixed distribution, is limited by the maximized likelihood for the NPML estimate of the mixing distribution for the original (one-parameter) exponential family distribution.

We illustrate these limitations with mixtures of negative binomials for the fabric fault data. The Poisson model using the log length variable had a deviance (relative to the saturated model) of 64.54 on 30 df, and fitting the negative binomial model, a gamma mixture of Poissons, reduced this to 52.09. If we now mix the negative binomial by introducing a new random effect w with marginal distribution $h(w \mid \xi)$ into its linear predictor, there is no conjugate form for this, so we will be limited to normal or other non-conjugate mixing requiring numerical integration. The marginal distribution of the negative binomial mixture is

$$m(y) = \int nb(y \mid \eta, r, w) h(w \mid \xi) \, dw,$$

where

$$nb(y \mid \eta, r, w) = \frac{\Gamma(y+r)}{\Gamma(r)y!} \left(\frac{e^{\eta+w}}{e^{\eta+w} + r} \right)^y \left(\frac{r}{e^{\eta+w} + r} \right)^r.$$

But since the negative binomial is itself the gamma mixture of Poissons, we have

$$m(y) = \int h(w \mid \xi) \left[\int P(y \mid \eta, z, w) g(z \mid r) \, dz \right] dw,$$

where

$$P(y \mid \eta, z, w) = \frac{e^{-e^{\eta+z+w}} e^{(\eta+z+w)y}}{y!}.$$

Changing variables to $u = z + w$ and z, with Jacobian 1, we have

$$m(y) = \int P(y \mid \eta, u) \left[\int h(u - z \mid \xi) g(z \mid r) \, dz \right] du$$

$$= \int P(y \mid \eta, u) k(u \mid \xi, r) \, du.$$

So the mixed negative binomial distribution is a mixed Poisson distribution with respect to a different mixing distribution k, the convolution of h and g. But for the single-parameter Poisson distribution, the minimized value of the deviance across *all* mixing distributions cannot be less than 49.37, the deviance for the NPML-estimated mixing distribution. So any compound negative binomial distribution for the fault data could only reduce the negative binomial deviance by at most 2.72 (from 52.09 to 49.37).

Such results greatly limit the possibilities for iterating mixing for any exponential family distribution—the NPML estimate will be the same for double mixing as for single mixing. An example is the *hierarchical mixture of experts* model (Jordan and Jacobs, 1994) in which the mixture of experts model is again mixed: each of the K mixture components is itself regarded as a mixture of L components with mixing probabilities π_{ikl}^* depending logistically on a parameter δ_{kl}^*. The marginal distribution of the response Y_i in such a model is then

$$m(y_i \mid \mathbf{x}_i, \delta_2, \ldots, \delta_K, \delta_{22}^*, \ldots, \delta_{2L}^*, \ldots, \delta_{K2}^*, \ldots, \delta_{KL}^*)$$

$$= \sum_{k=1}^{K} \pi_{ik}(\delta_k) \sum_{l=1}^{L} \pi_{ikl}^*(\delta_{kl}^*) f(y_i \mid \theta_{ikl})$$

$$= \sum_{k=1}^{K} \sum_{l=1}^{L} \pi_{ik}(\delta_k) \pi_{ikl}^*(\delta_{kl}^*) f(y_i \mid \theta_{ikl}),$$

where

$$\sum_{k=1}^{K} \sum_{l=1}^{L} \pi_{ik}(\delta_k) \pi_{ikl}^*(\delta_{kl}^*) = 1.$$

Write

$$\pi_{ik}(\delta_k)\pi^*_{ikl}(\delta^*_{kl}) = \pi^{**}_{im}(\gamma_m),$$

where m runs over the range $1, \ldots, M = KL$, with $\sum_{m=1}^{M} \pi^{**}_{im}(\gamma_m) = 1$, and γ is a parametric function of the δ and δ^* to be determined. Now

$$\begin{aligned}
\pi_{ik}(\delta_k)\pi^*_{ikl}(\delta^*_{kl}) &= \frac{e^{\delta_k x_i}}{\sum_{k=1}^{K} e^{\delta_k x_i}} \cdot \frac{e^{\delta^*_{kl} x_i}}{\sum_{l=1}^{L} e^{\delta^*_{kl} x_i}} \\
&= \frac{e^{(\delta_k + \delta^*_{kl}) x_i}}{\sum_{k=1}^{K} \sum_{l=1}^{L} e^{(\delta_k + \delta^*_{kl}) x_i}} \\
&= \frac{e^{\gamma_m x_i}}{\sum_{m=1}^{M} e^{\gamma_m x_i}},
\end{aligned}$$

where

$$\gamma_m = \gamma_{kl} = \delta_k + \delta^*_{kl}.$$

Thus, the marginal distribution of Y_i can be expressed as

$$m(y_i \mid x_i, \gamma_2, \ldots, \gamma_M) = \sum_{m=1}^{M} \pi^{**}_{im}(\gamma_m) f(y_i \mid \theta_{im}),$$

which is an M-component mixture with multinomial logit parameters regressing on x with coefficients γ_m.

Thus, the "hierarchical" mixture is equivalent to an M-component mixture of the original type, though with apparently many more components. Since there is an identifiable limit to the number of components, hierarchical mixing will not identify more components than simple mixing, and does not give greater generality.

9
Variance component models

9.1 Models with shared random effects

In this chapter we consider GLMs with *shared random effects* arising through variance component or repeated measures structure, for example, in two-stage sample designs, or longitudinal data. As we will see, the analysis of these models closely parallels that for overdispersion models, with slightly greater complexity in the EM algorithm. It is therefore quite straightforward to extend the GLIM modelling facilities to deal with multi-level models (two-level in our development). This possibility is somewhat surprising, since GLIM was never designed to fit multi-level models. The range of possible models in this category is also surprising.

We begin for simplicity of exposition with the two-level variance component model, for a two-stage random sample with upper- or second-level or *primary* sampling units (PSUs) indexed by $i = 1, \ldots, r$, and lower- or first-level or *secondary* sampling units (SSUs) indexed by j, sampled within each upper-level unit, where $j = 1, \ldots, n_i$. On each first-level unit we measure or record a response y_{ij}, and we have explanatory variables $\mathbf{x}$ which may be measured at both upper ($\mathbf{x}_{1i}$) and lower ($\mathbf{x}_{2ij}$) levels. We represent the distribution of the response Y by an exponential family member, with a link function and linear predictor involving the explanatory variables at both levels, and perhaps their cross-level interactions.

The common membership of the responses y_{ij} and $y_{ij'}$ in the same PSU implies, in most sampled populations, a greater homogeneity of these responses than of those in different PSUs. This homogeneity should be reflected in the model.

A natural way of representing this "common variation" is by adding a common unobserved random effect to the linear predictor for each lower-level unit in the same upper-level unit. Thus the common variation is modelled as an extra unobserved variable Z_i on the same scale as the linear predictor, but this variable is now shared between lower-level units in the same upper-level unit, rather than being unique to each observation as in the overdispersion model.

9.2 The normal/normal model

For the normal/normal model, we take the distribution of both Y and Z to be normal, with identity link: conditional on Z_i, the Y_{ij} are independently normal

$N(\mu_{ij}, \sigma^2)$, with

$$\mu_{ij} = \eta_{ij} = \boldsymbol{\beta}' \mathbf{x}_{ij} + \sigma_A Z_i,$$

$$\boldsymbol{\beta}' \mathbf{x}_{ij} = \boldsymbol{\beta}_1' \mathbf{x}_{1ij} + \boldsymbol{\beta}_2' \mathbf{x}_{2i},$$

$$Z_i \sim N(0, 1),$$

where σ_A^2 is the "among PSU" variance component, and σ^2 is the "within PSU" variance component. It follows immediately that marginally,

$$Y_{ij} \sim N(\boldsymbol{\beta}_1' \mathbf{x}_{1ij} + \boldsymbol{\beta}_2' \mathbf{x}_{2i}, \sigma^2 + \sigma_A^2),$$

$\text{Cov}[Y_{ij}, Y_{ij'}] = \sigma_A^2$ for $j \neq j'$, and $\text{Cov}[Y_{ij}, Y_{i'j'}] = 0$ for $i \neq i'$, so that correspondingly

$$\text{corr}[Y_{ij}, Y_{ij'}] = \rho = \frac{\sigma_A^2}{\sigma^2 + \sigma_A^2},$$

for $j \neq j'$, and $\text{corr}[Y_{ij}, Y_{i'j'}] = 0$ for $i \neq i'$.

The random effect induces an *intra-class* correlation between the lower-level responses on the same upper-level unit, and so *does* represent the greater homogeneity of the responses in the same upper-level unit. An equivalent form of the model is more familiar from *multi-level modelling* (e.g. Aitkin and Longford (1986), Goldstein (2003)):

$$Y_{ij} \mid Z_i \sim N(\boldsymbol{\beta}_1' \mathbf{x}_{1ij} + Z_i, \sigma^2), \quad j = 1, \ldots, n_i,$$

$$Z_i \sim N(\boldsymbol{\beta}_2' \mathbf{x}_{2i}, \sigma_A^2), \quad i = 1, \ldots, r.$$

In this form the role of the upper-level variables is clearer, in reducing or "explaining" the upper-level variance σ_A^2. It is an important aspect of these models that the parameter σ_A can appear either as a variance in the second form, or as a regression coefficient in the first. In fitting the model using an EM algorithm, the rate of convergence of the algorithm is affected by the size of σ_A in the second parametrization, but not in the first (Aitkin and Hinde, in Meng and van Dyk, 1997). In the normal model, the EM algorithm is not the most efficient algorithm for ML estimation, the scoring algorithm being the fastest (Longford, 1993), but in non-normal models the EM algorithm is very effective and relatively simple to program.

We now give the EM algorithm for the normal model using the first parametrization above. We maximize in the M-step the expected complete data log-likelihood, treating the Z_i as missing data. The complete data likelihood can be written as

$$L^* = \prod_{i=1}^{r} \left[\prod_{j=1}^{n_i} \frac{1}{\sqrt{2\pi}\sigma} \exp\left\{ -\frac{1}{2\sigma^2}(y_{ij} - \boldsymbol{\beta}'\mathbf{x}_{ij} - \sigma_A Z_i)^2 \right\} \right] \frac{1}{\sqrt{2\pi}} \exp\left\{ -\frac{1}{2}Z_i^2 \right\}.$$

The last term is the normal density of the Z_i, which does not involve any parameters, and so we may drop it from the likelihood. In the complete data model, Z_i is an additional explanatory variable, and the complete data log-likelihood involves the missing data only through Z_i and Z_i^2; in the E-step these values are replaced by their conditional expectations given all the observed data $\{y_{ij}\}$ and $\{\mathbf{x}_{ij}\}$: write

$$\mathrm{E}[Z_i \mid \{y_{ij}\}] = \tilde{z}_i,$$

$$\mathrm{E}[Z_i^2 \mid \{y_{ij}\}] = \mathrm{Var}[Z_i \mid \{y_{ij}\}] + E^2[Z_i \mid \{y_{ij}\}]$$

$$= V_i + \tilde{z}_i^2.$$

For the M-step, the normal equations for $\boldsymbol{\beta}$ and σ_A involve the cross-products of $\mathbf{x}$ with Z; in these equations Z_i is replaced by $\tilde{z}_i$, and the sum of squares $\sum Z_i^2$ is replaced by $\sum \tilde{z}_i^2 + \sum V_i$. The score equation for σ^2 is adjusted similarly.

For the E-step, it is easily seen that, given $\bar{Y}_{i.}$, Z_i is independent of $(Y_{ij} - \bar{Y}_{i.})$, where $\bar{Y}_{i.} = \sum_j Y_{ij}/n_i$. Since

$$\bar{Y}_{i.} \mid Z_i \sim N(\boldsymbol{\beta}'\bar{\mathbf{x}}_{i.} + \sigma_A Z_i, \sigma^2/n_i)$$

and $Z_i \sim N(0, 1)$, it follows easily that

$$Z_i \mid \bar{Y}_{i.} \sim N\left(\frac{\sigma_A}{\sigma_A^2 + \sigma^2/n_i}(\bar{Y}_{i.} - \boldsymbol{\beta}'\bar{\mathbf{x}}_{i.}), 1 - \frac{\sigma_A^2}{\sigma_A^2 + \sigma^2/n_i} \right).$$

The EM algorithm alternates between fitting the linear model to $\mathbf{x}$ and z, with z_i replaced by $\tilde{z}_i$ in the cross-products, and a diagonal loading on the SSP matrix of $\sum V_i$ corresponding to σ_A, and computing $\tilde{z}_i$ and V_i from the current parameter estimates.

The difficulty with other exponential family distributions is that the normal random effect model is not conjugate, and so numerical integration or other approximations are necessary. We give in Section 9.3 the same treatments of these models by Gaussian quadrature and NPML as for the overdispersion models of Chapter 8. Other approximate approaches to the variance component model have been extensively discussed, and we consider these in Section 9.4.

9.3   Exponential family two-level models

Restating the general model, we have a two-stage random sample y_{ij} with $i = 1, \ldots, r$, $j = 1, \ldots, n_i$, and $\sum n_i = n$, from an exponential family distribution $f(y \mid \theta)$ with canonical parameter θ and mean μ, and explanatory variables $X = (\mathbf{x}_{ij})$, related to μ through a link function $\eta_{ij} = g(\mu_{ij})$ with linear predictor $\eta_{ij} = \boldsymbol{\beta}' \mathbf{x}_{ij}$. Here the X matrix is understood to include both upper- and lower-level explanatory variables (and their interactions if any), with $\mathbf{x}_{ij} = \mathbf{x}_i$ for all j for an upper-level variable. Thus, the upper-level variable $\mathbf{x}_i$ is replicated n_i times for the n_i lower-level units in the i-th upper-level unit.

The random effect model has an unobserved common random effect Z_i for each lower-level unit in the i-th upper-level unit, the Z_i being initially assumed independently normally distributed $Z_i \sim N(0, 1)$, and conditionally on the Z_i, the Y_{ij} have independent GLMs with linear predictor $\eta_{ij} = \boldsymbol{\beta}' \mathbf{x}_{ij} + \sigma_A Z_i$. The random effect is modelled as acting on the same scale as the linear predictor; we again denote the standard deviation of the random effect by σ_A as in Section 9.2.

The likelihood is then

$$L(\boldsymbol{\beta}, \sigma_A) = \prod_{i=1}^{r} \left\{ \int \left[\prod_{j=1}^{n_i} f(y_{ij} \mid \boldsymbol{\beta}, \sigma_A, z_i) \right] \phi(z_i) \, \mathrm{d}z_i \right\},$$

where $\phi(z_i)$ is the standard normal density function.

The integral in the likelihood does not have a closed form except for $Y \mid z$ normal, and so for other response models we approximate it by Gaussian quadrature: we replace the integral over the normal Z_i by the finite sum over K Gaussian quadrature mass-points z_k with masses π_k. The likelihood is then

$$L(\boldsymbol{\beta}, \sigma_A) \doteq \prod_{i=1}^{r} \left\{ \sum_{k=1}^{K} \left[\prod_{j=1}^{n_i} f(y_{ij} \mid \boldsymbol{\beta}, \sigma_A, z_k) \right] \pi_k \right\}.$$

The likelihood is thus (approximately) the likelihood of a finite mixture of exponential family densities with known mixture proportions π_k at known mass-points z_k, with the linear predictor for the ij-th observation in the k-th mixture component being

$$\eta_{ijk} = \boldsymbol{\beta}' \mathbf{x}_{ij} + \sigma_A z_k.$$

Thus z_k becomes another observable variable in the regression, with regression coefficient σ_A.

The log-likelihood is

$$\ell(\boldsymbol{\beta}, \sigma_A) = \sum_i \log \left\{ \sum_k \pi_k m_{ik} \right\},$$

where for compactness we write

$$m_{ik} = \prod_j f_{ijk},$$

$$f_{ijk} = f(y_{ij} \mid \boldsymbol{\beta}, \sigma_A, z_k)$$
$$= \exp\{\theta_{ijk}y_{ij} - b(\theta_{ijk}) + c(y_{ij})\}$$

with

$$\mu_{ijk} = b'(\theta_{ijk}), \quad V_{ijk} = b''(\theta_{ijk}),$$

$$\eta_{ijk} = g(\mu_{ijk}) = \boldsymbol{\beta}'\mathbf{x}_{ij} + \sigma_A z_k,$$

$$g'_{ijk} = g'(\mu_{ijk}).$$

Then

$$\frac{\partial \ell}{\partial \boldsymbol{\beta}} = \sum_i \frac{\sum_k \pi_k m_{ik}(\partial \log m_{ik}/\partial \boldsymbol{\beta})}{\sum_k \pi_k m_{ik}} = \sum_i \sum_j \sum_k w_{ik}\mathbf{s}_{ijk}(\boldsymbol{\beta}),$$

where w_{ik} is the posterior probability that observation y_{ij} comes from component k:

$$w_{ik} = \frac{\pi_k m_{ik}}{\sum_\ell \pi_\ell m_{i\ell}}$$

and $\mathbf{s}_{ijk}(\boldsymbol{\beta})$ is the $\boldsymbol{\beta}$-component of the score for observation (ij) in component k:

$$\mathbf{s}_{ijk}(\boldsymbol{\beta}) = (y_{ij} - \mu_{ijk})\mathbf{x}_{ij}/V_{ijk}g'_{ijk}.$$

Similarly

$$s_{ijk}(\sigma_A) = (y_{ij} - \mu_{ijk})z_k/V_{ijk}g'_{ijk}.$$

Equating the score to zero gives likelihood equations which are simple weighted sums of those for an ordinary GLM with weights w_{ik}; alternately solving these equations for given weights w_{ik}, and updating these weights from the current parameter estimates, is an EM algorithm. The triple summation over (ij) and k is conveniently (though inefficiently) handled as in the simpler overdispersion case by expanding the data vectors to length Kn by replicating y and X K times, and the Gaussian quadrature variable z n times. Model fitting is then identical to that of a single sample of Kn observations with prior weight vector w. Initial estimates for the first E-step for $\boldsymbol{\beta}$ are conveniently obtained from the ordinary GLM fit, and for σ_A by arbitrary specification other than zero (e.g. $\sigma_A = 1$). A distinctive feature of the weights compared to those in the overdispersion model is that they are calculated for each upper-level unit in the E-step but applied to all lower-level units in this upper-level unit in the M-step.

9.4 Other approaches

A number of approaches have been suggested to avoid the full likelihood analysis above. First, the log-likelihood function can be approximated by a quadratic in the random effect, and standard computational methods for the normal variance component model can then be used, giving approximate ML or REML estimation (Laird and Ware, 1982). This approach has been implemented generally in slightly different forms by Breslow and Clayton (1993), McGilchrist (1994), Goldstein (2003) and Longford (1993), giving penalized quasi-likelihood (PQL) analyses. The success of the approximation depends on the closeness to normality (in the random effect) of the observed data likelihood, and may fail badly, for example, for binary response data (Rodriguez and Goldman, 1995).

Second, the integrals required in the E-step of the EM algorithm can be avoided by Laplace approximations (Steele, 1996) or by Monte Carlo integration (Walker, 1996; McCulloch, 1997).

Third, the problem can be circumvented by the generalized estimating equation (GEE) approach (Liang and Zeger, 1986; Diggle *et al.*, 2002). Here the *marginal* distribution of Y is assumed to be exponential family, and the repeated measures structure is represented by a covariance matrix model whose parameters are estimated by a form of QL (marginal quasi-likelihood (MQL)) which does not require a full parametric specification for the random effect distribution. This approach is widely used. It is formally inconsistent with the random effect model, since this model is always overdispersed because of the variance inflation from the random effect. However in Bernoulli models in which overdispersion cannot be identified, the analyses may be very similar (see Section 9.7.2 for an example).

Fourth, a fully Bayes approach can be followed, with the additional structure of a prior distribution on all the model parameters, and Markov Chain Monte Carlo methods can be used to obtain marginal posterior distributions of the parameters. Gelman *et al.* (1995) gave a detailed exposition of this approach, which is becoming increasingly popular with the widespread dissemination of Bayesian software.

A disadvantage of any approach using a specified parametric form for the "mixing" distribution of the unobserved random effects is the possible sensitivity of the conclusions to this specification. The influential paper by Heckman and Singer (1984) showed substantial changes in parameter estimates with quite small changes in mixing distribution specification; Davies (1987) showed similar effects. This difficulty can be avoided by non-parametric ML estimation of the mixing distribution in one-parameter exponential family distributions, concurrently with the structural model parameters. Clayton and Kaldor (1987) gave an example of this approach, in the simpler framework of a single-level overdispersion model.

9.5 NPML estimation of the masses and mass-points

As in Chapter 8, we may completely drop the parametric distributional assumption for the random effects, and estimate this distribution non-parametrically in single-parameter exponential family distributions. The analysis parallels that in Chapter 8: we treat the masses and mass-points as unknown parameters; the number K of mass-points is also unknown but is treated as fixed, and sequentially increased until the likelihood is maximized. Since the variance of the mixing distribution is a function of the unknown parameters, we absorb the scale parameter σ_A into the mass-point parameters z_k, with linear predictor

$$\eta_{ijk} = \boldsymbol{\beta}' \mathbf{x}_{ij} + z_k.$$

Thus z_k acts as an intercept parameter for the k-th component: it is estimated as in the overdispersion model by including a "component factor" in the model with K levels. One of the z_k parameters will be aliased with the intercept term β_0; alternatively the intercept can be removed from the model.

Differentiating the log-likelihood with respect to π_k, and using $\pi_K = 1 - \sum_1^{K-1} \pi_k$, we have directly

$$\frac{\partial \ell}{\partial \pi_k} = \sum_i \frac{m_{ik} - m_{iK}}{\sum_\ell \pi_\ell m_{i\ell}} = \sum_i \left\{ \frac{w_{ik}}{\pi_k} - \frac{w_{iK}}{\pi_K} \right\}.$$

Equating this to zero gives simply

$$\hat{\pi}_k = \sum_i w_{ik}/n,$$

the standard mixture ML result. The same EM algorithm applies with the additional calculation in each M-step of the estimate of π_k from the weights. Initial estimates of the z_k can be taken as the Gaussian quadrature values scaled by a factor depending on the standard deviation of Y. Hypothesis testing for regression model parameters (nested model comparisons) may be carried out via the LRT using differences of disparities in the usual way. The theoretical justification for this application of standard theory is given in Murphy and van der Vaart (2000).

9.6 Random coefficient models

The analysis can be extended to a class of random coefficient models as in Chapter 8. A frequent case of interest is when lower-level variables have slopes which vary across upper-level units. Consider a simple example with a single lower-level variable x_{1ij} whose coefficient β_1 varies across the upper-level units. We index it by $\beta_{1i} = \beta_1 + U_i$, where U_i represents variation about a "mean" β_1.

The remaining regression coefficients β_2 are fixed. Then conditional on U_i and Z_i, the regression model is

$$\eta_{ij} = \beta_1 x_{1ij} + \beta_2' x_{2ij} + Z_i + U_i x_{1ij},$$

while marginally Z_i and U_i have an unknown joint distribution $g(z, u)$. The likelihood is then

$$L(\beta) = \prod_i \left\{ \int \left[\prod_j f(y_{ij} \mid z_i, u_i) \right] g(z_i, u_i) \, dz_i \, du_i \right\}.$$

By estimating the joint distribution of z_i and u_i non-parametrically, we again obtain the NPML estimate (in the one-parameter exponential family) as a discrete distribution on a finite number of mass-points $(\hat{z}_k, \hat{u}_k)$ in the (z, u) plane, with an estimated mass $\hat{\pi}_k$ for the k-th component. As in Chapter 8, the likelihood is again that of a finite mixture, of regressions on x_1 with a different slope and intercept in each component of the mixture, and on $\mathbf{x}_2$ with the same regression coefficient in each component. The approach can be generalized to any number of explanatory variables, but the number of parameters in the model increases rapidly.

The fitting of these models parallels exactly that in Chapter 8: the random coefficients are handled in the computational implementation by including in the regression model, in addition to the main effect of the component factor, the interaction of this factor with the explanatory variable x_{1ij}. As in Chapter 8 the user provides both a `fixed` macro of the fixed-effect terms in the model, and a `random` macro of (lower-level) variables with random regression coefficients (the intercept term is always included).

Upper-level variable slopes may also be allowed to vary over upper-level units by including them in interactions with the random factor in the same way. This represents a form of generalized overdispersion at the upper level. However, incorporating both *lower-level* overdispersion *and* variance component structure requires two sets of random effects which are *both* modelled non-parametrically. This is beyond the scope of the present discussion. However, it is important to note that since compound Bernoulli models are unidentifiable, one set of random effects is sufficient for two-level Bernoulli models.

9.7   Variance component model fitting

To fit a variance component model, we use the file of macros in `allvc`, with the same `fixed` and `random` macros as for overdispersion modelling, and specify the number of mass-points as before.

The two-level structure has to be set by establishing the connection between lower- and upper-level units. This is done by using the `expand` macro with argument the variate containing the number of lower-level units in each upper-level unit. In the examples this variate is called `rep`. It is an essential requirement for the analysis, and obviates the need to have an upper-level identifier attached to each lower-level unit.

Gaussian quadrature is achieved by $use remz, NPML by $use remnp. The first use only of either of these macros requires three or four arguments:

(1) the y-variate;
(2) a vector of prior weights, of length the number of upper-level units r, which must be declared to be 1 if the observations are equally weighted; in the examples this variate is called wt;
(3) the rep variate, also of length r, used in the expand macro.

If the data are binomial, the fourth argument is the binomial denominator. The prior weight variable allows tabulated response data to be used.

We now consider several examples.

9.7.1 Children's height development
The first example is from Harrison and Brush (1990) also analysed in Goldstein (2003, pp. 132–3). The heights of 26 boys in Oxford were measured on nine occasions over two years, initially at age 12; the occasions were almost equally spaced in time through the school year. The data, given in the file height, are the height in cm, age expressed in months relative to an origin of 13 years, and months relative to an annual time origin of January 1. The heights at each actual age are shown in Fig. 9.1 for each boy. The points are joined by lines to identify the individual boys.

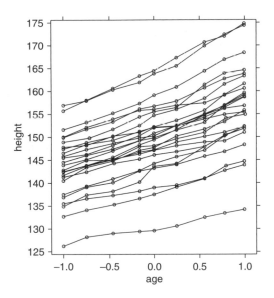

Fig. 9.1. Heights of 26 boys over two years

```
$input 'height' $
$var 234 boy $
$factor boy 26   $
$calc boy = %gl(26,9) $
$gstyle 20 col 1 sym 5 line 1 $
$var 26 style1 $calc style1 = 20 $
$calc tage = age + 13 $
$graph (s = 1 h = 'age' v = 'height' x = 11.9,14.1 y = 125,175)
        ht tage style1 boy $
```

The growth patterns are very similar though naturally there is variation in initial height, which increases with age.

We begin with a normal random effect model, fitted by 20-point Gaussian quadrature—we do not provide a separate macro for the normal model.

```
$input 'allvc' $
$var 26 wt rep $
$calc wt = 1 : rep = 9 $
$macro fixed age $endm
$macro random 1 $endm
$use expand rep $
$use mass20 $
$use remz ht wt rep $
```

	estimate	s.e.	parameter
1	6.526	0.8011	AGE(J_)
2	85.58	0.2970	Z_

scale parameter 62.78

```
MLE of sigma =    8.041
-2 log L = 1784.63 , deviance =  1784.63, after 6 iterations
```

The linear regression on age is very strong and the intra-class correlation is very high: $\sigma_A^2/(\sigma_A^2 + \sigma^2) = 0.991$! At this *growth spurt* period, is growth linear, or is it accelerating? We fit the quadratic term.

```
$calc a2 = age**2 $
$macro fixed age + a2 $endm
$use remz $
```

	estimate	s.e.	parameter
1	6.518	0.8025	AGE(J_)
2	0.7104	1.402	A2(J_)
3	85.41	0.4488	Z_

```
scale parameter 62.97

MLE of sigma =     8.036
-2 log L = 1784.35 , deviance =   1784.35, after 8 iterations
```

The deviance change is only 0.28. There is no evidence of curvature.

Is the assumption of a normal random effect distribution reasonable? With only 26 boys we do not have much information about it, but we try the discrete random effect model with linear age, with increasing numbers of mass-points.

```
$macro fixed age $endm
$use mass2 $use remnp ht wt rep $
$use mass3 $use remnp $
...
$use mass10 $use remnp $
```

The successive deviances and values of $\hat{\sigma}$ and K are shown in the table below.

K	Deviance	$\hat{\sigma}$
2	1466.33	5.16
3	1320.42	3.63
4	1212.63	2.81
5	1154.50	2.41
6	1054.80	1.92
7	1018.00	1.77
8	971.76	1.58
9	1018.00	1.77
10	932.46	1.44

Beyond $K = 7$ some mass-points are duplicated with the same total mass as in lower-dimension models, though the remaining mass is assigned differently over the other (different) mass-points.

This suggests that a 7-point discrete model is sufficiently complex. The deviance change from the 20-point Gaussian model is 766.63, which is very large, though a bootstrap test would be necessary to assess this formally. The within-boy standard deviation drops from 8.04 in the Gaussian model to 1.77 for the 7 mass-point discrete model. We consider the interpretation of this model below, but first consider a random slope model.

Is the linear growth trend consistent for all boys? We extend the 7-point model by allowing the linear trend to vary across boys.

```
$macro random age $endm
$use remnp$
          estimate        s.e.       parameter
     1      9.285        0.3584       AGE(J_)
```

```
        2          164.8        0.2318        MASS_(1)
        3          159.5        0.3279        MASS_(2)
        4          155.2        0.1466        MASS_(3)
        5          150.0        0.1095        MASS_(4)
        6          144.6        0.1462        MASS_(5)
        7          138.4        0.1893        MASS_(6)
        8          130.3        0.3279        MASS_(7)
        9         -0.614        0.6208        MASS_(2).AGE(J_)
       10         -2.260        0.4243        MASS_(3).AGE(J_)
       11         -2.840        0.3963        MASS_(4).AGE(J_)
       12         -3.265        0.4237        MASS_(5).AGE(J_)
       13         -4.077        0.4624        MASS_(6).AGE(J_)
       14         -5.562        0.6208        MASS_(7).AGE(J_)
scale parameter 0.9664


MLE of sigma =     1.595
mixture proportions :   0.0769   0.0385   0.1923   0.3451
                        0.1934   0.1154   0.0385
-2 log L =   970.66 , deviance =     970.66, after 8 iterations
```

The deviance decreases by 47.34 on 6 df, showing clear heterogeneity in slope. Examination of the slope differences shows that the age slope decreases smoothly with increasing mass-point, that is, with decreasing height at age 12. Smaller boys increase their heights less rapidly than taller boys. Before examining the fitted model, we check again for an acceleration term.

```
$macro fixed age + a2 $endm
$use remnp $
          estimate        s.e.        parameter
        1        9.277        0.3534        AGE(J_)
        2        0.736        0.1713        A2(J_)
        3      164.5        0.2396        MASS_(1)
        4      159.2        0.3312        MASS_(2)
        5      154.9        0.1613        MASS_(3)
        6      149.7        0.1296        MASS_(4)
        7      144.3        0.1611        MASS_(5)
        8      138.1        0.2000        MASS_(6)
        9      130.0        0.3311        MASS_(7)
       10       -0.614        0.6122        MASS_(2).AGE(J_)
       11       -2.261        0.4184        MASS_(3).AGE(J_)
       12       -2.842        0.3908        MASS_(4).AGE(J_)
       13       -3.265        0.4179        MASS_(5).AGE(J_)
       14       -4.081        0.4560        MASS_(6).AGE(J_)
       15       -5.562        0.6122        MASS_(7).AGE(J_)
scale parameter 0.9397


MLE of sigma =     1.572
mixture proportions :   0.0769   0.0385   0.1923   0.3452
```

```
                    0.1933   0.1154   0.0385
  -2 log L =   963.74 , deviance =   963.74, after 8 iterations
```

The deviance change of 6.92 on 1 df is large; with the random slope model we can now detect quadratic curvature in the regression.

Could this be a higher-order curvature? Adding the cubic term in age to the fixed macro gives a deviance change of only 0.62.

Is the quadratic curvature itself varying across boys? Adding the quadratic term in age to the random macro gives a deviance change of 1.13 on 6 df.

Is there a seasonal variation in growth? We define the four seasons represented by the four measurement times in each year (they occur in June/July, September, January, and April/May) by a four-level factor, and add this to the fixed part of the model. The deviance change is 0.33 on 3 df.

We conclude that the random slope but fixed quadratic model is appropriate. We examine the posterior probabilities of component membership for this model. We give the probabilities to 2 dp to save space, and suppress probabilities smaller than 0.01.

```
$tab the w_ mean for i_,k_ $
      K_      1      2      3      4      5      6      7
      I_
   1.000                           1
   2.000                                         1
   3.000                    1
   4.000      1
   5.000                           1
   6.000                                  1
   7.000                                  1
   8.000                           1
   9.000                                         1
  10.000                                                1
  11.000                           1
  12.000             1
  13.000             1
  14.000      1
  15.000                                  1
  16.000                           0.97   0.03
  17.000                                  1
  18.000                           1
  19.000      1
  20.000                                  1
  21.000                                  1
  22.000             1
  23.000                                  1
  24.000             1
  25.000                                         1
  26.000                                         1
```

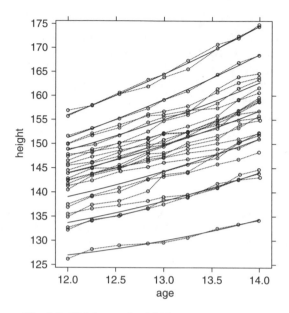

Fig. 9.2. Height trends of 26 boys over two years

With one exception, the boys are assigned to a single component with estimated probability 1.0. The random intercept and slope model classifies the boys into 7 homogeneous groups. The fitted models for each component are shown in Fig. 9.2, superimposed on the previous graph (though with a different graph symbol—the dashed lines are the observed heights, the solid curves are the fitted models for each component). The graph code is given below. The option r = n or refresh = no allows the second graph to be overlaid on the first. The postscript copy c = 'post' is needed to save the overlaid graphs.

```
$gstyle 21 col 1 line 2 sym 5 $
$var 26 style2 $calc style2 = 21 $
$var 26 style3 $calc style3 = 10 $
$assign lage = tage,tage,tage,tage,tage,tage,tage $
$tab the  fv mean for mass_,lage into mfv by mas_,tag $
$assign boyclass = 4,6,3,1,4,5,5,4,6,7,
4,3,3,2,5,4,5,4,1,4,4,3,4,3,6,6 $
$fac boyclass 7 : mas_ 7 $
$calc gboy = boyclass(boy) $
$factor gboy 7 $
$graph (s = 1 c = 'post' h = 'age' v = 'height' x = 11.9,14.1
        y = 125,175) mfv tag style3 mas_ $
```

```
$graph (s = 1 c = 'post' h = 'age' v = 'height' x = 11.9,14.1
        y = 125,175 r = n) ht tage style2 boy $
```

The increasing variability of heights with increasing age is now seen to be a combination of curvature and different slopes for the different groups.

9.7.2 School effects on pupil performance

Multi-level models are widely used in teacher and school effectiveness studies. A principal aim of these studies is to identify important teacher or school variables which affect pupil performance. The early use of multi-level models required normality assumptions for the school or teacher random effects. Since the numbers of schools, teachers or classes in such studies are much smaller than the number of pupils, the model assumption for these effects may be much more serious than for the pupil outcome variables. We illustrate this important point with the second example, a re-analysis of a study in Israeli schools by Zuzovsky and Aitkin (1990, 1991, 1994). We do not give the data. The study analysed was part of the second (1984) International Education Association study of science studies in Israeli elementary schools. A two-stage sample design provided complete data from 1587 pupils in 69 schools, one complete class being sampled from each school. This confounded class- and school-level effects; later studies sampled more than one class from each school. The outcome variable was the percentage of correct items on a 45-item science test. Explanatory variables at the pupil level were:

SEX (1 = male, 2 = female);
RCOMP, a reading comprehension test score (effective range 50–100) used as a surrogate for ability;
GEN, the number of generations the pupil's family had lived in Israel (0–3);
ORIGIN, a dummy variable for Ashkenazi (0) or Sephardic (1) origin; and
SES, family socioeconomic status index scaled to have mean 0 and variance 1.

Three of these variables were aggregated to the class level, giving (mean) class context variables CRCOMP, CORIGIN and CSES. Other variables at the class level were:

DISADVAN, the percentage of disadvantaged students in the class (0–100);
INSERV, a teacher input dummy variable taking the value 1 if the teacher had participated more than 10 hours per year in in-service training, and 0 otherwise;
TRAINING, a three-item composite index of the teacher's experience and preparation for science teaching;
REASON, a composite indicator of teacher's attitude measuring the extent to which the teacher preferred to foster reasoning in teaching science (1–3);
POSGROUP, a composite indicator of positive teacher attitudes towards group work (1–3);

LECTURE, the extent to which the teacher used a formal lecturing style in teaching science (1–4);

UNITS, the number of curriculum units studied in the class until the end of the fifth grade (1–16);

CLASSEXP, an aggregate pupil perception of the extent to which pupils do experiments in class (1–3);

DIVERSITY, teacher's report of a diversified and rich mode of instruction in the class (1–4);

ATMOSPHERE, a composite indicator of pupils' perception of a liberal atmosphere in class (free movement of pupils and absence of punishment) (1–3);

LABTIME, time allocated to laboratory work according to the teacher's report (1–4);

GROUPCHANGE, a dummy variable indicating whether the teacher changed the composition of pupil groups, reflecting flexible expectations of pupil progress (1 = no, 2 = yes).

The variables at the school level were:

ANIMALC, a dummy variable indicating whether the school had an animal corner (1) or not (0);

CURRICULUM, whether the new science curriculum had been introduced (1) or not (0);

CONDITIONS, the teacher assessment of whether the teaching conditions in school were adequate (1) or not (0).

Assuming a normal model for test score (the sum of 45 binary items) and for the school/teacher random effects, the two-level model was fitted with a full main-effect model for all the variables, using the analytic VARCL program of Longford (1987). This was reduced to a final model by backward elimination using a critical value of $t = 2.25$ for elimination. The full and final models are shown in Table 9.1.

All the pupil-level variables are important, with interpretations consistent with other Israeli school studies: girls perform less well than boys in science, children of Sephardic origin perform less well than those of Ashkenazi origin, and new migrants perform less well than native Israelis. At the upper level, classes with a high proportion of Sephardic children do less well than those with a low proportion, those with a liberal atmosphere perform better, as do those with changing group composition and an animal corner.

In a re-analysis, Aitkin (1996c) examined the sensitivity of the normal analysis through Gaussian quadrature with different numbers of mass-points, and non-parametric estimation with up to five mass-points. For the non-parametric estimation the deviances for successive numbers of mass-points were 12742.0

Table 9.1. Normal model estimates

Variable	Full model		Final model	
	Estimate	SE	Estimate	SE
1	24.0		25.5	
SEX	−3.70	0.65	−3.72	0.70
RCOMP	0.458	0.025	0.459	0.024
GEN	1.02	0.46	1.00	0.40
ORIGIN	−1.84	0.76	−1.85	0.70
SES	2.00	0.39	2.02	0.40
CORIGIN	−9.18	5.27	−8.10	2.94
CRCOMP	0.186	0.138		
CSES	1.13	2.71		
DISADVAN	0.063	0.054		
INSERV	1.24	1.63		
REASON	0.16	1.45		
LECTURE	1.47	1.04		
UNITS	0.302	0.313		
TRAINING	−3.79	2.21		
POSGROUP	−0.23	1.60		
CLASSEXP	3.71	2.45		
DIVERSITY	0.78	1.77		
ATMOSPHERE	10.2	4.71	10.4	4.58
LABTIME	0.64	1.15		
GROUPCHANGE	2.64	1.77	3.72	1.60
ANIMALC	3.49	1.59	4.33	1.49
CURRICULUM	0.72	1.63		
CONDITIONS	−1.10	1.70		
$\hat{\sigma}_A$	4.41		5.50	
$\hat{\sigma}$	12.23		12.21	
Deviance	12544.0		12556.5	

$(K = 1)$, 12560.8 $(K = 2)$, 12551.7 $(K = 3)$, 12539.0 $(K = 4)$, and 12539.0 $(K = 5)$. Parameter estimates for the full model are shown in Table 9.2 for the analytic normal, Gaussian quadrature and four-point non-parametric analysis.

Relative to the full-model analytic normal analysis, the four-point non-parametric analysis reduced the deviance by 5.0, which is not large. Relative to four-point Gaussian quadrature, however, the reduction is much larger—9.1 for the six extra parameters. This does not indicate a serious failure of the normal model, but a cause for concern is the substantial changes in many of the upper-level variable estimates, not just between the analytic normal and the non-parametric

Table 9.2. Normal, Gaussian quadrature and non-parametric estimates

Variable	Normal	GQ6	GQ4	GQ3	GQ2	NPML
1	24.0	26.7	26.6	30.2	25.6	27.1
SEX	−3.70	−3.55	−3.92	−3.81	−3.93	−3.86
RCOMP	0.458	0.457	0.459	0.469	0.471	0.460
GEN	1.02	0.82	0.98	1.03	1.20	1.01
ORIGIN	−1.84	−1.92	−2.03	−1.66	−2.71	−2.17
SES	2.00	1.88	1.94	1.94	2.19	2.00
CORIGIN	−9.18	−4.14	−3.63	0.02	−3.36	−3.61
CRCOMP	0.186	0.095	0.082	0.016	0.032	0.086
CSES	1.13	5.50	5.43	−2.96	0.46	4.96
DISADVAN	0.063	0.120	0.109	−0.060	−0.008	0.111
INSERV	1.24	3.35	2.33	5.36	2.43	0.63
REASON	0.16	−0.12	0.19	3.17	2.41	1.30
LECTURE	1.47	−0.20	0.23	1.44	−0.12	0.04
UNITS	0.302	−0.53	−0.77	−0.72	−0.30	−0.60
TRAINING	−3.79	−4.27	−4.55	−5.05	−3.22	−3.66
POSGROUP	−0.23	−1.39	−1.20	−2.71	−0.74	−0.68
CLASSEXP	3.71	−2.79	−4.64	−6.19	−4.50	−5.77
DIVERSITY	0.78	−1.45	0.09	−2.36	0.60	0.41
ATMOSPHERE	10.2	18.9	19.7	12.7	15.7	20.1
LABTIME	0.64	2.42	1.94	4.12	0.91	1.14
GROUPCHANGE	2.64	3.28	3.77	3.71	1.22	2.85
ANIMALC	3.49	−2.27	−0.91	−3.42	−0.35	0.56
CURRICULUM	0.72	−0.30	0.38	1.77	−1.37	1.04
CONDITIONS	−1.10	−0.26	0.93	3.44	2.03	0.75
$\hat{\sigma}_A$	4.41	7.37	6.16	6.01	5.77	6.21
$\hat{\sigma}$	12.23	12.10	12.12	12.11	12.31	12.11
Deviance	12544.0	12551.2	12548.1	12547.5	12561.23	12539.0

analysis, but among the analytic normal and Gaussian quadrature analyses as well. Backward elimination from the full non-parametric model with the same t-value criterion (though all final variables now have t-values of at least 2.70) results in a quite different final model, shown in Table 9.3 with the final analytic normal model.

Of particular note is that the class composition variable CORIGIN is replaced by the class composition variable CRCOMP—it is class ability rather than the proportion of Sephardic students which is important. Atmosphere and emphasis on reasoning are positively associated with success; the extent of class experiments and the number of units studied are negatively associated. GROUPCHANGE and ANIMALC are no longer important.

Why are these results so different? An examination of the four-point non-parametric estimate from the final model suggests a reason.

Table 9.3. Final models, normal and non-parametric

Variable	Normal		NPML	
	PE	SE	PE	SE
1	25.5		29.1	
SEX	−3.72	0.70	−3.78	0.63
RCOMP	0.459	0.024	0.467	0.025
GEN	1.00	0.40	1.21	0.33
ORIGIN	−1.85	0.70	−2.31	0.58
SES	2.02	0.40	2.08	0.34
CORIGIN	−8.10	2.94		
CRCOMP			0.149	0.052
CSES				
DISADVAN				
INSERV				
REASON			3.40	1.22
LECTURE				
UNITS			−0.54	0.20
TRAINING				
POSGROUP				
CLASSEXP			−5.67	2.04
DIVERSITY				
ATMOSPHERE	10.4	4.58	8.15	1.93
LABTIME				
GROUPCHANGE	3.72	1.60		
ANIMALC	4.33	1.49		
CURRICULUM				
CONDITIONS				
$\hat{\sigma}_A$	5.50		6.25	
$\hat{\sigma}$	12.21		12.20	
Deviance	12556.5		12556.7	

The estimated mass-points and masses are shown below, centred to an origin of zero. The standard deviation of the estimated distribution is 6.25. The estimated posterior probabilities of component membership are shown in Table 9.4, to 2 dp.

k	1	2	3	4
mass-pt	+11.4	+3.9	−6.4	−18.6
mass	0.074	0.509	0.403	0.014

Table 9.4. Probabilities of component membership

School/component	1	2	3	4	School/component	1	2	3	4
1		1			36		1		
2		1			37			1	
3			1		38		1		
4		0.11	0.89		39		0.98	0.02	
5			1		40				1
6			1		41	0.99	0.01		
7		1			42		0.04	0.96	
8		0.04	0.96		43			1	
9	0.03	0.97			44		0.94	0.06	
10			1		45			1	
11	0.47	0.53			46		1		
12			1		47			1	
13			1		48	0.01	0.99		
14		0.66	0.34		49	0.60	0.40		
15	0.82	0.18			50	1			
16	0.10	0.90			51	0.04	0.96		
17		1			52		1		
18			1		53			1	
19			1		54		1		
20			1		55			1	
21			1		56		1		
22	0.01	0.96	0.03		57		1		
23			1		58			1	
24		1			59	0.01	0.99		
25		1			60			1	
26		0.07	0.93		61		1		
27			1		62			1	
28		0.02	0.98		63		1		
29		1			64		1		
30		1			65	0.01	0.99		
31		1			66		0.98	0.02	
32		0.45	0.55		67	0.01	0.99		
33			1		68	0.97	0.03		
34		1			69		0.99	0.01	
35	0.01	0.92	0.07						

The fourth component has mass 0.014, and contains just one school—40—with probability 1. This component defines an *outlier*—this school does particularly badly, with a mean score 18.6 points below the average, and 29 points below the best. The other three components all contain several schools with high probability, and some others with appreciable probability. These remaining components represent general variation across schools.

The outlier school is removed in the non-parametric analysis since it is the only school in the fourth component. This removal has a substantial effect on the upper-level model, producing the changes in the importance of the upper-level variables. This change is accentuated by the large upper-level model, with 18 variables at this level for the 69 schools. The effects on the pupil-level estimates are much smaller as the sample size at this level is large.

The re-analysis provides a warning about the sensitivity of large upper-level models to outliers in the school sample; non-parametric estimation provides a routine check on this possibility, and if no substantial change is observed, a confirmation of the reasonableness of the normal random effect model assumption.

9.7.3 Multi-centre trial of beta-blockers

The third example is the 22-centre clinical trial of beta-blockers for reducing mortality after myocardial infarction, described by Yusuf *et al.* (1985) and analysed in detail in Gelman *et al.* (1995) and Aitkin (1999). The data are shown in Table 9.5, and are in the file betablok. In each centre r patients died out of n treated, under either the treatment T or control C condition. Studies involving multiple centres are common in drug trials, where an important issue is the *generalizability* of the treatment effect across different patient populations. The analysis of the multiple tables, or of multiple data sets of other kinds, is frequently called *meta-analysis*—*meta* in the sense of "at a higher level" than that of a single trial.

The outcomes are represented by a two-level model, with centres as the upper level and patients as the lower-level. There is only one explanatory variable, the treatment assignment at the lower-level. This is coded in the data file as 0 for the control and 1 for the beta-blocker treatment.

Table 9.5. 22-centre trial of beta-blockers

Centre	C		T		Centre	C		T	
	r	n	r	n		r	n	r	n
1	3	39	3	38	12	47	266	45	263
2	14	116	7	114	13	16	293	9	291
3	11	93	5	69	14	45	883	57	858
4	127	1520	102	1533	15	31	147	25	154
5	27	365	28	355	16	38	213	33	207
6	6	52	4	59	17	12	122	28	251
7	152	939	98	945	18	6	154	8	151
8	48	471	60	632	19	3	134	6	174
9	37	282	25	278	20	40	218	32	209
10	188	1921	138	1916	21	43	364	27	391
11	52	583	64	873	22	39	674	22	680

We input the data, and fit both normal and non-parametric models for Z.

```
$input 'betablok' : 'allvc' $
$calc treat = %gl(2,1) - 1 $
$macro fixed treat $endm
$macro random 1 $endm
$var 22 rep wt $calc rep = 2 : wt = 1 $
$error b n $
$use expand rep $
$use mass1 $use remz r wt rep n $
         estimate          s.e.       parameter
      1      -2.197       0.03359          1
      2      -0.2574      0.04942       TREAT(J_)
      3       0.000       aliased         Z_
   scale parameter 1.000


    -2 log L = 12179.74 , deviance =    305.76, after 2 iterations
```

The model fitted with one mass-point is identical to the fixed-effect model ignoring among-centre variation. We successively increase the number of mass-points, from 2–10 and 20. The parameter estimates, standard errors (expected complete data information-based) and the deviance are shown in Table 9.6.

The deviances vary considerably with small numbers of mass-points. The standard deviation σ_A also varies, but the treatment effect estimate $\hat{\beta}$ and its standard error are almost unchanged from the independence model, despite the very large change in deviance. To check the standard error calculation, we drop the treatment variable:

```
$use mass20 $
$macro fixed 1 $endm
$use remz $
         estimate          s.e.       parameter
      1      -2.196       0.02512          1
      2       0.6063      0.03793         Z_
   scale parameter 1.000


    -2 log L = 12003.31 , deviance =    129.33, after 19 iterations
```

The deviance change is 27.46, so the LRTS-based standard error is $0.2612/\sqrt{27.46} = 0.04985$, compared with 0.04976 from the GLIM output. There is almost no underestimation. The treatment produces a significant, though small, reduction in death risk compared to the control: the odds of death under the beta-blocker treatment are reduced by 23% relative to those under the control treatment.

Table 9.6. Gaussian quadrature for beta-blockers

K	α	SE	β	SE	σ_A	Deviance
1	−2.197	0.034	−0.257	0.049	0.000	305.76
2	−2.034	0.035	−0.257	0.050	0.366	148.06
3	−2.239	0.034	−0.258	0.050	0.360	103.55
4	−1.945	0.036	−0.260	0.050	0.398	114.34
5	−2.238	0.034	−0.258	0.050	0.455	102.34
6	−2.090	0.034	−0.262	0.050	0.359	105.05
7	−2.238	0.034	−0.259	0.050	0.528	103.60
8	−2.088	0.034	−0.262	0.050	0.410	101.92
9	−2.238	0.034	−0.259	0.050	0.588	105.35
10	−2.086	0.034	0.262	0.050	0.454	100.73
20	−2.074	0.034	−0.261	0.050	0.604	101.87

For the non-parametric estimation, we use remnp:

```
$use mass2 $use remnp r wt rep n $

          estimate        s.e.        parameter
     1     -0.2554      0.04967        TREAT(J_)
     2     -1.650       0.04939        MASS_(1)
     3     -2.394       0.03781        MASS_(2)
 scale parameter 1.000
 mixture proportions :  0.3025  0.6975
 -2 log L = 12019.22 , deviance =   145.24, after 5 iterations


 $use mass3 $use remnp $

          estimate        s.e.        parameter
     1     -0.2582      0.04974        TREAT(J_)
     2     -1.610       0.05136        MASS_(1)
     3     -2.250       0.03994        MASS_(2)
     4     -2.834       0.07369        MASS_(3)
 scale parameter 1.000
 mixture proportions :  0.2490  0.5118   0.2392
 -2 log L = 11975.27 , deviance =   101.29, after 7 iterations


$use mass4 $use remnp $

          estimate        s.e.        parameter
     1     -0.2582      0.04974        TREAT(J_)
     2     -2.025       86935.         MASS_(1)
     3     -1.610       0.05135        MASS_(2)
```

```
        4        -2.250       0.03994       MASS_(3)
        5        -2.834       0.07369       MASS_(4)
scale parameter 1.000
mixture proportions :  0.0000  0.2492  0.5116  0.2392
-2 log L = 11975.27 , deviance =   101.29, after 9 iterations
```

For the four mass-point model only three mass-points are needed—the first mass-point has no mass. The standard error for this mass-point estimate goes to ∞ because of the negligible weight attached to it. Further increases in K up to $K = 7$, by varying the scaling of the mass-points with tol_, leave the disparity unchanged, so the three mass-points appear to define the NPML estimate of the mixing distribution. However, this cannot be the case since the deviance for 10-point Gaussian quadrature is less than 101.29. We search further.

```
$use mass8 $use remnp $

              estimate          s.e.        parameter
        1      -0.2579        0.04977       TREAT(J_)
        2      -2.046          740.4        MASS_(1)
        3      -2.046          743.8        MASS_(2)
        4      -1.440         0.06903       MASS_(3)
        5      -1.786         0.06582       MASS_(4)
        6      -2.258         0.04035       MASS_(5)
        7      -2.975          0.1041       MASS_(6)
        8      -2.684         0.09935       MASS_(7)
        9      -2.046          437.3        MASS_(8)
scale parameter 1.000

mixture proportions :  0.0000  0.0000  0.1796  0.0978
                       0.4830  0.1647  0.0750  0.0000
-2 log L = 11966.94 , deviance =    92.96, after 29 iterations
```

Three mass-points are identical and have no mass (and very large standard errors), so this is a five-point estimate, with mass-points at -1.440, -1.786, -2.258, -2.684 and -2.975, and corresponding masses of 0.1796, 0.0978, 0.4830, 0.0750 and 0.1647. The masses are not increasing and decreasing smoothly, but are fairly symmetrical. This non-monotone behaviour makes the NPML estimate difficult to find from the Gaussian quadrature starting points with smaller K.

The treatment effect estimate is almost identical to that in the Gaussian quadrature analyses, but the standard deviation of the mass-point distribution is 1.148, much larger than that for the normal models, partly because of the relatively large masses at the extremes of the distribution.

Is there evidence of non-normal variation in the intercepts? The disparity (or deviance) change between the five-point Gaussian quadrature and the NPML analyses is 9.38 for the 8 additional parameters. This is not at all persuasive (by the usual χ^2_8 test)—there is no evidence that the normal assumption for the random effect is unreasonable.

We finally check on the constancy of the treatment effect, by fitting a random treatment model.

```
$macro fixed treat $endm
$macro random treat $endm
$use remnp $
          estimate        s.e.      parameter
     1      0.2703       0.1950      TREAT(J_)
     2     -2.424    17084552.       MASS_(1)
     3     -2.121        3228.       MASS_(2)
     4     -1.486       0.08908      MASS_(3)
     5     -1.684       0.08254      MASS_(4)
     6     -2.255       0.04500      MASS_(5)
     7     -2.895       0.1317       MASS_(6)
     8     -2.937       0.1451       MASS_(7)
     9      0.000       aliased      MASS_(8)
    10     -0.3951       4361.       MASS_(2).TREAT(J_)
    11     -0.4292      0.2342       MASS_(3).TREAT(J_)
    12     -0.7567      0.2332       MASS_(4).TREAT(J_)
    13     -0.5300      0.2058       MASS_(5).TREAT(J_)
    14     -0.7299      0.2845       MASS_(6).TREAT(J_)
    15      0.000       aliased      MASS_(7).TREAT(J_)
    16      0.000       aliased      MASS_(8).TREAT(J_)
  scale parameter 1.000

  mixture proportions :  0.0000  0.0000  0.1816  0.0869
                         0.4937  0.1581  0.0798  0.0000
  -2 log L = 11955.59 , deviance = 81.61, after 62 iterations
```

The deviance reduces to 81.61, again with five mass-points. This is a reduction in deviance of 11.35 for the four additional parameters. Since these interaction parameters do not affect the mixture structure, their zero null hypothesis values are internal to the parameter space, and the sample sizes are relatively large, the usual asymptotic χ^2 theory applies to the disparity or deviance difference. The P-value of 11.35 is 0.023 for χ^2_4—there is quite strong evidence for treatment heterogeneity.

The parameter estimates from mass 8 are aliased—the posterior weights for this component are so close to zero that the model terms are dropped. We examine the treatment variation across centres, and also examine the posterior probabilities of component membership for each centre. We table these probabilities below, rounded to 2 dp—they are difficult to read from the exponential form of the output. Since components 1, 2 and 8 have no mass, we omit these columns, and also omit probabilities less than 0.01. We adopt the convention that a centre is identified with a component if the posterior probability of the centre belonging to the component is at least 0.90.

```
$tabulate the w_ mean for i_,k_ $

   K_     3     4     5     6     7
   I_
    1   .02   .04   .76   .09   .10
    2         .05   .95
    3   .01   .08   .90   .01   .01
    4               1
    5               .96         .04
    6   .01   .08   .85   .03   .03
    7         1
    8               1
    9         .42   .58
   10               1
   11               1
   12   1
   13                     .98   .02
   14                           1
   15   .99   .01
   16   .99   .01
   17   .01   .19   .80
   18               .07   .48   .45
   19               .89   .11
   20   .98   .02
   21               1
   22                     1
```

The regressions in components 3–7 are given in Table 9.7.

Component 5, which contains 50% of the mass, is identified by the eight centres 2, 3, 4, 5, 8, 10, 11 and 21. This component has a treatment estimate of −0.260,

Table 9.7. Regressions in each component

k	β_{0k}	β_{1k}	π_k
3	−1.486	−0.159	0.1816
4	−1.684	−0.487	0.0869
5	−2.255	−0.260	0.4937
6	−2.895	−0.460	0.1581
7	−2.937	+0.270	0.0798

very close to the common value of -0.258 in the main-effect model. Components 4 and 6, containing 24% of the mass, and identified by centres 7, 13 and 22, have consistently larger treatment estimates, around -0.47, though the death rate under the control condition differs between these two components. Component 3 has a smaller treatment estimate, of -0.16. It is identified by centres 12, 15, 16 and 20.

Component 7 is identified only by centre 14 (note that its mass is 0.0798 which is not much larger than $1/22 = 0.0454$), which has an *increased* death risk under the beta-blocker treatment instead of a reduced risk. This centre has the fourth-largest trial.

Centres 1, 6, 9, 17, 18 and 19 are not identified with a single component. These centres have small or moderate size trials, which do not estimate the treatment effect very precisely. The large trials precisely identify single components; the small trials have posterior probabilities which are spread over these components.

Like centre 7, centres 17, 18 and 19 have positive instead of negative treatment estimates in the individual trials, but their trials are much smaller, so their estimates are more diffuse.

Thus we conclude that there is significant treatment variation across the centres. Half of the centres have consistent treatment effect estimates of -0.25, very close to the previous estimate of the *fixed* treatment effect. A quarter of the centres have *larger* treatment effect estimates of -0.5, and a smaller group of four centres have a smaller treatment effect estimate of -0.15. Centre 14 stands out as an "outlier" in having an *increase* in risk of $+0.27$ under the treatment instead of a reduction.

The analysis of this study by Aitkin (1999) used a three mass-point estimate for the intercept and slope variation which was incorrectly thought to be the NPML estimate. This analysis consequently found no significant evidence of variation in the treatment effect. Gelman *et al.* (1995) found a posterior median for the treatment effect of -0.25 with a posterior standard deviation of about 0.07. The estimated treatment effect distribution above has mean -0.251 and standard deviation 0.084 which are very similar, but our analysis identifies centre 14 as an outlier.

9.7.4   *Longitudinal study of obesity*

The fourth example is of data from the Muscatine, Iowa study of child development. A component of the Muscatine study was a longitudinal study of childhood obesity, discussed in Woolson and Clarke (1984). Part of the data from Woolson and Clarke were re-analysed by Fitzmaurice *et al.* (1994) and Aitkin (1999). These data were binary indicators of obesity on 1,014 children who were 7–9 years old in 1977, and were followed up in 1979 and 1981. Children were classified as obese if their weights were more than 210% of the population median weight for their gender and height; about 20% of children were classified as obese. The repeated binary response of interest y is whether the child is obese (1) or not (0) at each occasion. The sex of the child is also recorded as binary: male 0, female 1. Data on many

Table 9.8. Obesity status on three occasions

1977	1979	1981	Males	Females
0	0	0	150	154
0	0	1	15	14
0	1	0	8	13
0	1	1	8	19
1	0	0	8	2
1	0	1	9	6
1	1	0	7	6
1	1	1	20	21
—	1	1	13	8
—	1	0	3	1
—	0	1	2	4
—	0	0	42	47
1	—	1	3	4
1	—	0	1	0
0	—	1	6	16
0	—	0	16	3
1	1	—	11	11
1	0	—	1	1
0	1	—	3	3
0	0	—	38	25
—	—	1	14	13
—	—	0	55	39
—	1	—	4	5
—	0	—	33	23
1	—	—	7	7
0	—	—	45	47

children are incomplete, and only 460 children had complete data from all three occasions. Table 9.8 gives the child's obesity status at all three occasions for the 1,014 children with complete or incomplete data; the data are in the file obesity in a less compact form.

We input the data, with the rep and weight (wt) variables defined there. We follow previous analyses in recoding age to its linear and quadratic components a1 and a2.

```
$slen 108 $
$data y sex age $
$input 'obesity' $
$calc n = 1 $
   :  a1 = (age - 10)/2 : a2 = 3*a1**2 - 2 $
   : sa1 = sex*a1 : sa2 = sex*a2 $
```

```
$macro fixed sex + a1 + a2 + sa1 + sa2 $endm
$macro random 1 $endm
$use expand rep $
```

Fitzmaurice *et al.* modelled the *marginal* probability of response by a logistic
model with linear and quadratic age terms and their interactions with sex, and
saturated the covariance matrix between occasions. This is similar to the GEE
approach of Liang and Zeger but the parameters are estimated by full ML.
They analysed both the subset of children with complete data and the full
sample with both complete and incomplete data, and found substantial changes
in the conclusions, demonstrating the need for inclusion of all the data in the
analysis.

We repeat the analysis of the full sample with the random-effect model, with
the two-level structure of children and occasions within child. The intra-class
correlation structure on the logit scale is simple, though on the probability scale
this corresponds to a general correlation structure because of the non-linear
transformation.

```
$use mass1 $
$error b n $
$use mass1 $use remnp y wt rep n $
```

	estimate	s.e.	parameter
1	0.03388	0.1052	SEX(J_)
2	0.1458	0.09075	A1(J_)
3	0.01246	0.05226	A2(J_)
4	0.1128	0.1304	SA1(J_)
5	-0.08246	0.07339	SA2(J_)
6	-1.328	0.07400	MASS_(1)

```
scale parameter 1.000

mixture proportions :   1.000
-2 log L = 2271.02 , deviance =  2271.02, after 2 iterations
```

```
$use mass2 $use remnp $
```

	estimate	s.e.	parameter
1	0.02854	0.1592	SEX(J_)
2	0.3390	0.1400	A1(J_)
3	0.03134	0.07998	A2(J_)
4	0.3412	0.1952	SA1(J_)
5	-0.1907	0.1131	SA2(J_)
6	1.157	0.1303	MASS_(1)
7	-3.304	0.1549	MASS_(2)

```
scale parameter 1.000
```

```
  mixture proportions :   0.2399   0.7601
  -2 log L = 1892.09 , deviance =   1892.09, after 56 iterations


$use mass3 $use remnp $

            estimate          s.e.        parameter
      1       0.02375        0.1593        SEX(J_)
      2        0.3395        0.1401        A1(J_)
      3       0.03214       0.08008        A2(J_)
      4        0.3446        0.1955        SA1(J_)
      5       -0.1923        0.1132        SA2(J_)
      6        1.196         0.1322        MASS_(1)
      7       -2.904         0.1528        MASS_(2)
      8       -6.008         0.8820        MASS_(3)
  scale parameter 1.000


  mixture proportions :   0.2351   0.5480   0.2169
  -2 log L = 1892.06 , deviance =   1892.06, after 25 iterations


$use mass4 $use remnp $
  -- data list abolished
            estimate          s.e.        parameter
      1        0.1969        0.1611        SEX(J_)
      2        0.3277        0.1366        A1(J_)
      3       0.03263       0.07882        A2(J_)
      4        0.3827        0.1985        SA1(J_)
      5       -0.1985        0.1136        SA2(J_)
      6        2.510         0.2662        MASS_(1)
      7       -0.3389        0.1155        MASS_(2)
      8       -5.798         0.4745        MASS_(3)
      9       -7.577         2.879         MASS_(4)
  scale parameter 1.000


  mixture proportions :   0.1028   0.2561   0.5687   0.0724
  -2 log L = 1891.14 , deviance =   1891.14, after 23 iterations
```

The deviance changes slightly from two to four mass-points, as the large negative mass-points are driven further down the scale. This does not change the meaning of the model, and changes the other model parameters only slightly, apart from the sex effect. The large number of children who are not obese on all three occasions implies large masses at large negative values on the logit scale, but their precise locations have little effect on the fitted model. We proceed to simplify the model with two mass-points; we suppress the full output.

```
$use mass2 $use remnp $
$macro fixed sex + a1 + a2 + sa1 $endm
$use remnp $
$macro fixed sex + a1 + a2 $endm
$use remnp $
$macro fixed sex + a1 $endm
$use remnp $
$macro fixed a1 $endm
$use remnp $
$macro fixed 1 $endm
$use remnp $
```

Successive deviance changes for the omitted effects are 2.56 (sa2), 2.62 (sa1), 1.21 (a2), 0.00 (sex), 23.10 (a1). There is a linear age effect, the same for both sexes, and no sex difference. Parameter estimates for the a1 model are given below to 3 dp:

```
$macro fixed a1 $endmac
$use remnp $
```

	estimate	s.e.	parameter
1	0.505	0.101	A1(J_)
2	1.129	0.102	MASS_(1)
3	-3.301	0.132	MASS_(2)

scale parameter 1.000

mixture proportions : 0.242   0.758
-2 log L = 1898.49 , deviance = 1898.49, after 49 iterations

The LRTS-based standard error for a1 can be computed as $0.505/\sqrt{23.10} = 0.105$, only 4% larger than the GLIM figure.

The two-component mixture is identified by the "consistently obese" and "consistently non-obese" children, not surprisingly. The fitted probability of obesity is shown below with the response on each occasion (1 = obese).

Pattern	Obesity Probability	Pattern	Obesity Probability
0 0 0	.005	1 - 0	.621
- 0 0	.014	0 1 -	.706
0 - 0	.019	1 0 -	.706
0 0 -	.028	- - 1	.822
0 - -	.095	- 1 -	.871
- 0 -	.120	1 - -	.905

```
- - 0    .178        0 1 1    .972
1 0 0    .293        1 0 1    .972
0 1 0    .293        1 1 0    .972
0 0 1    .293        - 1 1    .990
- 0 1    .538        1 - 1    .993
- 1 0    .538        1 1 -    .995
0 - 1    .621        1 1 1    1.0
```

The fitted conditional (broken curves) and population-average (solid curve) models are shown in Fig. 9.3; the GLIM directives are given below:

```
$tab the %fv mean with ppp_ for j_ into mfv $
$ass age2 = age,age $
$sort sfp,sage2 %fv,age age $
$graph (s = 1 h = 'age' v = 'proportion' y = 0,1)
        sfp,mfv sage2,age 11,12,10 mass_, $
```

Table 9.9 gives the parameter estimates and standard errors for the full model, from both the Fitzmaurice *et al.* and the random effect model approaches, and Table 9.10 gives those for the final model (Fitzmaurice *et al.* did not give the estimates for the final linear model).

It is of interest that both ML analyses lead to the same models with nearly proportional parameter estimates and standard errors, though the models being

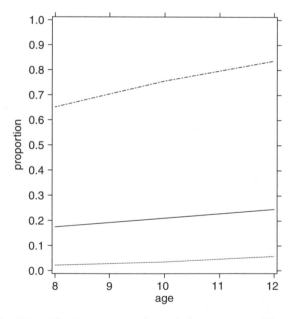

Fig. 9.3. Conditional (broken curves) and population-average (solid curve) obesity proportions

Table 9.9. Parameter estimates for full model

Parameter	Marginal ML estimate	SE	NPML estimate	SE
Intercept	−1.356	0.098	1.194	
GENDER	0.043	0.138	0.025	0.288
AGE(L)	0.142	0.063	0.339	0.150
AGE(Q)	0.014	0.035	0.032	0.088
GENDER.AGE(L)	0.162	0.096	0.345	0.209
GENDER.AGE(Q)	−0.089	0.049	−0.192	0.119
Deviance			1892.06	

Table 9.10. Parameter estimates for final model

Parameter	Marginal ML estimate	SE	NPML estimate	SE
Intercept	−1.321	0.094	1.151	
AGE(L)	0.220	0.059	0.507	0.105
AGE(Q)	−0.033	0.032	−0.065	0.059
Deviance			1897.27	
Intercept			1.165	
AGE(L)			0.507	0.105
Deviance			1898.50	

fitted are different: a marginal logit model in the first case and a conditional model in the second. The inflation of estimates and standard errors for the NPML analysis is a consequence of its much greater variance on the logit scale: $\pi^2/3 + \sigma_A^2 = 9.29$ compared with $\pi^2/3 = 3.29$ for the marginal logit model. The corresponding intra-class correlation is 0.65. Neuhaus and Jewell (1990) reported similar inflation in a paired study, and Neuhaus (1992) gave a general discussion of the relation between parameter estimates by these and other approaches.

Further discussion of these and other examples, and extensive references, can be found in Aitkin (1999). The data given in Woolson and Clarke (1984) are much more extensive, with five cohorts of children aged 5–7, 7–9, 9–11, 11–13 and 13–15 followed for four years, with obesity assessment every two years. The Muscatine study was even larger, with longer follow-up.

The data are in obesityfull, and the cohorts are identified by the weight variables wt6, wt8, wt10, wt12, wt14. We leave further analysis as an exercise for the reader.

9.8 Autoregressive random effect models

A limitation of the simple random effect model in Examples 2 and 3 is that it forces the correlation between any two repeated measures to be the same (at least on the underlying latent variable scale). It is a common experience however, that such correlations "decay" with increasing distance in time. For this reason models with other forms of correlation structure are widely used in longitudinal studies; a common one is the *autoregressive* model, in which the response on occasion j depends explicitly on responses at previous times. Autoregressive models are quite straightforward to fit, at least with complete longitudinal data, and provide a valuable generalization of the simple random effect model, but care is needed in the model specification when both random effect and autoregressive structure are modelled, because of a well-known difficulty called the *initial conditions problem* (see, e.g. Blundell and Bond, 1998; Aitkin and Alfó, 2003).

A simple example is the AR1 model, where the subscript t indexes time, for the response repeatedly measured within individual. Suppose for simplicity that each individual i is measured on T occasions, with an exponential family distribution for the response $Y_{i,t}$:

$$Y_{i,t} \mid y_{i,t-1}, z_i \sim ef(\eta_{i,t}), \quad t = 2, \ldots, T \text{ independently,}$$

where

$$\eta_{i,t} = \boldsymbol{\beta}'\mathbf{x}_{i,t} + \gamma y_{i,t-1} + z_i$$

and as before $z_i \sim \pi(z)$ for some random effect distribution π. Here γ is the autoregressive parameter. A difficulty with this model is that it does not specify the distribution of $Y_{i,1}$ at the first outcome. The difficulty could be avoided by conditioning on this outcome; it might appear that

$$f(y_{i,2}, \ldots, y_{i,T} \mid y_{i,1}) = \int \prod_{t=2}^{T} f(y_{i,t} \mid y_{i,t-1}, z_i)\pi(z_i)\,dz_i.$$

But individual i will have the same "propensity" random effect on the first measurement occasion as on the other occasions, and so $Y_{i,1}$ cannot be assumed independent of Z_i. In fact using Bayes' theorem

$$\pi(z_i \mid y_{i,1}) = f(y_{i,1} \mid z_i)\pi(z_i)/f(y_{i,1}).$$

and we could restate the model in the following way:

$$f(y_{i,1}, \ldots, y_{i,T}) = \int \prod_{t=2}^{T} f(y_{i,t} \mid y_{i,t-1}, z_i) f(y_{i,1} \mid z_i) \pi(z_i) \, dz_i$$

$$= \int \prod_{t=2}^{T} f(y_{i,t} \mid y_{i,t-1}, z_i) f(y_{i,1}) \pi(z_i \mid y_{i,1}) \, dz_i$$

$$= f(y_{i,1}) \int \prod_{t=2}^{T} f(y_{i,t} \mid y_{i,t-1}, z_i) \pi(z_i \mid y_{i,1}) \, dz_i,$$

or equivalently,

$$f(y_{i,2}, \ldots, y_{i,T} \mid y_{i,1}) = \int \prod_{t=2}^{T} f(y_{i,t} \mid y_{i,t-1}, z_i) \pi(z_i \mid y_{i,1}) \, dz_i.$$

The result is a random effect model with the $Y_{i,t}$ ($t = 2, \ldots, T$) conditioned on the initial outcome, but the distribution of Z_i has changed due to the conditioning, and depends on the initial outcome for each observation; it can no longer be considered the previous homogeneous distribution $\pi(z)$. This difficulty invalidates in general the simple approach of conditioning and using the standard random effect model; however for the conjugate normal/normal model for Y *and* the random effect Z, the conditioning on the first outcome still leaves a normal variance component model with a normal random effect, and this model can be re-parametrized to transfer the (conditional) mean of the normal random effect to the linear predictor for the response model, allowing the standard random effect analysis to be applied in the conditional model given the first outcome.

Aitkin and Alfó discussed alternative approaches to the conditioned model; a different solution is to complete the likelihood by specifying the initial outcome. The absence of a previous outcome means that the regression coefficients on the explanatory variables cannot in general be the same, nor can it be assumed that the variability of the random effect is unaffected by conditioning. We might therefore specify

$$Y_{i,1} \mid z_i \sim ef(\eta_{i,1}),$$

where

$$\eta_{i,1} = \delta' \mathbf{x}_{i,1} + \lambda z_i,$$

where λ is a scale factor.

The full random effect AR1 model can be fitted by re-structuring the data so that each outcome is an explanatory variable for the subsequent one; the first outcome is given a dummy zero variable as its "previous outcome". We define a dummy variable $t = 0$ for the first outcome and 1 for subsequent outcomes; the `fixed` macro then includes the previous outcome, the `t` variable and its interactions with the explanatory variables. The `random` macro includes the `t` variable, to allow for different variances of the random effect in the first unconditional model and the subsequent conditional models.

We illustrate with the subset of complete data from the Woolson and Clarke study (the missingness in the full data set greatly complicates this analysis, without providing much information about the autoregressive parameter). We extend the data structure as shown below:

```
! Complete data from Woolson and Clarke, restructured
! for AR1 model with dummy y_0 variable
$unit 48
$data x y sex age $
$read
! Complete data, males

0 0 0   8   0 0 0 10   0 0 0 12
0 0 0   8   0 0 0 10   0 1 0 12
0 0 0   8   0 1 0 10   1 0 0 12
0 0 0   8   0 1 0 10   1 1 0 12
0 1 0   8   1 0 0 10   0 0 0 12
0 1 0   8   1 0 0 10   0 1 0 12
0 1 0   8   1 1 0 10   1 0 0 12
0 1 0   8   1 1 0 10   1 1 0 12

! Complete data , females

0 0 1   8   0 0 1 10   0 0 1 12
0 0 1   8   0 0 1 10   0 1 1 12
0 0 1   8   0 1 1 10   1 0 1 12
0 0 1   8   0 1 1 10   1 1 1 12
0 1 1   8   1 0 1 10   0 0 1 12
0 1 1   8   1 0 1 10   0 1 1 12
0 1 1   8   1 1 1 10   1 0 1 12
0 1 1   8   1 1 1 10   1 1 1 12

$var 16 wt rep $
$calc n = 1 : rep = 3 $
$ass wt = 150 15  8   8 8 9 7 20
```

```
          154 14 13 19 2 6 6 21
$calc a1 = (age - 10)/2 : a2 = 3*a1**2 - 2 $
   : sa1 = sex*a1 : sa2 = sex*a2 $
   :  t = %gt(age,8) : ta1 = t*a1 $
  : tsex = t*sex : tsa1 = t*sa1 $
$macro fixed sex + t + x + tsex + ta1 + tsa1 $endm
$macro random t $endm
$use expand rep $
$use mass1 $
$use remnp y wt rep n $
          estimate        s.e.        parameter
     1      -1.414       0.1681        MASS_(1)
     2      -0.329       0.2486        SEX(J_)
     3      -0.895       0.2728        T(J_)
     4       2.739       0.1988        X(J_)
     5       1.006       0.3642        TSEX(J_)
     6       0.352       0.2725        TA1(J_)
     7      -0.748       0.3737        TSA1(J_)


scale parameter 1.000


mixture proportions :    1.000
-2 log L = 1197.37 , deviance =  1197.37, after 2 iterations
```

The one-mass-point model has no random effect, but the autoregression coefficient is very large. We increase the number of mass-points.

```
$use mass2 $use remnp $
-- data list abolished
          estimate        s.e.        parameter
     1       0.547       0.2488        MASS_(1)
     2      -7.305       2.214         MASS_(2)
     3      -0.545       0.3445        SEX(J_)
     4      -0.655       0.3697        T(J_)
     5       0.588       0.2575        X(J_)
     6       1.324       0.4648        TSEX(J_)
     7       0.462       0.3114        TA1(J_)
     8      -0.516       0.4389        TSA1(J_)
     9       4.292       2.234         MASS_(2).T(J_)
scale parameter 1.000


mixture proportions :    0.3025   0.6975
-2 log L = 1156.20 , deviance =  1156.20, after 65 iterations
```

```
$use mass3 $use remnp $
-- data list abolished
           estimate         s.e.        parameter
      1       0.733        0.2643        MASS_(1)
      2      -3.970        0.5827        MASS_(2)
      3      -7.234        3.683         MASS_(3)
      4      -0.631        0.3514        SEX(J_)
      5      -0.562        0.3924        T(J_)
      6       0.416        0.2720        X(J_)
      7       1.364        0.4736        TSEX(J_)
      8       0.468        0.3186        TA1(J_)
      9       0.000        aliased       TA2(J_)
     10      -0.492        0.4462        TSA1(J_)
     11       0.000        aliased       TSA2(J_)
     12       1.500        0.6606        MASS_(2).T(J_)
     13       1.312        3.921         MASS_(3).T(J_)
scale parameter 1.000
```

```
mixture proportions :  0.2751  0.4817  0.2432
-2 log L = 1156.20 , deviance =  1156.20, after 76 iterations
-- change to data values affects model
```

The deviance change is more than 40; even with autoregression on the previous outcome, there is a strong random effect dependence. The dependence on the previous outcome is greatly reduced, and there is little sign of different random effect variances. We simplify the random model.

```
$macro random 1 $endm
$use remnp $
           estimate         s.e.        ·parameter
      1       0.928        0.2613        MASS_(1)
      2      -2.830        0.2951        MASS_(2)
      3      -7.547        1.796         MASS_(3)
      4      -0.763        0.3456        SEX(J_)
      5       0.032        0.3745        T(J_)
      6      -0.187        0.2991        X(J_)
      7       1.360        0.4832        TSEX(J_)
      8       0.514        0.3400        TA1(J_)
      9       0.000        aliased       TA2(J_)
     10      -0.433        0.4712        TSA1(J_)
     11       0.000        aliased       TSA2(J_)
scale parameter 1.000
```

```
mixture proportions :   0.2433   0.4810   0.2757
-2 log L = 1157.23 , deviance =  1157.23, after 53 iterations
-- change to data values affects model
```

The deviance increases by only 1.03. We omit the t effect, followed by the dependence on the previous outcome.

```
$macro fixed sex + x + tsex + ta1 + ta2 + tsa1 + tsa2 $endm
$use remnp $
          estimate         s.e.        parameter
     1       0.928         0.2613       MASS_(1)
     2      -2.830         0.2951       MASS_(2)
     3      -7.547         1.796        MASS_(3)
     4      -0.763         0.3456       SEX(J_)
     5      -0.187         0.2991       X(J_)
     6       1.360         0.4832       TSEX(J_)
     7       0.562         0.5011       TA1(J_)
     8      -0.016         0.1873       TA2(J_)
     9      -0.433         0.4712       TSA1(J_)
    10       0.000         aliased      TSA2(J_)
scale parameter 1.000
```

```
mixture proportions :   0.2433   0.4810   0.2757
-2 log L = 1157.23 , deviance =  1157.23, after 53 iterations
```

```
$macro fixed sex + tsex + ta1 + ta2 + tsa1 + tsa2 $endm
$use remnp $
          estimate         s.e.        parameter
     1       0.815         0.2491       MASS_(1)
     2      -2.996         0.2784       MASS_(2)
     3      -5.260         0.7382       MASS_(3)
     4      -0.708         0.3376       SEX(J_)
     5       1.343         0.4770       TSEX(J_)
     6       0.426         0.4481       TA1(J_)
     7       0.029         0.1698       TA2(J_)
     8      -0.461         0.4668       TSA1(J_)
     9       0.000         aliased      TSA2(J_)
scale parameter 1.000
```

```
mixture proportions :   0.2512   0.5236   0.2252
-2 log L = 1157.37 , deviance =  1157.37, after 27 iterations
-- change to data values affects model
```

The omission of t gives no change in deviance, and of the previous outcome gives a deviance change of 0.14. Further reductions can be made in this model, but we conclude that the random effect model, rather than the autoregressive model, is the appropriate representation: given the random effect, no more complex dependence structure is necessary.

9.9 Latent variable models

Variance component models are a special case of more general *latent variable models* in which the covariance structure of a set of observed variables is expressed in terms of their joint *conditional independence* given a set of latent variables.

9.9.1 *The normal factor model*

Given a vector of continuous observable response variables $\mathbf{Y}' = (Y_1, \ldots, Y_r)$, and a vector of explanatory variables $\mathbf{x}' = (x_1, \ldots, x_p)$, we model the joint dependence of the responses, given the explanatory variables, by a set of *conditionally independent* regressions of the Y_j on a vector of *latent* or *factor* variables $\mathbf{Z}' = (Z_1, \ldots, Z_s)$:

$$\mathbf{Y} \mid \mathbf{Z} \sim N(\boldsymbol{\mu} + \Lambda \mathbf{Z}, \Psi),$$
$$\mathbf{Z} \sim N(\mathbf{0}, I),$$

where $\boldsymbol{\mu}$ is an r-vector of intercepts, Λ is an $r \times s$ matrix of *factor loadings* (regression coefficients) and Ψ is an $r \times r$ diagonal matrix of *specific variances*. The marginal distribution of the observable responses $\mathbf{Y}$ is then

$$\mathbf{Y} \sim N(\boldsymbol{\mu}, \Lambda \Lambda' + \Psi).$$

The zero mean and diagonal covariance matrix of $\mathbf{Z}$ appear restrictive, but a more general mean and covariance matrix are not identifiable. The independent factors can be *rotated* to any specific covariance structure, but this rotation is absorbed into the loading matrix, leaving the overall marginal distribution invariant.

The factor model can be fitted either by directly maximizing the observed data log-likelihood, or by using an EM algorithm (Rubin and Thayer, 1982), in which the latent variables Z_t are the unobserved data. We follow the latter approach.

In the complete data log-likelihood, the sufficient statistics are the sums involving the Z_t and the sums of squares and cross-products of the Z_t for the independent multiple regressions of each Y_j on $\mathbf{Z}$, and the residual sum of squares. These terms are replaced by their conditional expectations in the E-step, and the regression models are fitted separately in the M-step with appropriate adjustments to the SSP matrices, as in the variance component model.

The classical factor model does not contain any explanatory variables; these can be incorporated by a simple extension to the latent variable model:

$$Z_t \mid \mathbf{x} \sim N(\boldsymbol{\beta}_t'\mathbf{x}, 1), \quad t = 1, \ldots, s.$$

The general model can be fitted using an EM algorithm in the same way. In the complete data log-likelihood, the sufficient statistics now include the sums of squares and cross-products of the Z_t with the variables in $\mathbf{x}$. Here the representation of the model with independent factors allows the factor-explanatory variable regressions to be fitted independently as well.

The macros we have described for the variance component model can be adapted to fit factor models, by extending the two-level model to allow for different variances for each upper-level unit. Since we have not implemented this extension, we do not consider normal factor models further, though as shown below the closely related *item response models* which do not have a free variance parameter *can* be fitted directly using the variance component macros.

9.10   IRT models

A class of important models for examinees' responses to binary test items is based on *item response theory* (IRT), first developed by Lord (1952), Rasch (1960) and Birnbaum (in Lord and Novick, 1968) and extensively developed since by many psychometricians and statisticians (for a review, see van der Linden and Hambelton, 1997). We consider here only binary items, though ordered categorical response items are also possible.

The basic structure of such test data is a two-way array of binary responses Y_{ij}, indexed by person (examinee) i and test item j.

9.10.1   *The Rasch model*

The simplest IRT model (the "one-parameter" model) is the *Rasch* model (Rasch, 1960), developed and popularized later by Andersen (1973, 1980). The Rasch model is a main-effect logistic regression model with parameters for person i and item j:

$$Y_{ij} \sim b(1, p_{ij}) \text{ independently}, \quad i = 1, \ldots, n, \ j = 1, \ldots, r,$$
$$\text{logit } p_{ij} = \alpha_j + \theta_i,$$

where p_{ij} is the probability of a correct answer ($Y_{ij} = 1$) by person i on item j. The parameter α_j is the "easiness" of the j-th item, and θ_i is the "ability" of the i-th person.

If both sets of parameters are regarded as fixed effects, the model may be fitted directly as a Bernoulli GLM with n- and r-level factors for persons and items.

However, the ability estimates $\hat{\theta}_i$ are not consistent as $n \to \infty$ for fixed r, as their number $\to \infty$ as well.

This problem can be circumvented by conditioning: because of the simple model structure, as in the logistic model for 2×2 tables in Chapter 4, the marginal totals $Y_{+j} = \sum_i Y_{ij}$—the item totals (or proportions correct)—are sufficient for the item parameters α_j, and the $Y_{i+} = \sum_j Y_{ij}$—the total numbers of items correct—are sufficient for the ability parameters θ_i. The conditional distributions of $Y_{ij} \mid Y_{i+}$ and $Y_{ij} \mid Y_{+j}$ are therefore free of the parameters θ_i and α_j respectively, and so maximizing the conditional likelihoods in each case gives consistent estimates of the item easiness and person ability parameters. This model provides a very powerful justification for the use of the total test score as an estimate of person ability.

An alternative approach to the consistency difficulty is to model the ability parameters, for example by giving them a normal distribution:

$$Y_{ij} \mid \theta_i \sim b(1, p_{ij}) \text{ independently,}$$

$$\text{logit } p_{ij} = \alpha_j + \theta_i,$$

$$\theta_i \sim N(0, \sigma_A^2).$$

This model can be recognized immediately as a two-level variance component model, with persons as the upper level and item responses within person as the lower level. The items are different, rather than the same as in the obesity example, but the model structure is identical—items are differentiated by their easiness parameters, rather than through the age trend.

The model has an unidentifiability property as in the normal variance component model: the mean of the ability distribution is aliased with an item parameter: it is only *differences* of item parameters which are identified. A common convention in IRT is to constrain the item parameters to sum to zero, and include a non-zero ability mean in the model. We follow the zero mean model convention above. The random effect form of the model allows a simple change to the probit link function; this cannot be used in the conditional fixed-effects approach since it does not have sufficient statistics.

We give an example shortly, but first note that this model depends on the very strong assumption that the items differ only in easiness, or difficulty.

9.10.2 *The two-parameter model*

A more general model which weakens this assumption is the *two-parameter logit (2PL) model*, in which the items have different slope parameters as well as different intercept parameters:

$$Y_{ij} \sim b(1, p_{ij}),$$

$$\text{logit } p_{ij} = \alpha_j + \beta_j \theta_i.$$

The parameter β_j determines the discriminatory power of item j, in the sense that an item with a large value of β_j identifies ability over a narrower range than an item with a small value, as the probability of a correct answer changes more rapidly with small changes in ability. In this fixed-effect model, the parameters α_j, β_j and θ_i cannot be consistently estimated by conditioning, because the "sufficient" statistics now depend on the other model parameters. A distributional assumption for the ability parameters is essential; for example:

$$Y_{ij} \mid \theta_i \sim b(1, p_{ij}),$$
$$\text{logit } p_{ij} = \alpha_j + \beta_j \theta_i,$$
$$\theta_i \sim N(0, \sigma_A^2).$$

A probit link gives the *2PP model*.

The notational convention in IRT is unfortunately to reverse the usual notation for slopes and intercepts, to reverse the sign of the easiness parameter and to denote the resulting *item difficulty* parameter by β_j; the two-parameter logit model becomes

$$\text{logit } p_{ij} = \alpha_j(\theta_i - \beta_j)$$

and the one-parameter Rasch model becomes

$$\text{logit } p_{ij} = \theta_i - \beta_j.$$

The parameter α_j under this convention is called the *item discrimination* parameter. At the risk of further confusion, we maintain our conventional regression model notation for the parameters.

The model is the analogue for binary items of the normal factor model. It has an unidentifiability issue similar to that of the normal factor model: we could re-parametrize the model as

$$Y_{ij} \mid \theta_i^* \sim b(1, p_{ij}),$$
$$\text{logit } p_{ij} = \alpha_j + \beta_j^* \theta_i^*,$$
$$\theta_i^* \sim N(0, 1),$$

where $\theta_i^* = \theta_i / \sigma_A$, and $\beta_j^* = \sigma_A \beta_j$. Since θ_i is unobservable, the standard deviation of the ability distribution is unidentifiable if the regression parameters are unrestricted. However, this unidentifiability does not prevent the EM algorithm converging (one of its great assets). In the analysis below, the standard deviation is *not* fixed, so the β_j are determined only up to an arbitrary scale factor; common scalings are to fix the geometric mean of the β_j to be 1, or β_1 to be 1.

It is an important feature of the Rasch model that it does not suffer this unidentifiability: in (the random effect version of) this model σ is identifiable because
of the fixed equal slopes.

An interesting question is whether the distribution of ability in the 2PL model
is itself identifiable—is the normal distribution a convenient model assumption
which can be contradicted by the data, or would any other model give the same
estimates of the identifiable parameters? It is clear that the standard deviation of
the ability distribution is not identifiable, as noted above.

A deficiency of the two-parameter model is that it makes no provision for
guessing: a person with very low ability ($\theta_i \to -\infty$) has essentially zero probability of getting any item correct. This deficiency is addressed in the three-parameter
model.

9.10.3   *The three-parameter logit (3PL) model*

Guessing with multiple-choice items is modelled as a random choice amongst the
possible answers to the item. If the item has m possible answers, then a random
choice would suggest a probability of $1/m$ of a correct answer for a person regardless of ability. However, some answers for some items may be clearly incorrect
and so the probability of a correct answer by guessing may be different from $1/m$,
and is taken as an item-specific third parameter in the model.

The three-parameter model is a discrete mixture, with a probability γ_j of a
correct answer on item j for very low ability, increasing to 1 as ability $\to \infty$:

$$\Pr[Y_{ij} = 1 \mid \theta_i] = \gamma_j + (1 - \gamma_j)p_{ij},$$

$$\text{logit}\, p_{ij} = \alpha_j + \beta_j \theta_i,$$

$$\theta_i \sim N(0, 1).$$

The introduction of an additional mixture parameter further complicates both
the identification of the model parameters and the analysis of the data. Since
all the item response models are mixed Bernoulli models, these are identified only
by the common person ability across different items. This person ability does not
affect the guessing parameter, and so the guessing parameter is in general unidentifiable (van der Linden and Hambleton, 1997) without external information; this
may be provided by a prior distribution based on the plausibility of the alternatives
to the correct responses.

We now discuss an example.

9.10.4   *Example—The Law School Aptitude Test (LSAT)*

Bock and Aitkin (1981) illustrated the Gaussian quadrature analysis of the two-
parameter probit model with data from examinees on Sections 6 and 7 of the
LSAT. We use below the data from Section 7 which had five binary items, giving
32 possible response patterns, all of which were observed.

The data are given below, and are in the file `lsat`. A correct item is scored 1 and an incorrect one zero. The number with each pattern is *n*.

```
Item    n  Item    n  Item    n  Item    n
12345      12345      12345      12345
------------------------------------------
00000  12  01000  10  10000   7  11000    6
00001  19  01001   5  10001  39  11001   25
00010   1  01010   3  10010  11  11010    7
00011   7  01011   7  10011  34  11011   35
00100   3  01100   7  10100  14  11100   18
00101  19  01101  23  10101  51  11101  136
00110   3  01110   0  10110  15  11110   32
00111  17  01111  28  10111  90  11111  308
------------------------------------------
```

This section of the test was clearly quite easy (for this population of examinees): the total-score distribution is

```
Number of items correct  0   1   2    3    4    5
Number of examinees         12  40 104  205  321  308
```

We first estimate the item response probabilities directly from the sample means. The binary response data do not need to be read in, since they form a regular pattern.

```
$var 32 y1 y2 y3 y4 y5 i $
$calc y1 = %gl(2,16) - 1 : y2 = %gl(2,8) - 1 : y3 = %gl(2,4) - 1 $
    : y4 = %gl(2,2)  - 1 : y5 = %gl(2,1) - 1 :  i = %cu(1) $
$var 32 wt $
$data wt $
$read
12 19  1  7  3 19  3 17 10  5 3  7  7  23  8  28
 7 39 11 34 14 51 15 90  6 25 7 35 18 136 32 308
$tabulate the y1 mean with wt $
0.828
  : the y2 mean with wt $
0.658
  : the y3 mean with wt $
0.772
  : the y4 mean with wt $
```

```
0.606
  :  the y5 mean with wt $
0.843
```

Items 1 and 5 are very easy, item 4 the hardest. We repeat the Bock and Aitkin analysis of Section 7 with the two-parameter logit model, and fit the Rasch model first, as well.

```
$var 160 y n item $
$calc y(5*i-4) = y1(i) : y(5*i-3) = y2(i) : y(5*i-2) = y3(i) $
    :  y(5*i-1) = y4(i) : y(5*i) = y5(i) $
$var 32 rep $calc rep = 5 : n = 1 : item = %gl(5,1) $
$factor item 5 $
$input 'allvc' $
$error b n $
$unit 160 $
$macro fixed item $endm
$macro random 1 $endm
$use expand rep $
$use mass20 $use remz y wt rep n $
```

	estimate	s.e.	parameter
1	1.868	0.0924	ITEM(J_)(1)
2	0.791	0.0737	ITEM(J_)(2)
3	1.461	0.0834	ITEM(J_)(3)
4	0.521	0.0714	ITEM(J_)(4)
5	1.993	0.0956	ITEM(J_)(5)
6	1.010	0.0421	Z_

```
-2 log L = 5329.80 , deviance =  5329.80, after 23 iterations
```

The estimated standard deviation of ability is just over 1.0. How does the estimation of the ability distribution affect the estimated item parameters?

```
$use mass2 $use remnp y wt rep n $
```

	estimate	s.e.	parameter
1	2.385	0.0988	ITEM(J_)(1)
2	1.312	0.0783	ITEM(J_)(2)
3	1.981	0.0897	ITEM(J_)(3)
4	1.044	0.0750	ITEM(J_)(4)
5	2.509	0.1019	ITEM(J_)(5)
6	-1.826	0.0733	MASS_(2)

```
scale parameter 1.000
```

```
mixture proportions :   0.6857   0.3143
-2 log L = 5332.22 , deviance =   5332.22, after 67 iterations

$use mass3 $use remnp $

           estimate         s.e.        parameter
     1        3.640        0.1708        ITEM(J_)(1)
     2        2.564        0.1582        ITEM(J_)(2)
     3        3.233        0.1653        ITEM(J_)(3)
     4        2.295        0.1556        ITEM(J_)(4)
     5        3.765        0.1727        ITEM(J_)(5)
     6       -1.883        0.1523        MASS_(2)
     7       -3.510        0.1664        MASS_(3)
scale parameter 1.000


mixture proportions :   0.2011   0.6451   0.1539
-2 log L = 5330.62 , deviance =   5330.62, after 37 iterations

$use mass4 $use remnp $

           estimate         s.e.        parameter
     1        3.958        0.3463        ITEM(J_)(1)
     2        2.881        0.3402        ITEM(J_)(2)
     3        3.550        0.3437        ITEM(J_)(3)
     4        2.612        0.3390        ITEM(J_)(4)
     5        4.083        0.3473        ITEM(J_)(5)
     6       -1.435        0.3401        MASS_(2)
     7       -2.925        0.3384        MASS_(3)
     8       -4.655        0.3796        MASS_(4)
scale parameter 1.000


mixture proportions :   0.0508   0.4942   0.4178   0.0371
-2 log L = 5329.50 , deviance =   5329.50, after 8 iterations
```

The disparity remains constant, apart from small variations, as K is increased
further. Comparison of the item parameters is made difficult by the different
mass-point distributions. Since it is differences of item parameters which are identifiable, we give in Table 9.11 the parameter estimates centred on the first item
parameter, together with the corresponding mean and standard deviation of the
estimated ability distributions, with their masses and mass-points, for K up to 6.
The ability distributions are shown in Table 9.12. The item easiness parameters
are correlated +1.0 with the logistic transformations of the proportion correct on
the items.

Table 9.11. Item parameters for non-parametric and normal ability distributions

K	Item					μ	σ	Disparity
	1	2	3	4	5			
2	1.868	0.795	1.464	0.527	1.992	−0.057	0.848	5332.22
3	1.868	0.792	1.461	0.523	1.993	+0.017	1.104	5330.62
4	1.868	0.791	1.460	0.522	1.993	−0.014	0.985	5329.50
5	1.868	0.789	1.459	0.521	1.993	−0.020	1.015	5329.21
6	1.868	0.790	1.460	0.521	1.993	−0.030	1.002	5329.22
GQ20	1.868	0.791	1.461	0.522	1.993	0.000	1.011	5329.81

Table 9.12. Ability distributions

K	Mass-point mass					
2	0.631	−1.195				
	0.686	0.314				
3	1.638	−0.245	−1.872			
	0.201	0.645	0.154			
4	2.118	0.683	−0.807	−2.537		
	0.051	0.494	0.418	0.037		
5	2.306	1.270	0.002	−1.354	−5.140	
	0.012	0.252	0.510	0.220	0.006	
6	2.654	1.541	0.647	−0.701	−1.831	−6.021
	0.003	0.098	0.448	0.377	0.071	0.004

Although the discrete distributions differ in shape, they have essentially the same means and variances, and the item parameters are almost invariant; further, the disparities hardly change from the 20-point normal model. This is to be expected since at most about $K/2$ mass-points are identified with K items; the normal model for ability is not contradicted by the data. We conclude by computing the posterior mean of the ability distribution for each response pattern, using Gaussian quadrature. We will compare these means for the Rasch and 2PL models.

```
$tab the z_ mean with ww_ for j_ into rm $
```

We now fit the two-parameter model.

```
$macro random item $endm
$use mass20 $use remz $
```

```
            estimate            s.e.        parameter
    1        1.856           0.1040        ITEM(J_)(1)
    2        0.808           0.0770        ITEM(J_)(2)
    3        1.804           0.1141        ITEM(J_)(3)
    4        0.486           0.0694        ITEM(J_)(4)
    5        1.855           0.0998        ITEM(J_)(5)
    6        0.988           0.1023        Z_
    7        0.093           0.1360        ITEM(J_)(2).Z_
    8        0.718           0.1634        ITEM(J_)(3).Z_
    9       -0.223           0.1281        ITEM(J_)(4).Z_
   10       -0.252           0.1410        ITEM(J_)(5).Z_
scale parameter 1.000


-2 log L = 5317.62 , deviance =  5317.62, after 61 iterations

$tab the z_ mean with ww_ for j_ into tplm $
```

The estimated item intercepts and slopes are shown in Table 9.13.

The disparity change from the Rasch model is 12.18 on 4 df. The large-sample χ_4^2 distribution applies to the disparity change; there is strong evidence that the items are not equally discriminating. Item 3 is highly discriminating, while items 4 and 5 are less discriminating than items 1 and 2. The fitted logistic curves

$$p_j = \exp(\alpha_j + \beta_j z)/[1 + \exp(\alpha_j + \beta_j z)]$$

(the *item characteristic curves*) are shown in Fig. 9.4.

Item 5 is so easy (the probability of a correct answer is nearly 0.4 at $z = -3$), and so poorly discriminating, that it does not contribute much to the estimation of ability. Item 4 is much harder though also poorly discriminating.

The Rasch model uses only the six total-score patterns, while the 2PL model uses all 32 response patterns. However, the estimated ability posterior means are very similar (Fig. 9.5).

Table 9.13. Item intercepts and slopes

Item	Intercept	Slope
1	1.856	0.988
2	0.808	1.081
3	1.804	1.706
4	0.486	0.765
5	1.855	0.736

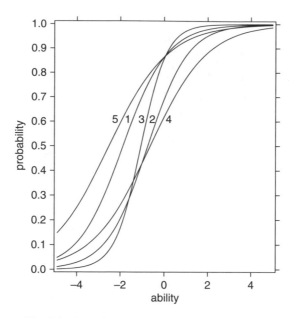

Fig. 9.4. Item characteristic curves, LSAT7 items

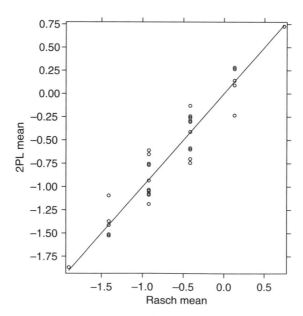

Fig. 9.5. Ability posterior means, LSAT7 items

```
$graph (s = 1 h = 'Rasch mean' v = '2PL mean') tplm,rm rm 5,10 $
```

The posterior means correlate +0.95. The posterior variances (not shown) are also similar.

9.10.5 Quadratic regressions

The item models are all *linear* in the ability variable $z\_$. However, there is no necessary restriction of these models to linearity—subject to identifiability, we may fit quadratic or other terms in $z\_$. These models cannot however be specified in the emz macro as the necessary quadratic term in $z\_$ cannot be specified as part of the model. The change required is very small—after $z\_$ is defined in the setup macro we define the quadratic term:

```
$calc  z2_ = z_**2 $
```

and the line in the emz macro

```
$fit (#fixed)(j_) + z_ + (#random)(j_).z_ - 1 $
```

is replaced by

```
$fit (#fixed)(j_) + z_ + z2_ + (#random)(j_).z_ - 1 $
```

For the 2PL model for the LSAT7 data, the resulting output from 20-point Gaussian quadrature is:

```
          estimate        s.e.       parameter
    1       1.887        0.0955       ITEM(J_)(1)
    2       0.809        0.0765       ITEM(J_)(2)
    3       1.478        0.0863       ITEM(J_)(3)
    4       0.540        0.0744       ITEM(J_)(4)
    5       2.012        0.0989       ITEM(J_)(5)
    6       0.987        0.0461       Z_
    7      -0.030        0.0331       Z2_
scale parameter 1.000

 -2 log L= 5329.69 , deviance =  5329.69, after 246 iterations
```

The deviance change of 0.12 clearly indicates that there is no evidence of quadratic regression—or that the quadratic term is unidentifiable. Without the information matrix we cannot easily distinguish these possibilities.

The two- and three-parameter models are widely used in large-scale testing. It is often important to include individual-level variables in the 2PL model, for example, to assess or account for the importance of sex, race and social class on outcomes. The inclusion of such variables in the model is straightforward—they correspond to upper-level variables in the usual two-level model.

9.11   Spatial dependence

A popular use of variance component models is in *small-area estimation*, in which geographical or other spatially-defined administrative regions or areas report counts of disease or accident cases, with explanatory variables for the cases and the regions, and we want to model the sources of variation over regions in the mean count (or the case proportion, if there is a population base for the count) while allowing for individual variations in risk. The small regions give individual area risk estimates with relatively large variability; the variance component model allows the among-region variability to improve the estimation of the local region risks. In biostatistical literature this is known as *disease mapping*. The first example in Aitkin (1999) is of this kind; it is taken from Tsutakawa (1985), who modelled the variation in lung cancer mortality rates for 84 cities in Missouri.

A limitation of the NPML analysis in this example, and generally in single- or two-level random effect models, is that it does not allow spatial dependence between neighbouring regions, as used for example in Clayton and Kaldor (1987), Breslow and Clayton (1993) and other authors. A popular extension (e.g. Besag *et al.*, 1991) of the simple random effect model for disease mapping is to include an additional spatial random effect for each region whose (conditional) mean value is equal to the average of the random effects for neighbouring regions (appropriately defined).

Care is needed in such models as the resulting joint distribution of all the random effects is likely to be singular unless it incorporates a regression parameter to reduce the very high intra-area correlation implied by the construction of the conditional means. This problem is usually handled by including an overdispersion effect as well for each region; this inflates the variance of the spatial random effects and reduces their correlation, but this is still determined by the size of the overdispersion variance component.

Initial spatial examination of the posterior means from the model *without* spatial dependence may be carried out to establish whether such dependence actually exists; as Clayton and Kaldor (1987) noted, "there is no a priori reason why geographic proximity should be reflected in correlated cancer rates". It should be noted that the simple variance component model provides consistent estimates of the regression model parameters in the presence of spatial correlation, but these are inefficient if real spatial dependence exists, and extensions of the NPML approach to spatial modelling would be very useful.

9.12   Multivariate correlated responses

A remarkable extension of the two-level model allows a very general analysis of multivariate responses linked by a latent structure. A simple example is the case of correlated bivariate Bernoulli responses, of which we gave an example in Chapter 5 (Section 5.9), with two signs of toxaemia in pregnant women. In Chapter 5 these were treated by contingency table analysis, but they can also be analysed through the two-level model: we simply allow different models for the two responses while maintaining their correlation through the common upper-level random effect. We do not give details here; the final model is very similar to that in Chapter 5 and does not lead to different conclusions.

A current important application of this approach is to survival studies in which a quality-of-life measure is also available, and the analysis needs to take into account the correlation between the survival event time and the quality of life measure. More general models could allow, for example, a normal response, a Poisson count, and a Bernoulli indicator to be linked through their conditional independence given a latent variable. Such extensions allow the development of very rich and complex model analyses.

9.13   Discreteness of the NPML estimate

The discrete nature of the NPML estimate may be found unattractive, if one believes *a priori* in the existence of a continuous mixing distribution. An alternative approach assuming a smooth mixing density was described by Davidian and Gallant (1993); it requires numerical quadrature with library optimization algorithms. Magder and Zeger (1996) took as a mixing distribution a finite mixture of normals, guaranteeing a continuous mixing distribution estimate. They found that the likelihood for the model with equal variances approaches a maximum as the common variance tends to zero (reflecting the optimality of the NPML estimate over *all* mixing distributions), and is very flat in the variance parameter away from zero, so the information about distributional shape of the mixing distribution is inherently limited.

A computational issue in current versions of GLIM is that only 6-figure accuracy can be obtained in the deviance and this may require relaxing or changing the convergence criterion (of successive deviance changes) in large data sets to prevent round-off error fluctuations in the value of the deviance.

References

Abramowitz, M. and Stegun, I.A. (eds) (1970) *Handbook of Mathematical Functions.* Dover, New York.

Aitkin, M.A. (1974) Simultaneous inference and choice of variable subsets in multiple regression. *Technometrics*, **16**, 221–7.

Aitkin, M.A. (1978) The analysis of unbalanced cross-classifications (with Discussion). *J. Roy. Statist. Soc. A*, **141**, 195–223.

Aitkin, M. (1980) Mixture applications of the EM algorithm in GLIM. In *Compstat 1980, Proceedings in Computational Statistics.* Physica-Verlag, Vienna, pp. 537–41.

Aitkin, M.A. (1981) A note on the regression analysis of censored data. *Technometrics*, **23**, 161–3.

Aitkin, M. (1987) Modelling variance heterogeneity in normal regression using GLIM. *J. Roy. Statist. Soc. C*, **36**, 332–9.

Aitkin, M. (1991) Posterior Bayes factors (with Discussion). *J. Roy. Statist. Soc. B*, **53**, 111–42.

Aitkin, M. (1994) An EM algorithm for overdispersion in generalized linear models. In *Proceedings of the 9th International Workshop on Statistical Modelling*, Exeter.

Aitkin, M. (1995) Probability model choice in single samples from exponential families using Poisson log-linear modelling, and model comparison using Bayes and posterior Bayes factors. *Statist. Comput.*, **5**, 113–20.

Aitkin, M. (1996a) A general maximum likelihood analysis of overdispersion in generalized linear models. *Statist. Comput.*, **6**, 251–62.

Aitkin, M. (1996b) A short history of a Vietnam War attitude survey. *Stats*, **17**, 1–9.

Aitkin, M. (1996c) Application of non-parametric maximum likelihood estimation in two-level variance component models for teacher effectiveness. In *Symposium on the Expansion Method*. Department of Economics, Odense University.

Aitkin M. (1997) The calibration of *P*-values, posterior Bayes factors and the AIC from the posterior distribution of the likelihood (with Discussion). *Statist. Comput.*, **7**, 253–72.

Aitkin, M. (1999) A general maximum likelihood analysis of variance components in generalized linear models. *Biometrics*, **55**, 117–28.

Aitkin, M. (2000) Review of *Computer-Assisted Analysis of Mixtures and Applications: Meta-Analysis, Disease Mapping and Others* (ed. D. Böhning), in *Biometrics*, **56**, 651–2.

Aitkin, M. (2001) Likelihood and Bayesian analysis of mixtures. *Statist. Model.*, **1**, 287–304.

Aitkin, M. and Alfo', M. (2003) Longitudinal analysis of repeated binary data using autoregressive and random effect modelling. *Statist. Model.*, **3**, 291–303.

Aitkin, M.A., Anderson, D. and Hinde, J. (1981) Statistical modelling of data on teaching styles (with Discussion). *J. Roy. Statist. Soc. A*, **144**, 419–61.

Aitkin, M., Boys, R. and Chadwick, T.J. (2005) Bayesian point null hypothesis testing via the posterior likelihood ratio. *Statist. Comput.*, (to appear).

Aitkin, M.A. and Clayton, D. (1980) The fitting of exponential, Weibull and extreme value distributions to complex censored survival data using GLIM. *Appl. Statist.*, **29**, 156–63.

Aitkin, M. and Foxall, R. (2003) Statistical modelling of artificial neural networks using the multi-layer perceptron. *Statist. Comput.*, **13**, 227–39.

Aitkin, M.A. and Francis, B. (1980) A GLIM macro for fitting the exponential or Weibull distribution to censored data. *GLIM Newslett.*, June, 19–25.

Aitkin, M.A. and Francis, B. (1982) Reader reaction: Interactive regression modelling. *Biometrics*, **38**, 511–13.

Aitkin, M. and Francis, B.J. (1995) Fitting overdispersed generalized linear models by nonparametric maximum likelihood. *GLIM Newslett.*, **25**, 37–45.

Aitkin, M.A., Laird, N. and Francis, B. (1983) A reanalysis of the Stanford heart transplant data (with Comments). *J. Amer. Statist. Assoc.*, **78**, 264–92.

Aitkin, M.A. and Longford, N.T. (1986) Statistical modelling issues in school effectiveness studies. *J. Roy. Statist. Soc. A*, **149**, 1–43.

Aitkin, M. and Rocci, R. (2002) A general maximum likelihood analysis of measurement error in generalized linear models. *Statist. Comput.*, **12**, 163–74.

Aitkin, M. and Tunnicliffe Wilson, G.T. (1980) Mixture models, outliers and the EM algorithm. *Technometrics*, **22**, 325–31.

Allen, D.M. (1971) Mean-square error of prediction as a criterion for selecting variables. *Technometrics*, **13**, 465–75.

Andersen, E.B. (1973) Conditional inference for multiple-choice questionnaires. *Brit. J. Math. Statist. Psych.*, **26**, 31–44.

Andersen, E.B. (1980) *Discrete Statistical Models with Social Science Applications*. North-Holland, Amsterdam.

Anderson, J.A. (1984) Regression and ordered categorical variables (with Discussion). *J. Roy. Statist. Soc. B*, **46**, 1–30.

Anderson, D.A. and Aitkin, M. (1985) Variance component models with binary response: interviewer variability. *J. Roy. Statist. Soc. B*, **47**, 203–10.

Anscombe, F.J. (1964) Normal likelihood functions. *Ann. Inst. Statist. Math.*, **26**, 1–19.

Ashford, J.R. (1959) An approach to the analysis of data from semi-quantal responses in biological response. *Biometrics*, **15**, 573–81.

Atkinson, A.C. (1981) Robustness, transformations and two graphical displays for outlying and influential observations in regression. *Biometrika*, **68**, 13–20.

Atkinson, A.C. (1982) Regression diagnostics, transformations and constructed variables (with Discussion). *J. Roy. Statist. Soc. B*, **44**, 1–36.

Atkinson, A.C. (1985) *Plots, Transformations and Regression. An Introduction to Graphical Methods of Diagnostic Regression Analysis*. Oxford.

Atkinson, C. and Polivy, J. (1976) Effects of delay, attack and retaliation on state depression and hostility. *J. Abnormal Psychol.*, **85**, 370–6.

Barnard, G.A. (1949) Statistical inference. *J. Roy. Statist. Soc. B*, **11**, 115–39.

Barnard, G.A., Jenkins, G.M. and Winsten, C.B. (1962) Likelihood inference and time series. *J. Roy. Statist. Soc. A*, **125**, 321–72.

Barnett, V. (1975) Probability plotting methods and order statistics. *Appl. Statist.*, **24**, 95–108.

Barnett, V. and Lewis, T. (1994) *Outliers in Statistical Data*, (2nd edn). Wiley, Chichester.

540   REFERENCES

Bartlett, M.S. (1957) A comment on D.V. Lindley's statistical paradox. *Biometrika*, **44**, 533–4.

Bartlett, R.H., Roloff, D.W., Cornell, R.G., Andrews, A.F., Dillon, P.W. and Zwischenberger, J.B. (1985) Extracorporeal circulation in neonatal respiratory failure: a prospective randomized study. *Pediatrics*, **76**, 479–87.

Baxter, L.A., Coutts, S. M. and Ross, G.A.F. (1980) Applications of linear models in motor insurance. In *Proceedings of the 21st International Congress of Actuaries, Zurich*, pp. 11–29.

Begg, C.B. (1990) On inferences from Wei's biased coin design for clinical trials (with Discussion). *Biometrika*, **77**, 467–84.

Belsley, D.A., Kuh, E. and Welsch, R.E. (1980) *Regression Diagnostics: Identifying Influential Data and Sources of Collinearity*. Wiley, New York.

Bennett, M.D. (1974) *The emotional response of husbands to suicide attempts by their wives*. Unpublished MD thesis, Sydney University.

Bennett, S. and Whitehead, J. (1981) Fitting logistic and log-logistic regression models to censored data using GLIM. *GLIM Newslett.*, **4**, 12–19.

Berger, J.O. and Wolpert, R. (1986) *The Likelihood Principle*. Institute of Mathematical Statistics, Hayward.

Besag, J., York, J. and Mollie, A. (1991). Bayesian image restoration, with two applications in spatial statistics. *Ann. Inst. Statist. Math.*, **43**, 1–59.

Birnbaum, A. (1962) On the foundations of statistical inference. *J. Amer. Statist. Assoc.*, **57**, 269–306.

Bishop, Y.M.M. (1969) Full contingency tables, logits and split contingency tables. *Biometrics*, **25**, 383–99.

Bishop, Y.M.M., Fienberg, S.E. and Holland, P.W. (1975) *Discrete Multivariate Analysis: Theory and practice*. MIT Press, Cambridge, Mass.

Bissell, A.F. (1972) A negative binomial model with varying element sizes. *Biometrika*, **59**, 435–41.

Blundell, R. and Bond, S. (1998) Initial conditions and moment restrictions in dynamic panel data models. *J. Economet.*, **68**, 115–32.

Bock, R.D. (1975) *Multivariate Statistical Methods in Behavioural Research*. McGraw-Hill, New York.

Bock, R.D. and Aitkin, M. (1981) Marginal maximum likelihood estimation of item parameters: an application of an EM algorithm. *Psychometrika*, **46**, 443–59.

Bock, R.D. and Yates, G. (1973) *MULTIQUAL: Log-Linear Analysis of Nominal or Ordinal Qualitative Data by the Method of Maximum Likelihood*. National Educational Resources, Chicago.

Böhning, D. (1999) *Computer-Assisted Analysis of Mixtures and Applications*. Chapman and Hall, Boca Raton.

Box, G.E.P. (1980) Sampling and Bayes inference in scientific modelling and robustness (with Discussion) *J. Roy. Statist. Soc. A*, **143**, 383–430.

Box, G.E.P. (1983) An apology for ecumenism in statistics. In *Scientific Inferences, Data Analysis and Robustness* (eds G.E.P. Box, T. Leonard and C.F. Wu). Academic Press, New York.

Box, G.E.P. and Cox, D.R. (1964) An analysis of transformations (with Discussion). *J. Roy. Statist. Soc. B*, **26**, 211–52.

Box, G.E.P. and Tidwell, P.W. (1962) Transformations of the independent variable. *Technometrics*, **4**, 531–50.

Breslow, N. (1974) Covariance analysis of censored survival data. *Biometrics*, **30**, 89–99.

Breslow, N.E. and Clayton, D.G. (1993) Approximate inference in generalized linear mixed models. *J. Amer. Statist. Assoc.*, **88**, 9–25.

Brown, P.J. (1982) Multivariate calibration (with Discussion). *J. Roy. Statist. Soc. B*, **44**, 287–321.

Brown, P.J., Stone, J. and Ord-Smith, C. (1983) Toxaemic signs during pregnancy. *Appl. Statist.*, **32**, 69–72.

Brownlee, K.A. (1965) *Statistical Theory and Methodology in Science and Engineering*, (2nd edn). Wiley, New York.

Butler, R. (1986) Predictive likelihood inference and applications (with Discussion). *J. Roy. Statist. Soc. B*, **48**, 1–38.

Carroll, R.J., Ruppert, D. and Stefanski, L.A. (1995) *Measurement Error in Nonlinear Models*. Chapman and Hall, London.

Chadwick, T.J. (2002) *A general Bayes theory of nested model comparisons*. Unpublished PhD thesis, University of Newcastle.

Chatfield, C. and Goodhart, G.J. (1970) The beta-binomial model for consumer purchasing behaviour. *Appl. Statist.*, **19**, 240–50.

Chernoff, H. (1954) On the distribution of the likelihood ratio. *Ann. Math. Statist.*, **25**, 573–8.

Clayton, D.G. and Hills, M. (1994) *Statistical Models in Epidemiology*. University Press, Oxford.

Clayton, D. and Kaldor, J. (1987) Empirical Bayes estimates of age-standardized relative risks for use in disease mapping. *Biometrics*, **43**, 671–81.

Collett, D. (2003) *Modelling Survival Data in Medical Research* (2nd edn). Chapman and Hall, Boca Raton.

Cook, R.D. (1977) Detection of influential observations in linear regression. *Technometrics*, **19**, 15–18.

Cook, R.D. and Weisberg, S. (1982) *Residuals and Influence in Regression*. Chapman and Hall, London.

Copas, J.B. (1983) Regression, prediction and shrinkage (with Discussion). *J. Roy. Statist. Soc. B*, **45**, 311–54.

Copas, J.B. (1997) Statistical inference for non-random samples. *J. Roy. Statist. Soc. B*, **59**, 55–95.

Cox, D.R. (1958) Some problems connected with statistical inference. *Ann. Math. Statist.*, **29**, 357.

Cox, D.R. (1972) Regression models and life tables (with Discussion). *J. Roy. Statist. Soc. B*, **34**, 187–220.

Cox, D.R. and Oakes, D. (1984) *Analysis of Survival Data*. Chapman and Hall, London.

Crouch, E.A.C. and Spiegelman, D. (1990) The evaluation of integrals of the form $\int_{-\infty}^{+\infty} f(t)\exp(-t^2)\,dt$: application to logistic-normal models. *J. Amer. Statist. Assoc.*, **85**, 464–9.

Crowder, M.J. (1978) Beta-binomial ANOVA for proportions. *Appl. Statist.*, **27**, 34–7.

Crowder, M.J. (1979) Inference about the intraclass correlation in the beta-binomial ANOVA for proportions. *J. Roy. Statist. Soc. B*, **41**, 230–4.

Crowder, M.J. (2001) *Classical Competing Risks*. Chapman and Hall, Boca Raton.

Crowley, J. and Hu, M. (1977) Covariance analysis of heart transplant survival data. *J. Amer. Statist. Assoc.*, **72**, 27–36.

Davidian, M. and Gallant, A.R. (1993) The nonlinear mixed effects model with a smooth random effects density. *Biometrika*, **80**, 475–88.

Davies, R.B. (1987) Mass point methods for dealing with nuisance parameters in longitudinal studies. In *Longitudinal Data Analysis* (ed. R. Crouchley). Avebury, Aldershot, Hants.

Day, N.E. (1969) Estimating the components of a mixture of two normal distributions. *Biometrika*, **56**, 463–74.

Dempster, A.P. (1974) The direct use of likelihood in significance testing. In *Proc. Conf. Foundational Questions in Statistical Inference* (eds O. Barndorff-Nielsen, P. Blaesild and G. Sihon), pp. 335–52.

Dempster, A.P. (1997) The direct use of likelihood in significance testing. *Statist. Comput.*, **7**, 247–52.

Dempster, A.P., Laird, N. M. and Rubin, D. B. (1977) Maximum likelihood from incomplete data via the EM algorithm (with Discussion). *J. Roy. Statist. Soc. B*, **39**, 1–38.

Dietz, E. and Böhning, D. (1995) Statistical inference based on a general model of unobserved heterogeneity. In *Statistical Modelling*. Springer-Verlag, New York.

Diggle, P.J., Heagerty, P., Liang, K.-Y. and Zeger, S.L. (2002) *Analysis of Longitudinal Data*. (2nd edn). Clarendon Press, Oxford.

Dodge, Y. (1999) The guinea pig of multiple regression. In *Robust Statistics, Data Analysis and Computer Intensive Methods: In honor of Peter Huber's 60th birthday*, Lecture notes in statistics, Vol. 109, pp. 91–117. Springer-Verlag, New York.

Draper, N.R. and Smith, H. (1998) *Applied Regression Analysis*. Wiley, New York.

Edwards, A.W.F. (1972) *Likelihood*. University Press, Cambridge.

Efron, B. (1986) Double exponential families and their use in generalized linear regression. *J. Amer. Statist. Assoc.*, **81**, 709–21.

Erickson, B.H. and Nosanchuk, T.A. (1979) *Understanding Data*. Open University Press, Milton Keynes.

Everitt, B.S. and Hand, D.J. (1981) *Finite Mixture Distributions*. Chapman and Hall, London.

Fay, R.E. (1996). Alternative paradigms for the analysis of imputed survey data. *J. Amer. Statist. Assoc.*, **91**, 490–8.

Feigl, P. and Zelen, M. (1965) Estimation of exponential probabilities with concomitant information. *Biometrics*, **21**, 826–38.

Filliben, J.J. (1975) The probability plot correlation coefficient test for normality. *Technometrics*, **17**, 111–17. Correction **17**, 520.

Finch, S.J., Mendell, N.R. and Thode, H.C. (1989) Probabilistic measures of adequacy of a numerical search for a global maximum. *J. Amer. Statist. Assoc.*, **84**, 1020–3.

Finney, D.J. (1947) The estimation from original records of the relationship between dose and quantal response. *Biometrika*, **34**, 320–34.

Fisher, R.A. (1935) The logic of inductive inference (with Discussion). *J. Roy. Statist. Soc.*, **98**, 39–54.

Fitzmaurice, G.M., Laird, N.M. and Lipsitz, S.R. (1994) Analysing incomplete longitudinal binary responses: a likelihood-based approach. *Biometrics*, **50**, 601–12.

Follman, D.A. and Lambert, D. (1989) Generalizing logistic regression by nonparametric mixing. *J. Amer. Statist. Assoc.*, **84**, 295–300.

Francis, B.J., Green, M. and Payne, C. (eds) (1993) *The GLIM System: Release 4 Manual.* Clarendon Press, Oxford.

Fuller, W.A. and Hidiroglou, M.A. (1978) Regression estimation after correcting for attenuation. *J. Amer. Statist. Assoc.*, **73**, 99–104.

Fuller, W.A. (1987) *Measurement Error Models.* Wiley, New York.

Gehan, E.A. (1965) A generalized Wilcoxon test for comparing arbitrarily singly-censored. *Biometrika*, **52**, 203–23.

Gelman, A., Carlin, J.B., Stern, H.S. and Rubin, D.B. (1995) *Bayesian Data Analysis.* Chapman and Hall, London.

Goldstein, H. (2003). *Multilevel Statistical Models* (3rd edn). Edward Arnold, London.

Goodman, L.A. (1979) Simple models for the analysis of ordered categorical data. *J. Amer. Statist Assoc*, **74**, 537–52.

Gottschalk, L.A. and Gleser, G.C. (1969) *The Measurement of Psychological States through the Content Analysis of Verbal Behavior.* University of California Press, Berkeley.

Haberman, S.J. (1974) Log-linear models for frequency tables with ordered classifications. *Biometrics*, **30**, 589–600.

Hall, P. and La Scala, B. (1990) Methodology and algorithms of empirical likelihood. *Int. Statist. Rev.*, **58**, 109–27.

Hartley, H.O. (1950) The maximum F-ratio as a short cut test for heterogeneity of variances. *Biometrika*, **37**, 308–12.

Heckman, J.J. and Singer, B. (1984) A method for minimizing the impact of distributional assumptions in econometric models of duration. *Econometrica*, **52**, 271–320.

Henderson, H.V. and Velleman, P.F. (1981) Building multiple regression models interactively. *Biometrics*, **37**, 391–411.

Higgins, J.E. and Koch, G.G. (1977) Variable selection and generalized chi-square analysis of categorical data applied to a large cross-sectional occupational health survey. *Int. Statist. Rev.*, **45**, 51–62.

Hinde, J. (1982) Compound Poisson regression models. In *GLIM82: Proceedings of the International Conference on Generalised Linear Models.* (ed. R. Gilchrist). Springer-Verlag, New York, pp. 109–21.

Hoaglin, D.C. and Welsch, R.E. (1978) The hat matrix in regression and ANOVA. *The Amer. Statist.*, **32**, 17–22. Corrigenda **32**, 146.

Hodges, J.L., Krech, D. and Crutchfield, R.S. (1975) *StatLab: An Empirical Introduction to Statistics.* McGraw-Hill Ryerson, Toronto.

Hollander, M., McKeague, I.W. and Yang, J. (1997) Likelihood ratio-based confidence bands for survival functions. *J. Amer. Statist. Assoc.*, **92**, 215–26.

Hope, A.C.A. (1968) A simplified Monte Carlo significance test procedure. *J. Roy. Statist. Soc. B*, **30**, 582–98.

Hosmer, D.W. and Lemeshow, S. (1999) *Applied Survival Analysis: Regression Modeling of Time to Event Data.* Wiley-Interscience, New York.

Huber, P.J. (1967) The behaviour of maximum likelihood estimates under non-standard conditions. In *Proceedings of the Fifth Berkeley Symposium in Mathematical Statistics and Probability, Vol 1, 221–33.* University of California Press, Berkeley CA.

Jacobs, R.A., Jordan, M.I., Nowlan, S.J. and Hinton, G.E. (1991) Adaptive mixtures of local experts. *Neural Comput.*, **3**, 79–87.

Jeffreys, H. (1961) *Theory of probability*. Clarendon Press, Oxford.

Johnson, P.O. and Neyman, J. (1936) Tests of certain linear hypotheses and their applications to some educational problems. *Statist. Res. Mem.*, **1**, 57–93.

Jones, K.A. (1975) Geographical contribution to the aetiology of chronic bronchitis. Unpublished B.Sc. dissertation, University of Southampton.

Jordan, M.I. and Jacobs, R.A. (1994) Hierarchical mixtures of experts and the EM algorithm. *Neural Comput.*, **6**, 181–214.

Jöreskog, K.G. and Sörbom, D. (1981) *LISREL V. Analysis of Linear Structural Relationships by Maximum Likelihood and Least Squares Methods*. Department of Statistics, University of Uppsala.

Jorgensen, B. (1997) *The Theory of Dispersion Models*. CRC Press, Boca Raton.

Kalbfleisch, J.D. and Prentice, R.L. (1980) *The Statistical Analysis of Failure Time Data*. Wiley, New York.

Kalbfleisch, J.D. and Sprott, D.A. (1970) Application of likelihood methods to models involving large numbers of parameters (with Discussion). *J. Roy. Statist. Soc. B*, **32**, 175–208.

Kaplan, E.L. and Meier, P. (1958) Non-parametric estimation from incomplete observations. *J. Amer. Statist. Assoc.*, **53**, 457–81.

Kass, R. and Raftery (1995) Bayes factors. *J. Amer. Statist. Assoc.*, **90**, 773–95.

Kiefer, J. and Wolfowitz, J. (1956) Consistency of the maximum likelihood estimator in the presence of infinitely many nuisance parameters. *Ann. Math. Statist.*, **27**, 887–906.

Ku, H.H. and Kullback, S. (1974) Loglinear models in contingency table analysis. *The Amer. Statist.*, **28**, 115–22.

Laird, N.M. (1978) Nonparametric maximum likelihood estimation of a mixing distribution. *J. Amer. Statist. Assoc.*, **73**, 805–11.

Laird, N.M. and Ware, J.H. (1982) Random-effects models for longitudinal data. *Biometrics*, **38**, 963–74.

Lange, K.L., Little, R.J.A. and Taylor, J.M.G. (1989) Robust statistical modeling using the *t* distribution. *J. Amer. Statist. Assoc.*, **84**, 881–96.

Larson, M.G. and Dinse, G.E. (1985) A mixture model for the regression analysis of competing risks data. *Appl. Statist.*, **34**, 201–11.

Lawless, J.F., Kalbfleisch, J.D. and Wild, C.J. (1999) Semiparametric methods for response-selective and missing data problems in regression. *J. Roy. Statist. Soc. B*, **61**, 413–38.

Lesaffre, E. and Spiessens, B. (2001) On the effect of the number of quadrature points in a logistic random-effects model: an example. *Appl. Statist.*, **50**, 325–35.

Liang, K.-Y. and Zeger, S.L. (1986) Longitudinal data analysis using generalized linear models. *Biometrika*, **73**, 13–22.

Lindsay, B.G. (1983) The geometry of mixture likelihoods, part I: a general theory. *Ann. Statist.*, **11**, 86–94.

Lindsay, B.G. (1995) *Mixture Models: Theory, Geometry and Applications*. Institute of Mathematical Statistics, Hayward.

Lindsey, J.K. (1974) Construction and comparison of statistical models. *J. Roy. Statist. Soc. B*, **36**, 418–25.

Lindsey, J.K. (1996) *Parametric Statistical Inference*. Clarendon Press, Oxford.

Lindsey, J.K. (1999) Some statistical heresies. *J. Roy. Statist. Soc. D*, **48**, 1–40.

Lindsey, J.K. and Mersch, G. (1992) Fitting and comparing probability distributions with log linear models. *Statist. Comput. Data Anal.*, **13**, 373–84.

Little, R.J.A. and Rubin, D.B. (1987) *Statistical Analysis with Missing Data.* John Wiley, New York.

Little, R.J.A. and Schluchter, M.D. (1985) Maximum likelihood estimation for mixed continuous and categorical data with missing values. *Biometrika*, **72**, 497–512.

Longford, N. (1987) A fast scoring algorithm for maximum likelihood estimation in unbalanced mixed models with nested random effects. *Biometrika*, **74**, 817–27.

Longford, N.T. (1993) *Random Coefficient Models.* Clarendon Press, Oxford.

Lord, F.M. (1952) A theory of test scores. *Psychometric Monographs*, No. 7.

Lord, F.M. and Novick, M.R. (1968) *Statistical Theories of Mental Test Scores.* Addison-Wesley, Reading Mass.

Louis, T.A. (1982) Finding the observed information matrix when using the EM algorithm. *J. Roy. Statist. Soc. B*, **44**, 226–33.

Magder, L.S. and Zeger, S.L. (1996) A smooth nonparametric estimate of a mixing distribution using mixtures of Gaussians. *J. Amer. Statist. Assoc.*, **91**, 1141–51.

Mantel, N. (1966) Evaluation of survival data and two new rank order statistics arising in its consideration. *Cancer Chemother. Reports*, **50**, 163–70.

Maritz, J.S. and Lwin, T. (1989) *Empirical Bayes Methods* (2nd edn). Chapman and Hall, London.

McCullagh, P. (1980) Regression models for ordinal data (with Discussion). *J. Roy. Statist. Soc. B*, **42**, 109–42.

McCullagh, P. and Nelder, J.A. (1989) *Generalized Linear Models.* Chapman and Hall, London.

McCulloch, C.E. (1997) Maximum likelihood algorithms for generalized linear mixed models. *J. Amer. Statist. Assoc.*, **92**, 162–70.

McGilchrist, C.A. (1994) Estimation in generalized mixed models. *J. Roy. Statist. Soc. B*, **56**, 61–9.

McLachlan, G.J. (1987) On bootstrapping the likelihood ratio test statistic for the number of components in a normal mixture. *Appl. Statist.*, **36**, 318–24.

McLachlan, G.J. and Basford, K.E. (1988) *Mixture Models: Inference and Applications to Clustering.* Marcel Dekker, New York.

McLachlan, G.J. and Krishnan, T. (1997) *The EM Algorithm and Extensions.* John Wiley, New York.

McLachlan, G.J. and Peel, D. (1997) On a resampling approach to choosing the number of components in normal mixture models. In *Computing Science and Statistics* **28** (eds L. Billard and N.I. Fisher). Interface Foundation, Fairfax Station, Virginia.

McLachlan, G.J. and Peel, D. (2000) *Finite Mixture Models.* John Wiley, New York.

Meng, X.L. and van Dyk, D.A. (1997) The EM algorithm–an old folk song sung to a fast new tune (with Discussion). *J. Roy. Statist. Soc. B*, **59**, 511–67.

Miller, R.G. and Halpern, J. (1982) Regression with censored data. *Biometrika*, **69**, 521–31.

Minder, Ch. E. and Whitney, J.B. (1975) A likelihood analysis of the linear calibration problem. *Technometrics*, **17**, 463–71.

Murphy, S.A. and van der Vaart, A.W. (2000) On profile likelihood (with Discussion). *J. Amer. Statist. Assoc.*, **95**, 449–85.

Nelder, J.A. (1977) A reformulation of linear models (with Discussion). *J. Roy. Statist. Soc. A*, **140**, 48–76.

Nelder, J.A. (1985) Quasi-likelihood and GLIM. In *Generalized Linear Models* (eds R. Gilchrist, B. Francis and J. Whittaker). Springer-Verlag, Berlin.

Nelder, J.A. and Wedderburn, R.W.M. (1972) Generalized linear models. *J. Roy. Statist. Soc. A*, **135**, 370–84.

Neuhaus, J.M. (1992) Statistical methods for longitudinal and clustered designs with binary responses. *Statist. Meth. Med. Res.*, **1**, 249–73.

Neuhaus, J.M. and Jewell, N.P. (1990) Some comments on Rosner's multiple logistic model for clustered data. *Biometrics*, **46**, 523–34.

Oakes, D. (1999) Direct calculation of the information matrix via the EM algorithm. *J. Roy. Statist. Soc. B*, **61**, 479–82.

Owen, A.B. (1988) Empirical likelihood ratio confidence intervals for a single functional. *Biometrika*, **75**, 237–49.

Owen, A.B. (1995) Nonparametric likelihood confidence bands for a distribution function. *J. Amer. Statist. Assoc.*, **90**, 516–21.

Owen, A.B. (2001) *Empirical Likelihood*. Chapman and Hall/CR, Boca Raton.

Patterson, H.D. and Thompson, R. (1971) Recovery of inter-block information when block sizes are unequal. *Biometrika*, **58**, 545–54.

Pfefferman, D. (1993) The role of sampling weights when modelling survey data. *Int. Statist. Rev.*, **61**, 317–37.

Postman, M., Huchra, J.P. and Geller, M.J. (1986) Probes of large-scale structures in the Corona Borealis region. *Astronom. J.*, **92**, 1238–47.

Potthoff, R.F. (1964) On the Johnson–Neyman technique and some extensions thereof. *Psychometrika*, **29**, 241–56.

Pregibon, D. (1981) Logistic regression diagnostics. *Ann. Statist.*, **9**, 705–24.

Prentice, R.L. (1973) Exponential survivals with censoring and explanatory variables. *Biometrika*, **60**, 279–88.

Prentice, R.L. and Gloeckler, L.A. (1978) Regression analysis of grouped survival data with application to breast cancer data. *Biometrics*, **34**, 57–67.

Racine, A., Grieve, A.P., Flühler, H. and Smith, A.F.M. (1986) Bayesian methods in practice: experiences in the pharmaceutical industry. *Appl. Statist.*, **35**, 93–120.

Rasch, G. (1960) *Probabilistic Models for Some Intelligence and Attainment Tests*. Danish Institute for Educational Research, Copenhagen.

Reiersöl, O. (1950) Identifiability of a linear relation between variables which are subject to error. *Econometrika*, **18**, 375–89.

Richardson, S. and Green, P.J. (1997) On Bayesian analysis of mixtures with an unknown number of components (with Discussion). *J. Roy. Statist. Soc. B*, **59**, 731–92.

Ridout, M.S. (1990) Non-convergence of Fisher's method of scoring–a simple example. GLIM Newsletter, 20.06.

Rodriguez, G. and Goldman, N. (1995) An assessment of estimation procedures for multilevel models with binary responses. *J. Roy. Statist. Soc. A*, **158**, 73–89.

Roeder, K. (1990) Density estimation with confidence sets exemplified by superclusters and voids in the galaxies. *J. Amer. Statist. Assoc.*, **85**, 617–24.

Roger, J.H. (1985) Using factors when fitting the scale parameter to Weibull and other survival regression models with censored data. *GLIM Newslett.*, **11**, 14–15.

Roger, J.H. and Peacock, S.B. (1982) Fitting the scale as a GLIM parameter for Weibull, extreme value, logistic and log-logistic regression models with censored data. *GLIM Newslett.*, **6**, 30–7.

Royall, R. (1997) *Statistical Evidence: A Likelihood Paradigm*. Chapman and Hall, London.

Rubin, D.B. (1976) Inference and missing data. *Biometrika*, **63**, 581–92.

Rubin, D.B. (1987) *Multiple Imputation for Nonresponse in Surveys*. Wiley, New York.

Rubin, D.B. and Thayer, D.T. (1982) EM algorithms for factor analysis. *Psychometrika*, **47**, 69–76.

Ryan, T., Joiner, B. and Ryan, B. (1976) *Minitab Students Handbook*. Duxbury Press, North Scituate, Mass.

Scallan, A., Gilchrist, R. and Green, M. (1984) Fitting parametric link functions in generalised linear models, *Comp. Stat. Data Anal.*, **2**, 37–49.

Schafer, D.W. (2001) Semiparametric maximum likelihood for measurement error regression. *Biometrics*, **57**, 53–61.

Schafer, J.L. (1997) *Analysis of Incomplete Multivariate Data*. Chapman and Hall, London.

Silvapulle, M.J. (1981) On the existence of maximum likelihood estimators for the binomial response model. *J. Roy. Statist. Soc. B*, **43**, 310–13.

Simpson, E.H. (1951) The interpretation of interaction in contingency tables. *J. Roy. Statist. Soc. B*, **13**, 238–41.

Skrondal, A. and Rabe-Hesketh, S. (2003). *Generalized Latent Variable Modeling: Multilevel, Longitudinal and Structural Equation Models*. Chapman and Hall/CRC Press, Boca Raton.

Smith, T.M.F. (1976) The foundations of survey sampling: a review (with Discussion) *J. Roy. Statist. Soc. A*, **139**, 183–204.

Smyth, G.K. (1986) Modelling the dispersion parameter in generalized linear models. In *Proc. Statist. Computing Section*. American Statistical Association, Alexandria, pp. 278–83.

Smyth, G.K. (1989) Generalized linear models with varying dispersion. *J. Roy. Statist. Soc. B*, **51**, 47–60.

Smyth, G.K. and Verbyla, A.P. (1999) Double generalized linear models: approximate REML and diagnostics. In *Proceedings of the 14th International Workshop on Statistical Modelling, Graz, July 19–23*, pp. 66–80.

Sprott, D.A. (1973) Normal likelihoods and their relation to large sample theory of estimation. *Biometrika*, **60**, 457–65.

Sprott, D.A. (1980) Maximum likelihood in small samples: estimation in the presence of nuisance parameters. *Biometrika*, **67**, 515–23.

Sprott, D.A. (2000) *Statistical Inference in Science*. Springer, New York.

Steele, B.M. (1996) A modified EM algorithm for estimation in generalized mixed models. *Biometrics*, **52**, 1295–310.

Stevens, S.S. (1946) On the theory of scales of measurement. *Science*, **103**, 677–80.

Stone, M. (1974) Cross-validatory choice and assessment of statistical predictions (with Discussion). *J. Roy. Statist. Soc. B*, **36**, 111–47.

Stroud, A.H. and Sechrest, D. (1966) *Gaussian Quadrature Formulas*. Prentice-Hall, Englewood Cliffs, New Jersey.

Stukel, T.A. (1988) Generalized logistic models. *J. Amer. Statist. Assoc.*, **83**, 426–31.

Tanner, M.A. (1996) *Tools for Statistical Inference: Methods for the Exploration of Posterior Distributions and Likelihood Functions* (3rd edn). Springer-Verlag, New York.

Tate, R.F. and Klett, G.W. (1959) Optimal confidence intervals for the variance of a normal distribution. *J. Amer. Statist. Assoc.*, **54**, 674–82.

Thode, H.C., Finch, S.J. and Mendell, N.R. (1988) Simulated percentage points for the null distribution of the likelihood ratio test for a mixture of two normals. *Biometrics*, **44**, 1195–201.

Thompson, R. and Baker, R.J. (1981) Composite link functions in generalized linear models. *Appl. Statist.*, **30**, 125–31.

Titterington, D.M., Smith, A.F.M. and Makov, U.E. (1985) *Statistical Analysis of Finite Mixture Distributions*. John Wiley, Chichester.

Tsutakawa, R.K. (1985) Estimation of cancer mortality rates: a Bayesian analysis of small frequencies. *Biometrics*, **41**, 69–79.

van der Linden, W.J. and Hambelton, R.K. (1997). *Handbook of Modern Item Response Theory*. Springer-Verlag, New York.

Verbeke, G. and Molenberghs, G. (2003) The use of score tests for inference on variance components. *Biometrics*, **59**, 254–62.

Walker, S. (1996) An EM algorithm for nonlinear random effects models. *Biometrics*, **52**, 934–44.

Wand, M.P. and Jones, M.C. (1994) *Kernel Smoothing*. CRC Press, Boca Raton.

Wechsler, D. (1949) *Wechsler Intelligence Scale for Children*. The Psychological corporation, New York.

Wedderburn, R.W.M. (1974) Quasi-likelihood functions, generalized linear models and the Gauss–Newton method. *Biometrika*, **61**, 439–47.

White, H. (1982) Maximum likelihood estimation of misspecified models. *Econometrica*, **50**, 1–25.

Whitehead, J. (1980) Fitting Cox's regression model to survival data using GLIM. *Appl. Statist.*, **29**, 268–75.

Wilson, E.B. and Hilferty, M.M. (1931) The distribution of chi-square. *Proc. Nat. Acad. Sci.*, **17**, 684–8.

Wolynetz, M.S. (1979) Maximum likelihood estimation in a linear model from confined and censored normal data. Algorithm AS139. *Appl. Statist.*, **28**, 195–206.

Woolson, R.F. and Clarke, W.R. (1984) Analysis of categorical incomplete longitudinal data. *J. Roy. Statist. Soc. A*, **147**, 87–99.

Wrigley, N. (1976) *Introduction to the Use of Logit Models in Geography*. CATMOG 10, University of East Anglia, Norwich: Geo. Abstracts Ltd.

Yates, F. (1984) Tests of significance for 2 × 2 contingency tables (with Discussion). *J. Roy. Statist. Soc. A*, **147**, 426–63.

Yusuf, S., Peto, R., Lewis, J., Collins, R. and Sleight, P. (1985) Beta blockade during and after myocardial infarction: an overview of the randomized trials. *Progr. Cardiovasc. Dis.*, **27**, 335–71.

Zelen, M. (1969) Play the winner rule and the controlled clinical trial. *J. Amer. Statist. Assoc.*, **64**, 131–46.

Zuzovsky, R. and Aitkin, M. (1990) Using a multi-level model and an indicator system in science education to assess the effect of school treatment on student achievement. *School Effect. School Improv.*, **1**, 121–38.

Zuzovsky, R. and Aitkin, M. (1991) Curricular change and science achievement in Israeli elementary schools. In *Pupils, Classrooms and Schools: International Studies of Schooling from a Multilevel Perspective* (eds D. Willms and S. Raudenbush). Academic Press, San Diego, pp. 25–36.

Zuzovsky, R. and Aitkin, M. (1994) A coupled process of conceptualizing a model of school effectiveness and developing an indicator system for monitoring effectiveness. *Tijdschrift voor Onderwijs Res.*, **19**, 65–81.

Dataset Index

Macros Index

Subject Index